Nonlinear Fiber Optics

Third Edition

OPTICS AND PHOTONICS

(formerly Quantum Electronics)

Series Editors

PAUL L. KELLEY
Tufts University
Medford, Massachusetts

IVAN P. KAMINOW
Lucent Technologies
Holmdel, New Jersey

GOVIND P. AGRAWAL
University of Rochester
Rochester, New York

Recently Published Books in the Series:

Jean-Claude Diels and Wolfgang Rudolph, *Ultrashort Laser Pulse Phenomena: Fundamentals, Techniques, and Applications on a Femtosecond Time Scale*
Eli Kapon, editor, *Semiconductor Lasers I: Fundamentals*
Eli Kapon, editor, *Semiconductor Lasers II: Materials and Structures*
P. C. Becker, N. A. Olsson, and J. R. Simpson, *Erbium-Doped Fiber Amplifiers: Fundamentals and Technology*
Raman Kashyap, *Fiber Bragg Gratings*
Katsunari Okamoto, *Fundamentals of Optical Waveguides*
Govind P. Agrawal, *Applications of Nonlinear Fiber Optics*

A complete list of titles in this series appears at the end of this volume.

Nonlinear Fiber Optics

Third Edition

GOVIND P. AGRAWAL

The Institute of Optics

University of Rochester

OPTICS AND PHOTONICS

ACADEMIC PRESS

A Harcourt Science and Technology Company

San Diego San Francisco New York Boston
London Sydney Tokyo

Academic Press
A Harcourt Science and Technology Company
525 B Street, Suite 1900, San Diego, California 92101-4495, USA
http://www.academicpress.com

Academic Press
Harcourt Place, 32 Jamestown Road, London NW1 7BY, UK
http://www.academicpress.com

Library of Congress Catalog Card Number: 00-111106

International Standard Book Number: 0-12-045143-3

PRINTED IN THE UNITED STATES OF AMERICA
00 01 02 03 04 05 ML 9 8 7 6 5 4 3 2 1

For Anne, Sipra, Caroline, and Claire

Contents

Preface

Since the publication of the first edition of this book in 1989, the field of *nonlinear fiber optics* has virtually exploded. A major factor behind such a tremendous growth was the advent of fiber amplifiers, made by doping silica or fluoride fibers with rare-earth ions such as erbium and neodymium. Such amplifiers revolutionized the design of fiber-optic communication systems, including those making use of optical solitons whose very existence stems from the presence of nonlinear effects in optical fibers. Optical amplifiers permit propagation of lightwave signals over thousands of kilometers as they can compensate for all losses encountered by the signal in the optical domain. At the same time, fiber amplifiers enable the use of massive wavelength-division multiplexing (WDM) and have led to the development of lightwave systems with capacities exceeding 1 Tb/s. Nonlinear fiber optics plays an increasingly important role in the design of such high-capacity lightwave systems. In fact, an understanding of various nonlinear effects occurring inside optical fibers is almost a prerequisite for a lightwave-system designer.

The third edition is intended to bring the book up-to-date so that it remains a unique source of comprehensive coverage on the subject of nonlinear fiber optics. An attempt was made to include recent research results on all topics relevant to the field of nonlinear fiber optics. Such an ambitious objective increased the size of the book to the extent that it was necessary to split it into two separate books. This book will continue to deal with the fundamental aspects of nonlinear fiber optics. A second book *Applications of Nonlinear Fiber Optics* is devoted to its applications; it is referred to as Part B in this text.

Nonlinear Fiber Optics, 3rd edition, retains most of the material that appeared in the first edition, with the exception of Chapter 6, which is now devoted to the polarization effects relevant for light propagation in optical fibers. Polarization issues have become increasingly more important, especially for high-speed lightwave systems for which the phenomenon of polarization-mode dispersion (PMD) has become a limiting factor. It is thus necessary that students learn about PMD and other polarization effects in a course devoted to nonlinear fiber optics.

The potential readership is likely to consist of senior undergraduate students, graduate students enrolled in the M. S. and Ph. D. degree programs, engineers and technicians involved with the telecommunication industry, and scientists working in the fields of fiber optics and optical communications. This revised edition should continue to be a useful text for graduate and senior-level courses dealing with nonlinear optics, fiber optics, or optical communications that are designed to provide mastery of the fundamental aspects. Some universities may even opt to offer a high-level graduate course devoted to solely nonlinear fiber optics. The problems provided at the end of each chapter should be useful to instructors of such a course.

Many individuals have contributed, either directly or indirectly, to the completion of the third edition. I am thankful to all of them, especially to my students whose curiosity led to several improvements. Several of my colleagues have helped me in preparing the third edition. I thank them for reading drafts and making helpful suggestions. I am grateful to many readers for their occasional feedback. Last, but not least, I thank my wife, Anne, and my daughters, Sipra, Caroline, and Claire, for understanding why I needed to spend many weekends on the book instead of spending time with them.

Govind P. Agrawal

Rochester, NY

Chapter 1

Introduction

This introductory chapter is intended to provide an overview of the fiber characteristics that are important for understanding the nonlinear effects discussed in later chapters. Section 1.1 provides a historical perspective on the progress in the field of fiber optics. Section 1.2 discusses various fiber properties such as optical loss, chromatic dispersion, and birefringence. Particular attention is paid to chromatic dispersion because of its importance in the study of nonlinear effects probed by using ultrashort optical pulses. Section 1.3 introduces various nonlinear effects resulting from the intensity dependence of the refractive index and stimulated inelastic scattering. Among the nonlinear effects that have been studied extensively using optical fibers as a nonlinear medium are self-phase modulation, cross-phase modulation, four-wave mixing, stimulated Raman scattering, and stimulated Brillouin scattering. Each of these effects is considered in detail in separate chapters. Section 1.4 gives an overview of how the text is organized for discussing such a wide variety of nonlinear effects in optical fibers.

1.1 Historical Perspective

Total internal reflection—the basic phenomenon responsible for guiding of light in optical fibers—is known from the nineteenth century. The reader is referred to a 1999 book for the interesting history behind the discovery of this phenomenon [1]. Although uncladded glass fibers were fabricated in the 1920s [2]–[4], the field of fiber optics was not born until the 1950s when the use of a cladding layer led to considerable improvement in the fiber charac-

1

teristics [5]–[8]. The idea that optical fibers would benefit from a dielectric cladding was not obvious and has a remarkable history [1].

The field of fiber optics developed rapidly during the 1960s, mainly for the purpose of image transmission through a bundle of glass fibers [9]. These early fibers were extremely lossy (loss >1000 dB/km) from the modern standard. However, the situation changed drastically in 1970 when, following an earlier suggestion [10], losses of silica fibers were reduced to below 20 dB/km [11]. Further progress in fabrication technology [12] resulted by 1979 in a loss of only 0.2 dB/km in the 1.55-μm wavelength region [13], a loss level limited mainly by the fundamental process of Rayleigh scattering.

The availability of low-loss silica fibers led not only to a revolution in the field of optical fiber communications [14]–[17] but also to the advent of the new field of nonlinear fiber optics. Stimulated Raman- and Brillouin-scattering processes in optical fibers were studied as early as 1972 [18]–[20]. This work stimulated the study of other nonlinear phenomena such as optically induced birefringence, parametric four-wave mixing, and self-phase modulation [21]–[25]. An important contribution was made in 1973 when it was suggested that optical fibers can support soliton-like pulses as a result of an interplay between the dispersive and nonlinear effects [26]. Optical solitons were observed in a 1980 experiment [27] and led to a number of advances during the 1980s in the generation and control of ultrashort optical pulses [28]–[32]. The decade of the 1980s also saw the development of pulse-compression and optical-switching techniques that exploited the nonlinear effects in fibers [33]–[40]. Pulses as short as 6 fs were generated by 1987 [41]. Several reviews and books cover the enormous progress made during the 1980s [42]–[52].

The field of nonlinear fiber optics continued to grow during the decade of the 1990s. A new dimension was added when optical fibers were doped with rare-earth elements and used to make amplifiers and lasers. Although fiber amplifiers were made as early as 1964 [53], it was only after 1987 that their development accelerated [54]. Erbium-doped fiber amplifiers attracted the most attention because they operate in the wavelength region near 1.55 μm and can be used for compensation of losses in fiber-optic lightwave systems [55], [56]. Such amplifiers were used for commercial applications beginning in 1995. Their use has led to a virtual revolution in the design of multichannel lightwave systems [14]–[17].

The advent of fiber amplifiers also fueled research on optical solitons [57]–[60] and led eventually to the concept of dispersion-managed solitons [61]–

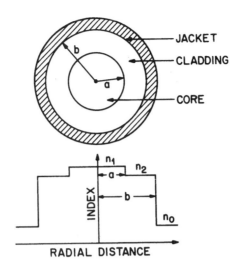

Figure 1.1 Schematic illustration of the cross section and the refractive-index profile of a step-index fiber.

[63]. In another development, fiber gratings, first made in 1978 [64], were developed during the 1990s to the point that they became an integral part of lightwave technology [65]. Nonlinear effects in fiber gratings and photonic-crystal fibers have attracted considerable attention since 1996 [66]–[71]. Clearly, the field of nonlinear fiber optics has grown considerably in the 1990s and is expected to do so during the twenty-first century. It has led to a number of advances important from the fundamental as well as the technological point of view. The interest in nonlinear fiber optics should continue in view of the development of the photonic-based technologies for information management.

1.2   Fiber Characteristics

In its simplest form, an optical fiber consists of a central glass core surrounded by a cladding layer whose refractive index n_2 is slightly lower than the core index n_1. Such fibers are generally referred to as step-index fibers to distinguish them from graded-index fibers in which the refractive index of the core decreases gradually from center to core boundary [72]–[74]. Figure 1.1 shows schematically the cross section and refractive-index profile of a step-index fiber. Two parameters that characterize an optical fiber are the relative core-cladding index difference

$$\Delta = \frac{n_1 - n_2}{n_1},$$ (1.2.1)

and the so-called V parameter defined as

$$V = k_0 a (n_1^2 - n_2^2)^{1/2},$$ (1.2.2)

where $k_0 = 2\pi/\lambda$, a is the core radius, and λ is the wavelength of light.

The V parameter determines the number of modes supported by the fiber. Fiber modes are discussed in Section 2.2, where it is shown that a step-index fiber supports a single mode if $V < 2.405$. Optical fibers designed to satisfy this condition are called single-mode fibers. The main difference between the single-mode and multimode fibers is the core size. The core radius a is typically 25–30 μm for multimode fibers. However, single-mode fibers with $\Delta \approx 0.003$ require a to be < 5 μm. The numerical value of the outer radius b is less critical as long as it is large enough to confine the fiber modes entirely. A standard value of $b = 62.5$ μm is commonly used for both single-mode and multimode fibers. Since nonlinear effects are mostly studied using single-mode fibers, the term optical fiber in this text refers to single-mode fibers unless noted otherwise.

1.2.1 Material and Fabrication

The material of choice for low-loss optical fibers is pure silica glass synthesized by fusing SiO_2 molecules. The refractive-index difference between the core and the cladding is realized by the selective use of dopants during the fabrication process. Dopants such as GeO_2 and P_2O_5 increase the refractive index of pure silica and are suitable for the core, while materials such as boron and fluorine are used for the cladding because they decrease the refractive index of silica. Additional dopants can be used depending on specific applications. For example, to make fiber amplifiers and lasers, the core of silica fibers is codoped with rare-earth ions using dopants such as $ErCl_3$ and Nd_2O_3. Similarly, Al_2O_3 is sometimes added to control the gain spectrum of fiber amplifiers.

The fabrication of optical fibers involves two stages [75]. In the first stage, a vapor-deposition method is used to make a cylindrical preform with the desired refractive-index profile and the relative core-cladding dimensions. A typical preform is 1-m long with 2-cm diameter. In the second stage, the preform

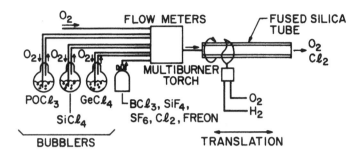

Figure 1.2 Schematic diagram of the MCVD process commonly used for fiber fabrication. (After Ref. [75].)

is drawn into a fiber using a precision-feed mechanism that feeds it into a furnace at a proper speed. During this process, the relative core-cladding dimensions are preserved. Both stages, preform fabrication and fiber drawing, involve sophisticated technology to ensure the uniformity of the core size and the index profile [75]–[77].

Several methods can be used for making a preform. The three commonly used methods are modified chemical vapor deposition (MCVD), outside vapor deposition (OVD), and vapor-phase axial deposition (VAD). Figure 1.2 shows a schematic diagram of the MCVD process. In this process, successive layers of SiO_2 are deposited on the inside of a fused silica tube by mixing the vapors of $SiCl_4$ and O_2 at a temperature of $\approx 1800°C$. To ensure uniformity, the multiburner torch is moved back and forth across the tube length. The refractive index of the cladding layers is controlled by adding fluorine to the tube. When a sufficient cladding thickness has been deposited with multiple passes of the torch, the vapors of $GeCl_4$ or $POCl_3$ are added to the vapor mixture to form the core. When all layers have been deposited, the torch temperature is raised to collapse the tube into a solid rod known as the preform.

This description is extremely brief and is intended to provide a general idea. The fabrication of optical fibers requires attention to a large number of technological details. The interested reader is referred to the extensive literature on this subject [75]–[77].

1.2.2 Fiber Losses

An important fiber parameter is a measure of power loss during transmission of optical signals inside the fiber. If P_0 is the power launched at the input of a

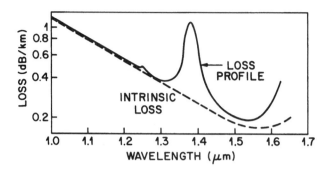

Figure 1.3 Measured loss spectrum of a single-mode silica fiber. Dashed curve shows the contribution resulting from Rayleigh scattering. (After Ref. [75].)

fiber of length L, the transmitted power P_T is given by

$$P_T = P_0 \exp(-\alpha L) \,, \qquad (1.2.3)$$

where the attenuation constant α is a measure of total fiber losses from all sources. It is customary to express α in units of dB/km using the relation (see Appendix A for an explanation of decibel units)

$$\alpha_{\text{dB}} = -\frac{10}{L} \log\left(\frac{P_T}{P_0}\right) = 4.343\alpha, \qquad (1.2.4)$$

where Eq. (1.2.3) was used to relate α_{dB} and α.

As one may expect, fiber losses depend on the wavelength of light. Figure 1.3 shows the loss spectrum of a silica fiber made by the MCVD process [75]. This fiber exhibits a minimum loss of about 0.2 dB/km near 1.55 μm. Losses are considerably higher at shorter wavelengths, reaching a level of a few dB/km in the visible region. Note, however, that even a 10-dB/km loss corresponds to an attenuation constant of only $\alpha \approx 2 \times 10^{-5}$ cm^{-1}, an incredibly low value compared to that of most other materials.

Several factors contribute to the loss spectrum of Fig. 1.3, with material absorption and Rayleigh scattering contributing dominantly. Silica glass has electronic resonances in the ultraviolet (UV) region and vibrational resonances in the far-infrared (FIR) region beyond 2 μm but absorbs little light in the wavelength region 0.5–2 μm. However, even a relatively small amount of impurities can lead to significant absorption in that wavelength window. From a practical point of view, the most important impurity affecting fiber loss is the

OH ion, which has a fundamental vibrational absorption peak at ≈ 2.73 μm. The overtones of this OH-absorption peak are responsible for the dominant peak seen in Fig. 1.3 near 1.4 μm and a smaller peak near 1.23 μm. Special precautions are taken during the fiber-fabrication process to ensure an OH-ion level of less than one part in one hundred million [75]. In state-of-the-art fibers, the peak near 1.4 μm can be reduced to below the 0.5-dB level. It virtually disappears in especially prepared fibers [78]. Such fibers with low losses in the entire 1.3–1.6 μm spectral region are useful for fiber-optic communications and were available commercially by the year 2000 (e.g., all-wave fiber).

Rayleigh scattering is a fundamental loss mechanism arising from density fluctuations frozen into the fused silica during manufacture. Resulting local fluctuations in the refractive index scatter light in all directions. The Rayleigh-scattering loss varies as λ^{-4} and is dominant at short wavelengths. As this loss is intrinsic to the fiber, it sets the ultimate limit on fiber loss. The intrinsic loss level (shown by a dashed line in Fig. 1.3) is estimated to be (in dB/km)

$$\alpha_R = C_R/\lambda^4, \qquad (1.2.5)$$

where the constant C_R is in the range 0.7–0.9 dB/(km-μm^4) depending on the constituents of the fiber core. As $\alpha_R = 0.12$–0.15 dB/km near $\lambda = 1.55$ μm, losses in silica fibers are dominated by Rayleigh scattering. In some glasses, α_R can be reduced to a level ~ 0.05 dB/km [79]. Such glasses may be useful for fabricating ultralow-loss fibers.

Among other factors that may contribute to losses are bending of fiber and scattering of light at the core-cladding interface [72]. Modern fibers exhibit a loss of ≈ 0.2 dB/km near 1.55 μm. Total loss of fiber cables used in optical communication systems is slightly larger (by ~ 0.03 dB/km) because of splice and cabling losses.

1.2.3 Chromatic Dispersion

When an electromagnetic wave interacts with the bound electrons of a dielectric, the medium response, in general, depends on the optical frequency ω. This property, referred to as chromatic dispersion, manifests through the frequency dependence of the refractive index $n(\omega)$. On a fundamental level, the origin of chromatic dispersion is related to the characteristic resonance frequencies at which the medium absorbs the electromagnetic radiation through oscillations of bound electrons. Far from the medium resonances, the refrac-

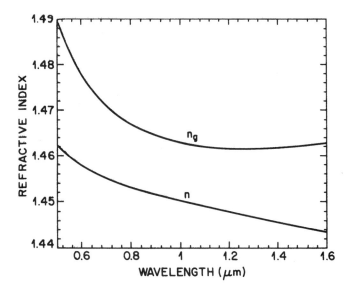

Figure 1.4 Variation of refractive index n and group index n_g with wavelength for fused silica.

tive index is well approximated by the Sellmeier equation [72]

$$n^2(\omega) = 1 + \sum_{j=1}^{m} \frac{B_j \omega_j^2}{\omega_j^2 - \omega^2}, \qquad (1.2.6)$$

where ω_j is the resonance frequency and B_j is the strength of jth resonance. The sum in Eq. (1.2.6) extends over all material resonances that contribute to the frequency range of interest. In the case of optical fibers, the parameters B_j and ω_j are obtained experimentally by fitting the measured dispersion curves [80] to Eq. (1.2.6) with $m = 3$ and depend on the core constituents [74]. For bulk-fused silica, these parameters are found to be [81] $B_1 = 0.6961663$, $B_2 = 0.4079426$, $B_3 = 0.8974794$, $\lambda_1 = 0.0684043$ μm, $\lambda_2 = 0.1162414$ μm, and $\lambda_3 = 9.896161$ μm, where $\lambda_j = 2\pi c / \omega_j$ and c is the speed of light in vacuum.

Fiber dispersion plays a critical role in propagation of short optical pulses because different spectral components associated with the pulse travel at different speeds given by $c/n(\omega)$. Even when the nonlinear effects are not important, dispersion-induced pulse broadening can be detrimental for optical communication systems. In the nonlinear regime, the combination of dispersion

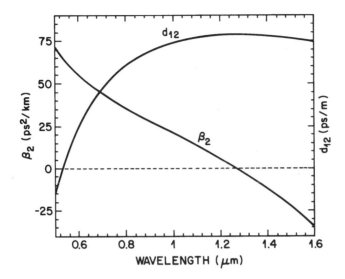

Figure 1.5 Variation of β_2 and d_{12} with wavelength for fused silica. The dispersion parameter $\beta_2 = 0$ near 1.27 μm.

and nonlinearity can result in a qualitatively different behavior, as discussed in later chapters. Mathematically, the effects of fiber dispersion are accounted for by expanding the mode-propagation constant β in a Taylor series about the frequency ω_0 at which the pulse spectrum is centered:

$$\beta(\omega) = n(\omega)\frac{\omega}{c} = \beta_0 + \beta_1(\omega - \omega_0) + \frac{1}{2}\beta_2(\omega - \omega_0)^2 + \cdots, \qquad (1.2.7)$$

where

$$\beta_m = \left(\frac{d^m\beta}{d\omega^m}\right)_{\omega=\omega_0} \qquad (m = 0, 1, 2, \ldots). \qquad (1.2.8)$$

The parameters β_1 and β_2 are related to the refractive index n and its derivatives through the relations

$$\beta_1 = \frac{1}{v_g} = \frac{n_g}{c} = \frac{1}{c}\left(n + \omega\frac{dn}{d\omega}\right), \qquad (1.2.9)$$

$$\beta_2 = \frac{1}{c}\left(2\frac{dn}{d\omega} + \omega\frac{d^2n}{d\omega^2}\right), \qquad (1.2.10)$$

where n_g is the group index and v_g is the group velocity. Physically speaking, the envelope of an optical pulse moves at the group velocity while the

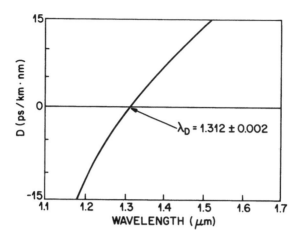

Figure 1.6 Measured variation of dispersion parameter D with wavelength for a single-mode fiber. (After Ref. [75].)

parameter β_2 represents dispersion of the group velocity and is responsible for pulse broadening. This phenomenon is known as the group-velocity dispersion (GVD), and β_2 is the GVD parameter.

Figures 1.4 and 1.5 show how n, n_g, and β_2 vary with wavelength λ in fused silica using Eqs. (1.2.6), (1.2.9), and (1.2.10). The most notable feature is that β_2 vanishes at a wavelength of about 1.27 μm and becomes negative for longer wavelengths. This wavelength is referred to as the zero-dispersion wavelength and is denoted as λ_D. However, note that dispersion does not vanish at $\lambda = \lambda_D$. Pulse propagation near this wavelength requires inclusion of the cubic term in Eq. (1.2.7). The coefficient β_3 appearing in that term is called the third-order-dispersion (TOD) parameter. Such higher-order dispersive effects can distort ultrashort optical pulses both in the linear [72] and nonlinear regimes [82]. Their inclusion is necessary only when the wavelength λ approaches λ_D to within a few nanometers.

The curves shown in Figs. 1.4 and 1.5 are for bulk-fused silica. The dispersive behavior of actual glass fibers deviates from that shown in these figures for the following two reasons. First, the fiber core may have small amounts of dopants such as GeO_2 and P_2O_5. Equation (1.2.6) in that case should be used with parameters appropriate to the amount of doping levels [74]. Second, because of dielectric waveguiding, the effective mode index is slightly lower than the material index $n(\omega)$ of the core, reduction itself being ω dependent [72]–

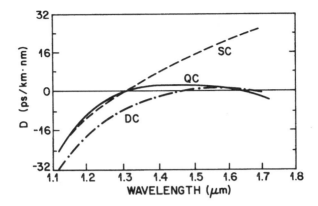

Figure 1.7 Variation of dispersion parameter D with wavelength for three kinds of fibers. Labels SC, DC, and QC stand for single-clad, double-clad, and quadruple-clad fibers, respectively. (After Ref. [84].)

[74]. This results in a waveguide contribution that must be added to the material contribution to obtain the total dispersion. Generally, the waveguide contribution to β_2 is relatively small except near the zero-dispersion wavelength λ_D where the two become comparable. The main effect of the waveguide contribution is to shift λ_D slightly toward longer wavelengths; $\lambda_D \approx 1.31$ μm for standard fibers. Figure 1.6 shows the measured total dispersion of a single-mode fiber [75]. The quantity plotted is the dispersion parameter D that is commonly used in the fiber-optics literature in place of β_2. It is related to β_2 by the relation

$$D = \frac{d\beta_1}{d\lambda} = -\frac{2\pi c}{\lambda^2}\beta_2 \approx \frac{\lambda}{c}\frac{d^2 n}{d\lambda^2}. \tag{1.2.11}$$

An interesting feature of the waveguide dispersion is that its contribution to D (or β_2) depends on fiber-design parameters such as core radius a and core-cladding index difference Δ. This feature can be used to shift the zero-dispersion wavelength λ_D in the vicinity of 1.55 μm where the fiber loss is minimum. Such dispersion-shifted fibers [83] have found applications in optical communication systems. They are available commercially and are known by names such as zero- and nonzero-dispersion-shifted fibers, depending on whether $D \approx 0$ at 1.55 μm or not. Those fibers in which GVD is shifted to the wavelength region beyond 1.6 μm exhibit a large positive value of β_2. They are called dispersion-compensating fibers (DCFs). The slope of the curve in Fig. 1.6 (called the dispersion slope) is related to the TOD parameter β_3.

Fibers with reduced slope have been developed in recent years for wavelength-division-multiplexing (WDM) applications.

It is possible to design dispersion-flattened optical fibers having low dispersion over a relatively large wavelength range 1.3–1.6 μm. This is achieved by using multiple cladding layers. Figure 1.7 shows the measured dispersion spectra for two such multiple-clad fibers having two (double-clad) and four (quadruple-clad) cladding layers around the core applications. For comparison, dispersion of a single-clad fiber is also shown by a dashed line. The quadruply clad fiber has low dispersion ($|D| \sim 1$ ps/km-nm) over a wide wavelength range extending from 1.25 to 1.65 μm. Waveguide dispersion can also be used to make fibers for which D varies along the fiber length. An example is provided by dispersion-decreasing fibers made by tapering the core diameter along the fiber length [85], [86].

Nonlinear effects in optical fibers can manifest qualitatively different behaviors depending on the sign of the GVD parameter. For wavelengths such that $\lambda < \lambda_D$, the fiber is said to exhibit normal dispersion as $\beta_2 > 0$ (see Fig. 1.5). In the normal-dispersion regime, high-frequency (blue-shifted) components of an optical pulse travel slower than low-frequency (red-shifted) components of the same pulse. By contrast, the opposite occurs in the anomalous-dispersion regime in which $\beta_2 < 0$. As seen in Fig. 1.5, silica fibers exhibit anomalous dispersion when the light wavelength exceeds the zero-dispersion wavelength ($\lambda > \lambda_D$). The anomalous-dispersion regime is of considerable interest for the study of nonlinear effects because it is in this regime that optical fibers support solitons through a balance between the dispersive and nonlinear effects.

An important feature of chromatic dispersion is that pulses at different wavelengths propagate at different speeds inside a fiber because of a mismatch in their group velocities. This feature leads to a walk-off effect that plays an important role in the description of the nonlinear phenomena involving two or more closely spaced optical pulses. More specifically, the nonlinear interaction between two optical pulses ceases to occur when the faster moving pulse completely walks through the slower moving pulse. This feature is governed by the walk-off parameter d_{12} defined as

$$d_{12} = \beta_1(\lambda_1) - \beta_1(\lambda_2) = v_g^{-1}(\lambda_1) - v_g^{-1}(\lambda_2), \qquad (1.2.12)$$

where λ_1 and λ_2 are the center wavelengths of two pulses and β_1 at these wavelengths is evaluated using Eq. (1.2.9). For pulses of width T_0, one can define the walk-off length L_W by the relation

$$L_W = T_0 / |d_{12}|. \qquad (1.2.13)$$

Figure 1.5 shows variation of d_{12} with λ_2 for fused silica using Eq. (1.2.12) with $\lambda_1 = 0.532$ μm. In the normal-dispersion regime ($\beta_2 > 0$), a longer-wavelength pulse travels faster, while the opposite occurs in the anomalous-dispersion region. For example, if a pulse at $\lambda_2 = 1.06$ μm copropagates with the pulse at $\lambda_1 = 0.532$ μm, it will separate from the shorter-wavelength pulse at a rate of about 80 ps/m. This corresponds to a walk-off length L_W of only 25 cm for $T_0 = 20$ ps. The group-velocity mismatch plays an important role for nonlinear effects involving cross-phase modulation [47].

1.2.4 Polarization-Mode Dispersion

As discussed in Chapter 2, even a single-mode fiber is not truly single mode because it can support two degenerate modes that are polarized in two orthogonal directions. Under ideal conditions (perfect cylindrical symmetry and stress-free fiber), a mode excited with its polarization in the x direction would not couple to the mode with the orthogonal y-polarization state. In real fibers, small departures from cylindrical symmetry because of random variations in core shape and stress-induced anisotropy result in a mixing of the two polarization states by breaking the mode degeneracy. Mathematically, the mode-propagation constant β becomes slightly different for the modes polarized in the x and y directions. This property is referred to as modal birefringence. The strength of modal birefringence is defined as [87]

$$B_m = \frac{|\beta_x - \beta_y|}{k_0} = |n_x - n_y|, \qquad (1.2.14)$$

where n_x and n_y are the modal refractive indices for the two orthogonally polarized states. For a given value of B_m, the two modes exchange their powers in a periodic fashion as they propagate inside the fiber with the period [87]

$$L_B = \frac{2\pi}{|\beta_x - \beta_y|} = \frac{\lambda}{B_m}. \qquad (1.2.15)$$

The length L_B is called the beat length. The axis along which the mode index is smaller is called the fast axis because the group velocity is larger for light propagating in that direction. For the same reason, the axis with the larger mode index is called the slow axis.

In standard optical fibers, B_m is not constant along the fiber but changes randomly because of fluctuations in the core shape and anisotropic stress. As a result, light launched into the fiber with a fixed state of polarization changes its polarization in a random fashion. This change in polarization is typically harmless for continuous-wave (CW) light because most photodetectors do not respond to polarization changes of the incident light. It becomes an issue for optical communication systems when short pulses are transmitted over long lengths [15]. If an input pulse excites both polarization components, the two components travel along the fiber at different speeds because of their different group velocities. The pulse becomes broader at the output end because group velocities change randomly in response to random changes in fiber birefringence (analogous to a random-walk problem). This phenomenon, referred to as polarization-mode dispersion (PMD), was studied extensively during the 1990s because of its importance for long-haul lightwave systems [88]–[98].

The extent of pulse broadening can be estimated from the time delay ΔT occurring between the two polarization components during propagation of an optical pulse. For a fiber of length L and constant birefringence B_m, ΔT is given by

$$\Delta T = \left| \frac{L}{v_{gx}} - \frac{L}{v_{gy}} \right| = L|\beta_{1x} - \beta_{1y}| = L\delta\beta_1, \qquad (1.2.16)$$

where $\delta\beta_1 = k_0(dB_m/d\omega)$ is related to fiber birefringence. Equation (1.2.16) cannot be used directly to estimate PMD for standard telecommunication fibers because of random changes in birefringence occurring along the fiber. These changes tend to equalize the propagation times for the two polarization components. In fact, PMD is characterized by the root-mean-square (RMS) value of ΔT obtained after averaging over random perturbations. The variance of ΔT is found to be [90]

$$\sigma_T^2 = \langle (\Delta T)^2 \rangle = 2(\Delta' l_c)^2 [\exp(-L/l_c) + L/l_c - 1], \qquad (1.2.17)$$

where Δ' is the intrinsic modal dispersion and the correlation length l_c is defined as the length over which two polarization components remain correlated; typical values of l_c are of the order of $10''$ m. For $L > 0.1$ km, we can use $l_c \ll L$ to find that

$$\sigma_T \approx \Delta' \sqrt{2l_c L} \equiv D_p \sqrt{L}, \qquad (1.2.18)$$

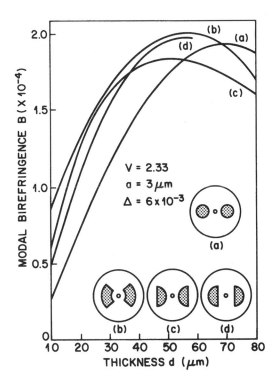

Figure 1.8 Variation of birefringence parameter B_m with thickness d of the stress-inducing element for four different polarization-preserving fibers. Different shapes of the stress-applying elements (shaded region) are shown in the inset. (After Ref. [101].)

where D_p is the PMD parameter. For most fibers, values of D_p are in the range 0.1–1 ps/$\sqrt{\text{km}}$. Because of its $\sqrt{L}$ dependence, PMD-induced pulse broadening is relatively small compared with the GVD effects. However, PMD becomes a limiting factor for high-speed communication systems designed to operate over long distances near the zero-dispersion wavelength of the fiber [92].

For some applications it is desirable that fibers transmit light without changing its state of polarization. Such fibers are called polarization-maintaining or polarization-preserving fibers [99]–[104]. A large amount of birefringence is introduced intentionally in these fibers through design modifications so that relatively small birefringence fluctuations are masked by it and do not affect the state of polarization significantly. One scheme breaks the cylindrical symmetry, making the fiber core elliptical in shape [104]. The degree of birefringence achieved by this technique is typically $\sim 10^{-6}$. An alternative scheme

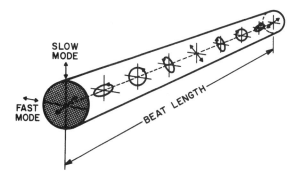

Figure 1.9 Evolution of state of polarization along a polarization-maintaining fiber when input signal is linearly polarized at 45° from the slow axis.

makes use of stress-induced birefringence and permits $B_m \sim 10^{-4}$. In a widely adopted design, two rods of borosilicate glass are inserted on the opposite sides of the fiber core at the preform stage. The resulting birefringence depends on the location and the thickness of the stress-inducing elements. Figure 1.8 shows how B_m varies with thickness d for four shapes of stress-inducing elements located at a distance of five times the core radius [101]. Values of $B_m \approx 2 \times 10^{-4}$ can be achieved for $d = 50$–60 μm. Such fibers are often named after the shape of the stress-inducing element, resulting in whimsical names such as "panda" and "bow-tie" fibers.

The use of polarization-maintaining fibers requires identification of the slow and fast axes before an optical signal can be launched into the fiber. Structural changes are often made to the fiber for this purpose. In one scheme, cladding is flattened in such a way that the flat surface is parallel to the slow axis of the fiber. Such a fiber is called the "D fiber" after the shape of the cladding [104] and makes axes identification relatively easy. When the polarization direction of the linearly polarized light coincides with the slow or the fast axis, the state of polarization remains unchanged during propagation. In contrast, if the polarization direction makes an angle with these axes, polarization changes continuously along the fiber in a periodic manner with a period equal to the beat length [see Eq. (1.2.15)]. Figure 1.9 shows schematically the evolution of polarization over one beat length of a birefringent fiber. The state of polarization changes over one-half of the beat length from linear to elliptic, elliptic to circular, circular to elliptic, and then back to linear but is rotated by 90° from the incident linear polarization. The process is repeated over the remaining half of the beat length such that the initial state is recovered at $z = L_B$ and its multiples. The beat length is typically ~ 1 m but can be as small as 1 cm for a strongly birefringent fiber with $B_m \sim 10^{-4}$.

1.3 Fiber Nonlinearities

The response of any dielectric to light becomes nonlinear for intense electromagnetic fields, and optical fibers are no exception. On a fundamental level, the origin of nonlinear response is related to anharmonic motion of bound electrons under the influence of an applied field. As a result, the total polarization **P** induced by electric dipoles is not linear in the electric field **E**, but satisfies the more general relation [105]–[108]

$$\mathbf{P} = \varepsilon_0 \left(\chi^{(1)} \cdot \mathbf{E} + \chi^{(2)} : \mathbf{EE} + \chi^{(3)} \vdots \mathbf{EEE} + \cdots \right), \qquad (1.3.1)$$

where ε_0 is the vacuum permittivity and $\chi^{(j)}$ ($j = 1, 2, \ldots$) is jth order susceptibility. In general, $\chi^{(j)}$ is a tensor of rank $j + 1$. The linear susceptibility $\chi^{(1)}$ represents the dominant contribution to **P**. Its effects are included through the refractive index n and the attenuation coefficient α discussed in Section 1.2. The second-order susceptibility $\chi^{(2)}$ is responsible for such nonlinear effects as second-harmonic generation and sum-frequency generation [106]. However, it is nonzero only for media that lack an inversion symmetry at the molecular level. As SiO_2 is a symmetric molecule, $\chi^{(2)}$ vanishes for silica glasses. As a result, optical fibers do not normally exhibit second-order nonlinear effects. Nonetheless, the electric-quadrupole and magnetic-dipole moments can generate weak second-order nonlinear effects. Defects or color centers inside the fiber core can also contribute to second-harmonic generation under certain conditions (see Chapter 10).

1.3.1 Nonlinear Refraction

The lowest-order nonlinear effects in optical fibers originate from the third-order susceptibility $\chi^{(3)}$, which is responsible for phenomena such as third-harmonic generation, four-wave mixing, and nonlinear refraction [106]. Unless special efforts are made to achieve phase matching, the nonlinear processes that involve generation of new frequencies (e.g. third-harmonic generation and four-wave mixing) are not efficient in optical fibers. Most of the nonlinear effects in optical fibers therefore originate from nonlinear refraction, a phenomenon referring to the intensity dependence of the refractive index. In its simplest form, the refractive index can be written as

$$\tilde{n}(\omega, |E|^2) = n(\omega) + n_2|E|^2, \qquad (1.3.2)$$

where $n(\omega)$ is the linear part given by Eq. (1.2.6), $|E|^2$ is the optical intensity inside the fiber, and n_2 is the nonlinear-index coefficient related to $\chi^{(3)}$ by the relation (see Section 2.3)

$$n_2 = \frac{3}{8n}\mathrm{Re}(\chi_{xxxx}^{(3)}), \qquad (1.3.3)$$

where Re stands for the real part and the optical field is assumed to be linearly polarized so that only one component $\chi_{xxxx}^{(3)}$ of the fourth-rank tensor contributes to the refractive index. The tensorial nature of $\chi^{(3)}$ can affect the polarization properties of optical beams through nonlinear birefringence. Such nonlinear effects are covered in Chapter 6.

The intensity dependence of the refractive index leads to a large number of interesting nonlinear effects; the two most widely studied are self-phase modulation (SPM) and cross-phase modulation (XPM). Self-phase modulation refers to the self-induced phase shift experienced by an optical field during its propagation in optical fibers. Its magnitude can be obtained by noting that the phase of an optical field changes by

$$\phi = \tilde{n}k_0L = (n + n_2|E|^2)k_0L, \qquad (1.3.4)$$

where $k_0 = 2\pi/\lambda$ and L is the fiber length. The intensity-dependent nonlinear phase shift $\phi_{\mathrm{NL}} = n_2k_0L|E|^2$ is due to SPM. Among other things, SPM is responsible for spectral broadening of ultrashort pulses [25] and formation of optical solitons in the anomalous-dispersion regime of fibers [26].

Cross-phase modulation refers to the nonlinear phase shift of an optical field induced by another field having a different wavelength, direction, or state of polarization. Its origin can be understood by noting that the total electric field $\mathbf{E}$ in Eq. (1.3.1) is given by

$$\mathbf{E} = \tfrac{1}{2}\hat{x}[E_1\exp(-i\omega_1 t) + E_2\exp(-i\omega_2 t) + \text{c.c.}], \qquad (1.3.5)$$

when two optical fields at frequencies ω_1 and ω_2, polarized along the x axis, propagate simultaneously inside the fiber. (The abbreviation c.c. stands for complex conjugate.) The nonlinear phase shift for the field at ω_1 is then given by

$$\phi_{NL} = n_2 k_0 L(|E_1|^2 + 2|E_2|^2), \qquad (1.3.6)$$

where we have neglected all terms that generate polarization at frequencies other than ω_1 and ω_2 because of their non-phase-matched character. The two terms on the right-hand side of Eq. (1.3.6) are due to SPM and XPM, respectively. An important feature of XPM is that, for equally intense optical fields of different wavelengths, the contribution of XPM to the nonlinear phase shift is twice that of SPM. Among other things, XPM is responsible for asymmetric spectral broadening of copropagating optical pulses. Chapters 6 and 7 discuss the XPM-related nonlinear effects.

1.3.2 Stimulated Inelastic Scattering

The nonlinear effects governed by the third-order susceptibility $\chi^{(3)}$ are elastic in the sense that no energy is exchanged between the electromagnetic field and the dielectric medium. A second class of nonlinear effects results from stimulated inelastic scattering in which the optical field transfers part of its energy to the nonlinear medium. Two important nonlinear effects in optical fibers fall in this category; both of them are related to vibrational excitation modes of silica. These phenomena, known as stimulated Raman scattering (SRS) and stimulated Brillouin scattering (SBS), were among the first nonlinear effects studied in optical fibers [18]–[20]. The main difference between the two is that optical phonons participate in SRS while acoustic phonons participate in SBS.

In a simple quantum-mechanical picture applicable to both SRS and SBS, a photon of the incident field (called the pump) is annihilated to create a photon at a lower frequency (belonging to the Stokes wave) and a phonon with the right energy and momentum to conserve the energy and the momentum. Of course, a higher-energy photon at the so-called anti-Stokes frequency can also be created if a phonon of right energy and momentum is available. Even though SRS and SBS are very similar in their origin, different dispersion relations for acoustic and optical phonons lead to some basic differences between the two. A fundamental difference is that SBS in optical fibers occurs only in the backward direction whereas SRS can occur in both directions.

Although a complete description of SRS and SBS in optical fibers is quite involved, the initial growth of the Stokes wave can be described by a simple relation. For SRS, this relation is given by

$$\frac{dI_s}{dz} = g_R I_p I_s,$$ (1.3.7)

where I_s is the Stokes intensity, I_p is the pump intensity, and g_R is the Raman-gain coefficient. A similar relation holds for SBS with g_R replaced by the Brillouin-gain coefficient g_B. Both g_R and g_B have been measured experimentally for silica fibers. The Raman-gain spectrum is found to be very broad, extending up to 30 THz [18]. The peak gain $g_R \approx 7 \times 10^{-14}$ m/W at pump wavelengths near 1.5 μm and occurs for the Stokes shift of ≈ 13 THz. In contrast, the Brillouin-gain spectrum is extremely narrow, with a bandwidth of <100 MHz. The peak value of Brillouin gain occurs for the Stokes shift of ~ 10 GHz for pump wavelengths near 1.5 μm. The peak gain is $\approx 6 \times 10^{-11}$ m/W for a narrow-bandwidth pump [19] and decreases by a factor of $\Delta v_p / \Delta v_B$ for a broad-bandwidth pump, where Δv_p is the pump bandwidth and Δv_B is the Brillouin-gain bandwidth.

An important feature of SRS and SBS is that they exhibit a threshold-like behavior, i.e., significant conversion of pump energy to Stokes energy occurs only when the pump intensity exceeds a certain threshold level. For SRS in a single-mode fiber with $\alpha L \gg 1$, the threshold pump intensity is given by [20]

$$I_p^{\text{th}} \approx 16(\alpha / g_R).$$ (1.3.8)

Typically $I_p^{\text{th}} \sim 10$ MW/cm^2, and SRS can be observed at a pump power ~ 1 W. A similar calculation for SBS shows that the threshold pump intensity is given by [20]

$$I_p^{\text{th}} \approx 21(\alpha / g_B).$$ (1.3.9)

As the Brillouin-gain coefficient g_B is larger by nearly three orders of magnitude compared with g_R, typical values of SBS threshold are ~ 1 mW. The nonlinear phenomena of SRS and SBS are discussed in Chapters 8 and 9, respectively.

1.3.3 Importance of Nonlinear Effects

Most measurements of the nonlinear-index coefficient n_2 in silica fibers yield a value in the range 2.2–3.4 $\times 10^{-20}$ m^2/W (see Appendix B), depending on both the core composition and whether the input polarization is preserved inside the fiber or not [109]. This value is small compared to most other nonlinear media by at least two orders of magnitude. Similarly, the measurements of Raman- and Brillouin-gain coefficients in silica fibers show that their values are smaller by two orders of magnitude or more compared with other common nonlinear media [47]. In spite of the intrinsically small values of the nonlinear coefficients in fused silica, the nonlinear effects in optical fibers can be observed at relatively low power levels. This is possible because of two important characteristics of single-mode fibers—a small spot size (mode diameter $< 10\ \mu$m) and extremely low loss (< 1 dB/km) in the wavelength range 1.0–1.6 μm.

A figure of merit for the efficiency of a nonlinear process in bulk media is the product IL_{eff} where I is the optical intensity and L_{eff} is the effective length of interaction region [110]. If light is focused to a spot of radius w_0, then $I = P/(\pi w_0^2)$, where P is the incident optical power. Clearly, I can be increased by focusing the light tightly to reduce w_0. However, this results in a smaller L_{eff} because the length of the focal region decreases with tight focusing. For a Gaussian beam, $L_{\mathrm{eff}} \sim \pi w_0^2/\lambda$, and the product

$$\left(IL_{\mathrm{eff}}\right)_{\mathrm{bulk}} = \left(\frac{P}{\pi w_0^2}\right)\frac{\pi w_0^2}{\lambda} = \frac{P}{\lambda} \tag{1.3.10}$$

is independent of the spot size w_0.

In single-mode fibers, spot size w_0 is determined by the core radius a. Furthermore, because of dielectric waveguiding, the same spot size can be maintained across the entire fiber length L. In this case, the interaction length L_{eff} is limited by the fiber loss α. Using $I(z) = I_0 \exp(-\alpha z)$, where $I_0 = P/(\pi w_0^2)$ and P is the optical power coupled into the fiber, the product $I L_{\mathrm{eff}}$ becomes

$$\left(IL_{\mathrm{eff}}\right)_{\mathrm{fiber}} = \int_0^L I(z)\exp(-\alpha z)\,dz = \frac{P}{\pi w_0^2 \alpha}[1 - \exp(-\alpha L)]. \tag{1.3.11}$$

A comparison of Eqs. (1.3.10) and (1.3.11) shows that, for sufficiently long fibers, the efficiency of a nonlinear process in optical fibers can be improved by a factor [110]

$$\frac{(IL_{\mathrm{eff}})_{\mathrm{fiber}}}{(IL_{\mathrm{eff}})_{\mathrm{bulk}}} = \frac{\lambda}{\pi w_0^2 \alpha}, \tag{1.3.12}$$

where $\alpha L \gg 1$ was assumed. In the visible region, the enhancement factor is $\sim 10^7$ for $\lambda = 0.53\ \mu$m, $w_0 = 2\ \mu$m, and $\alpha = 2.5 \times 10^{-5}$ cm^{-1} (10 dB/km). In the wavelength region near 1.55 μm ($\alpha = 0.2$ dB/km), the enhancement factor can approach 10^9. It is this tremendous enhancement in the efficiency of the nonlinear processes that makes silica fibers a suitable nonlinear medium for the observation of a wide variety of nonlinear effects at low power levels. Relatively weak nonlinearity of silica fibers becomes an issue in applications for which it is desirable to use a short fiber length (< 0.1 km). It is possible to make fibers by using nonlinear materials for which n_2 is larger than silica. Optical fibers made with lead silicate glasses have n_2 values larger by about a factor of ten [111]. Even larger values ($n_2 = 4.2 \times 10^{-18}$ m^2/W) have been measured in chalcogenide-glass fibers [112]. Such fibers are attracting considerable attention for making fiber devices such as amplifiers, tapers, switches, and gratings, and are likely to become important for nonlinear fiber optics [113]–[117].

1.4 Overview

This book is intended to provide a comprehensive account of the nonlinear phenomena in optical fibers. The field of nonlinear fiber optics has grown to the extent that its coverage requires two volumes. This volume covers fundamental aspects whereas a separate volume is devoted to device and system applications. Broadly speaking, Chapters 1–3 provide the background material and the mathematical tools needed for understanding the various nonlinear effects. Chapters 4–7 discuss the nonlinear effects that lead to spectral and temporal changes in an optical wave without changing its energy. Chapters 8–10 consider the nonlinear effects that generate new optical waves through an energy transfer from the incident waves.

Chapter 2 provides the mathematical framework needed for a theoretical understanding of the nonlinear phenomena in optical fibers. Starting from Maxwell's equations, the wave equation in a nonlinear dispersive medium is used to discuss the fiber modes and to obtain a basic propagation equation satisfied by the amplitude of the pulse envelope. The procedure emphasizes the various approximations made in the derivation of this equation. The numerical methods used to solve the basic propagation equation are then discussed with emphasis on the split-step Fourier method, also known as the beam-propagation method.

Chapter 3 focus on the dispersive effects that occur when the incident power and the fiber length are such that the nonlinear effects are negligible. The main effect of GVD is to broaden an optical pulse as it propagates through the fiber. Such dispersion-induced broadening is considered for several pulse shapes with particular attention paid to the effects of the frequency chirp imposed on the input pulse. The higher-order dispersive effects, important near the zero-dispersion wavelength of fibers, are also discussed.

Chapter 4 considers the nonlinear phenomenon of SPM occurring as a result of the intensity dependence of the refractive index. The main effect of SPM is to broaden the spectrum of optical pulses propagating through the fiber. The pulse shape is also affected if SPM and GVD act together to influence the optical pulse. The features of SPM-induced spectral broadening with and without the GVD effects are discussed in separate sections. The higher-order nonlinear and dispersive effects are also considered.

Chapter 5 is devoted to the study of optical solitons, a topic that has drawn considerable attention because of its fundamental nature as well as potential applications for optical fiber communications. The modulation instability is considered first to emphasize the importance of the interplay between the dispersive and nonlinear effects that can occur in the anomalous-GVD regime of optical fibers. The fundamental and higher-order solitons are then introduced together with the inverse scattering method used to solve the nonlinear Schrödinger equation. Dark solitons are also discussed briefly. The last section considers higher-order nonlinear and dispersive effects with emphasis on the soliton decay.

Chapters 6 and 7 focuses on the XPM effects occurring when two optical fields copropagate simultaneously and affect each other through the intensity dependence of the refractive index. The XPM-induced nonlinear coupling can occur not only when two beams of different wavelengths are incident on the fiber but also through the interaction between the orthogonally polarized components of a single beam in a birefringent fiber. The latter case is discussed first in Chapter 6 by considering the nonlinear phenomena such as the optical Kerr effect and birefringence-induced pulse shaping. Chapter 7 then focuses on the case in which two optical beams at different wavelengths are launched into the fiber. The XPM-induced coupling between the two beams can lead to modulation instability even in the normal-dispersion regime of the fiber. It can also lead to asymmetric spectral and temporal changes when the XPM effects are considered in combination with the SPM and GVD effects. The XPM-induced coupling between the counterpropagating waves is considered next with emphasis on its importance for fiber-optic gyroscopes.

Chapter 8 considers SRS, a nonlinear phenomenon in which the energy from a pump wave is transferred to a Stokes wave (downshifted by about 13 THz) as the pump wave propagates through the optical fiber. This happens only when the pump power exceeds a threshold level. The Raman gain and the Raman threshold in silica fibers are discussed first. Two separate sections then describe SRS for the case of a CW or quasi-CW pump and for the case of ultra-short pump pulses. In the latter case a combination of SPM, XPM, and GVD leads to new qualitative features. These features can be quite different depending on whether the pump and Raman pulses experience normal or anomalous GVD. The case of anomalous GVD is considered in the last section with emphasis on fiber-Raman soliton lasers. The applications of SRS to optical fiber communications are also discussed.

Chapter 9 is devoted to SBS, a nonlinear phenomenon that manifests in optical fibers in a way similar to SRS, but with important differences. Stimulated Brillouin scattering transfers a part of the pump energy to a counterpropagating Stokes wave, downshifted in frequency by only an amount ~ 10 GHz. Because of the small bandwidth (~ 10 MHz) associated with the Brillouin gain, SBS occurs efficiently only for a CW pump or pump pulses whose spectral width is smaller than the gain bandwidth. The characteristics of the Brillouin gain in silica fibers are discussed first. Chapter 9 then describes the theory of SBS by considering important features such as the Brillouin threshold, pump depletion, and gain saturation. The instabilities associated with SBS are also discussed. The experimental results on SBS are described with emphasis on fiber-Brillouin lasers and amplifiers. The last section is devoted to the implications of SBS for optical fiber communications.

Chapter 10 focuses on nonlinear parametric processes in which energy exchange among several optical waves occurs without an active participation of the nonlinear medium. Parametric processes occur efficiently only when a phase-matching condition is satisfied. This condition is relatively easy to satisfy for a nonlinear process known as four-wave mixing. The parametric gain associated with the four-wave-mixing process is obtained by considering nonlinear interaction among the four waves. The experimental results and the phase-matching techniques used to obtain them are discussed in detail. Parametric amplification is considered next together with its applications. The last two sections are devoted to second-harmonic generation in photosensitive fibers. The phenomenon of photosensitivity has attracted considerable attention during the 1990s because of its potential technological applications and is used routinely to make fiber gratings.

Problems

1.1 Calculate the propagation distance over which the injected optical power is reduced by a factor of two for three fibers with losses of 0.2, 20, and 2000 dB/km. Also calculate the attenuation constant α (in cm^{-1}) for the three fibers.

1.2 A single-mode fiber is measured to have $\lambda^2(d^2n/d\lambda^2) = 0.02$ at 0.8 μm. Calculate the dispersion parameters β_2 and D.

1.3 Calculate the numerical values of β_2 (in ps^2/km) and D [in ps/(km-nm)] at 1.5 μm when the modal index varies with wavelength as $n(\lambda) = 1.45 - s(\lambda - 1.3 \ \mu\text{m})^3$, where $s = 0.003 \ \mu\text{m}^{-3}$.

1.4 A 1-km-long single-mode fiber with the zero-dispersion wavelength at 1.4 μm is measured to have $D = 10$ ps/(km-nm) at 1.55 μm. Two pulses from Nd:YAG lasers operating at 1.06 and 1.32 μm are launched simultaneously into the fiber. Calculate the delay in the arrival time of the two pulses at the fiber output assuming that β_2 varies linearly with wavelength over the range 1.0–1.6 μm.

1.5 Equation (1.3.2) is often written in the alternate form $\tilde{n}(\omega, I) = n(\omega) + n_2' I$, where I is the optical intensity. What is the relationship between n_2 and n_2'? Use it to obtain the value of n_2 in units of m ^{2}/V^2 if $n_2' = 2.6 \times 10^{-20}$ m^2/W.

References

[1] J. Hecht, *City of Light* (Oxford University Press, New York, 1999).

[2] J. L. Baird, British Patent 285,738 (1928).

[3] C. W. Hansell, U. S. Patent 1,751,584 (1930).

[4] H. Lamm, Z. *Instrumenten.* **50**, 579 (1930).

[5] A. C. S. van Heel, *Nature* **173**, 39 (1954).

[6] H. H. Hopkins and N. S. Kapany, *Nature* **173**, 39 (1954); *Opt. Acta* **1**, 164 (1955).

[7] B. O'Brian, U. S. Patent 2,825,260 (1958).

[8] B. I. Hirschowitz, U. S. Patent 3,010,357 (1961).

[9] N. S. Kapany, *Fiber Optics: Principles and Applications* (Academic, New York, 1967).

[10] K. C. Kao and G. A. Hockham, *IEE Proc.* **113**, 1151 (1966).

[11] F. P. Kapron, D. B. Keck, and R. D. Maurer, *Appl. Phys. Lett.* **17**, 423 (1970).

[12] W. G. French, J. B. MacChesney, P. B. O'Connor, and G. W. Tasker, *Bell Syst. Tech. J.* **53**, 951 (1974).

[13] T. Miya, Y. Terunuma, T. Hosaka, and T. Miyashita, *Electron. Lett.* **15**, 106 (1979).

[14] I. P. Kaminow and T. L. Koch (eds.), *Optical Fiber Telecommunications III* (Academic Press, San Diego, CA, 1997).

[15] G. P. Agrawal, *Fiber-Optic Communication Systems*, 2nd ed. (Wiley, New York, 1997).

[16] R. Ramaswami and K. Sivarajan, *Optical Networks* (Morgan Kaufmann, Burlington, MA, 1998).

[17] G. Keiser, *Optical Fiber Communications*, 3rd ed. (McGraw-Hill, New York, 2000).

[18] R. H. Stolen, E. P. Ippen, and A. R. Tynes, *Appl. Phys. Lett.* **20**, 62 (1972).

[19] E. P. Ippen and R. H. Stolen, *Appl. Phys. Lett.* **21**, 539 (1972).

[20] R. G. Smith, *Appl. Opt.* **11**, 2489 (1972).

[21] R. H. Stolen and A. Ashkin, *Appl. Phys. Lett.* **22**, 294 (1973).

[22] R. H. Stolen, J. E. Bjorkholm, and A. Ashkin, *Appl. Phys. Lett.* **24**, 308 (1974).

[23] K. O. Hill, D. C. Johnson, B. S. Kawaski, and R. I. MacDonald, *J. Appl. Phys.* **49**, 5098 (1974).

[24] R. H. Stolen, *IEEE J. Quantum Electron.* **QE-11**, 100 (1975).

[25] R. H. Stolen and C. Lin, *Phys. Rev. A* **17**, 1448 (1978).

[26] A. Hasegawa and F. Tappert, *Appl. Phys. Lett.* **23**, 142 (1973).

[27] L. F. Mollenauer, R. H. Stolen, and J. P. Gordon, *Phys. Rev. Lett.* **45**, 1095 (1980).

[28] L. F. Mollenauer and R. H. Stolen, *Opt. Lett.* **9**, 13 (1984).

[29] L. F. Mollenauer, J. P. Gordon, and M. N. Islam, *IEEE J. Quantum Electron.* **QE-22**, 157 (1986).

[30] J. D. Kafka and T. Baer, *Opt. Lett.* **12**, 181 (1987).

[31] M. N. Islam, L. F. Mollenauer, R. H. Stolen, J. R. Simpson, and H. T. Shang, *Opt. Lett.* **12**, 814 (1987).

[32] A. S. Gouveia-Neto, A. S. L. Gomes, and J. R. Taylor, *Opt. Quantum Electron.* **20**, 165 (1988).

[33] H. Nakatsuka, D. Grischkowsky, and A. C. Balant, *Phys. Rev. Lett.* **47**, 910 (1981).

[34] C. V. Shank, R. L. Fork, R. Yen, R. H. Stolen, and W. J. Tomlinson, *Appl. Phys. Lett.* **40**, 761 (1982).

[35] B. Nikolaus and D. Grischkowsky, *Appl. Phys. Lett.* **42**, 1 (1983); *Appl. Phys. Lett.* **43**, 228 (1983).

[36] A. S. L. Gomes, A. S. Gouveia-Neto, and J. R. Taylor, *Opt. Quantum Electron.* **20**, 95 (1988).

[37] N. J. Doran and D. Wood, *Opt. Lett.* **13**, 56 (1988).

[38] M. C. Farries and D. N. Payne, *Appl. Phys. Lett.* **55**, 25 (1989).

[39] K. J. Blow, N. J. Doran, and B. K. Nayar, *Opt. Lett.* **14**, 754 (1989).

[40] M. N. Islam, E. R. Sunderman, R. H. Stolen, W. Pleibel, and J. R. Simpson, *Opt. Lett.* **14**, 811 (1989).

[41] R. L. Fork, C. H. Brito Cruz, P. C. Becker, and C. V. Shank, *Opt. Lett.* **12**, 483 (1987).

[42] H. G. Winful, in *Optical-Fiber Transmission*, E. E. Basch, ed. (Sams, Indianapolis, 1986).

[43] S. A. Akhmanov, V. A. Vysloukh, and A. S. Chirkin, *Sov. Phys. Usp.* **29**, 642 (1986).

[44] D. Cotter, *Opt. Quantum Electron.* **19**, 1 (1987); K. J. Blow and N. J. Doran, *IEE Proc.* **134**, Pt. J, 138 (1987).

[45] D. Marcuse, in *Optical Fiber Telecommunications II*, S. E. Miller and I. P. Kaminow, eds. (Academic Press, San Diego, CA, 1988).

[46] E. M. Dianov, P. V. Mamyshev, and A. M. Prokhorov, *Sov. J. Quantum Electron.* **15**, 1 (1988).

[47] R. R. Alfano (ed.), *The Supercontinuum Laser Source* (Springer-Verlag, New York, 1989).

[48] Y. S. Kivshar and B. A. Malomed, *Rev. Mod. Phys.* **61**, 763 (1989).

[49] A. Kumar, *Phys. Rep.* **187**, 63 (1990).

[50] J. R. Taylor (ed.), *Optical Solitons—Theory and Experiment* (Cambridge University Press, Cambridge, UK, 1992).

[51] G. P. Agrawal, in *Contemporary Nonlinear Optics*, G. P. Agrawal and R. W. Boyd, Eds. (Academic Press, San Diego, CA, 1992), Chap. 2.

[52] Y. Kodama and A. Hasegawa, in *Progress in Optics*, Vol. 30, E. Wolf, ed. (North-Holland, Amsterdam, 1992), Chap. 4.

[53] C. J. Koester and E. Snitzer, *Appl. Opt.* **3**, 1182 (1964).

[54] M. J. F. Digonnet, Ed., *Rare-Earth Doped Fiber Lasers and Amplifiers* (Dekker, New York, 1993).

[55] E. Desurvire, *Erbium-Doped Fiber Amplifiers* (Wiley, New York, 1994).

[56] P. C. Becker, N. A. Olsson, and J. R. Simpson, *Erbium-Doped Fiber Amplifiers: Fundamentals and Technology* (Academic Press, San Diego, CA, 1999).

[57] A. Hasegawa and Y. Kodama, *Solitons in Optical Communications* (Oxford University Press, New York, 1995).

[58] H. A. Haus and W. S. Wong, *Rev. Mod. Phys.* **68**, 423 (1996).

[59] R. J. Essiambre and G. P. Agrawal, in *Progress in Optics*, Vol. 37, E. Wolf, ed. (North-Holland, Amsterdam, 1997), Chap. 4.

[60] L. F. Mollenauer, J. P. Gordon, and P. V. Mamyshev, *Optical Fiber Telecommunications III*, I. P. Kaminow and T. L. Koch, eds. (Academic Press, San Diego, CA, 1997), Chap. 12.

[61] N. J. Smith, F. M. Knox, N. J. Doran, K. J. Blow, and I. Bennion, *Electron. Lett.* **32**, 54 (1996).

[62] I. Gabitov and S. K. Turitsyn, *JETP Lett.* **63**, 814 (1996).

[63] A. Hasegawa (ed.), *New Trends in Optical Soliton Transmission Systems* (Kluwer Academic, Boston, 1998).

[64] K. O. Hill, Y. Fujii, D. C. Johnson, and B. S. Kawasaki, *Appl. Phys. Lett.* **32**, 647 (1978).

[65] R. Kashyap, *Fiber Bragg Gratings* (Academic Press, San Diego, CA, 1999).

[66] C. M. de Sterke and J. E. Sipe, in *Progress in Optics*, Vol. 33, E. Wolf, ed. (North-Holland, Amsterdam, 1994), Chap. 3.

[67] C. M. de Sterke, B. J. Eggleton, and P. A. Krug, *IEEE J. Quantum Electron.* **15**, 1494 (1997).

[68] B. J. Eggleton, C. M. de Sterke, and R. E. Slusher, *J. Opt. Soc. Am. B* **14**, 2980 (1997).

[69] D. Taverner, N. G. R. Broderick, D. J. Richardson, R. I. Laming, and M. Isben, *Opt. Lett.* **23**, 259 (1998); *Opt. Lett.* **23**, 328 (1998).

[70] J. K. Ranka, R. S. Windeler, and A. J. Stentz, *Opt. Lett.* **25**, 25 (2000).

[71] W. J. Wadsworth, J. C. Knight, A. Ortigosa-Blanch, J. Arriaga, E. Silvestre, and P. S. J. Russell, *Electron. Lett.* **36**, 53 (2000).

[72] D. Marcuse, *Light Transmission Optics* (Van Nostrand Reinhold, New York, 1982), Chaps. 8 and 12.

[73] A. W. Snyder and J. D. Love, *Optical Waveguide Theory* (Chapman and Hall, London, 1983).

[74] M. J. Adams, *An Introduction to Optical Waveguides* (Wiley, New York, 1981), Chap. 7.

[75] T. Li (ed.), *Optical Fiber Communications: Fiber Fabrication*, Vol. 1 (Academic Press, San Diego, 1985).

[76] U. C. Paek, *J. Lightwave Technol.* **4**, 1048 (1986).

[77] B. J. Ainslie, *J. Lightwave Technol.* **9**, 220 (1991).

[78] K. Tsujikawa and M. Ohashi, *Electron. Lett.* **34**, 2057 (1998).

[79] M. Ohashi and K. Tsujikawa, *Opt. Fiber Technol.* **6**, 74 (2000).

[80] L. G. Cohen, *J. Lightwave Technol.* **LT-3**, 958 (1985).

[81] I. H. Malitson, *J. Opt. Soc. Am.* **55**, 1205 (1965).

[82] G. P. Agrawal and M. J. Potasek, *Phys. Rev. A* **33**, 1765 (1986).

[83] B. J. Ainslie and C. R. Day, *J. Lightwave Technol.* **LT-4**, 967 (1986).

[84] L. G. Cohen, W. L. Mammel, and S. J. Jang, *Electron. Lett.* **18**, 1023 (1982).

[85] V. A. Bogatyrjov, M. M. Bubnov, E. M. Dianov, and A. A. Sysoliatin, *Pure Appl. Opt.* **4**, 345 (1995).

[86] D. J. Richardson, R. P. Chamberlin, L. Dong, and D. N. Payne, *Electron. Lett.* **31**, 1681 (1995).

[87] I. P. Kaminow, *IEEE J. Quantum Electron.* **QE-17**, 15 (1981).

[88] C. D. Poole, *Opt. Lett.* **13**, 687 (1988); C. D. Poole, J. H. Winters, and J. A. Nagel, *Opt. Lett.* **16**, 372 (1991).

[89] N. Gisin, J. P. von der Weid, and J.-P. Pellaux *J. Lightwave Technol.* **9**, 821 (1991).

[90] G. J. Foschini and C. D. Poole, *J. Lightwave Technol.* **9**, 1439, (1991).

[91] J. Zhou and M. J. O'Mahony, *IEEE Photon. Technol. Lett.* **6**, 1265 (1994).

[92] E. Lichtman, *J. Lightwave Technol.* **13**, 898 (1995).

[93] Y. Suetsugu, T. Kato, and M. Nishimura, *IEEE Photon. Technol. Lett.* **7**, 887 (1995).

[94] P. K. A. Wai and C. R. Menyuk, *J. Lightwave Technol.* **14**, 148 (1996).

[95] D. Marcuse, C. R. Menyuk, and P. K. A. Wai, *J. Lightwave Technol.* **15**, 1735 (1997).

[96] H. Bulow, *IEEE Photon. Technol. Lett.* **10**, 696 (1998).

[97] C. Francia, F. Bruyere, D. Penninckx, and M. Chbat, *IEEE Photon. Technol. Lett.* **10**, 1739 (1998).

[98] D. Mahgerefteh and C. R. Menyuk, *IEEE Photon. Technol. Lett.* **11**, 340 (1999).

[99] D. N. Payne, A. J. Barlow, and J. J. R. Hansen, *IEEE J. Quantum Electron.* **QE-18**, 477 (1982).

[100] S. C. Rashleigh, *J. Lightwave Technol.* **LT-1**, 312 (1983).

[101] J. Noda, K. Okamoto, and Y. Sasaki, *J. Lightwave Technol.* **LT-4**, 1071 (1986).

[102] K. Tajima, M. Ohashi, and Y. Sasaki, *J. Lightwave Technol.* **7**, 1499 (1989).

[103] M. J. Messerly, J. R. Onstott, and R. C. Mikkelson, *J. Lightwave Technol.* **9**, 817 (1991).

[104] R. B. Dyott, *Elliptical Fiber Waveguides* (Artec House, Boston, 1995).

[105] N. Bloembergen, *Nonlinear Optics* (Benjamin, Reading, MA, 1977), Chap. 1.

[106] Y. R. Shen, *Principles of Nonlinear Optics* (Wiley, New York, 1984).

[107] P. N. Butcher and D. N. Cotter, *The Elements of Nonlinear Optics* (Cambridge University Press, Cambridge, UK, 1990).

[108] R. W. Boyd, *Nonlinear Optics* (Academic Press, San Diego, CA, 1992).

[109] G. P. Agrawal, in *Properties of Glass and Rare-Earth Doped Glasses for Optical Fibers*, D. Hewak, Ed. (IEE, London, 1998), pp. 17–21.

[110] E. P. Ippen, in *Laser Applications to Optics and Spectroscopy*, Vol. 2, S. F. Jacobs, M. Sargent III, J. F. Scott, and M. O. Scully, Eds. (Addison-Wesley, Reading, MA, 1975), Chap. 6.

[111] M. A. Newhouse, D. L. Weidman, and D. W. Hall, *Opt. Lett.* **15**, 1185 (1990).

[112] M. Asobe, T. Kanamori, and K. Kubodera, *IEEE Photon. Technol. Lett.* **4**, 362 (1992).

[113] M. Asobe, T. Kanamori, and K. Kubodera, *IEEE J. Quantum Electron.* **29**, 2325 (1993).

[114] M. Asobe, *Opt. Fiber Technol.* **3**, 142 (1997).

[115] A. Mori, Y. Ohishi, T. Kanamori, and S. Sudo, *Appl. Phys. Lett.* **70**, 1230 (1997).

[116] D. T. Schaafsma, J. A. Moon, J. S. Sanghera, and I. D. Aggarwal, *J. Lightwave Technol.* **15**, 2242 (1997).

[117] R. Mossadegh, J. S. Sanghera, D. Schaafsma, B. J. Cole, V. Q. Nguyen, P. E. Miklos, and I. D. Aggarwal, *J. Lightwave Technol.* **16**, 214 (1998).

Chapter 2

Pulse Propagation in Fibers

For an understanding of the nonlinear phenomena in optical fibers, it is necessary to consider the theory of electromagnetic wave propagation in dispersive nonlinear media. The objective of this chapter is to obtain a basic equation that governs propagation of optical pulses in single-mode fibers. Section 2.1 introduces Maxwell's equations and important concepts such as the linear and nonlinear parts of the induced polarization and the frequency-dependent dielectric constant. The concept of fiber modes is introduced in Section 2.2 where the single-mode condition is also discussed. Section 2.3 considers the theory of pulse propagation in nonlinear dispersive media in the slowly varying envelope approximation with the assumption that the spectral width of the pulse is much smaller than the frequency of the incident radiation. The numerical methods used to solve the resulting propagation equation are discussed in Section 2.4.

2.1 Maxwell's Equations

Like all electromagnetic phenomena, the propagation of optical fields in fibers is governed by Maxwell's equations. In the International System of Units (Système international d'unités or SI), these equations are [1]

$$\nabla \times \mathbf{E} = -\frac{\partial \mathbf{B}}{\partial t}, \tag{2.1.1}$$

$$\nabla \times \mathbf{H} = \mathbf{J} + \frac{\partial \mathbf{D}}{\partial t}, \tag{2.1.2}$$

$$\nabla \cdot \mathbf{D} = \rho_f, \tag{2.1.3}$$

31

$$\nabla \cdot \mathbf{B} = 0, \tag{2.1.4}$$

where $\mathbf{E}$ and $\mathbf{H}$ are electric and magnetic field vectors, respectively, and $\mathbf{D}$ and $\mathbf{B}$ are corresponding electric and magnetic flux densities. The current density vector $\mathbf{J}$ and the charge density ρ_f represent the sources for the electromagnetic field. In the absence of free charges in a medium such as optical fibers, $\mathbf{J} = 0$ and $\rho_f = 0$.

The flux densities $\mathbf{D}$ and $\mathbf{B}$ arise in response to the electric and magnetic fields $\mathbf{E}$ and $\mathbf{H}$ propagating inside the medium and are related to them through the constitutive relations given by [1]

$$\mathbf{D} = \varepsilon_0 \mathbf{E} + \mathbf{P}, \tag{2.1.5}$$

$$\mathbf{B} = \mu_0 \mathbf{H} + \mathbf{M}, \tag{2.1.6}$$

where ε_0 is the vacuum permittivity, μ_0 is the vacuum permeability, and $\mathbf{P}$ and $\mathbf{M}$ are the induced electric and magnetic polarizations. For a nonmagnetic medium such as optical fibers, $\mathbf{M} = 0$.

Maxwell's equations can be used to obtain the wave equation that describes light propagation in optical fibers. By taking the curl of Eq. (2.1.1) and using Eqs. (2.1.2), (2.1.5), and (2.1.6), one can eliminate $\mathbf{B}$ and $\mathbf{D}$ in favor of $\mathbf{E}$ and $\mathbf{P}$ and obtain

$$\nabla \times \nabla \times \mathbf{E} = -\frac{1}{c^2}\frac{\partial^2 \mathbf{E}}{\partial t^2} - \mu_0 \frac{\partial^2 \mathbf{P}}{\partial t^2}, \tag{2.1.7}$$

where c is the speed of light in vacuum and the relation $\mu_0 \varepsilon_0 = 1/c^2$ was used. To complete the description, a relation between the induced polarization $\mathbf{P}$ and the electric field $\mathbf{E}$ is needed. In general, the evaluation of $\mathbf{P}$ requires a quantum-mechanical approach. Although such an approach is often necessary when the optical frequency is near a medium resonance, a phenomenological relation of the form (1.3.1) can be used to relate $\mathbf{P}$ and $\mathbf{E}$ far from medium resonances. This is the case for optical fibers in the wavelength range 0.5– 2 μm that is of interest for the study of nonlinear effects. If we include only the third-order nonlinear effects governed by $\chi^{(3)}$, the induced polarization consists of two parts such that

$$\mathbf{P}(\mathbf{r},t) = \mathbf{P}_L(\mathbf{r},t) + \mathbf{P}_{\mathrm{NL}}(\mathbf{r},t), \tag{2.1.8}$$

where the linear part $\mathbf{P}_L$ and the nonlinear part $\mathbf{P}_{\mathrm{NL}}$ are related to the electric field by the general relations [2]–[4]

$$\mathbf{P}_L(\mathbf{r},t) = \varepsilon_0 \int_{-\infty}^{\infty} \chi^{(1)}(t-t') \cdot \mathbf{E}(\mathbf{r},t')\, dt', \tag{2.1.9}$$

$$\mathbf{P}_{NL}(\mathbf{r},t) = \varepsilon_0 \int\int\int_{-\infty}^{\infty} \chi^{(3)}(t-t_1,t-t_2,t-t_3) \vdots$$
$$\times \mathbf{E}(\mathbf{r},t_1)\mathbf{E}(\mathbf{r},t_2)\mathbf{E}(\mathbf{r},t_3)\,dt_1\,dt_2\,dt_3. \qquad (2.1.10)$$

These relations are valid in the electric-dipole approximation and assume that the medium response is local.

Equations (2.1.7)–(2.1.10) provide a general formalism for studying the third-order nonlinear effects in optical fibers. Because of their complexity, it is necessary to make several simplifying approximations. In a major simplification, the nonlinear polarization $\mathbf{P}_{NL}$ in Eq. (2.1.8) is treated as a small perturbation to the total induced polarization. This is justified because the nonlinear effects are relatively weak in silica fibers. The first step therefore consists of solving Eq. (2.1.7) with $\mathbf{P}_{NL} = 0$. Because Eq. (2.1.7) is then linear in $\mathbf{E}$, it is useful to write in the frequency domain as

$$\nabla \times \nabla \times \tilde{\mathbf{E}}(\mathbf{r},\omega) - \varepsilon(\omega)\frac{\omega^2}{c^2}\tilde{\mathbf{E}}(\mathbf{r},\omega) = 0, \qquad (2.1.11)$$

where $\tilde{\mathbf{E}}(\mathbf{r},\omega)$ is the Fourier transform of $\mathbf{E}(\mathbf{r},t)$ defined as

$$\tilde{\mathbf{E}}(\mathbf{r},\omega) = \int_{-\infty}^{\infty} \mathbf{E}(\mathbf{r},t)\exp(i\omega t)\,dt. \qquad (2.1.12)$$

The frequency-dependent dielectric constant appearing in Eq. (2.1.12) is defined as

$$\varepsilon(\omega) = 1 + \tilde{\chi}^{(1)}(\omega), \qquad (2.1.13)$$

where $\tilde{\chi}^{(1)}(\omega)$ is the Fourier transform of $\chi^{(1)}(t)$. As $\tilde{\chi}^{(1)}(\omega)$ is in general complex, so is $\varepsilon(\omega)$. Its real and imaginary parts can be related to the refractive index $n(\omega)$ and the absorption coefficient $\alpha(\omega)$ by using the definition

$$\varepsilon = (n + i\alpha c/2\omega)^2. \qquad (2.1.14)$$

From Eqs. (2.1.13) and (2.1.14), n and α are related to $\chi^{(1)}$ by the relations

$$n(\omega) = 1 + \tfrac{1}{2}\mathrm{Re}[\tilde{\chi}^{(1)}(\omega)], \qquad (2.1.15)$$
$$\alpha(\omega) = \frac{\omega}{nc}\mathrm{Im}[\tilde{\chi}^{(1)}(\omega)], \qquad (2.1.16)$$

where Re and Im stand for the real and imaginary parts, respectively. The frequency dependence of n and α has been discussed in Section 1.2.

Two further simplifications can be made before solving Eq. (2.1.11). First, because of low optical losses in fibers in the wavelength region of interest, the imaginary part of $\varepsilon(\omega)$ is small in comparison to the real part. Thus, we can replace $\varepsilon(\omega)$ by $n^2(\omega)$ in the following discussion of fiber modes and include fiber loss later in a perturbative manner. Second, as $n(\omega)$ is often independent of the spatial coordinates in both the core and the cladding of step-index fibers, one can use

$$\nabla \times \nabla \times \mathbf{E} \equiv \nabla(\nabla \cdot \mathbf{E}) - \nabla^2 \mathbf{E} = -\nabla^2 \mathbf{E}, \qquad (2.1.17)$$

where the relation $\nabla \cdot \mathbf{D} = \varepsilon \nabla \cdot \mathbf{E} = 0$ was used from Eq. (2.1.3). With these simplifications, Eq. (2.1.11) takes the form

$$\nabla^2 \tilde{\mathbf{E}} + n^2(\omega) \frac{\omega^2}{c^2} \tilde{\mathbf{E}} = 0. \qquad (2.1.18)$$

This equation is solved in the next section on fiber modes.

2.2   Fiber Modes

At any frequency ω, optical fibers can support a finite number of guided modes whose spatial distribution $\tilde{\mathbf{E}}(\mathbf{r}, \omega)$ is a solution of the wave equation (2.1.18) and satisfies all appropriate boundary conditions. In addition, the fiber can support a continuum of unguided radiation modes. Although the inclusion of radiation modes is crucial in problems involving transfer of power between bounded and radiation modes [5], they do not play an important role in the discussion of nonlinear effects. As fiber modes are covered in many textbooks [5]–[7], they are discussed only briefly in this section.

2.2.1   Eigenvalue Equation

Because of the cylindrical symmetry of fibers, it is useful to express the wave equation (2.1.18) in cylindrical coordinates ρ, ϕ, and z:

$$\frac{\partial^2 \tilde{\mathbf{E}}}{\partial \rho^2} + \frac{1}{\rho} \frac{\partial \tilde{\mathbf{E}}}{\partial \rho} + \frac{1}{\rho^2} \frac{\partial^2 \tilde{\mathbf{E}}}{\partial \phi^2} + \frac{\partial^2 \tilde{\mathbf{E}}}{\partial z^2} + n^2 k_0^2 \tilde{\mathbf{E}} = 0, \qquad (2.2.1)$$

where $k_0 = \omega/c = 2\pi/\lambda$ and $\tilde{\mathbf{E}}$ is the Fourier transform of the electric field $\mathbf{E}$, i.e.,

$$\mathbf{E}(\mathbf{r}, t) = \frac{1}{2\pi} \int_{-\infty}^{\infty} \tilde{\mathbf{E}}(\mathbf{r}, \omega) \exp(-i\omega t) \, d\omega. \qquad (2.2.2)$$

Similar relations exist for the magnetic field $\mathbf{H}(\mathbf{r},t)$. As $\mathbf{E}$ and $\mathbf{H}$ satisfy Maxwell's equations (2.1.1)–(2.1.4), only two components out of six are independent. It is customary to choose $\tilde{E}_z$ and $\tilde{H}_z$ as the independent components and express $\tilde{E}_\rho, \tilde{E}_\phi, \tilde{H}_\rho$, and $\tilde{H}_\phi$ in terms of $\tilde{E}_z$ and $\tilde{H}_z$. Both $\tilde{E}_z$ and $\tilde{H}_z$ satisfy Eq. (2.2.1). The wave equation for $\tilde{E}_z$ is easily solved by using the method of separation of variables, resulting in the following general form:

$$\tilde{E}_z(r,\omega) = A(\omega)F(\rho)\exp(\pm im\phi)\exp(i\beta z), \qquad (2.2.3)$$

where A is a normalization constant, β is the propagation constant, m is an integer, and $F(\rho)$ is the solution of

$$\frac{d^2F}{d\rho^2} + \frac{1}{\rho}\frac{dF}{d\rho} + \left(n^2 k_0^2 - \beta^2 - \frac{m^2}{\rho^2}\right)F = 0, \qquad (2.2.4)$$

where the refractive index $n = n_1$ for $\rho \leq a$ for a fiber of core radius a but takes the value n_2 outside the core ($\rho > a$).

Equation (2.2.4) is the well-known differential equation for Bessel functions. Its general solution inside the core can be written as

$$F(\rho) = C_1 J_m(\kappa\rho) + C_2 N_m(\kappa\rho), \qquad (2.2.5)$$

where J_m is the Bessel function, N_m is the Neumann function, and

$$\kappa = (n_1^2 k_0^2 - \beta^2)^{1/2}. \qquad (2.2.6)$$

The constants C_1 and C_2 are determined using the boundary conditions. As $N_m(\kappa\rho)$ has a singularity at $\rho = 0$, $C_2 = 0$ for a physically meaningful solution. The constant C_1 can be absorbed in A appearing in Eq. (2.2.3). Thus,

$$F(\rho) = J_m(\kappa\rho), \qquad \rho \leq a. \qquad (2.2.7)$$

In the cladding region ($\rho \geq a$), the solution $F(\rho)$ should be such that it decays exponentially for large ρ. The modified Bessel function K_m represents such a solution. Therefore,

$$F(\rho) = K_m(\gamma\rho), \qquad \rho \geq a, \qquad (2.2.8)$$

where

$$\gamma = (\beta^2 - n_2^2 k_0^2)^{1/2}. \qquad (2.2.9)$$

The same procedure can be followed to obtain the magnetic field component $\tilde{H}_z$. The boundary condition that the tangential components of $\tilde{E}$ and $\tilde{H}$ be continuous across the core-cladding interface requires that $\tilde{E}_z$, $\tilde{H}_z$, $\tilde{E}_\phi$, and $\tilde{H}_\phi$ be the same when $\rho = a$ is approached from inside or outside the core. The equality of these field components at $\rho = a$ leads to an eigenvalue equation whose solutions determine the propagation constant β for the fiber modes. Since the whole procedure is well known [5]–[7], we write the eigenvalue equation directly:

$$\left[\frac{J'_m(\kappa a)}{\kappa J_m(\kappa a)} + \frac{K'_m(\gamma a)}{\gamma K_m(\gamma a)} \right] \left[\frac{J'_m(\kappa a)}{\kappa J_m(\kappa a)} + \frac{n_2^2}{n_1^2} \frac{K'_m(\gamma a)}{\gamma K_m(\gamma a)} \right] = \left(\frac{m\beta k_0 (n_1^2 - n_2^2)}{a n_1 \kappa^2 \gamma^2} \right)^2,$$

(2.2.10)

where a prime denotes differentiation with respect to the argument and we used the important relation

$$\kappa^2 + \gamma^2 = (n_1^2 - n_2^2)k_0^2.$$

(2.2.11)

The eigenvalue equation (2.2.10) in general has several solutions for β for each integer value of m. It is customary to express these solutions by β_{mn}, where both m and n take integer values. Each eigenvalue β_{mn} corresponds to one specific mode supported by the fiber. The corresponding modal field distribution is obtained from Eq. (2.2.3). It turns out [5]–[7] that there are two types of fiber modes, designated as HE_{mn} and EH_{mn}. For $m = 0$, these modes are analogous to the transverse-electric (TE) and transverse-magnetic (TM) modes of a planar waveguide because the axial component of the electric field, or the magnetic field, vanishes. However, for $m > 0$, fiber modes become hybrid, i.e., all six components of the electromagnetic field are nonzero.

2.2.2 Single-Mode Condition

The number of modes supported by a specific fiber at a given wavelength depends on its design parameters, namely the core radius a and the core-cladding index difference $n_1 - n_2$. An important parameter for each mode is its cut-off frequency. This frequency is determined by the condition $\gamma = 0$. The value of κ when $\gamma = 0$ for a given mode determines the cut-off frequency from Eq. (2.2.11). It is useful to define a normalized frequency V by the relation

$$V = \kappa_c a = k_0 a (n_1^2 - n_2^2)^{1/2},$$

(2.2.12)

where κ_c is obtained from Eq. (2.2.11) by setting $\gamma = 0$.

The eigenvalue equation (2.2.10) can be used to determine the values of V at which different modes reach cut-off. The procedure is complicated, but has been described in many texts [5]–[7]. Since we are interested mainly in single-mode fibers, we limit the discussion to the cut-off condition that allows the fiber to support only one mode. A single-mode fiber supports only the HE_{11} mode, also referred to as the fundamental mode. All other modes are beyond cut-off if the parameter $V < V_c$, where V_c is the smallest solution of $J_0(V_c) = 0$ or $V_c \approx 2.405$. The actual value of V is a critical design parameter. Typically, microbending losses increase as V/V_c becomes small. In practice, therefore, fibers are designed such that V is close to V_c. The cut-off wavelength λ_c for single-mode fibers can be obtained by using $k_0 = 2\pi/\lambda_c$ and $V = 2.405$ in Eq. (2.2.12). For a typical value $n_1 - n_2 = 0.005$ for the index difference, $\lambda_c = 1.2 \ \mu$m for $a = 4 \ \mu$m, indicating that such a fiber supports a single mode only for $\lambda > 1.2 \ \mu$m. In practice, core radius should be below 2 μm for a fiber to support a single mode in the visible region.

2.2.3 Characteristics of the Fundamental Mode

The field distribution $\mathbf{E}(\mathbf{r}, t)$ corresponding to the HE_{11} mode has three non-zero components E_ρ, E_ϕ, and E_z, or in Cartesian coordinates E_x, E_y, and E_z. Among these, either E_x or E_y dominates. Thus, to a good degree of approximation, the fundamental fiber mode is linearly polarized in either x or y direction depending on whether E_x or E_y dominates. In this respect, even a single-mode fiber is not truly single mode because it can support two modes of orthogonal polarizations. The notation LP_{mn} is sometimes used to denote the linearly polarized modes, which are approximate solutions of Eq. (2.2.1). The fundamental mode HE_{11} corresponds to LP_{01} in this notation [6].

The two orthogonally polarized modes of a single-mode fiber are degenerate (i.e., they have the same propagation constant) under ideal conditions. In practice, irregularities such as random variations in the core shape and size along the fiber length break this degeneracy slightly, mix the two polarization components randomly, and scramble the polarization of the incident light as it propagates down the fiber. As discussed in Section 1.2.4, polarization-preserving fibers can maintain the linear polarization if the light is launched with its polarization along one of the principal axes of the fiber. Assuming that the incident light is polarized along a principal axis (chosen to coincide with the x axis), the electric field for the fundamental fiber mode HE_{11} is approximately given by

$$\tilde{\mathbf{E}}(\mathbf{r}, \omega) = \hat{x}\{A(\omega)F(x,y)\exp[i\beta(\omega)z]\}, \qquad (2.2.13)$$

where $A(\omega)$ is a normalization constant. The transverse distribution inside the core is found to be

$$F(x,y) = J_0(\kappa\rho), \qquad \rho \leq a, \qquad (2.2.14)$$

where $\rho = (x^2 + y^2)^{1/2}$ is the radial distance. Outside the fiber core, the field decays exponentially as [5]

$$F(x, y) = (a/\rho)^{1/2}J_0(\kappa a)\exp[-\gamma(\rho - a)], \qquad \rho \geq a, \qquad (2.2.15)$$

where $K_m(\gamma\rho)$ in Eq. (2.2.8) was approximated by the leading term in its asymptotic expansion and a constant factor was added to ensure the equality of $F(x, y)$ at $\rho = a$. The propagation constant $\beta(\omega)$ in Eq. (2.2.13) is obtained by solving the eigenvalue equation (2.2.10). Its frequency dependence results not only from the frequency dependence of n_1 and n_2 but also from the frequency dependence of κ. The former is referred to as material dispersion while the latter is called waveguide dispersion. As discussed in Section 1.3, material dispersion generally dominates unless the light wavelength is close to the zero-dispersion wavelength. The evaluation of $\beta(\omega)$ generally requires a numerical solution of Eq. (2.2.10) although approximate analytic expressions can be obtained in specific cases [5]. The effective mode index is related to β by $n_{\text{eff}} = \beta/k_0$.

As the use of modal distribution $F(x,y)$ given by Eqs. (2.2.14) and (2.2.15) is cumbersome in practice, the fundamental fiber mode is often approximated by a Gaussian distribution of the form

$$F(x,y) \approx \exp[-(x^2 + y^2)/w^2], \qquad (2.2.16)$$

where the width parameter w is determined by fitting the exact distribution to a Gaussian form or by following a variational procedure. Figure 2.1 shows the dependence of w/a on the fiber parameter V defined by Eq. (2.2.12). The comparison of the actual field distribution with the fitted Gaussian is also shown for a specific value $V = 2.4$. The quality of fit is generally quite good [8], particularly for V values in the neighborhood of 2. Figure 2.1 shows that $w \approx a$ for $V = 2$, indicating that the core radius provides a good estimate of w for telecommunication fibers for which $V \approx 2$. Note that w can be significantly larger than a for $V < 1.8$. The use of Gaussian approximation is of considerable practical value because of its relative simplicity.

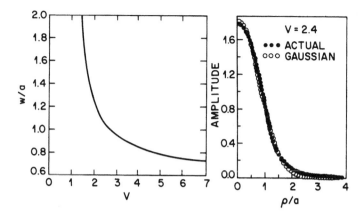

Figure 2.1 Variation of mode-width parameter w with V obtained by fitting the fundamental fiber mode to a Gaussian distribution. Traces on the right show the quality of fit for $V = 2.4$. (After Ref. [8].)

2.3 Pulse-Propagation Equation

The study of most nonlinear effects in optical fibers involves the use of short pulses with widths ranging from $\sim$10 ns to 10 fs. When such optical pulses propagate inside a fiber, both dispersive and nonlinear effects influence their shape and spectrum. In this section we derive a basic equation that governs propagation of optical pulses in nonlinear dispersive fibers. The starting point is the wave equation (2.1.7). By using Eqs. (2.1.8) and (2.1.17), it can be written in the form

$$\nabla^2 \mathbf{E} - \frac{1}{c^2} \frac{\partial^2 \mathbf{E}}{\partial t^2} = \mu_0 \frac{\partial^2 \mathbf{P}_L}{\partial t^2} + \mu_0 \frac{\partial^2 \mathbf{P}_{NL}}{\partial t^2}, \tag{2.3.1}$$

where the linear and nonlinear parts of the induced polarization are related to the electric field $\mathbf{E}(\mathbf{r}, t)$ through Eqs. (2.1.9) and (2.1.10), respectively.

2.3.1 Nonlinear Pulse Propagation

It is necessary to make several simplifying assumptions before solving Eq. (2.3.1). First, $\mathbf{P}_{NL}$ is treated as a small perturbation to $\mathbf{P}_L$. This is justified because nonlinear changes in the refractive index are $< 10^{-6}$ in practice. Second, the optical field is assumed to maintain its polarization along the fiber length so that a scalar approach is valid. This is not really the case, unless polarization-maintaining fibers are used, but the approximation works quite well in practice; it will be relaxed later in Chapter 6. Third, the optical field is assumed to be quasi-monochromatic, i.e., the pulse spectrum, centered at ω_0, is assumed to have a spectral width $\Delta\omega$ such that $\Delta\omega/\omega_0 \ll 1$. Since $\omega_0 \sim 10^{15}$ s^{-1}, the last assumption is valid for pulses as short as 0.1 ps. In the slowly varying envelope approximation adopted here, it is useful to separate the rapidly varying part of the electric field by writing it in the form

$$\mathbf{E}(\mathbf{r},t) = \tfrac{1}{2}\hat{x}[E(\mathbf{r},t)\exp(-i\omega_0 t) + \text{c.c.}], \qquad (2.3.2)$$

where $\hat{x}$ is the polarization unit vector, and $E(\mathbf{r},t)$ is a slowly varying function of time (relative to the optical period). The polarization components $\mathbf{P}_L$ and $\mathbf{P}_{NL}$ can also be expressed in a similar way by writing

$$\mathbf{P}_L(\mathbf{r},t) = \tfrac{1}{2}\hat{x}[P_L(\mathbf{r},t)\exp(-i\omega_0 t) + \text{c.c.}], \qquad (2.3.3)$$

$$\mathbf{P}_{NL}(\mathbf{r},t) = \tfrac{1}{2}\hat{x}[P_{NL}(\mathbf{r},t)\exp(-i\omega_0 t) + \text{c.c.}]. \qquad (2.3.4)$$

The linear component P_L can be obtained by substituting Eq. (2.3.3) in Eq. (2.1.9) and is given by

$$
\begin{aligned}
P_L(\mathbf{r},t) &= \varepsilon_0 \int_{-\infty}^{\infty} \chi_{xx}^{(1)}(t-t')E(\mathbf{r},t')\exp[i\omega_0(t-t')]\,dt' \\
&= \frac{\varepsilon_0}{2\pi} \int_{-\infty}^{\infty} \tilde{\chi}_{xx}^{(1)}(\omega)\tilde{E}(\mathbf{r},\omega-\omega_0)\exp[-i(\omega-\omega_0)t]\,d\omega, \quad (2.3.5)
\end{aligned}
$$

where $\tilde{E}(\mathbf{r},\omega)$ is the Fourier transform of $E(\mathbf{r},t)$ and is defined similarly to Eq. (2.1.12).

The nonlinear component $P_{NL}(\mathbf{r},t)$ is obtained by substituting Eq. (2.3.4) in Eq. (2.1.10). Considerable simplification occurs if the nonlinear response is assumed to be instantaneous so that the time dependence of $\chi^{(3)}$ in Eq. (2.1.10) is given by the product of three delta functions of the form $\delta(t-t_1)$. Equation (2.1.10) then reduces to

$$\mathbf{P}_{NL}(\mathbf{r},t) = \varepsilon_0 \chi^{(3)} \vdots \mathbf{E}(\mathbf{r},t)\mathbf{E}(\mathbf{r},t)\mathbf{E}(\mathbf{r},t). \qquad (2.3.6)$$

The assumption of instantaneous nonlinear response amounts to neglecting the contribution of molecular vibrations to $\chi^{(3)}$ (the Raman effect). In general, both electrons and nuclei respond to the optical field in a nonlinear manner. The nuclei response is inherently slower compared with the electronic response. For silica fibers the vibrational or Raman response occurs over a time scale 60–70 fs. Thus, Eq. (2.3.6) is approximately valid for pulse widths >1 ps. The Raman contribution is included later in this section.

When Eq. (2.3.2) is substituted in Eq. (2.3.6), $\mathbf{P}_{NL}(\mathbf{r},t)$ is found to have a term oscillating at ω_0 and another term oscillating at the third-harmonic frequency $3\omega_0$. The latter term requires phase matching and is generally negligible in optical fibers. By making use of Eq. (2.3.4), $P_{NL}(\mathbf{r},t)$ is given by

$$P_{NL}(\mathbf{r},t) \approx \varepsilon_0 \varepsilon_{NL} E(\mathbf{r},t), \tag{2.3.7}$$

where the nonlinear contribution to the dielectric constant is defined as

$$\varepsilon_{NL} = \tfrac{3}{4}\chi_{xxxx}^{(3)}|E(\mathbf{r},t)|^2. \tag{2.3.8}$$

To obtain the wave equation for the slowly varying amplitude $E(\mathbf{r},t)$, it is more convenient to work in the Fourier domain. This is generally not possible as Eq. (2.3.1) is nonlinear because of the intensity dependence of ε_{NL}. In one approach, ε_{NL} is treated as a constant during the derivation of the propagation equation [9], [10]. The approach is justified in view of the slowly varying envelope approximation and the perturbative nature of P_{NL}. Substituting Eqs. (2.3.2)–(2.3.4) in Eq. (2.3.1), the Fourier transform $\tilde{E}(r, \omega - \omega_0)$, defined as

$$\tilde{E}(\mathbf{r}, \omega - \omega_0) = \int_{-\infty}^{\infty} E(\mathbf{r},t) \exp[i(\omega - \omega_0)t]\,dt, \tag{2.3.9}$$

is found to satisfy the Helmholtz equation

$$\nabla^2 \tilde{E} + \varepsilon(\omega)k_0^2 \tilde{E} = 0, \tag{2.3.10}$$

where $k_0 = \omega/c$ and

$$\varepsilon(\omega) = 1 + \tilde{\chi}_{xx}^{(1)}(\omega) + \varepsilon_{NL} \tag{2.3.11}$$

is the dielectric constant whose nonlinear part ε_{NL} is given by Eq. (2.3.8). Similar to Eq. (2.1.14), the dielectric constant can be used to define the refractive index $\tilde{n}$ and the absorption coefficient $\tilde{\alpha}$. However, both $\tilde{n}$ and $\tilde{\alpha}$ become intensity dependent because of ε_{NL}. It is customary to introduce

$$\tilde{n} = n + n_2|E|^2, \qquad \tilde{\alpha} = \alpha + \alpha_2|E|^2. \tag{2.3.12}$$

Using $\varepsilon = (\tilde{n} + i\tilde{\alpha}/2k_0)^2$ and Eqs. (2.3.8) and (2.3.11), the nonlinear-index coefficient n_2 and the two-photon absorption coefficient α_2 are given by

$$n_2 = \frac{3}{8n}\mathrm{Re}(\chi_{xxxx}^{(3)}), \qquad \alpha_2 = \frac{3\omega_0}{4nc}\mathrm{Im}(\chi_{xxxx}^{(3)}). \tag{2.3.13}$$

The linear index n and the absorption coefficient α are related to the real and imaginary parts of $\tilde{\chi}_{xx}^{(1)}$ as in Eqs. (2.1.15) and (2.1.16). As α_2 is relatively small for silica fibers, it is often ignored. The parameter n_2 should not be confused with the cladding index of Section 2.2 even though the same notation has been used. From here onward, n_2 is a measure of the fiber nonlinearity.

Equation (2.3.10) can be solved by using the method of separation of variables. If we assume a solution of the form

$$\tilde{E}(\mathbf{r}, \omega - \omega_0) = F(x, y)\tilde{A}(z, \omega - \omega_0)\exp(i\beta_0 z), \qquad (2.3.14)$$

where $\tilde{A}(z, \omega)$ is a slowly varying function of z and β_0 is the wave number to be determined later, Eq. (2.3.10) leads to the following two equations for $F(x, y)$ and $\tilde{A}(z, \omega)$:

$$\frac{\partial^2 F}{\partial x^2} + \frac{\partial^2 F}{\partial y^2} + [\varepsilon(\omega)k_0^2 - \tilde{\beta}^2]F = 0, \qquad (2.3.15)$$

$$2i\beta_0\frac{\partial \tilde{A}}{\partial z} + (\tilde{\beta}^2 - \beta_0^2)\tilde{A} = 0. \qquad (2.3.16)$$

In obtaining Eq. (2.3.16), the second derivative $\partial^2\tilde{A}/\partial z^2$ was neglected since $\tilde{A}(z, \omega)$ is assumed to be a slowly varying function of z. The wave number $\tilde{\beta}$ is determined by solving the eigenvalue equation (2.3.15) for the fiber modes using a procedure similar to that used in Section 2.2. The dielectric constant $\varepsilon(\omega)$ in Eq. (2.3.15) can be approximated by

$$\varepsilon = (n + \Delta n)^2 \approx n^2 + 2n\Delta n, \qquad (2.3.17)$$

where Δn is a small perturbation given by

$$\Delta n = n_2|E|^2 + \frac{i\tilde{\alpha}}{2k_0}. \qquad (2.3.18)$$

Equation (2.3.15) can be solved using first-order perturbation theory [11]. We first replace ε with n^2 and obtain the modal distribution $F(x, y)$, and the corresponding wave number $\beta(\omega)$. For a single-mode fiber, $F(x, y)$ corresponds to the modal distribution of the fundamental fiber mode HE_{11} given by Eqs. (2.2.14) and (2.2.15), or by the Gaussian approximation (2.2.16). We then include the effect of Δn in Eq. (2.3.15). In the first-order perturbation theory, Δn does not affect the modal distribution $F(x, y)$. However, the eigenvalue $\tilde{\beta}$ becomes

$$\tilde{\beta}(\omega) = \beta(\omega) + \Delta\beta, \qquad (2.3.19)$$

where

$$\Delta\beta = \frac{k_0 \int\int_{-\infty}^{\infty} \Delta n |F(x,y)|^2 \, dxdy}{\int\int_{-\infty}^{\infty} |F(x,y)|^2 \, dxdy}. \tag{2.3.20}$$

This step completes the formal solution of Eq. (2.3.1) to the first order in perturbation $\mathbf{P}_{NL}$. Using Eqs. (2.3.2) and (2.3.12), the electric field $\mathbf{E}(\mathbf{r},t)$ can be written as

$$\mathbf{E}(\mathbf{r},t) = \tfrac{1}{2}\hat{x}\{F(x,y)A(z,t)\exp[i(\beta_0 z - \omega_0 t)] + \text{c.c.}\}, \tag{2.3.21}$$

where $A(z,t)$ is the slowly varying pulse envelope. The Fourier transform $\tilde{A}(z, \omega - \omega_0)$ of $A(z,t)$ satisfies Eq. (2.3.16), which can be written as

$$\frac{\partial \tilde{A}}{\partial z} = i[\beta(\omega) + \Delta\beta - \beta_0]\tilde{A}, \tag{2.3.22}$$

where we used Eq. (2.3.19) and approximated $\tilde{\beta}^2 - \beta_0^2$ by $2\beta_0(\tilde{\beta} - \beta_0)$. The physical meaning of this equation is clear. Each spectral component within the pulse envelope acquires, as it propagates down the fiber, a phase shift whose magnitude is both frequency and intensity dependent.

At this point, one can go back to the time domain by taking the inverse Fourier transform of Eq. (2.3.22), and obtain the propagation equation for $A(z,t)$. However, as an exact functional form of $\beta(\omega)$ is rarely known, it is useful to expand $\beta(\omega)$ in a Taylor series about the carrier frequency ω_0 as

$$\beta(\omega) = \beta_0 + (\omega - \omega_0)\beta_1 + \tfrac{1}{2}(\omega - \omega_0)^2\beta_2 + \tfrac{1}{6}(\omega - \omega_0)^3\beta_3 + \cdots, \tag{2.3.23}$$

where

$$\beta_m = \left(\frac{d^m \beta}{d\omega^m}\right)_{\omega = \omega_0} \qquad (m = 1, 2, \ldots). \tag{2.3.24}$$

The cubic and higher-order terms in this expansion are generally negligible if the spectral width $\Delta\omega \ll \omega_0$. Their neglect is consistent with the quasi-monochromatic assumption used in the derivation of Eq. (2.3.22). If $\beta_2 \approx 0$ for some specific values of ω_0 (in the vicinity of the zero-dispersion wavelength of the fiber, as discussed in Section 1.3.3), it may be necessary to include the cubic term. We substitute Eq. (2.3.23) in Eq. (2.3.22) and take the inverse Fourier transform by using

$$A(z,t) = \frac{1}{2\pi} \int_{-\infty}^{\infty} \tilde{A}(z, \omega - \omega_0) \exp[-i(\omega - \omega_0)t] \, d\omega. \tag{2.3.25}$$

During the Fourier-transform operation, $\omega - \omega_0$ is replaced by the differential operator $i(\partial/\partial t)$. The resulting equation for $A(z,t)$ becomes

$$\frac{\partial A}{\partial z} = -\beta_1 \frac{\partial A}{\partial t} - \frac{i\beta_2}{2}\frac{\partial^2 A}{\partial t^2} + i\Delta\beta A. \tag{2.3.26}$$

The term with $\Delta\beta$ includes the effect of fiber loss and nonlinearity. By using Eqs. (2.3.18) and (2.3.20), $\Delta\beta$ can be evaluated and substituted in (2.3.26). The result is

$$\frac{\partial A}{\partial z} + \beta_1 \frac{\partial A}{\partial t} + \frac{i\beta_2}{2}\frac{\partial^2 A}{\partial t^2} + \frac{\alpha}{2}A = i\gamma|A|^2 A, \tag{2.3.27}$$

where the nonlinear parameter γ is defined as

$$\gamma = \frac{n_2 \omega_0}{c A_{\text{eff}}}. \tag{2.3.28}$$

In obtaining Eq. (2.3.27) the pulse amplitude A is assumed to be normalized such that $|A|^2$ represents the optical power. The quantity $\gamma|A|^2$ is then measured in units of m^{-1} if n_2 is expressed in units of m^2/W (see Appendix B). The parameter A_{eff} is known as the effective core area and is defined as

$$A_{\text{eff}} = \frac{\left(\iint_{-\infty}^{\infty}|F(x,y)|^2 dxdy\right)^2}{\iint_{-\infty}^{\infty}|F(x,y)|^4 dxdy}. \tag{2.3.29}$$

Its evaluation requires the use of modal distribution $F(x,y)$ for the fundamental fiber mode. Clearly A_{eff} depends on fiber parameters such as the core radius and the core-cladding index difference. If $F(x,y)$ is approximated by a Gaussian distribution as in Eq. (2.2.16), $A_{\text{eff}} = \pi w^2$. The width parameter w depends on the fiber parameters and can be obtained using Fig. 2.1 and Eq. (2.2.12). Typically, A_{eff} can vary in the range 20–100 μm^2 in the 1.5-μm region depending on the fiber design. As a result, γ takes values in the range 1–10 W^{-1}/km if $n_2 \approx 2.6 \times 10^{-20}$ m^2/W is used (see Appendix B). In a large-effective-area fiber (LEAF), A_{eff} is increased intentionally to reduce the impact of fiber nonlinearity.

Equation (2.3.27) describes propagation of picosecond optical pulse in single-mode fibers. It is often referred to as the nonlinear Schrödinger (NLS) equation because it can be reduced to that form under certain conditions. It includes the effects of fiber losses through α, of chromatic dispersion through β_1 and β_2, and of fiber nonlinearity through γ. The physical significance of the parameters β_1 and β_2 has been discussed in Section 1.2.3. Briefly, the pulse envelope moves at the group velocity $v_g \equiv 1/\beta_1$ while the effects of group-velocity dispersion (GVD) are governed by β_2. The GVD parameter β_2 can be positive or negative depending on whether the wavelength λ is below or above the zero-dispersion wavelength λ_D of the fiber (see Fig. 1.5). In the anomalous-dispersion regime ($\lambda > \lambda_D$), β_2 is negative, and the fiber can support optical solitons. In standard silica fibers, $\beta_2 \sim 50$ ps^2/km in the visible region but becomes close to -20 ps^2/km near wavelengths ~ 1.5 μm, the change in sign occurring in the vicinity of 1.3 μm.

2.3.2 Higher-Order Nonlinear Effects

Although the propagation equation (2.3.27) has been successful in explaining a large number of nonlinear effects, it may need modification depending on the experimental conditions. For example, Eq. (2.3.27) does not include the effects of stimulated inelastic scattering such as SRS and SBS (see Section 1.3.2). If peak power of the incident pulse is above a threshold level, both SRS and SBS can transfer energy from the pulse to a new pulse, which may propagate in the same or the opposite direction. The two pulses interact with each other through the Raman or Brillouin gain and XPM. A similar situation occurs when two or more pulses at different wavelengths (separated by more than individual spectral widths) are incident on the fiber. Simultaneous propagation of multiple pulses is governed by a set of equations similar to Eq. (2.3.27), modified suitably to include the contributions of XPM and the Raman or Brillouin gain.

Equation (2.3.27) should also be modified for ultrashort optical pulses whose width is close to or < 1 ps [12]–[22]. The spectral width of such pulses becomes large enough that several approximations made in the derivation of Eq. (2.3.27) become questionable. The most important limitation turns out to be the neglect of the Raman effect. For pulses with a wide spectrum (> 0.1 THz), the Raman gain can amplify the low-frequency components of a pulse by transferring energy from the high-frequency components of the same pulse. This phenomenon is called intrapulse Raman scattering. As a result of it, the pulse spectrum shifts toward the low-frequency (red) side as the pulse propagates inside the fiber, a phenomenon referred to as the self-frequency shift [12]. The physical origin of this effect is related to the delayed nature of the Raman (vibrational) response [13]. Mathematically, Eq. (2.3.6) cannot be used in the derivation of Eq. (2.3.27); one must use the general form of the nonlinear polarization given in Eq. (2.1.10).

The starting point is again the wave equation (2.3.1). Equation (2.1.10) describes a wide variety of third-order nonlinear effects, and not all of them are relevant to our discussion. For example, nonlinear phenomena such as third-harmonic generation and four-wave mixing are unlikely to occur unless an appropriate phase-matching condition is satisfied (see Chapter 10). Nonresonant, incoherent (intensity-dependent) nonlinear effects can be included by assuming the following functional form for the third-order susceptibility [17]:

$$\chi^{(3)}(t-t_1,t-t_2,t-t_3) = \chi^{(3)}R(t-t_1)\delta(t-t_2)\delta(t-t_3), \qquad (2.3.30)$$

where $R(t)$ is the nonlinear response function normalized in a manner similar to the delta function, i.e., $\int_{-\infty}^{\infty} R(t)dt = 1$. By substituting Eq. (2.3.30) in Eq. (2.1.10) the nonlinear polarization is given by

$$\mathbf{P}_{NL}(\mathbf{r},t) = \varepsilon_0 \chi^{(3)} \mathbf{E}(\mathbf{r},t) \int_{-\infty}^{t} R(t-t_1)|\mathbf{E}(\mathbf{r},t_1)|^2 dt_1, \qquad (2.3.31)$$

where it is assumed that the electric field and the induced polarization vectors point along the same direction. The upper limit of integration in Eq. (2.3.31) extends only up to t because the response function $R(t-t_1)$ must be zero for $t_1 > t$ to ensure causality.

The analysis of Section 2.3.1 can still be used by working in the frequency domain. Using Eqs. (2.3.2)–(2.3.4), $\tilde{E}$ is found to satisfy [18]

$$\nabla^2 \tilde{E} + n^2(\omega)k_0^2 \tilde{E} = -ik_0\alpha + \chi^{(3)}\frac{\omega^2}{c^2}\int\int_{-\infty}^{\infty} \tilde{R}(\omega - \omega_1)$$
$$\times \tilde{E}(\omega_1,z)\tilde{E}(\omega_2,z)\tilde{E}^*(\omega_1 + \omega_2 - \omega,z)\,d\omega_1 d\omega_2, \qquad (2.3.32)$$

where $\tilde{R}(\omega)$ is the Fourier transform of $R(t)$. As before, one can treat the terms on the right-hand side as a small perturbation and first obtain the modal distribution by neglecting them. The effect of perturbation terms is to change the propagation constant for the fundamental mode by $\Delta\beta$ as in Eq. (2.3.19) but with a different expression for $\Delta\beta$. One can then define the slowly varying amplitude $A(z,t)$ as in Eq. (2.3.21) and obtain, after some algebra, the following equation for pulse evolution inside a single-mode fiber [18]:

$$\frac{\partial A}{\partial z} + \frac{\alpha}{2}A + \beta_1\frac{\partial A}{\partial t} + \frac{i\beta_2}{2}\frac{\partial^2 A}{\partial t^2} - \frac{\beta_3}{6}\frac{\partial^3 A}{\partial t^3}$$
$$= i\gamma\left(1 + \frac{i}{\omega_0}\frac{\partial}{\partial t}\right)\left(A(z,t)\int_{-\infty}^{\infty} R(t')|A(z,t-t')|^2 dt'\right), \qquad (2.3.33)$$

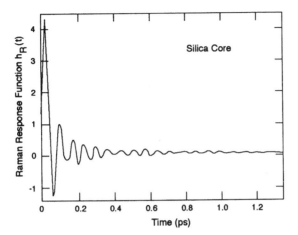

Figure 2.2 Temporal variation of the Raman response function $h_R(t)$ obtained by using the actual Raman-gain spectrum of silica fibers. (After Ref. [16].)

where γ is the nonlinear parameter as defined in Eq. (2.3.28). In general, the effective core area A_{eff} is also a function of ω because the mode distribution $F(x,y)$ is frequency dependent. However, the variation of A_{eff} over the pulse spectrum is typically negligible and can be included in a straightforward manner [19]. The time derivative appearing on the right-hand side of Eq. (2.3.33) results when Eq. (2.3.4) is used in Eq. (2.1.7) and the first-order time derivative of P_{NL} is retained in the analysis used for ultrashort pulses. This term is responsible for self-steepening and shock formation at a pulse edge and has been discussed extensively since 1967 [23]–[37]. This term also includes the nonlinear energy loss resulting from intrapulse Raman scattering. Equation (2.3.33) may be valid even when the slowly varying envelope approximation does not hold and can be used for pulses as short as a few optical cycles if enough higher-order dispersive terms are included [36], [37].

The response function $R(t)$ should include both the electronic and vibrational (Raman) contributions. Assuming that the electronic contribution is nearly instantaneous, the functional form of $R(t)$ can be written as [16]–[21]

$$R(t) = (1 - f_R)\delta(t) + f_R h_R(t), \qquad (2.3.34)$$

where f_R represents the fractional contribution of the delayed Raman response to nonlinear polarization P_{NL}. The Raman response function $h_R(t)$ is responsible for the Raman gain whose spectrum is given by

$$g_R(\Delta\omega) = \frac{\omega_0}{cn_0} f_R \chi^{(3)} \text{Im}[\tilde{h}_R(\Delta\omega)], \tag{2.3.35}$$

where $\Delta\omega = \omega - \omega_0$ and Im stands for the imaginary part. The real part of $\tilde{h}_R(\Delta\omega)$ can be obtained from the imaginary part by using the Kramers–Kronig relations [3]. The Fourier transform of $\tilde{h}_R(\Delta\omega)$ then provides the Raman response function $h_R(t)$. Figure 2.2 shows the temporal variation of $h_R(t)$ obtained by using the experimentally measured spectrum (see Fig. 8.1) of the Raman gain in silica fibers [16].

Attempts have been made to determine an approximate analytic form of the Raman response function. In view of the damped oscillations seen in Fig. 2.2, a useful form is [17]

$$h_R(t) = \frac{\tau_1^2 + \tau_2^2}{\tau_1 \tau_2^2} \exp(-t/\tau_2)\sin(t/\tau_1). \tag{2.3.36}$$

The parameters τ_1 and τ_2 are two adjustable parameters and are chosen to provide a good fit to the actual Raman-gain spectrum. Their appropriate values are $\tau_1 = 12.2$ fs and $\tau_2 = 32$ fs [17]. The fraction f_R can also be estimated from Eq. (2.3.35). By using the known numerical value of peak Raman gain f_R is estimated to be about 0.18 [16]–[18].

Equation (2.3.33) together with the response function $R(t)$ given by Eq. (2.3.34) governs evolution of ultrashort pulses in optical fibers. Its accuracy has been verified by showing that it preserves the number of photons during pulse evolution if fiber loss is ignored by setting $\alpha = 0$ [18]. The pulse energy is not conserved in the presence of intrapulse Raman scattering because a part of the pulse energy is absorbed by silica molecules. Equation (2.3.33) includes this source of nonlinear loss. It is easy to see that it reduces to the simpler equation obtained in Section 2.3.1 [Eq. (2.3.27)] for optical pulses much longer than the time scale of the Raman response function $h_R(t)$ because $R(t)$ for such pulses is replaced by the delta function $\delta(t)$. Noting that $h_R(t)$ becomes nearly zero for $t > 1$ ps (see Fig. 2.2), this replacement is valid for picosecond pulses having widths much greater than 1 ps. As the higher-order dispersion term (involving β_3) and the shock term (involving ω_0) are negligible for such pulses, Eq. (2.3.33) reduces to Eq. (2.3.27).

For pulses shorter than 5 ps but wide enough to contain many optical cycles (widths $\gg 10$ fs), we can simplify Eq. (2.3.33) using a Taylor-series expansion such that

$$|A(z, t - t')|^2 \approx |A(z,t)|^2 - t' \frac{\partial}{\partial t} |A(z,t)|^2. \tag{2.3.37}$$

This approximation is reasonable if the pulse envelope evolves slowly along the fiber. Defining the first moment of the nonlinear response function as

$$T_R \equiv \int_{-\infty}^{\infty} t R(t)\, dt = f_R \int_{-\infty}^{\infty} t\, h_R(t)\, dt = f_R \frac{d(\mathrm{Im}\, \tilde{h}_R)}{d(\Delta\omega)}\bigg|_{\Delta\omega = 0}, \tag{2.3.38}$$

and noting that $\int_0^{\infty} R(t)\, dt = 1$, Eq. (2.3.33) can be approximated by

$$\frac{\partial A}{\partial z} + \frac{\alpha}{2} A + \frac{i\beta_2}{2} \frac{\partial^2 A}{\partial T^2} - \frac{\beta_3}{6} \frac{\partial^3 A}{\partial T^3}$$
$$= i\gamma \left(|A|^2 A + \frac{i}{\omega_0} \frac{\partial}{\partial T} (|A|^2 A) - T_R A \frac{\partial |A|^2}{\partial T} \right), \tag{2.3.39}$$

where a frame of reference moving with the pulse at the group velocity v_g (the so-called retarded frame) is used by making the transformation

$$T = t - z/v_g \equiv t - \beta_1 z. \tag{2.3.40}$$

A second-order term involving the ratio T_R/ω_0 was neglected in arriving at Eq. (2.3.39) because of its smallness.

It is easy to identify the origin of the last three higher-order terms in Eq. (2.3.39). The term proportional to β_3 results from including the cubic term in the expansion of the propagation constant in Eq. (2.3.23). This term governs the effects of third-order dispersion and becomes important for ultrashort pulses because of their wide bandwidth [30]. The term proportional to ω_0^{-1} results from including the first derivative of P_{NL}. It is responsible for self-steepening and shock formation [23]–[37]. The last term proportional to T_R in Eq. (2.3.39) has its origin in the delayed Raman response, and is responsible for the self-frequency shift [12] induced by intrapulse Raman scattering. By using Eqs. (2.3.35) and (2.3.38), T_R can be related to the slope of the Raman gain spectrum [13] that is assumed to vary linearly with frequency in the vicinity of the carrier frequency ω_0. Its numerical value has recently been deduced experimentally [38], resulting in $T_R = 3$ fs at wavelengths ~ 1.55 μm. For pulses shorter than 1 ps, the Raman gain does not vary linearly over the entire pulse bandwidth, and the use of Eq. (2.3.39) becomes questionable for such short pulses. In practice, one may still be able to use it if T_R is treated as a fitting parameter.

For pulses of width $T_0 > 5$ ps, the parameters $(\omega_0 T_0)^{-1}$ and T_R/T_0 become so small (< 0.001) that the last two terms in Eq. (2.3.39) can be neglected. As the contribution of the third-order dispersion term is also quite small for such pulses (as long as the carrier wavelength is not too close to the zero-dispersion wavelength), one can use the simplified equation

$$i\frac{\partial A}{\partial z} + \frac{i\alpha}{2}A - \frac{\beta_2}{2}\frac{\partial^2 A}{\partial T^2} + \gamma|A|^2 A = 0. \tag{2.3.41}$$

This equation can also be obtained from Eq. (2.3.27) by using the transformation given in Eq. (2.3.40). In the special case of $\alpha = 0$, Eq. (2.3.41) is referred to as the NLS equation because it resembles the Schrödinger equation with a nonlinear potential term (variable z playing the role of time). To extend the analogy further, Eq. (2.3.39) is called the generalized (or extended) NLS equation. The NLS equation is a fundamental equation of nonlinear science and has been studied extensively in the context of solitons [39]–[46].

Equation (2.3.41) is the simplest nonlinear equation for studying third-order nonlinear effects in optical fibers. If the peak power associated with an optical pulse becomes so large that one needs to include the fifth and higher-order terms in Eq. (1.3.1), the NLS equation needs to be modified. A simple approach replaces the nonlinear parameter γ in Eq. (2.3.41) by

$$\gamma = \frac{\gamma_0 A}{1 + b_s|A|^2} \approx \gamma_0(1 - b_s|A|^2), \tag{2.3.42}$$

where b_s is the saturation parameter governing the power level at which the nonlinearity begins to saturate. For silica fibers, $b_s|A|^2 \ll 1$ in most practical situations, and one can use Eq. (2.3.41). If the peak intensity approaches 1 GW/cm^2, one can use the approximation $\gamma = \gamma_0(1 - b_s|A|^2)$ in Eq. (2.3.41). The resulting equation is often called the cubic-quintic (or quintic) NLS equation [45] because it contains terms involving both the third and fifth powers of the amplitude A. For the same reason, Eq. (2.3.41) is referred to as the cubic NLS equation. Fibers made by using materials with larger values of n_2 (such as silicate and chalcogenide fibers) are likely to exhibit the saturation effects at a lower peak-power level. Equation (2.3.42) may be more relevant for such fibers. It should also be useful for optical fibers whose core is doped with high-nonlinearity materials such as organic dyes [47] and semiconductors [48].

Equation (2.3.41) appears in optics in several different contexts. For example, the same equation holds for propagation of CW beams in planar waveguides when the variable T is interpreted as the spatial coordinate. The β_2

term in Eq. (2.3.41) then governs beam diffraction in the plane of the waveguide. This analogy between "diffraction in space" and "dispersion in time" is often exploited to advantage since the same equation governs the underlying physics.

2.4   Numerical Methods

The NLS equation [Eq. (2.3.39) or (2.3.41)] is a nonlinear partial differential equation that does not generally lend itself to analytic solutions except for some specific cases in which the inverse scattering method [39] can be employed. A numerical approach is therefore often necessary for an understanding of the nonlinear effects in optical fibers. A large number of numerical methods can be used for this purpose [49]–[64]. These can be classified into two broad categories known as: (i) the finite-difference methods; and (ii) the pseudospectral methods. Generally speaking, pseudospectral methods are faster by up to an order of magnitude to achieve the same accuracy [57]. The one method that has been used extensively to solve the pulse-propagation problem in nonlinear dispersive media is the split-step Fourier method [51], [52]. The relative speed of this method compared with most finite-difference schemes can be attributed in part to the use of the finite-Fourier-transform (FFT) algorithm [65]. This section describes various numerical techniques used to study the pulse-propagation problem in optical fibers with emphasis on the split-step Fourier method and its modifications.

2.4.1   Split-Step Fourier Method

To understand the philosophy behind the split-step Fourier method, it is useful to write Eq. (2.3.39) formally in the form

$$\frac{\partial A}{\partial z} = (\hat{D} + \hat{N})A, \qquad (2.4.1)$$

where $\hat{D}$ is a differential operator that accounts for dispersion and absorption in a linear medium and $\hat{N}$ is a nonlinear operator that governs the effect of fiber nonlinearities on pulse propagation. These operators are given by

$$\hat{D} = -\frac{i\beta_2}{2}\frac{\partial^2}{\partial T^2} + \frac{\beta_3}{6}\frac{\partial^3}{\partial T^3} - \frac{\alpha}{2}, \qquad (2.4.2)$$

$$\hat{N} = i\gamma \left(|A|^2 + \frac{i}{\omega_0} \frac{1}{A} \frac{\partial}{\partial T}(|A|^2 A) - T_R \frac{\partial |A|^2}{\partial T} \right). \tag{2.4.3}$$

In general, dispersion and nonlinearity act together along the length of the fiber. The split-step Fourier method obtains an approximate solution by assuming that in propagating the optical field over a small distance h, the dispersive and nonlinear effects can be pretended to act independently. More specifically, propagation from z to $z+h$ is carried out in two steps. In the first step, the nonlinearity acts alone, and $\hat{D} = 0$ in Eq. (2.4.1). In the second step, dispersion acts alone, and $\hat{N} = 0$ in Eq. (2.4.1). Mathematically,

$$A(z+h,T) \approx \exp(h\hat{D})\exp(h\hat{N})A(z,T). \tag{2.4.4}$$

The exponential operator $\exp(h\hat{D})$ can be evaluated in the Fourier domain using the prescription

$$\exp(h\hat{D})B(z,T) = F_T^{-1}\exp[h\hat{D}(i\omega)]F_T B(z,T), \tag{2.4.5}$$

where F_T denotes the Fourier-transform operation, $\hat{D}(i\omega)$ is obtained from Eq. (2.4.2) by replacing the differential operator $\partial/\partial T$ by $i\omega$, and ω is the frequency in the Fourier domain. As $\hat{D}(i\omega)$ is just a number in the Fourier space, the evaluation of Eq. (2.4.5) is straightforward. The use of the FFT algorithm [65] makes numerical evaluation of Eq. (2.4.5) relatively fast. It is for this reason that the split-step Fourier method can be faster by up to two orders of magnitude compared with most finite-difference schemes [57].

To estimate the accuracy of the split-step Fourier method, we note that a formally exact solution of Eq. (2.4.1) is given by

$$A(z+h,T) = \exp[h(\hat{D}+\hat{N})]A(z,T), \tag{2.4.6}$$

if $\hat{N}$ is assumed to be z independent. At this point, it is useful to recall the Baker–Hausdorff formula [66] for two noncommuting operators $\hat{a}$ and $\hat{b}$,

$$\exp(\hat{a})\exp(\hat{b}) = \exp\left(\hat{a}+\hat{b}+\frac{1}{2}[\hat{a},\hat{b}]+\frac{1}{12}[\hat{a}-\hat{b},[\hat{a},\hat{b}]]+\cdots\right), \tag{2.4.7}$$

where $[\hat{a},\hat{b}] = \hat{a}\hat{b} - \hat{b}\hat{a}$. A comparison of Eqs. (2.4.4) and (2.4.6) shows that the split-step Fourier method ignores the noncommutating nature of the operators $\hat{D}$ and $\hat{N}$. By using Eq. (2.4.7) with $\hat{a} = h\hat{D}$ and $\hat{b} = h\hat{N}$, the dominant error term is found to result from the single commutator $\frac{1}{2}h^2[\hat{D},\hat{N}]$. Thus, the split-step Fourier method is accurate to second order in the step size h.

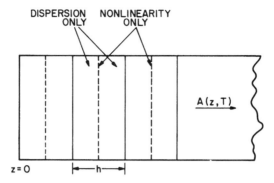

Figure 2.3 Schematic illustration of the symmetrized split-step Fourier method used for numerical simulations. Fiber length is divided into a large number of segments of width h. Within a segment, the effect of nonlinearity is included at the midplane shown by a dashed line.

The accuracy of the split-step Fourier method can be improved by adopting a different procedure to propagate the optical pulse over one segment from z to $z+h$. In this procedure Eq. (2.4.4) is replaced by

$$A(z+h,T) \approx \exp\left(\frac{h}{2}\hat{D}\right) \exp\left(\int_z^{z+h} \hat{N}(z')\,dz'\right) \exp\left(\frac{h}{2}\hat{D}\right) A(z,T). \quad (2.4.8)$$

The main difference is that the effect of nonlinearity is included in the middle of the segment rather than at the segment boundary. Because of the symmetric form of the exponential operators in Eq. (2.4.8), this scheme is known as the symmetrized split-step Fourier method [67]. The integral in the middle exponential is useful to include the z dependence of the nonlinear operator $\hat{N}$. If the step size h is small enough, it can be approximated by $\exp(h\hat{N})$, similar to Eq. (2.4.4). The most important advantage of using the symmetrized form of Eq. (2.4.8) is that the leading error term results from the double commutator in Eq. (2.4.7) and is of third order in the step size h. This can be verified by applying Eq. (2.4.7) twice in Eq. (2.4.8).

The accuracy of the split-step Fourier method can be further improved by evaluating the integral in Eq. (2.4.8) more accurately than approximating it by $h\hat{N}(z)$. A simple approach is to employ the trapezoidal rule and approximate the integral by [68]

$$\int_z^{z+h} \hat{N}(z')\,dz' \approx \frac{h}{2}[\hat{N}(z) + \hat{N}(z+h)]. \quad (2.4.9)$$

However, the implementation of Eq. (2.4.9) is not simple because $\hat{N}(z+h)$ is unknown at the midsegment located at $z+h/2$. It is necessary to follow an iterative procedure that is initiated by replacing $\hat{N}(z+h)$ by $\hat{N}(z)$. Equation (2.4.8) is then used to estimate $A(z+h,T)$ which in turn is used to calculate the new value of $\hat{N}(z+h)$. Although the iteration procedure is time-consuming, it can still reduce the overall computing time if the step size h can be increased because of the improved accuracy of the numerical algorithm. Two iterations are generally enough in practice.

The implementation of the split-step Fourier method is relatively straight-forward. As shown in Fig. 2.3, the fiber length is divided into a large number of segments that need not be spaced equally. The optical pulse is propagated from segment to segment using the prescription of Eq. (2.4.8). More specifically, the optical field $A(z,T)$ is first propagated for a distance $h/2$ with dispersion only using the FFT algorithm and Eq. (2.4.5). At the midplane $z+h/2$, the field is multiplied by a nonlinear term that represents the effect of nonlinearity over the whole segment length h. Finally, the field is propagated the remaining distance $h/2$ with dispersion only to obtain $A(z+h,T)$. In effect, the nonlinearity is assumed to be lumped at the midplane of each segment (dashed lines in Fig. 2.3).

The split-step Fourier method has been applied to a wide variety of optical problems including wave propagation in atmosphere [67], [68], graded-index fibers [69], semiconductor lasers [70], unstable resonators [71], and waveguide couplers [72], [73]. It is referred to as the beam-propagation method when applied to the propagation of CW optical beams in nonlinear media where dispersion is replaced by diffraction [69]–[73].

For the specific case of pulse propagation in optical fibers, the split-step Fourier method was first applied in 1973 [40]. Its use has become widespread since then [74]–[91] because of its fast execution compared with most finite-difference schemes [51]. Although the method is relatively straightforward to implement, it requires that step sizes in z and T be selected carefully to maintain the required accuracy. In particular, it is necessary to monitor the accuracy by calculating the conserved quantities such as the pulse energy (in the absence of absorption) along the fiber length. The optimum choice of step sizes depends on the complexity of the problem. Although a few guidelines are available [92]–[97], it may sometimes be necessary to repeat the calculation by reducing the step size to ensure the accuracy of numerical simulations. The time window should be wide enough to ensure that the pulse energy remains confined within the window. Typically, window size is 10–20 times the pulse width. In some problems, a part of the pulse energy may spread so rapidly that it may be difficult to prevent it from hitting the window boundary. This can lead to numerical instabilities as the energy reaching one edge of the window automatically reenters from the other edge (the use of the FFT algorithm implies periodic boundary conditions). It is common to use an "absorbing window" in which the radiation reaching window edges is artificially absorbed even though such an implementation does not preserve the pulse energy. In general, the split-step Fourier method is a powerful tool provided care is taken to ensure that it is used properly.

2.4.2 Finite-Difference Methods

Although the split-step Fourier method is commonly used for analyzing nonlinear effects in optical fibers, its use becomes quite time-consuming when the NLS equation is solved for simulating the performance of wavelength-division-multiplexed (WDM) lightwave systems. In such systems, the temporal resolution should be a small fraction of the entire bandwidth of the WDM signal. For a 100-channel system, the bandwidth approaches 10 THz, requiring a temporal resolution of ~ 10 fs. At the same time, the temporal window should be typically 1–10-ns wide, resulting in more than 10^5 mesh points in time domain. Even though each FFT operation is relatively fast, a large number of FFT operations on a large-size array leads to an overall computation time measured in hours (even days) on some of the fastest computers available in 1999. For this reason, there has been renewed interest in finite-difference methods in recent years.

Several different finite-difference schemes have been used to solve the NLS equations [57], [64]; some of the common ones are the Crank–Nicholson scheme and its variants, the hopscotch scheme and its variants, and the leapfrog method. A careful comparison of several finite-difference schemes with the split-step Fourier method shows that the latter is efficient only when the field amplitude varies slowly with time [64]. However, it is difficult to recommend a specific finite-difference scheme because the speed and accuracy depend to some extent on the number and form of the nonlinear terms included in the generalized NLS equation. A linearized Crank–Nicolson scheme can be faster by more than a factor of five under certain conditions.

Another situation in which finite-difference schemes are useful corresponds to propagation of ultrashort optical pulses whose width is so short that the pulse contains only a few optical cycles. The slowly varying envelope approximation does not always hold for such short pulses. In recent years attempts have been made to relax this approximation, and many new numerical techniques have been proposed [96]–[104]. Some of these techniques require the use of a finite-difference method in place of the split-step Fourier method. Finite-difference techniques for solving the paraxial wave equation have developed in parallel with the split-step Fourier method and are sometimes the method of choice. They can be extended beyond the validity of the paraxial approximation by using techniques such as the Lanczos orthogonalization [99] and the Padé approximation [104]. Other extensions include algorithms that can handle bidirectional beam propagation [103]. Most of these techniques have been developed in the context of beam propagation in planar waveguides, but they can be readily adopted for pulse propagation in optical fibers.

There are several limitations inherent in the use of the NLS equation for pulse propagation in optical fibers. The slowly varying envelope approximation has already been mentioned. Another one is related to the fact that backward propagating waves are totally ignored. If the fiber has a built-in index grating, a part of the pulse energy will be reflected backward because of Bragg diffraction. Such problems require simultaneous consideration of forward and backward propagating waves. The other major limitation is related to the neglect of the vector nature of the electromagnetic fields. In essence, polarization effects are completely ignored. As was seen in Section 1.2.4, optical fibers exhibit birefringence. The inclusion of the birefringence effects requires consideration of all components of electric and magnetic field vectors.

In the case of a linear medium, the algorithms that solve Maxwell's equations [Eqs. (2.1.1)–(2.1.4)] directly in the time domain by using finite-difference methods have been developed for many years [105]–[109]. Such algorithms have now been extended to the case of nonlinear media [110]–[115]. The delayed nature of nonlinear response was incorporated by using Eqs. (2.3.31) and (2.3.34) together with the functional form of the Raman response function given in Eq. (2.3.36). Conceptually, the main difference between the finite-difference time-domain (FDTD) method and the split-step Fourier method is that the former deals with all electromagnetic components without eliminating the carrier frequency ω_0 in contrast with what was done in Section 2.3 in deriving the NLS equation.

The FDTD method is certainly more accurate because it solves Maxwell's equations directly with a minimum number of approximations. However, improvement in accuracy is achieved only at the expense of a vast increase in the computational effort. This can be understood by noting that the time step needed to resolve the optical carrier is by necessity a fraction of the optical period and should often be < 1 fs. The step size along the fiber length is also required to be a fraction of the optical wavelength. It may be necessary to use this method for ultrashort pulses ($T_0 < 10$ fs) whose width is comparable to the optical period. In most applications of nonlinear fiber optics, pulses are much wider than the optical period, and Eq. (2.3.33) and its approximate forms such as Eq. (2.3.41) provide a reasonably accurate solution of the underlying Maxwell's equations.

Problems

2.1 Use Maxwell's equations to express the field components E_ρ, E_ϕ, H_ρ, and H_ϕ inside the fiber core in terms of E_z and H_z. Neglect the nonlinear part of the polarization in Eq. (2.1.8) for simplicity.

2.2 Derive eigenvalue equation (2.2.10) by matching the boundary conditions at the core-cladding interface of a step-index fiber. Consult References [5]–[7] if necessary.

2.3 Use the eigenvalue equation (2.2.10) to derive the single-mode condition in optical fibers.

2.4 A single-mode fiber has an index step of 0.005. Calculate the core radius if the fiber has a cut-off wavelength of 1 μm. Assume a core index of 1.45.

2.5 Derive an expression for the confinement factor Γ of single-mode fibers defined as the fraction of the total mode power contained inside the fiber core. Use the Gaussian approximation made in Eq. (2.2.16) for the fundamental fiber mode.

2.6 Estimate the full width at half maximum (FWHM) of the spot size associated with the fiber mode and the fraction of the mode power inside the core when the fiber of Problem 2.4 is used to transmit 1.3-μm light.

2.7 Derive Eq. (2.3.7) from Eq. (2.3.6). Explain the origin of the factor $\frac{3}{4}$ in the definition of ε_{NL}. Verify that Eq. (2.3.13) for n_2 follows from it.

2.8 Solve Eq. (2.3.15) by using perturbation theory to obtain the first-order correction for the propagation constant when ε_{NL} is small. Show that this correction is given by Eq. (2.3.20).

2.9 Show that Eq. (2.3.27) can be obtained by taking the Fourier transform indicated in Eq. (2.3.25) together with Eq. (2.3.22). Fill in all the missing steps.

2.10 Calculate the effective core area when the fiber of Problem 2.4 is used to transmit 1.3-μm light.

2.11 Take the Fourier transform of the Raman response function given by Eq. (2.3.36) and plot the real and imaginary parts as a function of frequency. What is the physical meaning of the resulting curves?

2.12 The Raman-gain spectrum of a fiber is approximated by a Lorentzian profile whose FWHM is 5 THz. The gain peak is located at 15 THz from the carrier frequency of the pulse. Derive an expression for the Raman response function of this fiber.

References

[1] P. Diament, *Wave Transmission and Fiber Optics* (Macmillan, New York, 1990), Chap. 3.

[2] Y. R. Shen, *Principles of Nonlinear Optics* (Wiley, New York, 1984), Chap. 1.

[3] M. Schubert and B. Wilhelmi, *Nonlinear Optics and Quantum Electronics* (Wiley, New York, 1986), Chap. 1.

[4] P. N. Butcher and D. N. Cotter, *The Elements of Nonlinear Optics* (Cambridge University Press, Cambridge, UK, 1990), Chap. 2.

[5] D. Marcuse, *Theory of Dielectric Optical Waveguides* (Academic Press, San Diego, CA, 1991), Chap. 2.

[6] A. W. Snyder and J. D. Love, *Optical Waveguide Theory* (Chapman and Hall, London, 1983), Chaps. 12–15.

[7] J. A. Buck, *Fundamentals of Optical Fibers* (Wiley, New York, 1995), Chap. 3.

[8] D. Marcuse, *J. Opt. Soc. Am.* **68**, 103 (1978).

[9] G. P. Agrawal, in *Supercontinuum Laser Source*, R. R. Alfano, ed. (Springer-Verlag, Heidelberg, 1989), Chap. 3.

[10] H. A. Haus, *Waves and Fields in Optoelectronics* (Prentice-Hall, Englewood Cliffs, 1984), Chap. 10.

[11] P. M. Morse and H. Feshbach, *Methods of Theoretical Physics* (McGraw-Hill, New York, 1953), Chap. 9.

[12] F. M. Mitschke and L. F. Mollenauer, *Opt. Lett.* **11**, 659 (1986).

[13] J. P. Gordon, *Opt. Lett.* **11**, 662 (1986).

[14] Y. Kodama and A. Hasegawa, *IEEE J. Quantum Electron.* **23**, 510 (1987).

[15] E. A. Golovchenko, E. M. Dianov, A. N. Pilipetskii, A. M. Prokhorov, and V. N. Serkin, *Sov. Phys. JETP. Lett.* **45**, 91 (1987).

[16] R. H. Stolen, J. P. Gordon, W. J. Tomlinson, and H. A. Haus, *J. Opt. Soc. Am. B* **6**, 1159 (1989).

[17] K. J. Blow and D. Wood, *IEEE J. Quantum Electron.* **25**, 2665 (1989).

[18] P. V. Mamyshev and S. V. Chernikov, *Opt. Lett.* **15**, 1076 (1990).

[19] S. V. Chernikov and P. V. Mamyshev, *J. Opt. Soc. Am. B* **8**, 1633 (1991).

[20] P. V. Mamyshev and S. V. Chernikov, *Sov. Lightwave Commun.* **2**, 97 (1992).

[21] R. H. Stolen and W. J. Tomlinson, *J. Opt. Soc. Am. B* **9**, 565 (1992).

[22] S. Blair and K. Wagner, *Opt. Quantum Electron.* **30**, 697 (1998).

[23] F. DeMartini, C. H. Townes, T. K. Gustafson, and P. L. Kelley, *Phys. Rev.* **164**, 312 (1967).

[24] D. Grischkowsky, E. Courtens, and J. A. Armstrong, *Phys. Rev. Lett.* **31**, 422 (1973).

[25] N. Tzoar and M. Jain, *Phys. Rev. A* **23**, 1266 (1981).

[26] D. Anderson and M. Lisak, *Phys. Rev. A* **27**, 1393 (1983).

[27] G. Yang and Y. R. Shen, *Opt. Lett.* **9**, 510 (1984).

[28] J. T. Manassah, M. A. Mustafa, R. A. Alfano, and P. P. Ho, *IEEE J. Quantum Electron.* **22**, 197 (1986).

[29] E. A. Golovchenko, E. M. Dianov, A. M. Prokhorov, and V. N. Serkin, *Sov. Phys. Dokl.* **31**, 494 (1986).

[30] E. Bourkoff, W. Zhao, R. I. Joseph, and D. N. Christodoulides, *Opt. Lett.* **12**, 272 (1987).

[31] K. Ohkuma, Y. H. Ichikawa, and Y. Abe, *Opt. Lett.* **12**, 516 (1987).

[32] Y. Kodama and K. Nozaki, *Opt. Lett.* **12** , 1038 (1987).

[33] P. Beaud, W. Hodel, B. Zysset, and H. P. Weber, *IEEE J. Quantum Electron.* **23**, 1938 (1987).

[34] A. B. Grudinin, E. M. Dianov, D. V. Korobkin, A. M. Prokhorov, V. N. Serkin, and D. V. Khaidarov, *Sov. Phys. JETP. Lett.* **46** , 221 (1987).

[35] K. Tai, A. Hasegawa, and N. Bekki, *Opt. Lett.* **13**, 392 (1988).

[36] S. Chi and S. Wang, *Opt. Quantum Electron.* **28**, 1351 (1996).

[37] T. Brabec and F. Krausz, *Phys. Rev. Lett.* **78**, 3282 (1997).

[38] A. K. Atieh, P. Myslinski, J. Chrostowski, and P. Galko, *J. Lightwave Technol.* **17**, 216 (1999).

[39] V. E. Zakharov and A. B. Shabat, *Sov. Phys. JETP* **34**, 62 (1972).

[40] A. Hasegawa and F. Tappert, *Appl. Phys. Lett.* **23**, 142 (1973).

[41] R. K. Dodd, J. C. Eilbeck, J. D. Gibbon, and H. C. Morris, *Solitons and Nonlinear Wave Equations* (Academic Press, New York, 1984).

[42] M. J. Ablowitz, *Nonlinear Evolution Equations and Inverse Scattering* Cambridge University Press, New York, 1991).

[43] G. L. Lamb, Jr., *Elements of Soliton Theory* (Dover, New York, 1994).

[44] C. H. Gu (ed.), *Soliton Theory and its Applications* (Springer-Verlag, New York, 1995).

[45] N. N. Akhmediev and A. Ankiewicz, *Solitons* (Chapman and Hall, New York, 1997).

[46] C. Sulem, *Nonlinear Schrödinger Equation; Self-Focusing and Wave Collapse* (Springer-Verlag, New York, 1999).

[47] G. D. Peng, Z. Xiong, and P. L. Chu, *Opt. Fiber Technol.* **5**, 242 (1999).

[48] B. J. Inslie, H. P. Girdlestone, D. Cotter, *Electron. Lett.* **23**, 405 (1987).

[49] V. I. Karpman and E. M. Krushkal, *Sov. Phys. JETP* **28**, 277 (1969).

[50] N. Yajima and A. Outi, *Prog. Theor. Phys.* **45**, 1997 (1971).

[51] R. H. Hardin and F. D. Tappert, *SIAM Rev. Chronicle* **15**, 423 (1973).

[52] R. A. Fisher and W. K. Bischel, *Appl. Phys. Lett.* **23**, 661 (1973); *J. Appl. Phys.* **46**, 4921 (1975).

[53] M. J. Ablowitz and J. F. Ladik, *Stud. Appl. Math.* **55**, 213 (1976).

[54] I. S. Greig and J. L. Morris, *J. Comput. Phys.* **20**, 60 (1976).

[55] B. Fornberg and G. B. Whitham, *Phil. Trans. Roy. Soc.* **289**, 373 (1978).

[56] M. Delfour, M. Fortin, and G. Payre, *J. Comput. Phys.* **44**, 277 (1981).

[57] T. R. Taha and M. J. Ablowitz, *J. Comput. Phys.* **55**, 203 (1984).

[58] D. Pathria and J. L. Morris, *J. Comput. Phys.* **87**, 108 (1990).

[59] L. R. Watkins and Y. R. Zhou, *J. Lightwave Technol.* **12**, 1536 (1994).

[60] M. S. Ismail, *Int. J. Comput. Math.* **62**, 101 (1996).

[61] K. V. Peddanarappagari and M. Brandt-Pearce, *J. Lightwave Technol.* **15**, 2232 (1997); *J. Lightwave Technol.* **16**, 2046 (1998).

[62] E. H. Twizell, A. G. Bratsos, and J. C. Newby, *Math. Comput. Simul.* **43**, 67 (1997).

[63] W. P. Zeng, *J. Comput. Math.* **17**, 133 (1999).

[64] Q. Chang, E. Jia, and W. Sun, *J. Comput. Phys.* **148**, 397 (1999).

[65] J. W. Cooley and J. W. Tukey, *Math. Comput.* **19**, 297 (1965).

[66] G. H. Weiss and A. A. Maradudin, *J. Math. Phys.* **3**, 771 (1962).

[67] J. A. Fleck, J. R. Morris, and M. D. Feit, *Appl. Phys.* **10**, 129 (1976).

[68] M. Lax, J. H. Batteh, and G. P. Agrawal, *J. Appl. Phys.* **52**, 109 (1981).

[69] M. D. Feit and J. A. Fleck, *Appl. Opt.* **17**, 3990 (1978); *Appl. Opt.* **18**, 2843 (1979).

[70] G. P. Agrawal, *J. Appl. Phys.* **56**, 3100 (1984); *J. Lightwave Technol.* **2**, 537 (1984).

[71] M. Lax, G. P. Agrawal, M. Belic,B. J. Coffey, and W. H. Louisell, *J. Opt. Soc. Am. A* **2**, 732 (1985).

[72] B. Hermansson, D. Yevick, and P. Danielsen, *IEEE J. Quantum Electron.* **19**, 1246 (1983).

[73] L. Thylen, E. M. Wright, G. I. Stegeman, C. T. Seaton, and J. V. Moloney, *Opt. Lett.* **11**, 739 (1986).

[74] D. Yevick and B. Hermansson, *Opt. Commun.* **47**, 101 (1983).

[75] V. A. Vysloukh, *Sov. J. Quantum Electron.* **13**, 1113 (1983).

[76] K. J. Blow, N. J. Doran, and E. Cummins, *Opt. Commun.* **48**, 181 (1983).

[77] A. Hasegawa, *Opt. Lett.* **9**, 288 (1984).

[78] H. E. Lassen, F. Mengel, B. Tromberg, N. C. Albertson, and P. L. Christiansen, *Opt. Lett.* **10**, 34 (1985).

[79] G. P. Agrawal and M. J. Potasek, *Phys. Rev. A* **33**, 1765 (1986).

[80] M. J. Potasek, G. P. Agrawal, and S. C. Pinault, *J. Opt. Soc. Am. B* **3**, 205 (1986).

[81] P. K. A. Wai, C. R. Menyuk, Y. C. Lee, and H. H. Chen, *Opt. Lett.* **11**, 464 (1986); C. R. Menyuk, *Opt. Lett.* **12**, 614 (1987).

[82] E. M. Dianov, A. M. Prokhorov, and V. N. Serkin, *Opt. Lett.* **11**, 168 (1986).

[83] V. A. Vysloukh and T. A. Matveeva, *Sov. J. Quantum Electron.* **17**, 498 (1987).

[84] D. Krökel, N. J. Halas, G. Giuliani, and D. Grischkowsky, *Phys. Rev. Lett.* **60**, 29 (1988).

[85] G. P. Agrawal, *Phys. Rev. A* **44**, 7493 (1991).

[86] V. V. Afanasjev, V. N. Serkin, and V. A. Vysloukh, *Sov. Lightwave Commun.* **2**, 35 (1992).

[87] M. Matsumoto and A. Hasegawa, *Opt. Lett.* **18**, 897 (1993).

[88] K. C. Chan and H. F. Liu, *Opt. Lett.* **18** , 1150 (1993).

[89] H. Ghafourishiraz, P. Shum, and M. Nagata, *IEEE J. Quantum Electron.* **31**, 190 (1995).

[90] M. Margalit and M. Orenstein, *Opt. Commun.* **124**, 475 (1996).

[91] C. Francia, *IEEE Photon. Technol. Lett.* **11**, 69 (1999).

[92] J. Van Roey, J. van der Donk, and P. E. Lagasse, *J. Opt. Soc. Am.* **71**, 803 (1981).

[93] L. Thylen, *Opt. Quantum Electron.* **15**, 433 (1983).

[94] J. Saijonmaa and D. Yevick, *J. Opt. Soc. Am.* **73**, 1785 (1983).

[95] D. Yevick and B. Hermansson, *J. Appl. Phys.* **59**, 1769 (1986); *IEEE J. Quantum Electron.* **25**, 221 (1989).

[96] M. D. Feit and J. A. Fleck, *J. Opt. Soc. Am. B* **5**, 633 (1988).

[97] D. Yevick and M. Glassner, *Opt. Lett.* **15**, 174 (1990).

[98] D. Yevick and B. Hermansson, *IEEE J. Quantum Electron.* **26**, 109 (1990).

[99] R. Ratowsky and J. A. Fleck, *Opt. Lett.* **16**, 787 (1991).

[100] J. Gerdes and R. Pregla, *J. Opt. Soc. Am. B* **8**, 389 (1991).

[101] G. R. Hadley, *Opt. Lett.* **16**, 624 (1991); *IEEE J. Quantum Electron.* **28**, 363 (1992).

[102] G. R. Hadley, *Opt. Lett.* **17**, 1426 (1992); *Opt. Lett.* **17**, 1743 (1992).

[103] P. Kaczmarski and P. E. Lagasse, *Electron. Lett.* **24**, 675 (1988).

[104] Y. Chung and N. Dagli, *IEEE J. Quantum Electron.* **26**, 1335 (1990); *IEEE J. Quantum Electron.* **27**, 2296 (1991).

[105] K. S. Yee, *IEEE Trans. Antennas Propag.* **14**, 302 (1966).

[106] A. Taflove, *Wave Motion* **10**, 547 (1988).

[107] C. F. Lee, R. T. Shin, and J. A. Kong, in *Progress in Electromagnetic Research*, J. A. Kong, ed. (Elsevier, New York, 1991), Chap. 11.

[108] S. Maeda, T. Kashiwa, and I. Fukai, *IEEE Trans. Microwave Theory Tech.* **39**, 2154 (1991).

[109] R. M. Joseph, S. C. Hagness, and A. Taflove, *Opt. Lett.* **16**, 1412 (1991).

[110] P. M. Goorjian and A. Taflove, *Opt. Lett.* **16**, 180 (1992).

[111] P. M. Goorjian, A. Taflove, R. M. Joseph, and S. C. Hagness, *IEEE J. Quantum Electron.* **28**, 2416 (1992).

[112] R. M. Joseph, P. M. Goorjian, and A. Taflove, *Opt. Lett.* **17**, 491 (1993).

[113] R. W. Ziolkowski and J. B. Judkins, *J. Opt. Soc. Am. B* **10**, 186 (1993).

[114] M. Zoboli, F. Di Pasquale, and S. Selleri, *Opt. Commun.* **97**, 11 (1993).

[115] P. M. Goorjian and Y. Silberberg, *J. Opt. Soc. Am. B* **14**, 3523 (1997).

Chapter 3

Group-Velocity Dispersion

The preceding chapter discussed how the combined effects of group-velocity dispersion (GVD) and self-phase modulation (SPM) on optical pulses propagating inside a fiber can be studied by solving a pulse-propagation equation. Before considering the general case, it is instructive to study the effects of GVD alone. This chapter considers the pulse-propagation problem by treating fibers as a linear optical medium. In Section 3.1 we discuss the conditions under which the GVD effects dominate over the nonlinear effects by introducing two length scales associated with GVD and SPM. Dispersion-induced broadening of optical pulses is considered in Section 3.2 for several specific pulse shapes, including Gaussian and 'sech' pulses. The effects of initial frequency chirping are also discussed in this section. Section 3.3 is devoted to the effects of third-order dispersion on pulse broadening. An analytic theory capable of predicting dispersive broadening for pulses of arbitrary shapes is also given in this section. Section 3.4 discusses how the GVD limits the performance of optical communication systems and how the technique of dispersion management can be used to combat them in practice.

3.1 Different Propagation Regimes

In Section 2.3 we obtained the nonlinear Schrödinger (NLS) equation that governs propagation of optical pulses inside single-mode fibers. For pulse widths >5 ps, one can use Eq. (2.3.41) given by

$$i\frac{\partial A}{\partial z} = -\frac{i\alpha}{2}A + \frac{\beta_2}{2}\frac{\partial^2 A}{\partial T^2} - \gamma|A|^2 A, \tag{3.1.1}$$

where A is the slowly varying amplitude of the pulse envelope and T is measured in a frame of reference moving with the pulse at the group velocity v_g ($T = t - z/v_g$). The three terms on the right-hand side of Eq. (3.1.1) govern, respectively, the effects of fiber losses, dispersion, and nonlinearity on pulses propagating inside optical fibers. Depending on the initial width T_0 and the peak power P_0 of the incident pulse, either dispersive or nonlinear effects may dominate along the fiber. It is useful to introduce two length scales, known as the dispersion length L_D and the nonlinear length L_{NL} [1]–[3]. Depending on the relative magnitudes of L_D, L_{NL}, and the fiber length L, pulses can evolve quite differently.

Let us introduce a time scale normalized to the input pulse width T_0 as

$$\tau = \frac{T}{T_0} = \frac{t - z/v_g}{T_0}. \tag{3.1.2}$$

At the same time, we introduce a normalized amplitude U as

$$A(z, \tau) = \sqrt{P_0} \exp(-\alpha z/2) U(z, \tau), \tag{3.1.3}$$

where P_0 is the peak power of the incident pulse. The exponential factor in Eq. (3.1.3) accounts for fiber losses. By using Eqs. (3.1.1)–(3.1.3), $U(z, \tau)$ is found to satisfy

$$i\frac{\partial U}{\partial z} = \frac{\text{sgn}(\beta_2)}{2L_D} \frac{\partial^2 U}{\partial \tau^2} - \frac{\exp(-\alpha z)}{L_{NL}} |U|^2 U, \tag{3.1.4}$$

where $\text{sgn}(\beta_2) = \pm 1$ depending on the sign of the GVD parameter β_2 and

$$L_D = \frac{T_0^2}{|\beta_2|}, \qquad L_{NL} = \frac{1}{\gamma P_0}. \tag{3.1.5}$$

The dispersion length L_D and the nonlinear length L_{NL} provide the length scales over which dispersive or nonlinear effects become important for pulse evolution. Depending on the relative magnitudes of L, L_D, and L_{NL}, the propagation behavior can be classified in the following four categories.

When fiber length L is such that $L \ll L_{NL}$ and $L \ll L_D$, neither dispersive nor nonlinear effects play a significant role during pulse propagation. This can be seen by noting that both terms on the right-hand side of Eq. (3.1.4) can be neglected in that case. (It is assumed that the pulse has a smooth temporal profile so that $\partial^2 U/\partial \tau^2 \sim 1$.) As a result, $U(z, \tau) = U(0, \tau)$, i.e., the pulse maintains its shape during propagation. The fiber plays a passive role in this regime and acts as a mere transporter of optical pulses (except for reducing the pulse energy because of fiber losses). This regime is useful for optical communication systems. For $L \sim 50$ km, L_D and L_{NL} should be larger than 500 km for distortion-free transmission. One can estimate T_0 and P_0 from Eq. (3.1.5) for given values of the fiber parameters β_2 and γ. At $\lambda = 1.55$ μm, $|\beta_2| \approx 20$ ps^2/km, and $\gamma \approx 3$ W^{-1}km^{-1} for standard telecommunication fibers. The use of these values in Eq. (3.1.5) shows that the dispersive and nonlinear effects are negligible for $L < 50$ km if $T_0 > 100$ ps and $P_0 \sim 1$ mW. However, L_D and L_{NL} become smaller as pulses become shorter and more intense. For example, L_D and L_{NL} are ~ 100 m for $T_0 \sim 1$ ps and $P_0 \sim 1$ W. For such optical pulses, both the dispersive and nonlinear effects need to be included if fiber length exceeds a few meters.

When the fiber length is such that $L \ll L_{NL}$ but $L \sim L_D$, the last term in Eq. (3.1.4) is negligible compared to the other two. The pulse evolution is then governed by GVD, and the nonlinear effects play a relatively minor role. The effect of GVD on propagation of optical pulses is discussed in this chapter. The dispersion-dominant regime is applicable whenever the fiber and pulse parameters are such that

$$\frac{L_D}{L_{NL}} = \frac{\gamma P_0 T_0^2}{|\beta_2|} \ll 1. \tag{3.1.6}$$

As a rough estimate, $P_0 \ll 1$ W for 1-ps pulses if we use typical values for the fiber parameters γ and $|\beta_2|$ at $\lambda = 1.55$ μm.

When the fiber length L is such that $L \ll L_D$ but $L \sim L_{NL}$, the dispersion term in Eq. (3.1.4) is negligible compared to the nonlinear term (as long as the pulse has a smooth temporal profile such that $\partial^2 U/\partial \tau^2 \sim 1$). In that case, pulse evolution in the fiber is governed by SPM that leads to spectral broadening of the pulse. This phenomenon is considered in Chapter 4. The nonlinearity-dominant regime is applicable whenever

$$\frac{L_D}{L_{NL}} = \frac{\gamma P_0 T_0^2}{|\beta_2|} \gg 1. \tag{3.1.7}$$

This condition is readily satisfied for relatively wide pulses ($T_0 > 100$ ps) with a peak power $P_0 \sim 1$ W. Note that SPM can lead to pulse shaping in the presence of weak GVD effects. If the pulse develops a sharp leading or trailing edge, the dispersion term may become important even when Eq. (3.1.7) is initially satisfied.

When the fiber length L is longer or comparable to both L_D and L_{NL}, dispersion and nonlinearity act together as the pulse propagates along the fiber. The interplay of the GVD and SPM effects can lead to a qualitatively different behavior compared with that expected from GVD or SPM alone. In the anomalous-dispersion regime ($\beta_2 < 0$), the fiber can support solitons. In the normal-dispersion regime ($\beta_2 > 0$), the GVD and SPM effects can be used for pulse compression. Equation (3.1.4) is extremely helpful in understanding pulse evolution in optical fibers when both dispersive and nonlinear effects should be taken into account. However, this chapter is devoted to the linear regime, and the following discussion is applicable to pulses whose parameters satisfy Eq. (3.1.6).

3.2   Dispersion-Induced Pulse Broadening

The effect of GVD on optical pulses propagating in a linear dispersive medium [4]–[17] are studied by setting $\gamma = 0$ in Eq. (3.1.1). If we define the normalized amplitude $U(z, T)$ according to Eq. (3.1.3), $U(z, T)$ satisfies the following linear partial differential equation:

$$i\frac{\partial U}{\partial z} = \frac{\beta_2}{2}\frac{\partial^2 U}{\partial T^2}. \tag{3.2.1}$$

This equation is similar to the paraxial wave equation that governs diffraction of CW light and becomes identical to it when diffraction occurs in only one transverse direction and β_2 is replaced by $-\lambda/(2\pi)$, where λ is the wavelength of light. For this reason, the dispersion-induced temporal effects have a close analogy with the diffraction-induced spatial effects [2].

Equation (3.2.1) is readily solved by using the Fourier-transform method. If $\tilde{U}(z, \omega)$ is the Fourier transform of $U(z, T)$ such that

$$U(z, T) = \frac{1}{2\pi}\int_{-\infty}^{\infty}\tilde{U}(z, \omega)\exp(-i\omega T)\,d\omega, \tag{3.2.2}$$

then it satisfies an ordinary differential equation

$$i\frac{\partial \tilde{U}}{\partial z} = -\tfrac{1}{2}\beta_2\omega^2\tilde{U}, \tag{3.2.3}$$

whose solution is given by

$$\tilde{U}(z, \omega) = \tilde{U}(0, \omega)\exp\left(\frac{i}{2}\beta_2\omega^2 z\right). \tag{3.2.4}$$

Equation (3.2.4) shows that GVD changes the phase of each spectral component of the pulse by an amount that depends on both the frequency and the propagated distance. Even though such phase changes do not affect the pulse spectrum, they can modify the pulse shape. By substituting Eq. (3.2.4) in Eq. (3.2.2), the general solution of Eq. (3.2.1) is given by

$$U(z,T) = \frac{1}{2\pi} \int_{-\infty}^{\infty} \tilde{U}(0,\omega) \exp\left(\frac{i}{2}\beta_2\omega^2 z - i\omega T\right) d\omega, \qquad (3.2.5)$$

where $\tilde{U}(0,\omega)$ is the Fourier transform of the incident field at $z = 0$ and is obtained using

$$\tilde{U}(0,\omega) = \int_{-\infty}^{\infty} U(0,T)\exp(i\omega T)\,dT. \qquad (3.2.6)$$

Equations (3.2.5) and (3.2.6) can be used for input pulses of arbitrary shapes.

3.2.1 Gaussian Pulses

As a simple example, consider the case of a Gaussian pulse for which the incident field is of the form [8]

$$U(0,T) = \exp\left(-\frac{T^2}{2T_0^2}\right), \qquad (3.2.7)$$

where T_0 is the half-width (at $1/e$-intensity point) introduced in Section 3.1. In practice, it is customary to use the full width at half maximum (FWHM) in place of T_0. For a Gaussian pulse, the two are related as

$$T_{\text{FWHM}} = 2(\ln 2)^{1/2}T_0 \approx 1.665T_0. \qquad (3.2.8)$$

By using Eqs. (3.2.5)–(3.2.7) and carrying out the integration, the amplitude at any point z along the fiber is given by

$$U(z,T) = \frac{T_0}{(T_0^2 - i\beta_2 z)^{1/2}} \exp\left(-\frac{T^2}{2(T_0^2 - i\beta_2 z)}\right). \qquad (3.2.9)$$

Thus, a Gaussian pulse maintains its shape on propagation but its width T_1 increases with z as

$$T_1(z) = T_0[1 + (z/L_D)^2]^{1/2}, \qquad (3.2.10)$$

where the dispersion length $L_D = T_0^2/|\beta_2|$. Equation (3.2.10) shows how GVD broadens a Gaussian pulse. The extent of broadening is governed by the dispersion length L_D. For a given fiber length, short pulses broaden more because

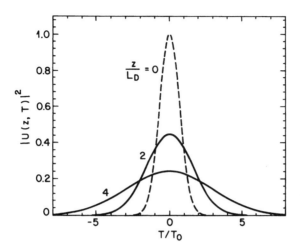

Figure 3.1 Dispersion-induced broadening of a Gaussian pulse inside a fiber at $z = 2L_D$ and $z = 4L_D$. Dashed curve shows the incident pulse at $z = 0$.

of a smaller dispersion length. At $z = L_D$, a Gaussian pulse broadens by a factor of $\sqrt{2}$. Figure 3.1 shows the extent of dispersion-induced broadening for Gaussian pulses by plotting $|U(z,T)|^2$ at $z = 0$, $2L_D$, and $4L_D$.

A comparison of Eqs. (3.2.7) and (3.2.9) shows that although the incident pulse is unchirped (with no phase modulation), the transmitted pulse becomes chirped. This can be seen clearly by writing $U(z,T)$ in the form

$$U(z,T) = |U(z,T)| \exp[i\phi(z,T)], \tag{3.2.11}$$

where

$$\phi(z,T) = -\frac{\mathrm{sgn}(\beta_2)(z/L_D)}{1+(z/L_D)^2} \frac{T^2}{T_0^2} + \frac{1}{2} \tan^{-1}\left(\frac{z}{L_D}\right). \tag{3.2.12}$$

The time dependence of the phase $\phi(z,T)$ implies that the instantaneous frequency differs across the pulse from the central frequency ω_0. The difference $\delta\omega$ is just the time derivative $-\partial\phi/\partial T$ [the minus sign is due to the choice $\exp(-i\omega_0 t)$ in Eq. (2.3.2)] and is given by

$$\delta\omega(T) = -\frac{\partial\phi}{\partial T} = \frac{\mathrm{sgn}(\beta_2)(2z/L_D)}{1+(z/L_D)^2} \frac{T}{T_0^2}. \tag{3.2.13}$$

Equation (3.2.13) shows that the frequency changes linearly across the pulse, i.e., a fiber imposes linear frequency chirp on the pulse. The chirp $\delta\omega$ depends on the sign of β_2. In the normal-dispersion regime ($\beta_2 > 0$), $\delta\omega$ is negative at the leading edge ($T < 0$) and increases linearly across the pulse; the opposite occurs in the anomalous-dispersion regime ($\beta_2 > 0$).

Dispersion-induced pulse broadening can be understood by recalling from Section 1.3 that different frequency components of a pulse travel at slightly different speeds along the fiber because of GVD. More specifically, red components travel faster than blue components in the normal-dispersion regime ($\beta_2 > 0$), while the opposite occurs in the anomalous-dispersion regime ($\beta_2 < 0$). The pulse can maintain its width only if all spectral components arrive together. Any time delay in the arrival of different spectral components leads to pulse broadening.

3.2.2 Chirped Gaussian Pulses

For an initially unchirped Gaussian pulse, Eq. (3.2.10) shows that dispersion-induced broadening of the pulse does not depend on the sign of the GVD parameter β_2. Thus, for a given value of the dispersion length L_D, the pulse broadens by the same amount in the normal- and anomalous-dispersion regimes of the fiber. This behavior changes if the Gaussian pulse has an initial frequency chirp [9]. In the case of linearly chirped Gaussian pulses, the incident field can be written as [compare with Eq. (3.2.7)]

$$U(0,T) = \exp\left(-\frac{(1+iC)}{2}\frac{T^2}{T_0^2}\right), \qquad (3.2.14)$$

where C is a chirp parameter. By using Eq. (3.2.11) one finds that the instantaneous frequency increases linearly from the leading to the trailing edge (up-chirp) for $C > 0$ while the opposite occurs (down-chirp) for $C < 0$. It is common to refer to the chirp as positive or negative depending on whether C is positive or negative.

The numerical value of C can be estimated from the spectral width of the Gaussian pulse. By substituting Eq. (3.2.14) in Eq. (3.2.6), $\tilde{U}(0,\omega)$ is given by

$$\tilde{U}(0,\omega) = \left(\frac{2\pi T_0^2}{1+iC}\right)^{1/2} \exp\left(-\frac{\omega^2 T_0^2}{2(1+iC)}\right). \qquad (3.2.15)$$

The spectral half-width (at $1/e$-intensity point) from Eq. (3.2.15) is given by

$$\Delta\omega = (1+C^2)^{1/2}/T_0. \qquad (3.2.16)$$

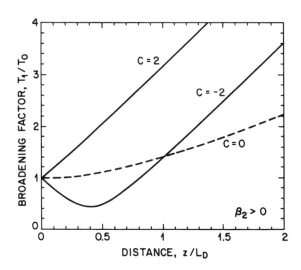

Figure 3.2 Broadening factor for a chirped Gaussian pulse as a function of distance. Dashed curve corresponds to the case of an unchirped Gaussian pulse. For $\beta_2 < 0$, the same curves are obtained if the sign of C is reversed.

In the absence of frequency chirp ($C = 0$), the spectral width is transform-limited and satisfies the relation $\Delta\omega T_0 = 1$. The spectral width is enhanced by a factor of $(1 + C^2)^{1/2}$ in the presence of linear chirp. Equation (3.2.16) can be used to estimate $|C|$ from measurements of $\Delta\omega$ and T_0.

To obtain the transmitted field, $\tilde{U}(0, \omega)$ from Eq. (3.2.15) is substituted in Eq. (3.2.5). The integration can be carried out analytically with the result

$$U(z, T) = \frac{T_0}{[T_0^2 - i\beta_2 z(1 + iC)]^{1/2}} \exp\left(-\frac{(1 + iC)T^2}{2[T_0^2 - i\beta_2 z(1 + iC)]}\right). \quad (3.2.17)$$

Thus, even a chirped Gaussian pulse maintains its Gaussian shape on propagation. The width T_1 after propagating a distance z is related to the initial width T_0 by the relation [9]

$$\frac{T_1}{T_0} = \left[\left(1 + \frac{C\beta_2 z}{T_0^2}\right)^2 + \left(\frac{\beta_2 z}{T_0^2}\right)^2\right]^{1/2}. \quad (3.2.18)$$

This equation shows that broadening depends on the relative signs of the GVD parameter β_2 and the chirp parameter C. Whereas a Gaussian pulse broadens monotonically with z if $\beta_2 C > 0$, it goes through an initial narrowing stage when $\beta_2 C < 0$. Figure 3.2 shows this behavior by plotting the broadening factor T_1/T_0 as a function of z/L_D for $C = 2$. In the case $\beta_2 C < 0$, the pulse width becomes minimum at a distance

$$z_{\min} = \frac{|C|}{1+C^2} L_D. \tag{3.2.19}$$

The minimum value of the pulse width at $z = z_{\min}$ is given by

$$T_1^{\min} = \frac{T_0}{(1+C^2)^{1/2}}. \tag{3.2.20}$$

By using Eqs. (3.2.16) and (3.2.10) one finds that at $z = z_{\min}$ the pulse width is Fourier-transform-limited because $\Delta\omega T_1^{\min} = 1$.

Initial narrowing of the pulse for the case $\beta_2 C < 0$ can be understood by referring to Eq. (3.2.13), which shows the dispersion-induced chirp imposed on an initially unchirped Gaussian pulse. When the pulse is initially chirped and the condition $\beta_2 C < 0$ is satisfied, the dispersion-induced chirp is in opposite direction to that of the initial chirp. As a result the net chirp is reduced, leading to pulse narrowing. The minimum pulse width occurs at a point at which the two chirps cancel each other. With a further increase in the propagation distance, the dispersion-induced chirp starts to dominate over the initial chirp, and the pulse begins to broaden. The net chirp as a function of z can be obtained from Eq. (3.2.17) by using Eqs. (3.2.11) and (3.2.13); it shows the qualitative behavior discussed in the preceding.

3.2.3 Hyperbolic-Secant Pulses

Although pulses emitted from many lasers can be approximated by a Gaussian shape, it is necessary to consider other pulse shapes. Of particular interest is the hyperbolic-secant pulse shape that occurs naturally in the context of optical solitons and pulses emitted from some mode-locked lasers. The optical field associated with such pulses often takes the form

$$U(0,T) = \operatorname{sech}\left(\frac{T}{T_0}\right) \exp\left(-\frac{iCT^2}{2T_0^2}\right), \tag{3.2.21}$$

where the chirp parameter C controls the initial chirp similarly to that of Eq. (3.2.14). The transmitted field $U(z,T)$ is obtained by using Eqs. (3.2.5), (3.2.6), and (3.2.21). Unfortunately, it is not easy to evaluate the integral in Eq. (3.2.5)

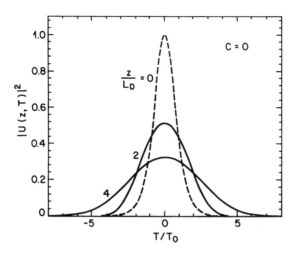

Figure 3.3 Pulse shapes at $z = 2L_D$ and $z = 4L_D$ of a pulse whose shape at $z = 0$ (dashed curve) is described by a "sech" profile. Compare with Fig. 3.1 where the case of a Gaussian pulse is shown.

in a closed form for non-Gaussian pulse shapes. Figure 3.3 shows the transmitted pulse shapes calculated numerically at $z = 2L_D$ and $z = 4L_D$ for the case of unchirped pulses ($C = 0$). A comparison of Figs. 3.1 and 3.3 shows that the qualitative features of dispersion-induced broadening are nearly identical for the Gaussian and "sech" pulses. Note that T_0 appearing in Eq. (3.2.21) is not the FWHM but is related to it by

$$T_{\mathrm{FWHM}} = 2\ln(1 + \sqrt{2})T_0 \approx 1.763\,T_0. \tag{3.2.22}$$

This relation should be used if the comparison is made on the basis of FWHM. The same relation for a Gaussian pulse is given in Eq. (3.2.8).

3.2.4 Super-Gaussian Pulses

So far we have considered pulse shapes with relatively broad leading and trailing edges. As one may expect, dispersion-induced broadening is sensitive to pulse edge steepness. In general, a pulse with steeper leading and trailing edges broadens more rapidly with propagation simply because such a pulse has a wider spectrum to start with. Pulses emitted by directly modulated semiconductor lasers fall in this category and cannot generally be approximated by a Gaussian pulse. A super-Gaussian shape can be used to model the effects of

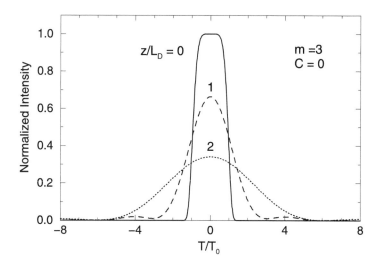

Figure 3.4 Pulse shapes at $z = L_D$ and $z = 2L_D$ of a pulse whose shape at $z = 0$ (dashed curve) is described by a super-Gaussian profile. Compare with Fig. 3.1 where the case of a Gaussian pulse is shown.

steep leading and trailing edges on dispersion-induced pulse broadening. For a super-Gaussian pulse, Eq. (3.2.14) is generalized to take the form [16]

$$U(0,T) = \exp\left[-\frac{1+iC}{2}\left(\frac{T}{T_0}\right)^{2m}\right], \tag{3.2.23}$$

where the parameter m controls the degree of edge sharpness. For $m = 1$ we recover the case of chirped Gaussian pulses. For larger value of m, the pulse becomes square shaped with sharper leading and trailing edges. If the rise time T_r is defined as the duration during which the intensity increases from 10 to 90% of its peak value, it is related to the parameter m as

$$T_r = (\ln 9)\frac{T_0}{2m} \approx \frac{T_0}{m}. \tag{3.2.24}$$

Thus the parameter m can be determined from the measurements of T_r and T_0.

Figure 3.4 shows the pulse shapes at $z = 0$, L_D, and $2L_D$ in the case of an unchirped super-Gaussian pulse ($C = 0$) with $m = 3$. It should be compared with Fig. 3.1 where the case of a Gaussian pulse ($m = 1$) is shown. The differences between the two can be attributed to the steeper leading and trailing edges associated with a super-Gaussian pulse. Whereas the Gaussian

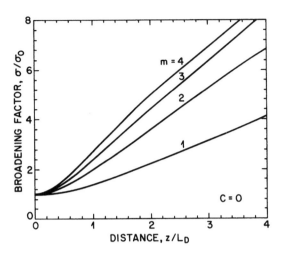

Figure 3.5 Variation of broadening factor σ/σ_0 with distance for several super-Gaussian pulses with different values of m. The case $m = 1$ corresponds to a Gaussian pulse. Pulse edges become steeper with increasing values of m.

pulse maintains its shape during propagation, the super-Gaussian pulse not only broadens at a faster rate but also distorts in shape. Enhanced broadening of a super-Gaussian pulse can be understood by noting that its spectrum is wider than that of a Gaussian pulse because of steeper leading and trailing edges. As the GVD-induced delay of each frequency component is directly related to its separation from the central frequency ω_0, a wider spectrum results in a faster rate of pulse broadening.

For complicated pulse shapes such as those seen in Fig. 3.4, the FWHM is not a true measure of the pulse width. The width of such pulses is more accurately described by the root-mean-square (RMS) width σ defined as [8]

$$\sigma = [\langle T^2 \rangle - \langle T \rangle^2]^{1/2}, \tag{3.2.25}$$

where

$$\langle T^p \rangle = \frac{\int_{-\infty}^{\infty} T^p |U(z,T)|^2 \, dT}{\int_{-\infty}^{\infty} |U(z,T)|^2 \, dT}. \tag{3.2.26}$$

To see how pulse broadening depends on the steepness of pulse edges, Fig. 3.5 shows the broadening factor σ/σ_0 of super-Gaussian pulses as a function of the propagation distance for values of m ranging from 1 to 4. Here σ_0 is the initial RMS width of the pulse at $z = 0$. The case $m = 1$ corresponds to a

Gaussian pulse; the pulse edges become increasingly steeper for larger values of m. Noting from Eq. (3.2.24) that the rise time is inversely proportional to m, it is evident that a pulse with a shorter rise time broadens faster. The curves in Fig. 3.5 are drawn for the case of initially unchirped pulses ($C = 0$). When the pulses are initially chirped, the magnitude of pulse broadening depends on the sign of the product $\beta_2 C$. The qualitative behavior is similar to that shown in Fig. 3.2 for the case of a Gaussian pulse ($m = 1$). In particular, even super-Gaussian pulses exhibit initial narrowing when $\beta_2 C < 0$. It is possible to evaluate the broadening factor analytically using Eqs. (3.2.5) and (3.2.23)–(3.2.26) with the result [17]

$$\frac{\sigma}{\sigma_0} = \left[1 + \frac{\Gamma(1/2m)}{\Gamma(3/2m)} \frac{C\beta_2 z}{T_0^2} + m^2(1+C^2) \frac{\Gamma(2 - 1/2m)}{\Gamma(3/2m)} \left(\frac{\beta_2 z}{T_0^2} \right)^2 \right]^{1/2},$$

(3.2.27)

where Γ is the gamma function. For a Gaussian pulse ($m = 1$) the broadening factor reduces to that given in Eq. (3.2.18).

3.2.5 Experimental Results

The initial compression of chirped pulses has been observed experimentally using pulses emitted from a directly modulated semiconductor laser. In one experiment [10], the incident pulse at a wavelength of 1.54 μm was positively chirped ($C > 0$). It compressed by about a factor of 5 after propagating 104 km in the anomalous-GVD regime of a fiber with $\beta_2 \approx -20$ ps^2/km. In another experiment, the semiconductor laser emitted a negatively chirped pulse ($C < 0$) at a wavelength of 1.21 μm [11]. After propagating a distance of 1.5 km in the normal-dispersion regime ($\beta_2 = 15$ ps^2/km), the pulse width decreased from 190 to 150 ps. When the fiber length was increased to 6 km, the pulse width increased to 300 ps, in agreement with the qualitative behavior shown in Fig. 3.2. In a different experiment much shorter optical pulses (initial FWHM ≈ 26 ps) at 1.3 μm were obtained from a distributed-feedback (DFB) semiconductor laser by using the gain-switching technique [15]. As the pulses were negatively chirped ($C < 0$), a dispersion-shifted fiber was employed with a positive GVD at 1.3 μm ($\beta_2 \approx 12$ ps^2/km). The pulse compressed by a factor of three after propagating inside a 4.8-km-long fiber and then started to broaden with a further increase in the fiber length.

Compression of chirped picosecond pulses through GVD in optical fibers has been used to advantage in some experiments in which a gain-switched

DFB semiconductor laser was used as a source of solitons [18]–[21]. Even though a relatively broad optical pulse (duration 20–40 ps) emitted from such lasers is far from being transform limited, its passage through a fiber of optimized length with positive GVD produces compressed optical pulses that are nearly transform limited. In a 1989 demonstration of this technique [20], 14-ps pulses were obtained at the 3-GHz repetition rate by passing the gain-switched pulse through a polarization-preserving, dispersion-shifted, 3.7-km-long optical fiber with $\beta_2 = 23$ ps^2/km at the 1.55-μm operating wavelength. In another experiment, a narrowband optical filter was used to control the spectral width of the gain-switched pulse before its compression [21]. An erbium-doped fiber amplifier then amplified and compressed the pulse simultaneously. It was possible to generate nearly transform-limited 17-ps optical pulses at repetition rates of 6–24 GHz. Pulses as short as 3 ps were obtained by 1990 with this technique [22].

In a related method, amplification of picosecond pulses in a semiconductor laser amplifier produces optical pulses chirped such that they can be compressed by using optical fibers with anomalous GVD [23]–[25]. The method is useful in the wavelength region near 1.5 μm because silica fibers commonly exhibit anomalous GVD in that spectral region. The technique was demonstrated in 1989 by using 40-ps input pulses obtained from a 1.52-μm mode-locked semiconductor laser [23]. The pulse was first amplified in a semiconductor laser amplifier and then compressed by about a factor of two by propagating it through an 18-km-long fiber with $\beta_2 = -18$ ps^2/km. Such a compression mechanism was useful for transmitting a 16-Gb/s signal over 70 km of standard telecommunication fiber [24].

3.3   Third-Order Dispersion

The dispersion-induced pulse broadening discussed in Section 3.2 is due to the lowest-order GVD term proportional to β_2 in Eq. (2.3.23). Although the contribution of this term dominates in most cases of practical interest, it is sometimes necessary to include the third-order term proportional to β_3 in this expansion. For example, if the pulse wavelength nearly coincides with the zero-dispersion wavelength λ_D, $\beta_2 \approx 0$; the β_3 term then provides the dominant contribution to the GVD effects [6]. For ultrashort pulses (width $T_0 < 1$ ps), it is necessary to include the β_3 term even when $\beta_2 \neq 0$ because the expansion parameter $\Delta\omega/\omega_0$ is no longer small enough to justify the truncation of the expansion in Eq. (2.3.23) after the β_2 term.

This section considers the dispersive effects by including both β_2 and β_3 terms while still neglecting the nonlinear effects. The appropriate propagation equation for the amplitude $A(z,T)$ is obtained from Eq. (2.3.39) after setting $\gamma = 0$. Using Eq. (3.1.3), $U(z,T)$ satisfies the following equation:

$$i\frac{\partial U}{\partial z} = \frac{\beta_2}{2}\frac{\partial^2 U}{\partial T^2} + \frac{i\beta_3}{6}\frac{\partial^3 U}{\partial T^3}. \tag{3.3.1}$$

This equation can also be solved by using the Fourier technique of Section 3.2. In place of Eq. (3.2.5) the transmitted field is obtained from

$$U(z,T) = \frac{1}{2\pi}\int_{-\infty}^{\infty} \tilde{U}(0,\omega)\exp\left(\frac{i}{2}\beta_2\omega^2 z + \frac{i}{6}\beta_3\omega^3 z - i\omega T\right)d\omega, \tag{3.3.2}$$

where the Fourier transform $\tilde{U}(0,\omega)$ of the incident field is given by Eq. (3.2.6). Equation (3.3.2) can be used to study the effect of higher-order dispersion if the incident field $U(0,T)$ is specified. In particular, one can consider Gaussian, super-Gaussian, or hyperbolic-secant pulses in a manner analogous to Section 3.2. An analytic solution in terms of the Airy functions can be obtained for Gaussian pulses [6].

3.3.1 Changes in Pulse Shape

As one may expect, pulse evolution along the fiber depends on the relative magnitudes of β_2 and β_3, which in turn depend on the deviation of the optical wavelength λ_0 from λ_D. At $\lambda_0 = \lambda_D$, $\beta_2 = 0$, and typically $\beta_3 \approx 0.1$ ps³/km. However, $|\beta_2| \approx 1$ ps²/km even when λ_0 differs from λ_D by as little as 10 nm. In order to compare the relative importance of the β_2 and β_3 terms in Eq. (3.3.1), it is useful to introduce a dispersion length associated with the third-order dispersion (TOD) term as

$$L'_D = T_0^3/|\beta_3|. \tag{3.3.3}$$

The TOD effects play a significant role only if $L'_D \leq L_D$ or $T_0|\beta_2/\beta_3| \leq 1$. For a 100-ps pulse, this condition implies that $\beta_2 < 10^{-3}$ ps²/km when $\beta_3 = 0.1$ ps³/km. Such low values of β_2 are realized only if λ_0 and λ_D differ by $< 10^{-2}$ nm! In practice, it is difficult to match λ_0 and λ_D to such an accuracy, and the contribution of β_3 is generally negligible compared with that of

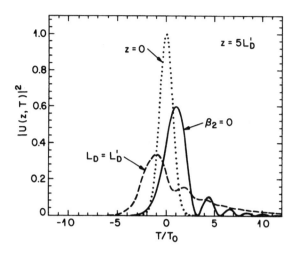

Figure 3.6 Pulse shapes at $z = 5L'_D$ of an initially Gaussian pulse at $z = 0$ (dotted curve) in the presence of higher-order dispersion. Solid curve is for the case of $\lambda_0 = \lambda_D$. Dashed curve shows the effect of finite β_2 in the case of $L_D = L'_D$.

β_2. This was indeed the case in the experiments in which 1.32-μm picosecond pulses were propagated over a few-kilometer-long fiber [26], [27],. The situation changes completely for ultrashort pulses with widths in the femtosecond range. For example, β_2 can be as large as 1 ps^2/km for $T_0 = 0.1$ ps before the contribution of β_3 becomes negligible. As $L'_D \sim 10$ m for such values of T_0, the effect of TOD can be studied experimentally by propagating 100-fs pulses across a few-meter-long fiber.

Figure 3.6 shows the pulse shapes at $z = 5L'_D$ for an initially unchirped Gaussian pulse [$C = 0$ in Eq. (3.2.14)] for $\beta_2 = 0$ (solid curve) and for a value of β_2 such that $L_D = L'_D$ (dashed curve). Whereas a Gaussian pulse remains Gaussian when only the β_2 term in Eq. (3.3.1) contributes to GVD (Fig. 3.1), the TOD distorts the pulse such that it becomes asymmetric with an oscillatory structure near one of its edges. In the case of positive β_3 shown in Fig. 3.6, oscillations appear near the trailing edge of the pulse. When β_3 is negative, it is the leading edge of the pulse that develops oscillations. When $\beta_2 = 0$, oscillations are deep, with intensity dropping to zero between successive oscillations. However, these oscillations damp significantly even for relatively small values of β_2. For the case $L_D = L'_D$ shown in Fig. 3.6 ($\beta_2 = \beta_3/T_0$), oscillations have nearly disappeared, and the pulse has a long tail on the trailing side. For larger values of β_2 such that $L_D \ll L'_D$, the pulse shape becomes nearly Gaussian as

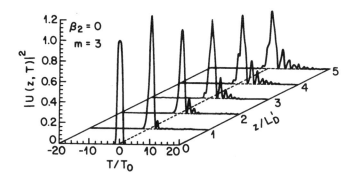

Figure 3.7 Evolution of a super-Gaussian pulse with $m = 3$ along the fiber length for the case of $\beta_2 = 0$ and $\beta_3 > 0$. Higher-order dispersion is responsible for the oscillatory structure near the trailing edge of the pulse.

the TOD plays a relatively minor role.

Equation (3.3.2) can be used to study pulse evolution for other pulse shapes (with or without chirp). By way of an example, Fig. 3.7 shows evolution of an unchirped super-Gaussian pulse at the zero-dispersion wavelength ($\beta_2 = 0$) with $C = 0$ and $m = 3$ in Eq. (3.2.23). It is clear that pulse shapes can vary widely depending on the initial conditions. In practice, one is often interested in the extent of dispersion-induced broadening rather than details of pulse shapes. As the FWHM is not a true measure of the width of pulses shown in Figs. 3.6 and 3.7, we use the RMS width σ defined in Eq. (3.2.25). In the case of Gaussian pulses, it is possible to obtain a simple analytic expression of σ that includes the effects of β_2, β_3, and the initial chirp C on dispersion broadening [9].

3.3.2 Broadening Factor

To calculate σ from Eq. (3.2.25), we need to find the nth moment $\langle T^n \rangle$ of T using Eq. (3.2.26). As the Fourier transform $\tilde{U}(z, \omega)$ of $U(z, T)$ is known from Eq. (3.3.2), it is useful to evaluate $\langle T^n \rangle$ in the frequency domain. By using the Fourier transform $\tilde{I}(z, \omega)$ of the pulse intensity $|U(z, T)|^2$,

$$\tilde{I}(z, \omega) = \int_{-\infty}^{\infty} |U(z, T)|^2 \exp(i\omega T) \, dT, \tag{3.3.4}$$

and differentiating it n times, we obtain

$$\lim_{\omega \to 0} \frac{\partial^n}{\partial \omega^n} \tilde{I}(z, \omega) = (i)^n \int_{-\infty}^{\infty} T^n |U(z, T)|^2 \, dT. \tag{3.3.5}$$

Using Eq. (3.3.5) in Eq. (3.2.26) we find that

$$\langle T^n \rangle = \frac{(-i)^n}{N_c} \lim_{\omega \to 0} \frac{\partial^n}{\partial \omega^n} \tilde{I}(z, \omega), \tag{3.3.6}$$

where the normalization constant

$$N_c = \int_{-\infty}^{\infty} |U(z, T)|^2 \, dT \equiv \int_{-\infty}^{\infty} |U(0, T)|^2 \, dT. \tag{3.3.7}$$

From the convolution theorem

$$\tilde{I}(z, \omega) = \int_{-\infty}^{\infty} \tilde{U}(z, \omega - \omega') \tilde{U}^*(z, \omega') \, d\omega'. \tag{3.3.8}$$

Performing the differentiation and limit operations indicated in Eq. (3.3.6), we obtain

$$\langle T^n \rangle = \frac{(i)^n}{N_c} \int_{-\infty}^{\infty} \tilde{U}^*(z, \omega) \frac{\partial^n}{\partial \omega^n} \tilde{U}(z, \omega) \, d\omega. \tag{3.3.9}$$

In the case of a chirped Gaussian pulse $\tilde{U}(z, \omega)$ can be obtained from Eqs. (3.2.15) and (3.3.2) and is given by

$$\tilde{U}(z, \omega) = \left(\frac{2\pi T_0^2}{1 + iC} \right)^{1/2} \exp \left[\frac{i\omega^2}{2} \left(\beta_2 z + \frac{iT_0^2}{1 + iC} \right) + \frac{i}{6} \beta_3 \omega^3 z \right]. \tag{3.3.10}$$

If we differentiate Eq. (3.3.10) two times and substitute the result in Eq. (3.3.9), we find that the integration over ω can be performed analytically. Both $\langle T \rangle$ and $\langle T^2 \rangle$ can be obtained by this procedure. Using the resulting expressions in Eq. (3.2.25), we obtain [9]

$$\frac{\sigma}{\sigma_0} = \left[\left(1 + \frac{C\beta_2 z}{2\sigma_0^2} \right)^2 + \left(\frac{\beta_2 z}{2\sigma_0^2} \right)^2 + (1 + C^2)^2 \frac{1}{2} \left(\frac{\beta_3 z}{4\sigma_0^3} \right)^2 \right]^{1/2}, \tag{3.3.11}$$

where σ_0 is the initial RMS width of the chirped Gaussian pulse ($\sigma_0 = T_0/\sqrt{2}$). As expected, Eq. (3.3.11) reduces to Eq. (3.2.18) for $\beta_3 = 0$.

Equation (3.3.11) can be used to draw several interesting conclusions. In general, both β_2 and β_3 contribute to pulse broadening. However, the dependence of their contributions on the chirp parameter C is qualitatively different.

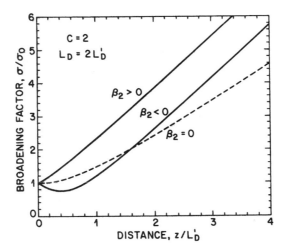

Figure 3.8 Variation of broadening factor with propagated distance for a chirped Gaussian pulse in the vicinity of λ_D such that $L_D = 2L_D'$. Dashed curve corresponds to the case of $\lambda_0 = \lambda_D$ so that L_D is infinite ($\beta_2 = 0$).

Whereas the contribution of β_2 depends on the sign of $\beta_2 C$, the contribution of β_3 is independent of the sign of both β_3 and C. Thus, in contrast to the behavior shown in Fig. 3.2, a chirped pulse propagating exactly at the zero-dispersion wavelength never experiences width contraction. However, even small departures from the exact zero-dispersion wavelength can lead to initial pulse contraction. This behavior is illustrated in Fig. 3.8 where the broadening factor σ/σ_0 is plotted as a function of propagation distance for $C = 2$ and $L_D = 2L_D'$. Dashed curve shows for comparison broadening expected when $\beta_2 = 0$. In the anomalous-dispersion regime the contribution of β_2 can counteract the β_3 contribution in such a way that dispersive broadening is less than that expected when $\beta_2 = 0$ for $z \sim L_D'$. For large values of z such that $z \gg L_D/|C|$, Eq. (3.3.11) can be approximated by

$$\sigma/\sigma_0 = (1+C^2)^{1/2}[1 + (L_D/2L_D')^2]^{1/2}(z/L_D), \tag{3.3.12}$$

where we have used Eqs. (3.1.5) and (3.3.3). The linear dependence of the RMS pulse width on the propagation distance for large z is a general feature that holds for arbitrary pulse shapes, as discussed in the next section.

Equation (3.3.11) can be generalized to include the effects of a finite source bandwidth [9]. Spontaneous emission in any light source produces amplitude and phase fluctuations that manifest as a finite bandwidth $\delta\omega$ of the source spectrum centered at ω_0 [28]. If the source bandwidth $\delta\omega$ is much smaller than the pulse bandwidth $\Delta\omega$ [given by Eq. (3.2.16) for chirped Gaussian pulses], its effect on the pulse broadening can be neglected. However, for many light sources used in optical communications [such as light-emitting diodes (LEDs) and multimode semiconductor lasers] this condition is not satisfied, and it becomes necessary to include the effect of source bandwidth. In the case of a Gaussian spectrum, the generalized form of Eq. (3.3.11) is given by [9]

$$\frac{\sigma^2}{\sigma_0^2} = \left(1 + \frac{C\beta_2 z}{2\sigma_0^2}\right)^2 + (1 + V_\omega^2)\left(\frac{\beta_2 z}{2\sigma_0^2}\right)^2 + (1 + C^2 + V_\omega^2)^2 \frac{1}{2}\left(\frac{\beta_3 z}{4\sigma_0^3}\right)^2,$$

(3.3.13)

where $V_\omega = 2\sigma_\omega\sigma_0$ and σ_ω is the RMS width of the Gaussian source spectrum. This equation describes broadening of chirped Gaussian pulses in a linear dispersive medium under quite general conditions. It can be used to discuss the effect of GVD on the performance of fiber-optic communication systems.

3.3.3   Arbitrary-Shape Pulses

The formal similarity of Eq. (3.2.1) with the Schrödinger equation can be exploited to obtain an analytic expression of the RMS width for pulses of arbitrary shape while including the third- and higher-order dispersive effects [29]. For this purpose, we write Eq. (3.3.1) in an operator form as

$$i\frac{\partial U}{\partial z} = \hat{H}U,$$

(3.3.14)

where the operator $\hat{H}$ includes, in its general form, dispersive effects to all orders and is given by

$$\hat{H} = -\sum_{n=2}^{\infty} \frac{i^n}{n!}\left(\frac{\partial}{\partial T}\right)^n = \frac{\beta_2}{2}\frac{\partial^2}{\partial T^2} + \frac{i\beta_3}{6}\frac{\partial^3}{\partial T^3} + \cdots.$$

(3.3.15)

Using Eq. (3.2.26) and assuming that $U(z, T)$ is normalized such that $\int_{-\infty}^{\infty} |U|^2 dT = 1$, the first and second moments of T are found to evolve with z as

$$\frac{d\langle T\rangle}{dz} = i\langle[\hat{H}, T]\rangle,$$

(3.3.16)

$$\frac{d\langle T^2\rangle}{dz} = -\langle[\hat{H}, [\hat{H}, T]]\rangle,$$

(3.3.17)

where $[\hat{H}, T] \equiv \hat{H}T - T\hat{H}$ stands for the commutator.

Equations (3.3.16) and (3.3.17) can be integrated analytically and result in the following general expressions [29]:

$$\langle T \rangle = a_0 + a_1 z, \tag{3.3.18}$$
$$\langle T^2 \rangle = b_0 + b_1 z + b_2 z^2, \tag{3.3.19}$$

where the coefficients depend only on the incident field $U_0(T) \equiv U(0, T)$ and are defined as

$$a_0 = \int_{-\infty}^{\infty} U_0^*(T) T U_0(T) \, dT, \tag{3.3.20}$$

$$a_1 = i \int_{-\infty}^{\infty} U_0^*(T) [\hat{H}, T] U_0(T) \, dT, \tag{3.3.21}$$

$$b_0 = \int_{-\infty}^{\infty} U_0^*(T) T^2 U_0(T) \, dT, \tag{3.3.22}$$

$$b_1 = i \int_{-\infty}^{\infty} U_0^*(T)) [\hat{H}, T^2] U_0(T) \, dT, \tag{3.3.23}$$

$$b_2 = -\frac{1}{2} \int_{-\infty}^{\infty} U_0^*(T) [\hat{H}, [\hat{H}, T^2]] U_0(T) \, dT. \tag{3.3.24}$$

Physically, $\langle T \rangle$ governs asymmetry of pulse shape while $\langle T^2 \rangle$ is a measure of pulse broadening. Higher-order moments $\langle T^3 \rangle$ and $\langle T^4 \rangle$ can also be calculated by this technique and govern the skewness and kurtosis of the intensity profile, respectively. For initially symmetric pulses, $a_0 = 0$. If the effects of third- and higher-order dispersion are negligible it is easy to show that a_1 is also zero. With $\langle T \rangle = 0$ in that case, the pulse retains its symmetric nature during its transmission through optical fibers when β_2 dominates. Note that $\sigma^2 \equiv \langle T^2 \rangle - \langle T \rangle^2$ varies quadratically along the fiber length for pulses of arbitrary shape and chirp even when third- and higher-order dispersive effects are included.

As a simple example, consider the case of an unchirped 'sech' pulse discussed in Section 3.2.3 numerically and retain only the effects of GVD ($\beta_m = 0$ for $m > 2$). Using $U_0(T) = (2T_0)^{-1/2} \text{sech}(t/T_0)$ in Eqs. (3.3.20)–(3.3.24) one can show that $a_0 = a_1 = b_1 = 0$ while

$$b_0 = (\pi^2/12) T_0^2, \qquad b_2 = \beta_2^2/(3T_0^2). \tag{3.3.25}$$

Noting that $\sigma_0^2 = b_0$ and $\sigma^2 = b_0 + b_2 z^2$, the broadening factor becomes

$$\frac{\sigma}{\sigma_0} = \left[1 + \left(\frac{\pi \beta_2 z}{6 \sigma_0^2} \right)^2 \right]^{1/2}, \qquad (3.3.26)$$

where $\sigma_0 = (\pi / \sqrt{12}) T_0$ is the RMS width of the input pulse. This result should be compared with Eq. (3.3.11) obtained for a Gaussian pulse after setting $C = 0$ and $\beta_3 = 0$. Noting that $\pi/6 \approx 0.52$, one can conclude that a "sech" pulse broadens almost at the same rate and exhibits the same qualitative behavior as a Gaussian pulse when the comparison is made on the basis of their RMS widths.

The preceding analysis can readily be extended to chirped pulses. For a chirped Gaussian pulse, all integrals in Eqs. (3.3.20)–(3.3.24) can be evaluated in a closed form, and one recovers Eq. (3.3.11) for the broadening factor. For a super-Gaussian pulse Eq. (3.2.27) is obtained when third-order dispersion is neglected. It is possible to obtain σ/σ_0 in a closed form for a super-Gaussian pulse even when both β_2 and β_3 are finite but the resulting expression is quite complex [17].

The effect of third-order dispersion is to make the intensity profile asymmetric and introduce a long oscillating tail similar to that seen in Fig. 3.6. The quantity $\langle T \rangle$ provides a simple measure of this asymmetry. If we consider again the example of a 'sech' pulse, we find that $\langle T \rangle$ is zero initially but changes linearly with z at a rate given by $a_1 = \beta_3/(6T_0^2)$. The same behavior occurs for a Gaussian pulse but $\langle T \rangle$ changes at a different rate. These results are in agreement with the numerically calculated pulse shapes in Fig. 3.6. As seen there, pulse develops a long tail on the trailing edge for positive values of β_3, resulting in $\langle T \rangle > 0$.

The most important conclusion that one can draw from Eqs. (3.3.19) and (3.3.26) is that, for a long fiber whose length $L \gg L_D$, the GVD-induced pulse broadening scales as L/L_D irrespective of the pulse shape. As the dispersion length $L_D \equiv T_0^2/|\beta_2|$ scales as T_0^2, it decreases rapidly as pulses become shorter. As an example, $L_D = 100$ km for pulses with $T_0 = 10$ ps launched into a dispersion-shifted fiber having $|\beta_2| = 1$ ps²/km but becomes only 1 km if pulse width is reduced to $T_0 = 1$ ps. Such a pulse will broaden by a factor ~100 in a 100-km-long fiber. Because L can exceed thousands of kilometers for fiber-optic communication systems designed to transmit information over transoceanic distances, it is evident that GVD-induced pulse broadening limits the performance of most lightwave systems. The next section is devoted to the

GVD-induced limitations and the dispersion-management schemes developed to overcome them in practice.

3.3.4 Ultrashort-Pulse Measurements

As the GVD and TOD effects can change the shape and width of ultrashort pulses considerably, one should consider how such pulses can be measured experimentally. For pulses broader than 100 ps, pulse characteristics can be measured directly by using a high-speed photodetector. Streak cameras can be used for pulses as short as 1 ps. However, most of them work best in the visible spectral region and cannot be used at wavelengths near 1.55 μm.

A common technique for characterizing ultrashort optical pulses is based on the nonlinear phenomenon of second-harmonic generation. In this method, known as the autocorrelation technique, the pulse is sent through a nonlinear crystal together with a delayed (or advanced) replica of its own [30]. A second-harmonic signal is generated inside the crystal only when two pulses overlap in time. Measuring the second-harmonic power as a function of the delay time produces an autocorrelation trace. The width of this trace is related to the width of the original pulse. The exact relationship between the two widths depends on the pulse shape. If pulse shape is known *a priori*, or it can be inferred indirectly, the autocorrelation trace provides an accurate measurement of the pulse width. This technique can measure widths down to a few femtoseconds but provides little information on details of the pulse shape. In fact, an autocorrelation trace is always symmetric even when the pulse shape is known to be asymmetric. The use of cross correlation, a technique in which an ultrashort pulse of known shape and width is combined with the original pulse inside a second-harmonic crystal, solves this problem to some extent. The auto- and cross-correlation techniques can also make use of other nonlinear effects such as third-harmonic generation [31] and two-photon absorption [32]. All such methods, however, record intensity correlation and cannot provide any information on the phase or chirp variations across the pulse.

An interesting technique, called frequency-resolved optical gating (FROG) and developed during the 1990s, solves this problem quite nicely [33]–[37]. It not only can measure the pulse shape but can also provide information on how the optical phase and the frequency chirp vary across the pulse. The technique works by recording a series of spectrally resolved autocorrelation traces and uses them to deduce the intensity and phase profiles associated with the pulse. It has been used to characterize pulse propagation in optical fibers with considerable success [35]–[37]. A related technique, known as time-resolved optical gating (TROG), has been introduced recently [38]. In this method, the pulse is passed through a dispersive medium (e.g., an optical fiber) whose GVD can be varied over a certain range, and a number of autocorrelation traces are recorded for different GVD values. Both the intensity and phase profiles can be deduced from such autocorrelation traces.

3.4 Dispersion Management

In a fiber-optic communication system, information is transmitted over a fiber by using a coded sequence of optical pulses whose width is determined by the bit rate B of the system. Dispersion-induced broadening of pulses is undesirable as it interferes with the detection process and leads to errors if the pulse spreads outside its allocated bit slot ($T_B = 1/B$). Clearly, GVD limits the bit rate B for a fixed transmission distance L [39]. The dispersion problem becomes quite serious when optical amplifiers are used to compensate for fiber losses because L can exceed thousands of kilometers for long-haul systems. A useful measure of the information-transmission capacity is the bit rate–distance product BL. This section discusses how the BL product is limited by fiber dispersion and how it can be improved by using the technique of dispersion management.

3.4.1 GVD-Induced Limitations

Consider first the case in which pulse broadening is dominated by the large spectral width σ_ω of the source. For a Gaussian pulse, the broadening factor can be obtained from Eq. (3.3.13). Assuming that the contribution of the β_3 term is negligible together with $C = 0$ and $V_\omega \gg 1$, the RMS pulse width σ is given by

$$\sigma = [\sigma_0^2 + (\beta_2 L \sigma_\omega)^2]^{1/2} = [\sigma_0^2 + (DL\sigma_\lambda)^2]^{1/2}, \qquad (3.4.1)$$

where L is the fiber-link length and σ_λ is the RMS spectral width of the source in wavelength units. The dispersion parameter D is related to the GVD parameter β_2 as indicated in Eq. (1.2.11).

One can relate σ to the bit rate B by using the criterion that the broadened pulse should remain confined to its own bit slot ($T_B = 1/B$). A commonly used criterion is $4\sigma < T_B$ [39]; for a Gaussian pulse, at least 95% of the pulse energy remains within the bit slot when this condition is satisfied. The limiting bit rate is obtained using $4B\sigma < 1$. Assuming $\sigma_0 \ll \sigma$, this condition becomes

$$BL|D|\sigma_\lambda < 1/4. \tag{3.4.2}$$

As an illustration, consider the case of multimode semiconductor lasers [28] for which $\sigma_\lambda \approx 2$ nm. If the system is operating near $\lambda = 1.55$ μm using standard fibers, $D \approx 16$ ps/(km-nm). With these parameter values, Eq. (3.4.2) requires $BL < 8$ (Gb/s)-km. For a 100-km-long fiber, GVD restricts the bit rate to relatively low values of only 80 Mb/s. However, if the system is designed to operate near the zero-dispersion wavelength (occurring near 1.3 μm) such that $|D| < 1$ ps/(km-nm), the BL product increases to beyond 100 (Gb/s)-km.

Modern fiber-optic communication systems operating near 1.55 μm reduce the GVD effects using dispersion-shifted fibers designed such that the minimum-loss wavelength and the zero-dispersion wavelengths nearly coincide. At the same time, they use lasers designed to operate in a single longitudinal mode such that the source spectral width is well below 100 MHz [28]. Under such conditions, $V_\omega \ll 1$ in Eq. (3.3.13). If we neglect the β_3 term and set $C = 0$, Eq. (3.3.13) can be approximated by

$$\sigma = [\sigma_0^2 + (\beta_2 L/2\sigma_0)^2]^{1/2}. \tag{3.4.3}$$

A comparison with Eq. (3.4.1) reveals a major difference: Dispersion-induced broadening now depends on the initial width σ_0. In fact, σ can be minimized by choosing an optimum value of σ_0. The minimum value of σ is found to occur for $\sigma_0 = (|\beta_2|L/2)^{1/2}$ and is given by $\sigma = (|\beta_2|L)^{1/2}$. The limiting bit rate is obtained by using $4B\sigma < 1$ or the condition

$$B(|\beta_2|L)^{1/2} < 1/4. \tag{3.4.4}$$

The main difference from Eq. (3.4.2) is that B scales as $L^{-1/2}$ rather than L^{-1}. Figure 3.9 compares the decrease in the bit rate with increasing L by choosing $D = 16$ ps/(km-nm) and $\sigma_\lambda = 0$, 1, and 5 nm. Equation (3.4.4) was used in the case $\sigma_\lambda = 0$.

For a lightwave system operating exactly at the zero-dispersion wavelength, $\beta_2 = 0$ in Eq. (3.3.13). Assuming $V_\omega \ll 1$ and $C = 0$, the pulse width is given by

$$\sigma = [\sigma_0^2 + \tfrac{1}{2}(\beta_3 L/4\sigma_0^2)^2]^{1/2}. \tag{3.4.5}$$

Similar to the case of Eq. (3.4.3), σ can be minimized by optimizing the input pulse width σ_0. The minimum value of σ_0 is found to occur for $\sigma_0 =$

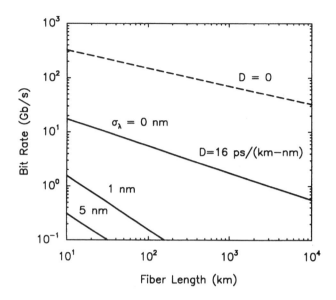

Figure 3.9 Limiting bit rate as a function of the fiber length for σ_λ = 0, 1 and 5 nm. The case σ_λ = 0 corresponds to an optical source whose spectral width is much smaller than the bit rate. Dashed line shows the case β_2 = 0.

$(|\beta_3|L/4)^{1/3}$. The limiting bit rate is obtained by using the condition $4B\sigma < 1$ and is given by [39]

$$B(|\beta_3|L)^{1/3} < 0.324. \tag{3.4.6}$$

The dispersive effects are most forgiving in this case. For a typical value β_3 = 0.1 ps³/km, the bit rate can be as large as 150 Gb/s for L = 100 km. It decreases to only 70 Gb/s even when L increases by a factor of 10 because of the $L^{-1/3}$ dependence of the bit rate on the fiber length. The dashed line in Fig. 3.9 shows this dependence using Eq. (3.4.6) with β_3 = 0.1 ps³/km. Clearly, the performance of a lightwave system can be considerably improved by operating it close to the zero-dispersion wavelength of the fiber.

3.4.2 Dispersion Compensation

Even though operation at the zero-dispersion wavelength is most desirable from the standpoint of pulse broadening, other considerations may preclude such a design. For example, at most one channel can be located at the zero-dispersion wavelength in a wavelength-division-multiplexed (WDM) system.

Moreover, as discussed in Chapter 10, strong four-wave mixing occurring when GVD is relatively low forces WDM systems to operate away from the zero-dispersion wavelength so that each channel has a finite value of β_2. Of course, GVD-induced pulse broadening then becomes of serious concern. The technique of dispersion management provides a solution to this dilemma. It consists of combining fibers with different characteristics such that the average GVD of the entire fiber link is quite low while the GVD of each fiber section is chosen to be large enough to make the four-wave-mixing effects negligible [40]. In practice, a periodic dispersion map is used with a period equal to the amplifier spacing (typically 50–100 km). Amplifiers compensate for accumulated fiber losses in each section. Between each pair of amplifiers, just two kinds of fibers, with opposite signs of β_2, are combined to reduce the average dispersion to a small value. When the average GVD is set to zero, dispersion is totally compensated.

Such a dispersion-compensation technique takes advantage of the linear nature of Eq. (3.2.1). The basic idea can be understood from Eq. (3.2.5) representing the general solution of Eq. (3.2.1). For a dispersion map consisting of two fiber segments, Eq. (3.2.5) becomes

$$U(L_m,t) = \frac{1}{2\pi} \int_{-\infty}^{\infty} \tilde{U}(0,\omega) \exp\left[\frac{i}{2}\omega^2(\beta_{21}L_1 + \beta_{22}L_2) - i\omega t\right] d\omega, \quad (3.4.7)$$

where $L_m = L_1 + L_2$ is the dispersion-map period, and β_{2j} is the GVD parameter of the fiber segment of length L_j ($j = 1, 2$). By using $D_j = -(2\pi c/\lambda^2)\beta_{2j}$, the condition for dispersion compensation can be written as

$$D_1 L_1 + D_2 L_2 = 0. \quad (3.4.8)$$

As $A(L_m,t) = A(0,t)$ when Eq. (3.4.8) is satisfied, the pulse recovers its initial width after each map period even though pulse width can change significantly within each period.

Equation (3.4.8) can be satisfied in several different ways. If two segments are of equal lengths ($L_1 = L_2$), the two fibers should have $D_1 = -D_2$. Fibers with equal and opposite values of GVD can be made by shifting the zero-dispersion wavelength appropriately during the manufacturing stage. However, a large quantity of standard fiber is already installed in existing lightwave systems. Because this fiber has anomalous GVD with $D \approx 16$ ps/(km-nm), its dispersion can be compensated by using a relatively short segment of dispersion-compensating fiber (DCF), designed to have 'normal' GVD with values of $D > -100$ ps/(km-nm).

The idea of using a DCF has been around since 1980 [41]. However, it was only after the advent of optical amplifiers in 1990 that the development of DCFs accelerated in pace [42]–[51]. There are two basic approaches to designing DCFs. In one approach, the DCF supports a single mode, but it is designed with a relatively small value of the fiber parameter V. As discussed in Section 2.2, the fundamental mode is weakly confined when $V \approx 1$. Because a large fraction of the mode propagates inside the cladding layer where the refractive index is smaller, the waveguide contribution to the GVD is quite different for such fibers, resulting in $D \sim -100$ ps/(km-nm). A depressed-cladding design is often used in practice.

Single-mode DCFs suffer from several problems. First, 1 km of DCF compensates dispersion for only 8–10 km of standard fiber. Second, DCF losses are relatively high in the 1.55-μm wavelength region ($\alpha \approx 0.5$ dB/km). Third, because of a relatively small mode diameter, the optical intensity is larger at a given input power, resulting in enhanced nonlinear effects. Most of the problems associated with a single-mode DCF can be solved to some extent by using a two-mode fiber designed with values of V such that the higher-order mode is near cutoff ($V \approx 2.5$). Such fibers have almost the same loss as the single-mode fiber but can be designed such that the dispersion parameter D for the higher-order mode has large negative values [44]–[46]. Indeed, values of D as large as -770 ps/(km-nm) have been measured for elliptical-core fibers [46]. A 1-km length of such a DCF can compensate the GVD of a 40-km-long fiber, adding relatively little to the total link loss. An added advantage of the two-mode DCF is that it allows for broadband dispersion compensation [44]. However, its use requires a mode-conversion device capable of transferring radiation from the fundamental to the higher-order mode supported by the DCF. Several such all-fiber devices have been developed [52]–[54]. As an alternative, a chirped fiber grating can be used for dispersion compensation [55].

3.4.3 Compensation of Third-Order Dispersion

When the bit rate of a single channel exceeds 100 Gb/s, one must use ultrashort pulses (width $\sim$1 ps) in each bit slot. For such short optical pulses, the pulse spectrum becomes broad enough that it is difficult to compensate GVD over the entire bandwidth of the pulse (because of the frequency dependence of β_2). The simplest solution to this problem is provided by fibers, or other devices, designed such that both β_2 and β_3 are compensated simultaneously [56]–[70]. The necessary conditions for designing such fibers can be obtained from Eq. (3.3.2). For a fiber link containing two different fibers of lengths L_1 and L_2, the conditions for broadband dispersion compensation are given by

$$\beta_{21}L_1 + \beta_{22}L_2 = 0 \qquad \text{and} \qquad \beta_{31}L_1 + \beta_{32}L_2 = 0, \qquad (3.4.9)$$

where β_{2j} and β_{3j} are the GVD and TOD parameters for fiber of length L_j ($j = 1, 2$). It is generally difficult to satisfy both conditions simultaneously over a wide wavelength range. However, for a 1-ps pulse, it is sufficient to satisfy Eq. (3.4.9) over a 4–5 nm bandwidth. This requirement is easily met for DCFs [56], especially designed with negative values of β_3 (sometimes called reverse-dispersion fibers). Fiber gratings, liquid-crystal modulators, and other devices can also be used for this purpose [62]–[68].

Several experiments have demonstrated signal transmission over distances ~100 km at high bit rates (100 Gb/s or more) using simultaneous compensation of both GVD and TOD. In a 1996 experiment, a 100-Gb/s signal was transmitted over 560 km with 80-km amplifier spacing [57]. In a later experiment, bit rate was extended to 400 Gb/s by using 0.98-ps optical pulses within the 2.5-ps time slot [58]. Without compensation of the TOD, the pulse broadened to 2.3 ps after 40 km and exhibited a long oscillatory tail extending over 5–6 ps (see Fig. 3.6). With partial compensation of TOD, the oscillatory tail disappeared and the pulse width reduced to 1.6 ps. In another experiment [59], a planar lightwave circuit was designed to have a dispersion slope of -15.8 ps/nm^2 over a 170-GHz bandwidth. It was used to compensate the TOD over 300 km of a dispersion-shifted fiber for which $\beta_3 \approx 0.05$ ps/(km-nm^2) at the operating wavelength. The dispersion compensator eliminated the long oscillatory tail and reduced the width of the main peak from 4.6 to 3.8 ps. The increase in pulse width from its input value of 2.6 ps can be attributed to the PMD effects.

The dispersion-compensation technique has also been used for femtosecond optical pulses. For a pulse with $T_0 = 0.1$ ps, the TOD length L_D' is only 10 m for a typical value $\beta_3 = 0.1$ ps^3/km. Such pulses cannot propagate more than a few meters before becoming severely distorted even when β_2 is compensated fully so that its average value is zero. Nonetheless, a 0.5-ps pulse ($T_0 \approx 0.3$ ps) was transmitted over 2.5 km of fiber using a 445-m-long DCF with $\beta_2 \approx 98$ ps^2/km and $\beta_3 \approx -0.5$ ps^3/km [60]. The output pulse was slightly distorted because β_3 could not be fully compensated. In a later experiment, a

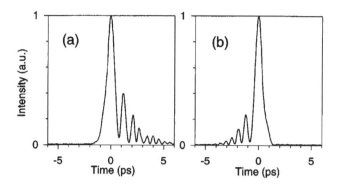

Figure 3.10 Experimentally observed shapes of a 0.5-ps input pulse at the output of a 2.5-km GVD-compensated fiber link. The value of β_3 was changed from 0.124 (left) to -0.076 ps^3/km (right) using a liquid-crystal modulator. (After Ref. [62].)

liquid-crystal modulator was used to compensate for the residual β_3 [62], and the pulse remained nearly unchanged after propagating over 2.5 km of the dispersion-compensated fiber link. In a 1999 experiment [66], the use of the same technique with a different DCF (length 1.5 km) permitted transmission of a 0.4-ps pulse ($T_0 \approx 0.25$ ps) over 10.6 km of fiber with little distortion in the pulse shape. The main advantage of a liquid-crystal modulator is that it acts as a programmable pulse shaper. It can even be used to enhance the TOD effects artificially. As an example, Fig. 3.10 shows the pulse shapes at the output of a 2.5-km GVD-compensated fiber link when the effective value of β_3 changes from 0.124 to -0.076 ps^3/km [62]. The observed pulse shapes are in agreement with those predicted by Eq. (3.3.2) as long as the nonlinear effects remain negligible.

When both β_2 and β_3 are nearly compensated, propagation of femtosecond optical pulses is limited by the fourth-order dispersive effects governed by the parameter β_4. In a 1999 experiment, the combination of a DCF and a frequency-resolved, programmable, dispersion compensator compensated β_2, β_3, and β_4 simultaneously over a 30-nm-wide wavelength range [67]. This scheme allowed transmission of a 0.2-ps pulse train with 22-nm bandwidth over a distance of 85 km. In a later experiment, 0.25-ps pulses could be transmitted over 139 km when dispersion up to fourth order was compensated using a DCF with a negative slope [70]. Input pulses were prechirped appropriately with a phase modulator. On the system level, a single high-speed channel at 640 Gb/s (obtained through time-division multiplexing) has been transmitted over 92 km by compensating β_2 and β_3 over the entire link consisting of standard, dispersion-shifted, and reverse-dispersion fibers [69].

Problems

3.1 A dispersion-shifted fiber is measured to have $D = 2$ ps/(km-nm) at 1.55 μm. It has an effective core area of 40 μm^2. Calculate the dispersion and nonlinear lengths when (i) 10-ps pulses with 100-mW peak power and (ii) 1-ps pulses with 1-W peak power are launched into the fiber. Are nonlinear effects important in both cases?

3.2 A chirped Gaussian pulse is well described by Eq. (3.2.14) with $C = 5$ and $T_0 = 50$ ps. Determine the temporal and spectral widths (FWHM) of this pulse.

3.3 Prove that for an unchirped Gaussian pulse of arbitrary width, the product $\Delta v \Delta t$ approximately equals 0.44, where Δt and Δv are the temporal and spectral widths (both measured as FWHM), respectively.

3.4 Repeat Problem 3.3 for an unchirped "sech" pulse and prove that $\Delta v \Delta t$ approximately equals 0.315.

3.5 Starting with Eq. (3.2.23), derive an expression for the RMS width of a super-Gaussian pulse.

3.6 Show that a chirped Gaussian pulse is compressed initially inside a single-mode fiber when $\beta_2 C < 0$. Derive expressions for the minimum width and the fiber length at which the minimum occurs.

3.7 Evaluate the integral in Eq. (3.3.2) numerically for an unchirped Gaussian pulse with 1-ps width (FWHM) assuming $\beta_2 = 0$ and $\beta_3 = 0.1$ ps^3/km. Plot the pulse shapes for $L = 2$ and 4 km. What happens to the pulse shape if the sign of β_3 is reversed or the input pulse is chirped?

3.8 Calculate the RMS width of an unchirped Gaussian pulse using Eqs. (3.3.18)–(3.3.24). Retain the β_2 and β_3 terms in Eq. (3.3.15) but neglect all others.

3.9 Estimate the limiting bit rate for a 60-km single-mode fiber link at 1.3- and 1.55-μm wavelengths assuming transform-limited 50-ps (FWHM) input pulses. Assume $\beta_2 = 0$ and -20 ps^2/km and $\beta_3 = 0.1$ and 0 ps^3/km at 1.3 and 1.55 μm wavelengths, respectively.

3.10 An optical communication system is operating with chirped Gaussian input pulses. Assume $\beta_3 = 0$ and $V_\omega \ll 1$ in Eq. (3.3.13) and obtain a condition on the bit rate in terms of the parameters C, β_2, and L.

References

[1] I. N. Sisakyan and A. B. Shvartsburg, *Sov. J. Quantum Electron.* **14**, 1146 (1984).

[2] S. A. Akhmanov, V. A. Vysloukh, and A. S. Chirkin, *Optics of Femtosecond Laser Pulses* (American Institute of Physics, New York, 1992), Chap. 1.

[3] G. P. Agrawal, in *Supercontinuum Laser Source*, R. R. Alfano, ed. (Springer-Verlag, Heidelberg, 1989), Chap. 3.

[4] C. G. B. Garrett and D. E. McCumber, *Phys. Rev. A* **1**,305 (1970).

[5] H. G. Unger, *Arch. Elecktron. Uebertragungstech.* **31**, 518 (1977).

[6] M. Miyagi and S. Nishida, *Appl. Opt.* **18**, 678 (1979).

[7] D. Gloge, *Electron. Lett.* **15**, 686 (1979).

[8] D. Marcuse, *Appl. Opt.* **19**, 1653 (1980).

[9] D. Marcuse, *Appl. Opt.* **20**, 3573 (1981).

[10] K. Iwashita, K. Nakagawa, Y. Nakano, and Y. Suzuki, *Electron. Lett.* **18**, 873 (1982).

[11] C. Lin and A. Tomita, *Electron. Lett.* **19**, 837 (1983).

[12] D. Anderson, M. Lisak, and P. Anderson, *Opt. Lett.* **10**, 134 (1985).

[13] F. Koyama and Y. Suematsu, *IEEE J. Quantum Electron.* **21**, 292 (1985).

[14] K. Tajima and K. Washio, *Opt. Lett.* **10**, 460 (1985).

[15] A. Takada, T. Sugie, and M. Saruwatari, *Electron. Lett.* **21**, 969 (1985).

[16] G. P. Agrawal and M. J. Potasek, *Opt. Lett.* **11**, 318 (1986).

[17] D. Anderson and M. Lisak, *Opt. Lett.* **11**, 569 (1986).

[18] A. Takada, T. Sugie, and M. Saruwatari, *J. Lightwave Technol.* **5**, 1525 (1987).

[19] K. Iwatsuki, A. Takada, and M. Saruwatari, *Electron. Lett.* **24**, 1572 (1988).

[20] K. Iwatsuki, A. Takada, S. Nishi, and M. Saruwatari, *Electron. Lett.* **25**, 1003 (1989).

[21] M. Nakazawa, K. Suzuki, and Y. Kimura, *Opt. Lett.* **15** 588 (1990).

[22] R. T. Hawkins, *Electron. Lett.* **26**, 292 (1990).

[23] G. P. Agrawal and N. A. Olsson, *Opt. Lett.* **14**, 500 (1989).

[24] N. A. Olsson, G. P. Agrawal, and K. W. Wecht, *Electron. Lett.* **25**, 603 (1989).

[25] G. P. Agrawal and N. A. Olsson, *IEEE J. Quantum Electron.* **25**, 2297 (1989).

[26] D. M. Bloom, L. F. Mollenauer, C. Lin, D. W. Taylor, and A. M. DelGaudio, *Opt. Lett.* **4**, 297 (1979).

[27] C. Lin, A. Tomita, A. R. Tynes, P. F. Glodis, and D. L. Philen, *Electron. Lett.* **18**, 882 (1982).

[28] G. P. Agrawal and N. K. Dutta, *Semiconductor Lasers*, 2nd ed. (Van Nostrand Reinhold, New York, 1993), Chap. 6.

[29] D. Anderson and M. Lisak, *Phys. Rev. A* **35**, 184 (1987).

[30] J. C. Diels, *Ultrashort Laser Pulse Phenomena* (Academic Press, San Diego, 1996).

[31] D. Meshulach, Y. Barad, and Y. Silberberg, *J. Opt. Soc. Am. B* **14**, 2122 (1997).

[32] D. T. Reid, W. Sibbett, J. M. Dudley, L. P. Barry, B. C. Thomsen, and J. D. Harvey, *Appl. Opt.* **37**, 8142 (1998).

[33] K. W. DeLong, D. N. Fittinghoff, and R. Trebino, *IEEE J. Quantum Electron.* **32**, 1253 (1996).

[34] D. J. Kane, *IEEE J. Quantum Electron.* **35**, 421 (1999).

[35] J. M. Dudley, L. P. Barry, P. G. Bollond, J. D. Harvey, and R. Leonhardt, *Opt. Fiber Technol.* **4**, 237 (1998).

[36] J. M. Dudley, L. P. Barry, J. D. Harvey, M. D. Thomson, B. C. Thomsen, P. G. Bollond, and R. Leonhardt, *IEEE J. Quantum Electron.* **35**, 441 (1999).

[37] F. G. Omenetto, B. P. Luce, D. Yarotski, and A. J. Taylor, *Opt. Lett.* **24**, 1392 (1999).

[38] R. G. M. P. Koumans and A. Yariv, *IEEE J. Quantum Electron.* **36**, 137 (2000); *IEEE Photon. Technol. Lett.* **12**, 666 (2000).

[39] G. P. Agrawal, *Fiber-Optic Communication Systems* (Wiley, New York, 1997), Chap. 2.

[40] A. H. Gnauck and R. M. Jopson, *Optical Fiber Telecommunications III*, I. P. Kaminow and T. L. Koch, eds. (Academic Press, San Diego, CA, 1997), Chap. 7.

[41] C. Lin, H. Kogelnik, and L. G. Cohen, *Opt. Lett.* **5**, 476 (1980).

[42] D. S. Larner and V. A. Bhagavatula, *Electron. Lett.* **21**, 1171 (1985).

[43] A. M. Vengsarkar and W. A. Reed, *Opt. Lett.* **18**, 924 (1993).

[44] C. D. Poole, J. M. Wiesenfeld, A. R. McCormick, and K. T. Nelson, *Opt. Lett.* **17**, 985 (1992).

[45] C. D. Poole, J. M. Wiesenfeld, and D. J. DiGiovanni, *IEEE Photon. Technol. Lett.* **5**, 194 (1993).

[46] C. D. Poole, J. M. Wiesenfeld, D. J. DiGiovanni, and A. M. Vengsarkar, *J. Lightwave Technol.* **12** 1746 (1994).

[47] A. J. Antos and D. K. Smith, *J. Lightwave Technol.* **12**, 1739 (1994).

[48] K. Thyagarajan, V. K. Varshney, P. Palai, A. K. Ghatak, and I. C. Goel, *IEEE Photon. Technol. Lett.* **8**, 1510 (1996).

[49] A. Goel and R. K. Shevgaonkar, *IEEE Photon. Technol. Lett.* **8**, 1668 (1996).

[50] M. Onishi, T. Kashiwada, Y. Ishiguro, Y. Koyano, M. Nishimura, and H. Kanamori, *Fiber Integ. Opt.* **16**, 277 (1997).

[51] M. J. Yadlowsky, E. M. Deliso, and V. L. daSilva, *Proc. IEEE* **85**, 1765 (1997)

[52] R. C. Youngquist, J. L. Brooks, and H. J. Shaw, *Opt. Lett.* **9**, 177 (1984).

[53] J. N. Blake, B. Y. Kim, and H. J. Shaw, *Opt. Lett.* **11**, 177 (1986).

[54] C. D. Poole, C. D. Townsend, and K. T. Nelson, *J. Lightwave Technol.* **9**, 598 (1991).

[55] R. Kashyap, *Fiber Bragg Gratings* (Academic Press, San Diego, CA, 1999).

[56] C. C. Chang, A. M. Weiner, A. M. Vengsarakar, and D. W. Peckham, *Opt. Lett.* **21**, 1141 (1996).

[57] S. Kawanishi, H. Takara, O. Kamatani, T. Morioka, and M. Saruwatari, *Electron. Lett.* **32**, 470 (1996).

[58] S. Kawanishi, H. Takara,, T. Morioka, O. Kamatani, K. Takiguchi, T. Kitoh, and M. Saruwatari, *Electron. Lett.* **32**, 916 (1996).

[59] K. Takiguchi, S. Kawanishi, H. Takara, K. Okamoto, and Y. Ohmori, *Electron. Lett.* **32**, 755 (1996).

[60] C. C. Chang and A. M. Weiner, *IEEE J. Quantum Electron.* **33**, 1455 (1997).

[61] M. Durkin, M. Ibsen, M. J. Cole, and R. I. Laming, *Electron. Lett.* **23**, 1891 (1997).

[62] C. C. Chang, H. P. Sardesai, and A. M. Weiner, *Opt. Lett.* **23**, 283 (1998).

[63] K. Takiguchi, S. Kawanishi, H. Takara, A. Himeno, and K. Hattori, *J. Lightwave Technol.* **16**, 1647 (1998).

[64] T. Imai, T. Komukai, and M. Nakazawa, *Electron. Lett.* **34**, 2422 (1998).

[65] H. Tsuda, K. Okamoto, T. Ishii, K. Naganuma, Y. Inoue, H. Takenouchi, and T. Kurokawa, *IEEE Photon. Technol. Lett.* **11**, 569 (1999).

[66] S. Shen and A. M. Weiner, *IEEE Photon. Technol. Lett.* **11**, 827 (1999).

[67] F. Futami, K. Taira, K. Kikuchi, and A. Suzuki, *Electron. Lett.* **35**, 2221 (1999).

[68] T. Komukai, T. Inui, and M. Nakazawa, *IEEE J. Quantum Electron.* **36**, 409 (2000).

[69] T. Yamamoto, E. Yoshida, K. R. Tamura, K. Yonenaga, and M. Nakazawa, *IEEE Photon. Technol. Lett.* **12**, 353 (2000).

[70] M. D. Pelusi, F. Futami, K. Kikuchi, and A. Suzuki, *IEEE Photon. Technol. Lett.* **12**, 795 (2000).

Chapter 4

Self-Phase Modulation

An interesting manifestation of the intensity dependence of the refractive index in nonlinear optical media occurs through self-phase modulation (SPM), a phenomenon that leads to spectral broadening of optical pulses [1]–[9]. SPM is the temporal analog of self-focusing. Indeed, it was first observed in 1967 in the context of transient self-focusing of optical pulses propagating in a CS_2-filled cell [1]. By 1970, SPM had been observed in solids and glasses by using picosecond pulses. The earliest observation of SPM in optical fibers was made with a fiber whose core was filled with CS_2 [7]. This work led to a systematic study of SPM in a silica-core fiber [9]. This chapter considers SPM as a simple example of the nonlinear optical effects that can occur in optical fibers. Section 4.1 is devoted to the case of pure SPM by neglecting the GVD effects. The effects of GVD on SPM are discussed in Section 4.2 with particular emphasis on the SPM-induced frequency chirp. Section 4.3 extends the results to include the higher-order nonlinear effects such as self-steepening.

4.1 SPM-Induced Spectral Broadening

A general description of SPM in optical fibers requires numerical solutions of the pulse-propagation equation (2.3.39) obtained in Section 2.3. The simpler equation (2.3.41) can be used for pulse widths $T_0 > 5$ ps. A further simplification occurs if the effect of GVD on SPM is negligible so that the β_2 term in Eq. (2.3.41) can be set to zero. The conditions under which GVD can be ignored were discussed in Section 3.1 by introducing the length scales L_D and L_{NL} [see Eq. (3.1.5)]. In general, the pulse width and the peak power should

97

be such that $L_D \gg L > L_{NL}$ for a fiber of length L. Equation (3.1.7) shows that the GVD effects are negligible for relatively wide pulses ($T_0 > 100$ ps) with a large peak power ($P_0 > 1$ W).

4.1.1   Nonlinear Phase Shift

In terms of the normalized amplitude $U(z, T)$ defined as in Eq. (3.1.3), the pulse-propagation equation (3.1.4), in the limit $\beta_2 = 0$, becomes

$$\frac{\partial U}{\partial z} = \frac{ie^{-\alpha z}}{L_{NL}} |U|^2 U, \tag{4.1.1}$$

where α accounts for fiber losses. The nonlinear length is defined as

$$L_{NL} = (\gamma P_0)^{-1}, \tag{4.1.2}$$

where P_0 is the peak power and γ is related to the nonlinear-index coefficient n_2 as in Eq. (2.3.28). Equation (4.1.1) can be solved substituting $U = V \exp(i\phi_{NL})$ and equating the real and imaginary parts so that

$$\frac{\partial V}{\partial z} = 0; \qquad \frac{\partial \phi_{NL}}{\partial z} = \frac{e^{-\alpha z}}{L_{NL}} V^2. \tag{4.1.3}$$

As the amplitude V does not change along the fiber length L, the phase equation can be integrated analytically to obtain the general solution

$$U(L, T) = U(0, T) \exp[i\phi_{NL}(L, T)], \tag{4.1.4}$$

where $U(0, T)$ is the field amplitude at $z = 0$ and

$$\phi_{NL}(L, T) = |U(0, T)|^2 (L_{eff}/L_{NL}), \tag{4.1.5}$$

with the effective length L_{eff} defined as

$$L_{eff} = [1 - \exp(-\alpha L)]/\alpha. \tag{4.1.6}$$

Equation (4.1.4) shows that SPM gives rise to an intensity-dependent phase shift but the pulse shape remains unaffected. The nonlinear phase shift ϕ_{NL} in Eq. (4.1.5) increases with fiber length L. The quantity L_{eff} plays the role of an effective length that is smaller than L because of fiber losses. In the absence of fiber losses, $\alpha = 0$, and $L_{eff} = L$. The maximum phase shift ϕ_{max} occurs at the pulse center located at $T = 0$. With U normalized such that $|U(0,0)| = 1$, it is given by

$$\phi_{max} = L_{eff}/L_{NL} = \gamma P_0 L_{eff}. \qquad (4.1.7)$$

The physical meaning of the nonlinear length L_{NL} is clear from Eq. (4.1.7)—it is the effective propagation distance at which $\phi_{max} = 1$. If we use a typical value $\gamma = 2$ $W^{-1}km^{-1}$ in the 1.55-μm wavelength region, $L_{NL} = 50$ km at a power level $P_0 = 10$ mW and decreases inversely with an increase in P_0.

The SPM-induced spectral broadening is a consequence of the time dependence of ϕ_{NL}. This can be understood by noting that a temporally varying phase implies that the instantaneous optical frequency differs across the pulse from its central value ω_0. The difference $\delta\omega$ is given by

$$\delta\omega(T) = -\frac{\partial\phi_{NL}}{\partial T} = -\left(\frac{L_{eff}}{L_{NL}}\right)\frac{\partial}{\partial T}|U(0,T)|^2, \qquad (4.1.8)$$

where the minus sign is due to the choice of the factor $\exp(-i\omega_0 t)$ in Eq. (2.3.2). The time dependence of $\delta\omega$ is referred to as frequency chirping. The chirp induced by SPM increases in magnitude with the propagated distance. In other words, new frequency components are generated continuously as the pulse propagates down the fiber. These SPM-generated frequency components broaden the spectrum over its initial width at $z = 0$.

The extent of spectral broadening depends on the pulse shape. Consider, for example, the case of a super-Gaussian pulse with the incident field $U(0,T)$ given by Eq. (3.2.23). The SPM-induced chirp $\delta\omega(T)$ for such a pulse is

$$\delta\omega(T) = \frac{2m}{T_0}\frac{L_{eff}}{L_{NL}}\left(\frac{T}{T_0}\right)^{2m-1}\exp\left[-\left(\frac{T}{T_0}\right)^{2m}\right], \qquad (4.1.9)$$

where $m = 1$ for a Gaussian pulse. For larger values of m, the incident pulse becomes nearly rectangular with increasingly steeper leading and trailing edges. Figure 4.1 shows variation of the nonlinear phase shift ϕ_{NL} and the induced frequency chirp $\delta\omega$ across the pulse at $L_{eff} = L_{NL}$ in the cases of a Gaussian pulse ($m = 1$) and a super-Gaussian pulse ($m = 3$). As ϕ_{NL} is directly proportional to $|U(0,T)|^2$ in Eq. (4.1.5), its temporal variation is identical to that of the pulse intensity. The temporal variation of the induced chirp $\delta\omega$ has several interesting features. First, $\delta\omega$ is negative near the leading edge (red shift) and becomes positive near the trailing edge (blue shift) of the pulse. Second, the chirp is linear and positive (up-chirp) over a large central region of the

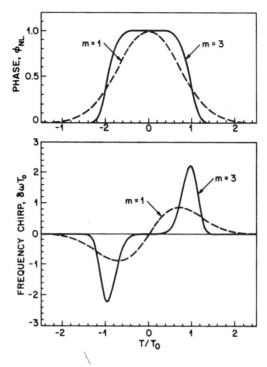

Figure 4.1 Temporal variation of SPM-induced phase shift ϕ_{NL} and frequency chirp $\delta\omega$ for Gaussian (dashed curve) and super-Gaussian (solid curve) pulses.

Gaussian pulse. Third, the chirp is considerably larger for pulses with steeper leading and trailing edges. Fourth, super-Gaussian pulses behave differently than Gaussian pulses because the chirp occurs only near pulse edges and does not vary in a linear fashion.

4.1.2   Changes in Pulse Spectra

An estimate of the magnitude of SPM-induced spectral broadening can be obtained from the peak value of $\delta\omega$ in Fig. 4.1. More quantitatively, we can calculate the peak value by maximizing $\delta\omega(T)$ from Eq. (4.1.9). By setting its time derivative to zero, the maximum value of $\delta\omega$ is given by

$$\delta\omega_{max} = \frac{mf(m)}{T_0}\phi_{max}, \qquad (4.1.10)$$

where ϕ_{max} is given in Eq. (4.1.7) and $f(m)$ is defined as

$$f(m) = 2\left(1 - \frac{1}{2m}\right)^{1-1/2m} \exp\left[-\left(1 - \frac{1}{2m}\right)\right]. \tag{4.1.11}$$

The numerical value of f depends on m only slightly; $f = 0.86$ for $m = 1$ and tends toward 0.74 for large values of m. To obtain the broadening factor, the width parameter T_0 should be related to the initial spectral width $\Delta\omega_0$ of the pulse. For an unchirped Gaussian pulse, $\Delta\omega_0 = T_0^{-1}$ from Eq. (3.2.16), where $\Delta\omega_0$ is the $1/e$ half-width. Equation (4.1.10) then becomes (with $m = 1$)

$$\delta\omega_{max} = 0.86 \, \Delta\omega_0 \phi_{max}, \tag{4.1.12}$$

showing that the spectral broadening factor is approximately given by the numerical value of the maximum phase shift ϕ_{max}. In the case of a super-Gaussian pulse, it is difficult to estimate $\Delta\omega_0$ because its spectrum is not Gaussian. However, if we use Eq. (3.2.24) to obtain the rise time, $T_r = T_0/m$, and assume that $\Delta\omega_0$ approximately equals T_r^{-1}, Eq. (4.1.10) shows that the broadening factor of a super-Gaussian pulse is also approximately given by ϕ_{max}. With $\phi_{max} \sim 100$ possible for intense pulses or long fibers, SPM can broaden the spectrum considerably. In the case of intense ultrashort pulses, the broadened spectrum can extend over 100 THz or more, especially when SPM is accompanied by other nonlinear processes such as stimulated Raman scattering and four-wave mixing. Such an extreme spectral broadening is referred to as supercontinuum [4].

The actual shape of the pulse spectrum $S(\omega)$ is obtained by taking the Fourier transform of Eq. (4.1.4). Using $S(\omega) = |\tilde{U}(L,\omega)|^2$, we obtain

$$S(\omega) = \left|\int_{-\infty}^{\infty} U(0,T) \exp[i\phi_{NL}(L,T) + i(\omega - \omega_0)T]dT\right|^2. \tag{4.1.13}$$

In general, the spectrum depends not only on the pulse shape but also on the initial chirp imposed on the pulse. Figure 4.2 shows the spectra of an unchirped Gaussian pulse for several values of the maximum phase shift ϕ_{max}. For a given fiber length, ϕ_{max} increases linearly with peak power P_0 according to Eq. (4.1.7). Thus, spectral evolution seen in Fig. 4.2 can be observed experimentally by increasing the peak power. Figure 4.3 shows the experimentally observed spectra [9] of nearly Gaussian pulses ($T_0 \approx 90$ ps), obtained from an argon-ion laser, at the output of a 99-m-long fiber with 3.35-μm core diameter (parameter $V = 2.53$). The experimental spectra are also labeled with

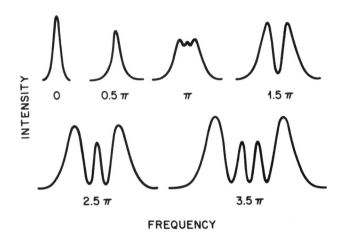

Figure 4.2 SPM-broadened spectra for an unchirped Gaussian pulse. Spectra are labeled by the maximum nonlinear phase shift ϕ_{max}. (After Ref. [9].)

ϕ_{max} and should be compared with the calculated spectra of Fig. 4.2. Slight asymmetry seen in the experimental traces can be attributed to the asymmetric shape of the incident pulse [9]. The overall agreement between theory and the experiment is remarkably good.

The most notable feature of Figs. 4.2 and 4.3 is that SPM-induced spectral broadening is accompanied by an oscillatory structure covering the entire frequency range. In general, the spectrum consists of many peaks, and the outermost peaks are the most intense. The number of peaks depends on ϕ_{max} and increases linearly with it. The origin of the oscillatory structure can be understood by referring to Fig. 4.1 where the time dependence of the SPM-induced frequency chirp is shown. In general, the same chirp occurs at two values of T, showing that the pulse has the same instantaneous frequency at two distinct points. Qualitatively speaking, these two points represent two waves of the same frequency but different phases that can interfere constructively or destructively depending on their relative phase difference. The multipeak structure in the pulse spectrum is a result of such interference [1]. Mathematically, the Fourier integral in Eq. (4.1.13) gets dominant contributions at the two values of T at which the chirp is the same. These contributions, being complex quantities, may add up in phase or out of phase. Indeed, one can use the method of stationary phase to obtain an analytic expression of $S(\omega)$ that is valid for large values of ϕ_{max}. This expression shows that the number of peaks

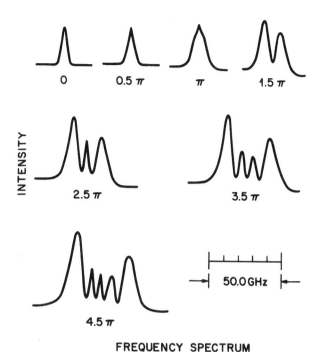

FREQUENCY SPECTRUM

Figure 4.3 Experimentally observed spectra for a nearly Gaussian pulse at the output of a 99-m-long fiber. Spectra are labeled by the maximum phase shift ϕ_{max} related linearly to the peak power. (After Ref. [9].)

M in the SPM-broadened spectrum is given approximately by the relation [3]

$$\phi_{max} \approx (M - \tfrac{1}{2})\pi. \qquad (4.1.14)$$

Equation (4.1.14) together with Eq. (4.1.12) can be used to estimate the initial spectral width $\Delta\omega_0$ or the pulse width T_0 if the pulse is unchirped [6]. The method is accurate however only if $\phi_{max} \gg 1$. To obtain a more accurate measure of spectral broadening, one should use the RMS spectral width $\Delta\omega_{rms}$ defined as

$$\Delta\omega_{rms}^2 = \langle(\omega - \omega_0)^2\rangle - \langle(\omega - \omega_0)\rangle^2, \qquad (4.1.15)$$

where the angle brackets denote an average over the SPM-broadened spectrum given in Eq. (4.1.13). More specifically,

$$\langle(\omega - \omega_0)\rangle^n = \frac{\int_{-\infty}^{\infty}(\omega - \omega_0)^n S(\omega)\, d\omega}{\int_{-\infty}^{\infty} S(\omega)\, d\omega}. \qquad (4.1.16)$$

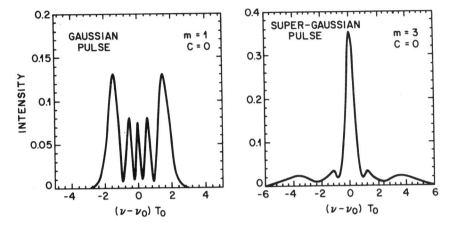

Figure 4.4 Comparison of SPM-broadened spectra for unchirped Gaussian and super-Gaussian pulses at a peak power corresponding to $\phi_{max} = 4.5\pi$.

Using a procedure similar to that of Section 3.3, the spectral broadening factor for a Gaussian pulse is given by [10]

$$\frac{\Delta\omega_{rms}}{\Delta\omega_0} = \left(1 + \frac{4}{3\sqrt{3}}\phi_{max}^2\right)^{1/2}, \qquad (4.1.17)$$

where $\Delta\omega_0$ is the initial RMS spectral width of the pulse.

4.1.3 Effect of Pulse Shape and Initial Chirp

As mentioned before, the shape of the SPM-broadened spectrum depends on the pulse shape and on the initial chirp if the input pulse is chirped [11]. Figure 4.4 compares the pulse spectra for Gaussian ($m = 1$) and super-Gaussian ($m = 3$) pulses obtained using Eq. (3.2.23) in Eq. (4.1.13) and performing the integration numerically. In both cases, input pulses are assumed to be unchirped ($C = 0$). The fiber length and the peak power are chosen such that $\phi_{max} = 4.5\pi$. The qualitative differences between the two spectra can be understood by referring to Fig. 4.1, where the SPM-induced chirp is shown for the Gaussian and super-Gaussian pulses. The spectral range is about three times larger for the super-Gaussian pulse because the maximum chirp from Eq. (4.1.10) is about three times larger in that case. Even though both spectra in Fig. 4.4 exhibit five peaks, in agreement with Eq. (4.1.14), most of the

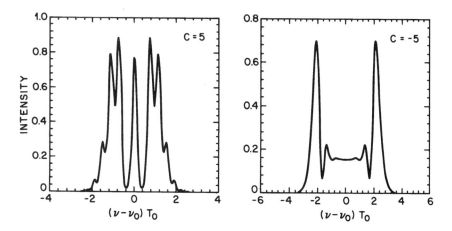

Figure 4.5 Effect of initial frequency chirp on SPM-broadened spectra of a chirped Gaussian pulse for $C = 5$ and $C = -5$. The two spectra should be compared with the left spectrum in Fig. 4.4 where $C = 0$. In all cases $\phi_{\max} = 4.5\pi$.

energy remains in the central peak for the super-Gaussian pulse. This is so because the chirp is nearly zero over the central region in Fig. 4.1 for such a pulse as a consequence of the nearly uniform intensity of a super-Gaussian pulse for $|T| < T_0$. The frequency chirp occurs mainly near the leading and trailing edges. As these edges become steeper, the tails in Fig. 4.4 extend over a longer frequency range but, at the same time, carry less energy because chirping occurs over a small time duration.

An initial frequency chirp can also lead to drastic changes in the SPM-broadened pulse spectrum. This is illustrated in Fig. 4.5, which shows the spectra of a Gaussian pulse with positive and negative chirps [$C = \pm 5$ in Eq. (3.2.23)] under conditions identical to those of Fig. 4.4, i.e., $\phi_{\max} = 4.5\pi$. A comparison of these spectra with the spectrum of the unchirped Gaussian pulse (left plot in Fig. 4.4) shows how the initial chirp leads to qualitative changes in SPM-induced spectral broadening. A positive chirp increases the number of spectral peaks while the opposite occurs in the case of a negative chirp. This can be understood by noting that the SPM-induced frequency chirp is linear and positive (frequency increases with increasing T) over the central portion of a Gaussian pulse (see Fig. 4.1). Thus, it adds with the initial chirp for $C > 0$, resulting in an enhanced oscillatory structure. In the case of $C < 0$, the two chirp contributions are of opposite signs except near the pulse edges. The outermost peaks in Fig. 4.5 for $C = -5$ are due to the residual chirp near the leading and trailing edges.

For negative values of the chirp parameter C, pulse spectrum at the fiber output can become narrower than that of initially unchirped pulses. Such a spectral narrowing has been seen experimentally using 100-fs pulses (emitted from a mode-locked Ti:sapphire laser operating near 0.8 μm) and chirping them with a prism pair before launching them into a 48-cm-long fiber [11]. The 10.6-nm spectral width of input pulses was nearly unchanged at low peak powers but became progressively smaller as the peak power was increased. It reduced to 3.1 nm at a 1.6-kW peak power. The output spectral width also changed with the fiber length at a given peak power and exhibited a minimum value of 2.7 nm for a fiber length of 28 cm at the 1-kW peak power. The spectrum rebroadened for longer fibers. These results can be understood qualitatively by noting that the spectrum narrows as long as the SPM-induced chirp compensates the initial chirp. For a quantitative modeling of the experimental data it is necessary to include the effects of GVD for 100-fs pulses used in the experiment. This issue is covered in Section 4.2.

4.1.4 Effect of Partial Coherence

In the preceding discussion, SPM-induced spectral broadening occurs only for optical pulses because, as seen in Eq. (4.1.5), the nonlinear phase shift mimics temporal variations of the pulse shape. Indeed, the SPM-induced chirp in Eq. (4.1.8) vanishes for continuous-wave (CW) radiation, implying that a CW beam would not experience any spectral broadening in optical fibers. This conclusion, however, is a consequence of an implicit assumption that the input optical field is perfectly coherent. In practice, all optical beams are only partially coherent. The degree of coherence for laser beams is large enough that the effects of partial coherence are negligible in most cases of practical interest. For example, SPM-induced spectral broadening of optical pulses is relatively unaffected by the partial temporal coherence of the laser source as long as the coherence time T_c of the laser beam is much larger than the pulse width T_0.

When the coherence time becomes shorter than the pulse width, effects of partial coherence must be included [12]–[18]. In the case of a CW beam, SPM can lead to spectral broadening during its propagation inside an optical fiber. The physical reason behind such broadening can be understood by noting that partially coherent light exhibits both intensity and phase fluctuations. The

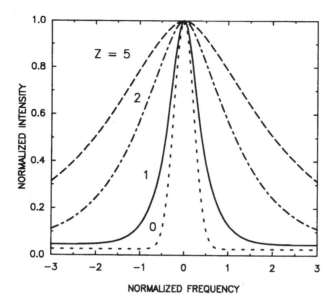

Figure 4.6 SPM-induced spectral broadening of a partially coherent CW beam for several values of Z. The curve marked $Z = 0$ shows the input Gaussian spectrum.

SPM converts intensity fluctuations into additional phase fluctuations [see Eq. (4.1.5)] and broadens the optical spectrum. Alternatively, SPM reduces the coherence time T_c as the CW beam propagates inside the fiber, making it less and less coherent.

The optical spectrum of partially coherent light at the fiber output is obtained using the Wiener–Khintchine theorem [19]:

$$S(\omega) = \int_{-\infty}^{\infty} \Gamma(z, \tau) \exp(i\omega\tau)\, d\tau, \qquad (4.1.18)$$

where the coherence function $\Gamma(z, \tau)$ is defined as

$$\Gamma(z, \tau) = \langle U^*(z, T) U(z, T + \tau) \rangle. \qquad (4.1.19)$$

The optical field $U(z, T)$ inside the fiber at a distance z is known from Eq. (4.1.4). The angle brackets denote an ensemble average over fluctuations in the input field $U(0, T)$. The statistical properties of $U(0, T)$ depend on the optical source and are generally quite different for laser and nonlaser sources.

The average in Eq. (4.1.19) can be performed analytically for thermal sources because both the real and imaginary parts of $U(0, T)$ follow a Gaussian distribution for such a source. Even though the laser light used commonly in nonlinear-optics experiments is far from being thermal, it is instructive to consider the case of a thermal field. The coherence function of Eq. (4.1.19) in that specific case is found to evolve as [13]

$$\Gamma(Z, \tau) = \Gamma(0, \tau)[1 + Z^2(1 - |\Gamma(0, \tau)|^2)]^{-2}, \qquad (4.1.20)$$

where $Z = L_{\text{eff}}/L_{\text{NL}}$ is the normalized propagation distance. For a perfectly coherent field, $\Gamma(0, \tau) = 1$. Equation (4.1.20) shows that such a field remains perfectly coherent on propagation. In contrast, partially coherent light becomes progressively less coherent as it travels inside the fiber. Such a coherence degradation can be understood by noting that SPM converts intensity fluctuations into additional phase fluctuations, making light less coherent.

The spectrum is obtained by substituting Eq. (4.1.20) in Eq. (4.1.18). The integral can be performed analytically in some specific cases [12], but in general requires numerical evaluation (through the FFT algorithm, for example). As an example, Fig. 4.6 shows the optical spectra at several propagation distances assuming a Gaussian form for the input coherence function,

$$\Gamma(0, \tau) = \exp[-(\tau^2/2T_c^2)], \qquad (4.1.21)$$

where T_c is the coherence time of the input field. As expected, shortening of the coherence time is accompanied by SPM-induced spectral broadening. Little broadening occurs until light has propagated a distance equal to the nonlinear length L_{NL}, but the spectrum broadens by about a factor of 8 at $Z = 5$. The spectral shape is quite different qualitatively compared with those seen in Fig. 4.2 for the case of a completely coherent pulse. In particular, note the absence of a multipeak structure.

One may ask how the SPM-broadened spectrum of an optical pulse is affected by the partial coherence of the optical source. Numerical simulations show that each peak of the multipeak structure seen in Fig. 4.2 begins to broaden when the coherence time becomes comparable to or shorter than the pulse width. As a result, individual peaks begin to merge together. In the limit of very short coherence time, the multipeak structure disappears altogether, and spectral broadening has features similar to those seen in Fig. 4.6. The SPM-induced coherence degradation and the associated spectral broadening has been observed experimentally by using stimulated Raman scattering (see Chapter 8) as a source of partially coherent light [14].

4.2 Effect of Group-Velocity Dispersion

The SPM effects discussed in Section 4.1 describe the propagation behavior realistically only for relatively long pulses ($T_0 > 100$ ps) for which the dispersion length L_D is much larger compared with both the fiber length L and the nonlinear length L_{NL}. As pulses become shorter and the dispersion length becomes comparable to the fiber length, it becomes necessary to consider the combined effects of GVD and SPM [8]. New qualitative features arise from an interplay between GVD and SPM. In the anomalous-dispersion regime of an optical fiber, the two phenomena can cooperate in such a way that the pulse propagates as an optical soliton (Chapter 5). In the normal-dispersion regime, the combined effects of GVD and SPM can be used for pulse compression. This section considers the temporal and spectral changes that occur when the effects of GVD are included in the description of SPM [20]–[31].

4.2.1 Pulse Evolution

The starting point is the nonlinear Schrödinger (NLS) equation (2.3.41) or Eq. (3.1.4). The later equation can be written in a normalized form as

$$i\frac{\partial U}{\partial \xi} = \text{sgn}(\beta_2)\frac{1}{2}\frac{\partial^2 U}{\partial \tau^2} - N^2 e^{-\alpha z}|U|^2 U, \qquad (4.2.1)$$

where ξ and τ represent the normalized distance and time variables defined as

$$\xi = z/L_D, \qquad \tau = T/T_0, \qquad (4.2.2)$$

and the parameter N is introduced by using

$$N^2 = \frac{L_D}{L_{NL}} \equiv \frac{\gamma P_0 T_0^2}{|\beta_2|}. \qquad (4.2.3)$$

The physical significance of N will become clear in Chapter 5 where the integer values of N are found to be related to the soliton order. The practical significance of the parameter N is that solutions of Eq. (4.2.1) obtained for a specific N value are applicable to many practical situations through the scaling law of Eq. (4.2.3). For example, if $N = 1$ for $T_0 = 1$ ps and $P_0 = 1$ W, the calculated results apply equally well for $T_0 = 10$ ps and $P_0 = 10$ mW or $T_0 = 0.1$ ps and $P_0 = 100$ W. As evident from Eq. (4.2.3), N governs the relative importance of the SPM and GVD effects on pulse evolution along the fiber. Dispersion

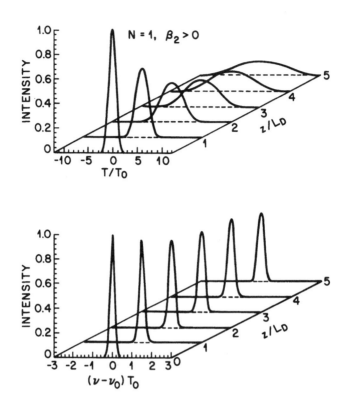

Figure 4.7 Evolution of pulse shapes (upper plot) and optical spectra (lower plot) over a distance of $5L_D$ for an initially unchirped Gaussian pulse propagating in the normal-dispersion regime of the fiber ($\beta_2 > 0$) with parameters such that $N = 1$.

dominates for $N \ll 1$ while SPM dominates for $N \gg 1$. For values of $N \sim 1$, both SPM and GVD play an equally important role during pulse evolution. In Eq. (4.2.1), $\mathrm{sgn}(\beta_2) = \pm 1$ depending on whether GVD is normal ($\beta_2 > 0$) or anomalous ($\beta_2 < 0$). The split-step Fourier method of Section 2.4 can be used to solve Eq. (4.2.1) numerically.

Figure 4.7 shows evolution of the shape and the spectrum of an initially unchirped Gaussian pulse in the normal-dispersion regime of a fiber using $N = 1$ and $\alpha = 0$. The qualitative behavior is quite different from that expected when either GVD or SPM dominates. In particular, the pulse broadens much more rapidly compared with the $N = 0$ case (no SPM). This can be understood by noting that SPM generates new frequency components that are red-shifted near the leading edge and blue-shifted near the trailing edge of the

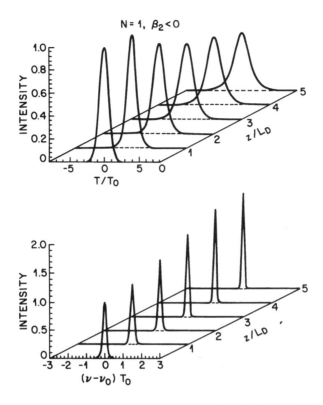

Figure 4.8 Evolution of pulse shapes (upper plot) and optical spectra (lower plot) under conditions identical to those of Fig. 4.7 except that the Gaussian pulse propagates in the anomalous-dispersion regime ($\beta_2 < 0$).

pulse. As the red components travel faster than the blue components in the normal-dispersion regime, SPM leads to an enhanced rate of pulse broadening compared with that expected from GVD alone. This in turn affects spectral broadening as the SPM-induced phase shift ϕ_{NL} becomes less than that occurring if the pulse shape were to remain unchanged. Indeed, $\phi_{max} = 5$ at $z = 5L_D$, and a two-peak spectrum is expected in the absence of GVD. The single-peak spectrum for $z/L_D = 5$ in Fig. 4.7 implies that the effective ϕ_{max} is below π because of pulse broadening.

The situation is different for pulses propagating in the anomalous-dispersion regime of the fiber. Figure 4.8 shows the pulse shapes and spectra under conditions identical to those of Fig. 4.7 except that the sign of the GVD parameter has been reversed ($\beta_2 < 0$). The pulse broadens initially at a rate much lower

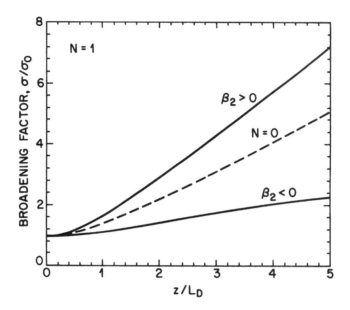

Figure 4.9 Broadening factor of Gaussian pulses in the cases of normal ($\beta_2 > 0$) and anomalous ($\beta_2 < 0$) GVD. The parameter $N = 1$ in both cases. Dashed curve shows for comparison the broadening expected in the absence of SPM ($N = 0$).

than that expected in the absence of SPM and then appears to reach a steady state for $z > 4L_D$. At the same time, the spectrum narrows rather than exhibiting broadening expected by SPM in the absence of GVD. This behavior can be understood by noting that the SPM-induced chirp given by Eq. (4.1.9) is positive while the dispersion-induced chirp given by Eq. (3.2.13) is negative for $\beta_2 < 0$. The two chirp contributions nearly cancel each other along the center portion of the Gaussian pulse when $L_D = L_{NL}$ ($N = 1$). Pulse shape adjusts itself during propagation to make such cancelation as complete as possible. Thus, GVD and SPM cooperate with each other to maintain a chirp-free pulse. The preceding scenario corresponds to soliton evolution; initial broadening of the Gaussian pulse occurs because the Gaussian profile is not the characteristic shape associated with a fundamental soliton. Indeed, if the input pulse is chosen to be a "sech" pulse [Eq. (3.2.21) with $C = 0$], both its shape and spectrum remain unchanged during propagation. When the input pulse deviates from a 'sech' shape, the combination of GVD and SPM affects the pulse in such a way that it evolves to become a 'sech' pulse, as seen in Fig. 4.8. This aspect is discussed in detail in Chapter 5.

4.2.2 Broadening Factor

Figures 4.7 and 4.8 show that the main effect of SPM is to alter the broadening rate imposed on the pulse by the GVD alone. Figure 4.9 shows the broadening factor σ/σ_0 as a function of z/L_D for $N = 1$ when unchirped Gaussian pulses are launched into the fiber. Here σ is the RMS width defined by Eq. (3.2.25) and σ_0 is its initial value. The dashed line shows for comparison the broadening factor in the absence of SPM ($N = 0$). The SPM enhances the broadening rate in the normal-dispersion regime and decreases it in the anomalous-dispersion regime. The slower broadening rate for $\beta_2 < 0$ is useful for 1.55-μm optical communication systems for which $\beta_2 \approx -20$ ps^2/km when standard fibers (the zero-dispersion wavelength near 1.3-μm) are used. The performance of such systems is dispersion limited to the extent that the bit rate–distance product BL is typically below 100 (Gb/s)-km for chirped pulses with $C = -5$. It has been shown that the BL product can be nearly doubled by increasing the peak power in the range 20–30 mW [26]. This enhancement is due to the SPM-induced pulse narrowing seen in Fig. 4.8 for the case $\beta_2 < 0$.

It is generally necessary to solve Eq. (4.2.1) numerically to study the combined effects of GVD and SPM. However, even an approximate analytic expression for the pulse width would be useful to see the functional dependence of the broadening rate on various physical parameters. Several approaches have been used to solve the NLS equation approximately [32]–[39]. A variational approach was used as early as 1983. It assumes that the pulse maintains a certain shape during propagation inside the fiber while its width or chirp can change with z. In the case of a Gaussian pulse of the form of Eq. (3.2.14), the parameters T_0 and C are allowed to vary with z. Their evolution equations can be obtained using the variational principle [32] or the path-integral formulation [33]. This method is quite powerful because it provides physical insight in the evolution behavior even for initially chirped pulses. However, its validity is limited to values of $N < 1$ for which the pulse shape does not change drastically. This approach is also useful for solitons as discussed in Chapter 5.

In a different approach [35], the NLS equation is first solved by neglecting the GVD effects. The result is used as the initial condition, and Eq. (4.2.1) is solved again by neglecting the SPM effects. The approach is similar to the split-step Fourier method of Section 2.4 except that the step size is equal to the fiber length. The RMS pulse width can be calculated analytically by following the method discussed in Section 3.3. In the case of an unchirped Gaussian pulse incident at the input end of a fiber of length L, the broadening factor is given by [35]

$$\frac{\sigma}{\sigma_0} = \left[1 + \sqrt{2}\phi_{\max}\frac{L}{L_D} + \left(1 + \frac{4}{3\sqrt{3}}\phi_{\max}^2\right)\frac{L^2}{L_D^2}\right]^{1/2}, \qquad (4.2.4)$$

where $\phi_{\max}$ is the SPM-induced maximum phase shift given by Eq. (4.1.7). This expression is fairly accurate for $\phi_{\max} < 1$.

In another approach, Eq. (4.2.1) is solved in the frequency domain [36]. Such a spectral approach shows that SPM can be viewed as a four-wave-mixing process [22] in which two photons at pump frequencies are annihilated to create two photons at frequencies shifted toward the blue and red sides. These newly created spectral components result in SPM-induced spectral broadening of the pulse. The oscillatory structure of the SPM spectra is due to the phase-matching requirement of four-wave mixing (see Chapter 10). Although in general the equation describing evolution of the spectral components should be solved numerically, it can be solved analytically in some cases if the pulse shape is assumed not to change significantly.

Another method that has been used with success studies evolution of the pth moment $\langle T^p \rangle$ as defined in Eq. (3.2.26), using the NLS equation (3.1.1) or (4.2.1). If the fiber loss is neglected assuming that it is compensated using optical amplifiers and Eq. (3.1.1) is first multiplied by $T^p A^*$ and then integrated over T, $\langle T^p \rangle$ evolves as [37]

$$\frac{d\langle T^p \rangle}{dz} = \frac{i\beta_2}{2W}\int_{-\infty}^{\infty} T^p\left(A^*\frac{\partial^2 A}{\partial T^2} - A\frac{\partial^2 A^*}{\partial T^2}\right)dT, \qquad (4.2.5)$$

where $W = \int_{-\infty}^{\infty}|A(z,T)|^2\,dT$ represents the pulse energy that does not change along the fiber in the absence of losses. The integral on the right-hand side of Eq. (4.2.5) can be evaluated in a closed form if we use Eq. (4.1.4) as the approximate solution for $A(z,T)$. This is a major simplification because it amounts to assuming that the pulse shape does not change along the fiber.

With the preceding simplification, evolution of the RMS width of the pulse, defined as $\sigma = [\langle T^2 \rangle - \langle T \rangle^2]^{1/2}$, is governed by

$$\frac{d\sigma}{dz} = \frac{\gamma\beta_2 S_p}{2\sigma}z, \qquad (4.2.6)$$

where S_p depends on both the shape and the peak power of the input pulse and is defined as

$$S_p = \frac{1}{W}\int_{-\infty}^{\infty}|A(0,T)|^4\,dT. \qquad (4.2.7)$$

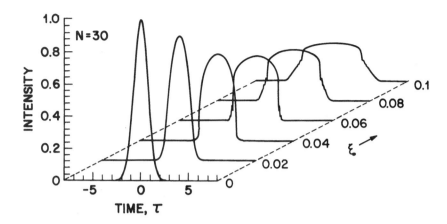

Figure 4.10 Evolution of an initially unchirped Gaussian pulse for $N = 30$ at $z/L_D = 0.1$ in the normal-dispersion regime of an optical fiber.

Equation (4.2.6) can be integrated readily to obtain

$$\sigma^2(z) = \sigma_0^2 + \tfrac{1}{2}\gamma\beta_2 S_p z^2, \tag{4.2.8}$$

where σ_0 is the RMS width of the input pulse at $z = 0$. This remarkably simple result can be used for any pulse shape after S_p is calculated from Eq. (4.2.7). It provides a reasonably accurate estimate of the pulse width when its predictions are compared with the numerical results such as those shown in Fig. 4.9. A variant of the moment method has been used to include the effects of fiber losses and to predict not only the pulse width but also the spectral width and the frequency chirp [38]. The moment method can also be used for dispersion-managed lightwave systems in which optical amplifiers are used periodically for compensating fiber losses [39]. See Chapter 7 of Part B for further details.

4.2.3 Optical Wave Breaking

Equation (4.2.1) suggests that the effects of SPM should dominate over those of GVD for values of $N \gg 1$, at least during the initial stages of pulse evolution. In fact, by introducing a new distance variable as $Z = N^2 \xi = z/L_{NL}$, Eq. (4.2.1) can be written as

$$i\frac{\partial U}{\partial Z} - \frac{d}{2}\frac{\partial^2 U}{\partial\tau^2} + |U|^2 U, \tag{4.2.9}$$

where fiber losses are neglected and $d = \beta_2/(\gamma P_0 T_0^2)$ is a small parameter. Using the transformation

$$U(z,T) = \sqrt{\rho(z,T)} \exp\left(i \int_0^T v(z,T) \, dT \right), \qquad (4.2.10)$$

in Eq. (4.2.9), the pulse-propagation problem reduces approximately to a fluid-dynamics problem in which the variables ρ and v play, respectively, the role of density and velocity of a fluid [40]. In the optical case, these variables represent the power and chirp profiles of the pulse. For a square-shape pulse, the pulse-propagation problem becomes identical to the one related to "breaking of a dam" and can be solved analytically. This solution is useful for lightwave systems using the NRZ format and provides considerable physical insight [41]–[43].

The approximate solution, although useful, does not account for a phenomenon termed optical wave breaking [44]–[50]. It turns out that GVD cannot be treated as a small perturbation even when N is large. The reason is that, because of a large amount of the SPM-induced frequency chirp imposed on the pulse, even weak dispersive effects lead to significant pulse shaping. In the case of normal dispersion ($\beta_2 > 0$), the pulse becomes nearly rectangular with relatively sharp leading and trailing edges and is accompanied by a linear chirp across its entire width [20]. It is this linear chirp that can be used to compress the pulse by passing it through a dispersive delay line.

The GVD-induced pulse shaping has another effect on pulse evolution. It increases the importance of GVD because the second derivative in Eq. (4.2.1) becomes large near the pulse edges. As a consequence, the pulse develops a fine structure near its edges. Figure 4.10 shows pulse evolution for $N = 30$ for the case of an initially unchirped Gaussian pulse. The oscillatory structure near pulse edges is present already at $z/L_D = 0.06$. Further increase in z leads to broadening of the pulse tails. Figure 4.11 shows the pulse shape and the spectrum at $z/L_D = 0.08$. The noteworthy feature is that rapid oscillations near pulse edges are always accompanied by the sidelobes in the spectrum. The central multipeak part of the spectrum is also considerably modified by GVD. In particular, the minima are not as deep as expected from SPM alone.

The physical origin of temporal oscillations near the pulse edges is related to optical wave breaking [44]. Both GVD and SPM impose frequency chirp on the pulse as it travels down the fiber. However, as seen from Eqs. (3.2.13) and (4.1.9), although the GVD-induced chirp is linear with time, the SPM-induced chirp is far from being linear across the entire pulse. Because of the nonlinear

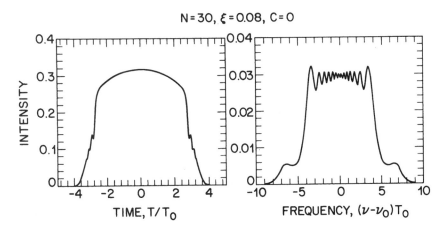

Figure 4.11 Shape and spectrum of an initially unchirped ($C = 0$) Gaussian pulse at $z/L_D = 0.08$. All parameters are identical to those of Fig. 4.10. Spectral sidelobes and temporal structure near pulse edges are due to optical wave breaking.

nature of the composite chirp, different parts of the pulse propagate at different speeds [49]. In particular, in the case of normal GVD ($\beta_2 > 0$), the red-shifted light near the leading edge travels faster and overtakes the unshifted light in the forward tail of the pulse. The opposite occurs for the blue-shifted light near the trailing edge. In both cases, the leading and trailing regions of the pulse contain light at two different frequencies that interfere. Oscillations near the pulse edges in Fig. 4.10 are a result of such interference.

The phenomenon of optical wave breaking can also be understood as a four-wave-mixing process (see Section 10.1). Nonlinear mixing of two different frequencies ω_1 and ω_2 in the pulse tails creates new frequencies at $2\omega_1 - \omega_2$ and $2\omega_2 - \omega_1$. The spectral sidelobes in Fig. 4.11 represent these new frequency components. Temporal oscillations near pulse edges and the spectral sidelobes are manifestations of the same phenomenon. It is interesting to note that optical wave breaking does not occur in the case of anomalous GVD. The reason is that the red-shifted part of the pulse cannot take over the fast-moving forward tail. Instead, the energy in the pulse tail spreads out, and the pulse acquires a pedestal [49].

The results shown in Figs. 4.10 and 4.11 are obtained for an unchirped pulse ($C = 0$). Pulses emitted from practical laser sources are often chirped and may follow quite a different evolution pattern depending on the sign and magnitude of the chirp parameter C [46]. Figure 4.12 shows the pulse shape

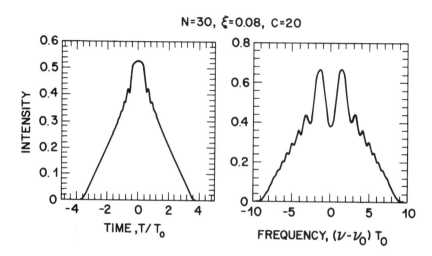

Figure 4.12 Pulse shape and spectrum under conditions identical to those of Fig. 4.11 except that the input Gaussian pulse is chirped with $C = 20$.

and the spectrum under conditions identical to those of Fig. 4.11 except for the chirp parameter, which has a value $C = 20$. A comparison of the two figures illustrates how much an initial chirp can modify the propagation behavior. For an initially chirped pulse, the shape becomes nearly triangular rather than rectangular. At the same time, the spectrum exhibits an oscillatory structure in the wings while the central SPM-like structure (seen in Fig. 4.11 for the case of an unchirped pulse) has almost disappeared. These changes in the pulse shape and spectrum can be understood qualitatively by recalling that a positive initial chirp adds to the SPM-induced chirp. As a result, optical wave breaking sets in earlier for chirped pulses. Pulse evolution is also sensitive to fiber losses. For an actual comparison between theory and experiment it is necessary to include both the chirp and losses in numerical simulations.

4.2.4   Experimental Results

The combined effects of GVD and SPM in optical fibers were first observed in an experiment in which 5.5-ps (FWHM) pulses from a mode-locked dye laser (at 587 nm) were propagated through a 70-m fiber [20]. For an input peak power of 10 W ($N \approx 7$), output pulses were nearly rectangular and had a positive linear chirp. The pulse shape was deduced from autocorrelation measurements as pulses were too short to be measured directly (see Section 3.3.4).

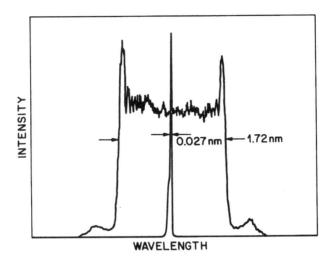

Figure 4.13 Output spectrum of 35-ps input pulses showing SPM-induced spectral broadening. Initial pulse spectrum is also shown for comparison. (After Ref. [44].)

In a later experiment, much wider pulses (FWHM ≈ 150 ps) from a Nd:YAG laser operating at 1.06 μm were transmitted through a 20-km-long fiber [23]. As the peak power of the input pulses was increased from 1 to 40 W (corresponding to N in the range 20–150), the output pulses broadened, became nearly rectangular and then developed substructure near its edges, resulting in an evolution pattern similar to that shown in Fig. 4.10. For such long fibers, it is necessary to include fiber losses. The experimental results were indeed in good agreement with the predictions of Eq. (4.2.1).

The evidence of optical wave breaking was seen in an experiment in which 35-ps (FWHM) pulses at 532 nm (from a frequency-doubled Nd:YAG laser) with peak powers of 235 W were propagated through a 93.5-m-long polarization-maintaining fiber [44]. Figure 4.13 shows the experimentally observed spectrum of the output pulses. Even though $N \approx 173$ in this experiment, the formal similarity with the spectrum shown in Fig. 4.11 is evident. In fact, the phenomenon of optical wave breaking was discovered in an attempt to explain the presence of the sidelobes in Fig. 4.13. In a 1988 experiment [47], the frequency chirp across the pulse was directly measured by using a combination of a streak camera and a spectrograph. The spectral sidelobes associated with the optical wave breaking were indeed found to be correlated with the generation of new frequencies near the pulse edges. In a later experiment [48], rapid oscillations across the leading and trailing edges of the optical pulse were directly observed by using a cross-correlation technique that permitted subpicosecond resolution. The experimental results were in excellent agreement with the predictions of Eq. (4.2.1).

4.2.5 Effect of Third-Order Dispersion

If the optical wavelength λ_0 nearly coincides with the zero-dispersion wavelength λ_D so that $\beta_2 \approx 0$, it is necessary to include the effects of third-order dispersion (TOD) on SPM-induced spectral broadening [51]–[60]. The pulse-propagation equation is obtained from Eq. (2.3.34) by setting $\beta_2 = 0$ and neglecting the higher-order nonlinear terms. If we introduce the dispersion length L_D' from Eq. (3.3.3) and define $\xi' = z/L_D'$ as the normalized distance, we obtain

$$i\frac{\partial U}{\partial \xi'} = \text{sgn}(\beta_3)\frac{i}{6}\frac{\partial^3 U}{\partial \tau^3} - \bar{N}^2 e^{-\alpha z}|U|^2 U, \qquad (4.2.11)$$

where

$$\bar{N}^2 = \frac{L_D'}{L_{\text{NL}}} = \frac{\gamma P_0 T_0^3}{|\beta_3|}. \qquad (4.2.12)$$

Similar to Eq. (4.2.1), the parameter $\bar{N}$ governs the relative importance of the GVD and SPM effects during pulse evolution; GVD dominates for $\bar{N} \ll 1$ while SPM dominates for $\bar{N} \gg 1$. Equation (4.2.11) can be solved numerically using the split-step Fourier method of Section 2.4. In the following discussion we assume $\beta_3 > 0$ and neglect fiber losses by setting $\alpha = 0$.

Figure 4.14 shows the shape and the spectrum of an initially unchirped Gaussian pulse at $\xi' = 5$ for the case $\bar{N} = 1$. The pulse shape should be compared with that shown in Fig. 3.6 where SPM effects were absent ($\bar{N} = 0$). The effect of SPM is to increase the number of oscillations seen near the trailing edge of the pulse. At the same time, the intensity does not become zero at the oscillation minima. The effect of GVD on the spectrum is also evident in Fig. 4.14. In the absence of GVD, a symmetric two-peak spectrum is expected (similar to the one shown in Fig. 4.2 for the case $\phi_{\text{max}} = 1.5\pi$) since $\phi_{\text{max}} = 5$ for the parameter values used in Fig. 4.14. The effect of GVD is to introduce spectral asymmetry without affecting the two-peak structure. This behavior is in sharp contrast with the one shown in Fig. 4.6 for the normal-dispersion case where GVD hindered splitting of the spectrum.

Pulse evolution exhibits qualitatively different features for large values of N. As an example, Fig. 4.15 shows the shape and spectrum of an initially

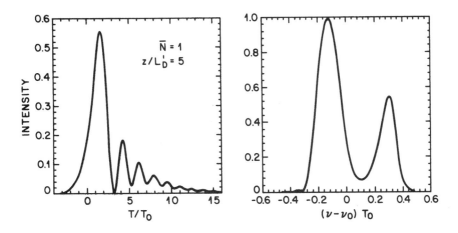

Figure 4.14 Pulse shape and spectrum of unchirped Gaussian pulses propagating exactly at the zero-dispersion wavelength with $\bar{N} = 1$ and $z = 5L'_D$.

unchirped Gaussian pulse at $\xi' = 0.1$ for the case $\bar{N} = 10$. The pulse develops an oscillatory structure with deep modulation. Because of rapid temporal variations, the third derivative in Eq. (4.2.11) becomes large locally, and the GVD effects become more important as the pulse propagates inside the fiber. The most noteworthy feature of the spectrum is that the pulse energy becomes concentrated in two spectral bands, a feature common for all values of $\bar{N} \geq 1$. As one of the spectral bands lies in the anomalous-dispersion regime, the pulse energy in that band can form a soliton [59]. The energy in the other spectral band, lying in the normal-dispersion regime of the fiber, disperses with propagation. The soliton-related features are discussed later in Chapter 5. The important point to note is that, because of SPM-induced spectral broadening, the pulse does not really propagate at the zero-dispersion wavelength even if $\beta_2 \approx 0$ initially. In effect, the pulse creates its own β_2 through SPM. Roughly speaking, the effective value of β_2 is given by

$$|\beta_2| \approx \beta_3 |\delta \omega_{max}/2\pi|, \qquad (4.2.13)$$

where $\delta \omega_{max}$ is the maximum chirp given by Eq. (4.1.10). Physically, β_2 is determined by the position of the dominant outermost spectral peaks in the SPM-broadened spectrum.

In dispersion-managed fiber links, β_2 is large locally but nearly vanishes on average. The effects of TOD play an important role in such links, especially for short optical pulses [61]. The spectral and temporal evolution depends on

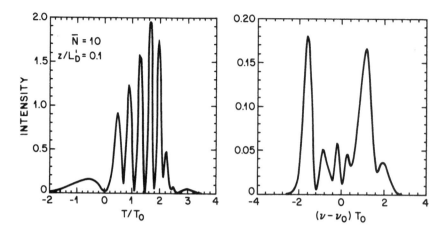

Figure 4.15 Pulse shape and spectrum under conditions identical to those of Fig. 4.14 except that $\bar{N} = 10$ and $z/L'_D = 0.1$.

whether the dispersion-compensating fiber (DCF) is placed before or after the the standard fiber (pre- or postcompensation). In the case of postcompensation, the pulse develops an oscillating tail because of TOD and exhibits spectral narrowing. These features have been seen experimentally by transmitting 0.4-ps pulses over a 2.5-km-long dispersion-compensated fiber link.

4.3   Higher-Order Nonlinear Effects

The discussion of SPM so far is based on the simplified propagation equation (2.3.41). For ultrashort optical pulses ($T_0 < 1$ ps), it is necessary to include the higher-order nonlinear effects through Eq. (2.3.39). If Eq. (3.1.3) is used to define the normalized amplitude U, this equation takes the form

$$\frac{\partial U}{\partial z} + i\frac{\text{sgn}(\beta_2)}{2L_D}\frac{\partial^2 U}{\partial \tau^2} = \frac{\text{sgn}(\beta_3)}{6\,L'_D}\frac{\partial^3 U}{\partial \tau^3}$$
$$+ i\frac{e^{-\alpha z}}{L_{\text{NL}}}\left(|U|^2 U + is\frac{\partial}{\partial \tau}(|U|^2 U) - \tau_R U\frac{\partial |U|^2}{\partial \tau}\right), \qquad (4.3.1)$$

where L_D, L'_D, and L_{NL} are the three length scales defined as

$$L_D = \frac{T_0^2}{|\beta_2|}, \qquad L'_D = \frac{T_0^3}{|\beta_3|}, \qquad L_{\text{NL}} = \frac{1}{\gamma P_0}. \qquad (4.3.2)$$

The parameters s and τ_R govern the effects of self-steepening and intrapulse Raman scattering, respectively, and are defined as

$$s = \frac{1}{\omega_0 T_0}, \qquad \tau_R = \frac{T_R}{T_0}. \tag{4.3.3}$$

Both of these effects are quite small for picosecond pulses but must be considered for ultrashort pulses with $T_0 < 0.1$ ps.

4.3.1 Self-Steepening

Self-steepening results from the intensity dependence of the group velocity [62]–[65]. Its effects on SPM were first considered in liquid nonlinear media [2] and later extended to optical fibers [66]–[70]. Self-steepening leads to an asymmetry in the SPM-broadened spectra of ultrashort pulses [71]–[75].

Before solving Eq. (4.3.1) numerically, it is instructive to consider the dispersionless case by setting $\beta_2 = \beta_3 = 0$. Equation (4.3.1) can be solved analytically in this specific case if we also set $\tau_R = 0$ [64]. Defining a normalized distance as $Z = z/L_{NL}$ and neglecting fiber losses ($\alpha = 0$), Eq. (4.3.1) becomes

$$\frac{\partial U}{\partial Z} + s\frac{\partial}{\partial \tau}(|U|^2 U) = i|U|^2 U. \tag{4.3.4}$$

Using $U = \sqrt{I}\,\exp(i\phi)$ in Eq. (4.3.4) and separating the real and imaginary parts, we obtain the following two equations:

$$\frac{\partial I}{\partial Z} + 3sI\frac{\partial I}{\partial \tau} = 0, \tag{4.3.5}$$

$$\frac{\partial \phi}{\partial Z} + sI\frac{\partial \phi}{\partial \tau} = I. \tag{4.3.6}$$

Since the intensity equation (4.3.5) is decoupled from the phase equation (4.3.6), it can be solved easily using the method of characteristics. Its general solution is given by [66]

$$I(Z, \tau) = f(\tau - 3sIZ), \tag{4.3.7}$$

where we used the initial condition $I(0, \tau) = f(\tau)$, where $f(\tau)$ describes the pulse shape at $z = 0$. Equation (4.3.7) shows that each point τ moves along a straight line from its initial value, and the slope of the line is intensity dependent. This feature leads to pulse distortion. As an example, consider the case of a Gaussian pulse for which

$$I(0, \tau) \equiv f(\tau) = \exp(-\tau^2). \tag{4.3.8}$$

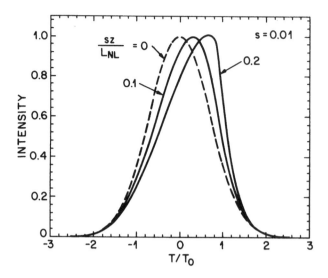

Figure 4.16 Self-steepening of a Gaussian pulse in the dispersionless case. Dashed curve shows the input pulse shape at $z = 0$.

From Eq. (4.3.7), the pulse shape at a distance Z is obtained by using

$$I(Z, \tau) = \exp[-(\tau - 3sIZ)^2]. \qquad (4.3.9)$$

The implicit relation for $I(Z, \tau)$ should be solved for each τ to obtain the pulse shape at a given value of Z. Figure 4.16 shows the calculated pulse shapes at $sZ = 0.1$ and 0.2 for $s = 0.01$. As the pulse propagates inside the fiber, it becomes asymmetric, with its peak shifting toward the trailing edge. As a result, the trailing edge becomes steeper and steeper with increasing Z. Physically, the group velocity of the pulse is intensity dependent such that the peak moves at a lower speed than the wings.

Self-steepening of the pulse eventually creates an optical shock, analogous to the development of an acoustic shock on the leading edge of a sound wave [64]. The distance at which the shock is formed is obtained from Eq. (4.3.9) by requiring that $\partial I / \partial \tau$ be infinite at the shock location. It is given by [67]

$$z_s = \left(\frac{e}{2}\right)^{1/2} \frac{L_{NL}}{3s} \approx 0.39(L_{NL}/s). \qquad (4.3.10)$$

A similar relation holds for a "sech" pulse with only a slight change in the numerical coefficient (0.43 in place of 0.39). For picosecond pulses with $T_0 = 1$

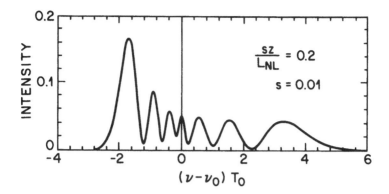

Figure 4.17 Spectrum of a Gaussian pulse at a distance $z = 0.2L_{NL}/s$, where $s = 0.01$ and L_{NL} is the nonlinear length. Self-steepening is responsible for the asymmetry in the SPM-broadened spectrum. The effects of GVD are neglected.

ps and $P_0 \sim 1$ W, the shock occurs at a distance $z_s \sim 100$ km. However, for femtosecond pulses with $T_0 < 100$ fs and $P_0 > 1$ kW, z_s becomes < 1 m. As a result, significant self-steepening of the pulse can occur in a few-centimeter-long fiber. Optical shocks with an infinitely sharp trailing edge never occur in practice because of the GVD; as the pulse edge becomes steeper, the disper-sive terms in Eq. (4.3.1) become increasingly more important and cannot be ignored. The shock distance z_s is also affected by fiber losses α. In the dispersionless case, fiber losses delay the formation of optical shocks; if $\alpha z_s > 1$, the shock does not develop at all [67].

Self-steepening also affects SPM-induced spectral broadening. In the dispersionless case, $\phi(z, \tau)$ is obtained by solving Eq. (4.3.6). It can then be used to calculate the spectrum using

$$S(\omega) = \left| \int_{-\infty}^{\infty} [I(z, \tau)]^{1/2} \exp[i\phi(z, \tau) + i(\omega - \omega_0)\tau] d\tau \right|^2. \qquad (4.3.11)$$

Figure 4.17 shows the calculated spectrum at $sz/L_{NL} = 0.2$ for $s = 0.01$. The most notable feature is spectral asymmetry— the red-shifted peaks are more intense than blue-shifted peaks. The other notable feature is that SPM-induced spectral broadening is larger on the blue side (called the anti-Stokes side in the terminology used for stimulated Raman scattering) than the red side (or the Stokes side). Both of these features can be understood qualitatively from the changes in the pulse shape induced by self-steepening. The spectrum is asymmetric simply because pulse shape is asymmetric. A steeper trailing edge

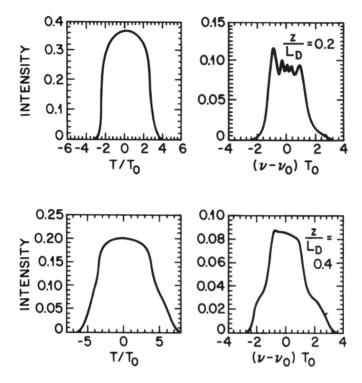

Figure 4.18 Pulse shapes and spectra at $z/L_D = 0.2$ (upper row) and 0.4 (lower row) for a Gaussian pulse propagating in the normal-dispersion regime of the fiber. The other parameters are $\alpha = 0$, $\beta_3 = 0$, $s = 0.01$, and $N = 10$.

of the pulse implies larger spectral broadening on the blue side as SPM generates blue components near the trailing edge (see Fig. 4.1). In the absence of self-steepening ($s = 0$), a symmetric six-peak spectrum is expected because $\phi_{max} \approx 6.4\pi$ for the parameter values used in Fig. 4.17. Self-steepening stretches the blue portion. The amplitude of the high-frequency peaks decreases because the same energy is distributed over a wider spectral range.

4.3.2   Effect of GVD on Optical Shocks

The spectral features seen in Fig. 4.17 are considerably affected by GVD, which cannot be ignored when short optical pulses propagate inside silica fibers [76]–[83]. The pulse evolution in this case is studied by solving Eq. (4.3.1) numerically. Figure 4.18 shows the pulse shapes and the spectra at

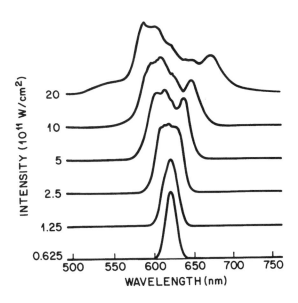

Figure 4.19 Experimentally observed spectra of 40-fs input pulses at the output of a 7-mm-long fiber. Spectra are labeled by the peak intensity of input pulses. The top spectrum corresponds to $N \approx 7.7$. (After Ref. [76].)

$z/L_D = 0.2$ and 0.4 in the case of an initially unchirped Gaussian pulse propagating with normal dispersion ($\beta_2 > 0$) and $\beta_3 = 0$. The parameter N defined in Eq. (4.2.3) is taken to be 10, resulting in $L_D = 100L_{NL}$. In the absence of GVD ($\beta_2 = 0$), the pulse shape and the spectrum shown in the upper row of Fig. 4.18 reduce to those shown in Figs. 4.16 and 4.17 in the case of $sz/L_{NL} = 0.2$. A direct comparison shows that both the shape and spectrum are significantly affected by GVD even though the propagation distance is only a fraction of the dispersion length ($z/L_D = 0.2$). The lower row of Fig. 4.18 shows the pulse shape and spectrum at $z/L_D = 0.4$; the qualitative changes induced by GVD are self-evident. For this value of z/L_D, the propagation distance z exceeds the shock distance z_s given by Eq. (4.3.10). It is the GVD that dissipates the shock by broadening the steepened trailing edge, a feature clearly seen in the asymmetric pulse shapes of Fig. 4.18. Although the pulse spectra do not exhibit deep oscillations (seen in Fig. 4.17 for the dispersionless case), the longer tail on the blue side is a manifestation of self-steepening. With a further increase in the propagation distance z, the pulse continues to broaden while the spectrum remains nearly unchanged.

The effect of self-steepening on pulse evolution has been seen experimentally in liquids and solids as a larger spectral broadening on the blue side compared with that on the red side [4]. In these early experiments, GVD played a relatively minor role, and the spectral structure similar to that of Fig. 4.17 was observed. In the case of optical fibers, the GVD effects are strong enough that the spectra similar to those of Fig. 4.18 are expected to occur in practice. In an experiment on pulse compression [76], 40-fs optical pulses at 620 nm were propagated over a 7-mm-long fiber. Figure 4.19 shows the experimentally observed spectra at the fiber output for several values of peak intensities. The spectrum broadens asymmetrically with a longer tail on the blue side than on the red side. This feature is due to self-steepening. In this experiment, the self-steepening parameter $s \approx 0.026$, and the dispersion length $L_D \approx 1$ cm if we use $T_0 = 24$ fs (corresponding to a FWHM of 40 fs for a Gaussian pulse). Assuming an effective core area of 10 μm^2, the peak power corresponding to the top trace of Fig. 4.19 is about 200 kW. This value results in a nonlinear length $L_{NL} \approx 0.16$ mm and $N \approx 7.7$. Equation (4.3.1) can be used to simulate the experiment by using these parameter values. Inclusion of the β_3 term is generally necessary to reproduce the detailed features of the experimentally observed spectra of Fig. 4.19 [69]. Similar conclusions were reached in another experiment in which asymmetric spectral broadening of 55-fs pulses from a 620-nm dye laser was observed in a 11-mm-long optical fiber [77].

4.3.3   Intrapulse Raman Scattering

The discussion so far has neglected the last term in Eq. (4.3.1) that is responsible for intrapulse Raman scattering. In the case of optical fibers, this term becomes quite important for ultrashort optical pulses ($T_0 < 1$ ps) and should be included in modeling pulse evolution of such short pulses in optical fibers [79]–[83]. The effects of intrapulse Raman scattering are most dramatic in the context of solitons, where they lead to new phenomena such as decay and self-frequency shift of solitons (see Chapter 5). However, even in the case of normal GVD, the inclusion of both self-steepening and intrapulse Raman scattering is essential for an agreement between theory and experiments.

Figure 4.20(a) shows the experimentally recorded pulse spectrum after 109-fs 'sech' pulses ($T_0 \approx 60$ fs) with 7.4 kW peak power were sent through a 6-m-long fiber [83]. The fiber had $\beta_2 \approx 4$ ps^2/km and $\beta_3 \approx 0.06$ ps^3/km at the 1260-nm wavelength used in this experiment. The three traces b–d show the prediction of Eq. (4.3.1) under three different conditions. Both self-steepening and intrapulse Raman scattering were neglected in the trace b and included in

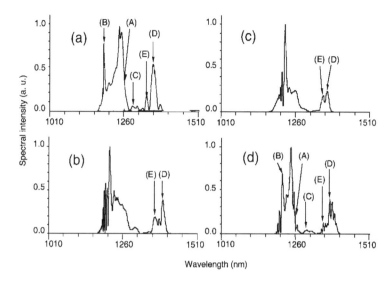

Figure 4.20 Experimental spectrum of 109-fs input pulses at the output of a 6-m-long fiber (a) and predictions of the generalized NLS equation with (b) $s = \tau_R = 0$, (c) $s = 0$, and (d) both s and τ_R nonzero. Letters (A)–(E) mark different spectral features observed experimentally. (After Ref. [83].)

the trace d, while only the latter was included in the trace c. All experimental features, marked as (A)–(E), were reproduced only when both higher-order nonlinear effects were included in the model. Inclusion of the fourth-order dispersion was also necessary for a good agreement. Even the predicted pulse shapes were in agreement with the cross-correlation traces obtained experimentally.

The SPM and other nonlinear effects such as stimulated Raman scattering and four-wave mixing, occurring simultaneously inside optical fibers, can broaden the spectrum of an ultrashort pulse so much that it may extend over 100 nm or more. Such extreme spectral broadening is called supercontinuum, a phenomenon that attracted considerable attention during the 1990s because of its potential applications [84]–[93]. Pulse spectra extending over as much as 300 nm have been generated using various types of optical fibers. Among other applications, supercontinuum is useful for WDM lightwave systems because its spectral filtering can provide an optical source capable of emitting synchronized pulse trains at hundreds of wavelengths simultaneously [86]–[88].

Problems

4.1 A 1.06-μm Q-switched Nd:YAG laser emits unchirped Gaussian pulse with 1-nJ energy and 100-ps width (FWHM). Pulses are transmitted through a 1-km-long fiber having a loss of 3 dB/km and an effective core area of 20 μm^2. Calculate the maximum values of the nonlinear phase shift and the frequency chirp at the fiber output.

4.2 Use a fast-Fourier-transform algorithm to calculate the spectrum of the chirped output pulse of Problem 4.1. Does the number of spectral peaks agree with the prediction of Eq. (4.1.14)?

4.3 Repeat Problem 4.1 for a hyperbolic-secant pulse.

4.4 Determine the shape, width, and peak power of the optical pulse that will produce a linear chirp at the rate of 1 GHz/ps over a 100-ps region when transmitted through the fiber of Problem 4.1.

4.5 Calculate numerically the SPM-broadened spectra of a super-Gaussian pulse ($m = 3$) for $C = -5$, 0, and 5. Assume a peak power such that $\phi_{max} = 4.5\pi$. Compare your spectra with those shown in Fig. 4.5 and comment on the main qualitative differences.

4.6 Use the split-step Fourier method of Section 2.4 for solving Eq. (4.2.1) numerically. Generate curves similar to those shown in Figs. 4.7 and 4.8 for a 'sech' pulse using $N = 1$ and $\alpha = 0$. Compare your results with the Gaussian-pulse case and discuss the differences qualitatively.

4.7 Use the computer program developed for Problem 4.7 to study numerically optical wave breaking for an unchirped super-Gaussian pulse with $m = 3$ by using $N = 30$ and $\alpha = 0$. Compare your results with those shown in Figs. 4.10 and 4.11 for a Gaussian pulse.

4.8 Show that the solution (4.3.9) is indeed the solution of Eq. (4.3.4) for an input Gaussian pulse. Calculate the phase profile $\phi(Z, \tau)$ at $sZ = 0.2$ analytically (if possible) or numerically.

References

[1] F. Shimizu, *Phys. Rev. Lett.* **19**, 1097 (1967).

[2] T. K. Gustafson, J. P. Taran, H. A. Haus, J. R. Lifsitz, and P. L. Kelley, *Phys. Rev.* **177**, 306 (1969).

[3] R. Cubeddu, R. Polloni, C. A. Sacchi, and O. Svelto, *Phys. Rev. A* **2**, 1955 (1970).

[4] R. R. Alfano and S. L. Shapiro, *Phys. Rev. Lett.* **24**, 592 (1970); *Phys. Rev. Lett.* **24**, 1217 (1970).

[5] Y. R. Shen and M. M. T. Loy, *Phys. Rev. A* **3**, 2099 (1971).

[6] C. H. Lin and T. K. Gustafson, *IEEE J. Quantum Electron.* **8**, 429 (1972).

[7] E. P. Ippen, C. V. Shank, and T. K. Gustafson, *Appl. Phys. Lett.* **24**, 190 (1974).

[8] R. A. Fisher and W. K. Bischel, *J. Appl. Phys.* **46**, 4921 (1975).

[9] R. H. Stolen and C. Lin, *Phys. Rev. A* **17**, 1448 (1978).

[10] S. C. Pinault and M. J. Potasek, *J. Opt. Soc. Am. B* **2**, 1318 (1985).

[11] M. Oberthaler and R. A. Höpfel, *Appl. Phys. Lett.* **63**, 1017 (1993).

[12] J. T. Manassah, *Opt. Lett.* **15**, 329 (1990); *Opt. Lett.* **16**, 1638 (1991).

[13] B. Gross and J. T. Manassah, *Opt. Lett.* **16**, 1835 (1991).

[14] M. T. de Araujo, H. R. da Cruz, and A. S. Gouveia-Neto, *J. Opt. Soc. Am. B* **8**, 2094 (1991).

[15] H. R. da Cruz, J. M. Hickmann, and A. S. Gouveia-Neto, *Phys. Rev. A* **45**, 8268 (1992).

[16] J. N. Elgin, *Opt. Lett.* **18**, 10 (1993); *Phys. Rev. A* **47**, 4331 (1993).

[17] S. Cavalcanti, G. P. Agrawal, and M. Yu, *Phys. Rev. A* **51**, 4086 (1995).

[18] J. Garnier, L. Videau, C. Gouédard, and A. Migus, *J. Opt. Soc. Am. B* **15**, 2773 (1998).

[19] L. Mandel and E. Wolf, *Optical Coherence and Quantum Optics* (Cambridge University Press, New York, 1995).

[20] H. Nakatsuka, D. Grischkowsky, and A. C. Balant, *Phys. Rev. Lett.* **47**, 910 (1981).

[21] D. Grischkowsky and A. C. Balant, *Appl. Phys. Lett.* **41**, 1 (1982).

[22] J. Botineau and R. H. Stolen, *J. Opt. Soc. Am.* **72**, 1592 (1982).

[23] B. P. Nelson, D. Cotter, K. J. Blow, and N. J. Doran, *Opt. Commun.* **48**, 292 (1983).

[24] W. J. Tomlinson, R. H. Stolen, and C. V. Shank, *J. Opt. Soc. Am. B* **1**, 139 (1984).

[25] I. N. Sisakyan and A. B. Shvartsburg, *Sov. J. Quantum Electron.* **14**, 1146 (1984).

[26] M. J. Potasek and G. P. Agrawal, *Electron. Lett.* **22**, 759 (1986).

[27] A. Kumar and M. S. Sodha, *Electron. Lett.* **23**, 275 (1987).

[28] M. J. Potasek and G. P. Agrawal, *Phys. Rev. A* **36**, 3862 (1987).

[29] J. M. Hickmann, J. F. Martino-Filho, and A. S. L. Gomes, *Opt. Commun.* **84**, 327 (1991).

[30] A. Kumar, *Phys. Rev. A* **44**, 2130 (1991).

[31] P. Weidner and A. Penzkofer, *Opt. Quantum Electron.* **25**, 1 (1993).

[32] D. Anderson, *Phys. Rev. A* **27**, 3135 (1983).

[33] A. M. Fattakhov and A. S. Chirkin, *Sov. J. Quantum Electron.* **14**, 1556 (1984).

[34] D. Anderson, *IEE Proc.* **132**, Pt. J, 122 (1985).

[35] M. J. Potasek, G. P. Agrawal, and S. C. Pinault, *J. Opt. Soc. Am. B* **3**, 205 (1986).

[36] C. Pask and A. Vatarescu, *J. Opt. Soc. Am. B* **3**, 1018 (1986).

[37] D. Marcuse, *J. Lightwave Technol.* **10**, 17 (1992).

[38] P. A. Bélanger and N. Bélanger, *Opt. Commun.* **117**, 56 (1995).

[39] Q. Yu and C. Fan, *IEEE J. Quantum Electron.* **15**, 444 (1997).

[40] Y. Kodama and S. Wabnitz, *Opt. Lett.* **20**, 2291 (1995).

[41] Y. Kodama and S. Wabnitz, *Electron. Lett.* **31**, 1761 (1995).

[42] Y. Kodama, Wabnitz, and K. Tanaka *Opt. Lett.* **21**, 719 (1996).

[43] A. M. Kamchatnov and H. Steudel, *Opt. Commun.* **162**, 162 (1999).

[44] W. J. Tomlinson, R. H. Stolen, and A. M. Johnson, *Opt. Lett.* **10**, 457 (1985).

[45] A. M. Johnson and W. M. Simpson, *J. Opt. Soc. Am. B* **2**, 619 (1985).

[46] H. E. Lassen, F. Mengel, B. Tromborg, N. C. Albertsen, and P. L. Christiansen, *Opt. Lett.* **10**, 34 (1985).

[47] J.-P. Hamaide and P. Emplit, *Electron. Lett.* **24**, 818 (1988).

[48] J. E. Rothenbeg, *J. Opt. Soc. Am. B* **6**, 2392 (1989); *Opt. Lett.* **16**, 18 (1991).

[49] D. Anderson, M. Desaix, M. Lisak, and M. L. Quiroga-Teixeiro, *J. Opt. Soc. Am. B* **9**, 1358 (1992).

[50] D. Anderson, M. Desaix, M. Karlsson, M. Lisak, and M. L. Quiroga-Teixeiro, *J. Opt. Soc. Am. B* **10**, 1185 (1993).

[51] K. J. Blow, N. J. Doran, and E. Cummins, *Opt. Commun.* **48**, 181 (1983).

[52] V. A. Vysloukh, *Sov. J. Quantum Electron.* **13**, 1113 (1983).

[53] G. P. Agrawal and M. J. Potasek, *Phys. Rev. A* **33**, 1765 (1986).

[54] P. K. A. Wai, C. R. Menyuk, Y. C. Lee, and H. H. Chen, *Opt. Lett.* **11**, 464 (1986).

[55] G. R. Boyer and X. F. Carlotti, *Opt. Commun.* **60**, 18 (1986); *Phys. Rev. A* **38**, 5140 (1988).

[56] P. K. A. Wai, C. R. Menyuk, H. H. Chen, and Y. C. Lee, *Opt. Lett.* **12**, 628 (1987).

[57] A. S. Gouveia-Neto, M. E. Faldon, and J. R. Taylor, *Opt. Lett.* **13**, 770 (1988).

[58] S. Wen and S. Chi, *Opt. Quantum Electron.* **21**, 335 (1989).

[59] P. K. A. Wai, H. H. Chen, and Y. C. Lee, *Phys. Rev. A* **41**, 426 (1990).

[60] J. N. Elgin, *Opt. Lett.* **15**, 1409 (1992).

[61] S. Shen, C. C. Chang, H. P. Sardesai, V. Binjrajka, and A. M. Weiner, *IEEE J. Quantum Electron.* **17**, 452 (1999).

[62] L. A. Ostrovskii, *Sov. Phys. JETP* **24**, 797 (1967).

[63] R. J. Jonek and R. Landauer, *Phys. Lett.* **24A**, 228 (1967).

[64] F. DeMartini, C. H. Townes, T. K. Gustafson, and P. L. Kelley, *Phys. Rev.* **164**, 312 (1967).

[65] D. Grischkowsky, E. Courtens, and J. A. Armstrong, *Phys. Rev. Lett.* **31**, 422 (1973).

[66] N. Tzoar and M. Jain, *Phys. Rev. A* **23**, 1266 (1981).

[67] D. Anderson and M. Lisak, *Phys. Rev. A* **27**, 1393 (1983).

[68] E. A. Golovchenko, E. M. Dianov, A. M. Prokhorov, and V. N. Serkin, *JETP Lett.* **42**, 87 (1985); *Sov. Phys. Dokl.* **31**, 494 (1986).

[69] E. Bourkoff, W. Zhao, R. L. Joseph, and D. N. Christoulides, *Opt. Lett.* **12**, 272 (1987); *Opt. Commun.* **62**, 284 (1987).

[70] W. Zhao and E. Bourkoff, *IEEE J. Quantum Electron.* **24**, 365 (1988).

[71] R. L. Fork, C. V. Shank, C. Herlimann, R. Yen, and W. J. Tomlinson, *Opt. Lett.* **8**, 1 (1983).

[72] G. Yang and Y. R. Shen, *Opt. Lett.* **9**, 510 (1984).

[73] J. T. Manassah, M. A. Mustafa, R. R. Alfano, and P. P. Ho, *Phys. Lett.* **113A**, 242 (1985); *IEEE J. Quantum Electron.* **22**, 197 (1986).

[74] D. Mestdagh and M. Haelterman, *Opt. Commun.* **61**, 291 (1987).

[75] B. R. Suydam, in *Supercontinuum Laser Source*, R. R. Alfano, ed. (Springer-Verlag, New York, 1989), Chap. 6.

[76] W. H. Knox, R. L. Fork, M. C. Downer, R. H. Stolen, and C. V. Shank, *Appl. Phys. Lett.* **46**, 1120 (1985).

[77] G. R. Boyer and M. Franco, *Opt. Lett.* **14**, 465 (1989).

[78] J. R. de Oliveira, M. A. de Moura, J. M. Hickmann, and A. S. L. Gomes, *J. Opt. Soc. Am. B* **9**, 2025 (1992).

[79] A. B. Grudinin, E. M. Dianov, D. V. Korobkin, A. M. Prokhorov, V. N. Serkin, and D. V. Khaidarov, *JETP Lett.* **46**, 221 (1987).

[80] W. Hodel and H. P. Weber, *Opt. Lett.* **12**, 924 (1987).

[81] V. Yanosky and F. Wise, *Opt. Lett.* **19**, 1547 (1994).

[82] C. Headley and G. P. Agrawal, *J. Opt. Soc. Am. B* **13**, 2170 (1996).

[83] G. Boyer, *Opt. Lett.* **24**, 945 (1999).

[84] R. R. Alfano (ed.), *Supercontinuum Laser Source*, (Springer-Verlag, New York, 1989).

[85] B. Gross and J. T. Manassah, *J. Opt. Soc. Am. B* **9**, 1813 (1992).

[86] T. Morioka, K. Mori, and M. Saruwatari, *Electron. Lett.* **29**, 862 (1993).

[87] T. Morioka, K. Uchiyama, S. Kawanishi, S. Suzuki, and M. Saruwatari, *Electron. Lett.* **31**, 1064 (1995).

[88] M. C. Nuss, W. H. Knox, and U. Koren, *Electron. Lett.* **32**, 1311 (1996).

[89] M. Guy, S. V. Chernikov, and J. R. Taylor, *IEEE Photon. Technol. Lett.* **9**, 1017 (1997).

[90] T. Okuno, M. Onishi, and M. Nishimura, *IEEE Photon. Technol. Lett.* **10**, 72 (1998).

[91] Y. Takushima, F. Futami, and K. Kikuchi, *IEEE Photon. Technol. Lett.* **10**, 1560 (1998).

[92] M. Nakazawa, K. Tamura, H. Kubota, and E. Yoshida, *Opt. Fiber Technol.* **4**, 215 (1998).

[93] G. A. Nowak, J. Kim, and M. N. Islam, *Appl. Opt.* **38**, 7364 (1999).

Chapter 5

Optical Solitons

A fascinating manifestation of the fiber nonlinearity occurs through optical solitons, formed as a result of the interplay between the dispersive and nonlinear effects. The word *soliton* refers to special kinds of wave packets that can propagate undistorted over long distances. Solitons have been discovered in many branches of physics. In the context of optical fibers, not only are solitons of fundamental interest but they have also found practical applications in the field of fiber-optic communications. This chapter is devoted to the study of pulse propagation in optical fibers in the regime in which both the group-velocity dispersion (GVD) and self-phase modulation (SPM) are equally important and must be considered simultaneously.

The chapter is organized as follows. Section 5.1 considers the phenomenon of modulation instability and shows that propagation of a continuous-wave (CW) beam inside optical fibers is inherently unstable because of the nonlinear phenomenon of SPM and leads to formation of a pulse train in the anomalous-dispersion regime of optical fibers. Section 5.2 discusses the inverse-scattering method and uses it to obtain soliton solutions of the underlying wave-propagation equation. The properties of the fundamental and higher-order solitons are considered in this section. Section 5.3 is devoted to other kinds of solitons forming in optical fibers, with emphasis on dark solitons. Section 5.4 considers the effects of external perturbations on solitons. Perturbations discussed include fiber losses, amplification of solitons, and noise introduced by optical amplifiers. Higher-order nonlinear effects such as self-steepening and intrapulse Raman scattering are the focus of Section 5.5.

135

5.1   Modulation Instability

Many nonlinear systems exhibit an instability that leads to modulation of the steady state as a result of an interplay between the nonlinear and dispersive effects [1]–[30]. This phenomenon is referred to as the *modulation instability* and was studied during the 1960s in such diverse fields as fluid dynamics [2]–[4], nonlinear optics [5]–[7] and plasma physics [8]–[11]. In the context of optical fibers, modulation instability requires anomalous dispersion and manifests itself as breakup of the CW or quasi-CW radiation into a train of ultrashort pulses. This section discusses modulation instability in optical fibers as an introduction to soliton theory.

5.1.1   Linear Stability Analysis

Consider propagation of CW light inside an optical fiber. The starting point is the simplified propagation equation (2.3.41). If fiber losses are ignored, this equation takes the form

$$i\frac{\partial A}{\partial z} = \frac{\beta_2}{2}\frac{\partial^2 A}{\partial T^2} - \gamma|A|^2 A, \tag{5.1.1}$$

and is referred to as the nonlinear Schrödinger (NLS) equation in the soliton literature. As discussed in Section 2.3, $A(z,T)$ represents the amplitude of the pulse envelope, β_2 is the GVD parameter, and the nonlinear parameter γ is responsible for SPM. In the case of CW radiation, the amplitude A is independent of T at the input end of the fiber at $z = 0$. Assuming that $A(z,T)$ remains time independent during propagation inside the fiber, Eq. (5.1.1) is readily solved to obtain the steady-state solution

$$\bar{A} = \sqrt{P_0}\exp(i\phi_{\mathrm{NL}}), \tag{5.1.2}$$

where P_0 is the incident power and $\phi_{\mathrm{NL}} = \gamma P_0 z$ is the nonlinear phase shift induced by SPM. Equation (5.1.2) implies that CW light should propagate through the fiber unchanged except for acquiring a power-dependent phase shift (and for reduction in power in the presence of fiber losses).

Before reaching this conclusion, however, we must ask whether the steady-state solution (5.1.2) is stable against small perturbations. To answer this question, we perturb the steady state slightly such that

$$A = (\sqrt{P_0} + a)\exp(i\phi_{\mathrm{NL}}) \tag{5.1.3}$$

and examine evolution of the perturbation $a(z, T)$ using a linear stability analysis. Substituting Eq. (5.1.3) in Eq. (5.1.1) and linearizing in a, we obtain

$$i\frac{\partial a}{\partial z} = \frac{\beta_2}{2}\frac{\partial^2 a}{\partial T^2} - \gamma P_0(a + a^*). \tag{5.1.4}$$

This linear equation can be solved easily in the frequency domain. However, because of the a^* term, the Fourier components at frequencies Ω and $-\Omega$ are coupled. Thus, we should consider its solution in the form

$$a(z, T) = a_1 \exp[i(Kz - \Omega T)] + a_2 \exp[-i(Kz - \Omega T)], \tag{5.1.5}$$

where K and Ω are the wave number and the frequency of perturbation, respectively. Equations (5.1.4) and (5.1.5) provide a set of two homogeneous equations for a_1 and a_2. This set has a nontrivial solution only when K and Ω satisfy the following dispersion relation

$$K = \pm\tfrac{1}{2}|\beta_2\Omega|[\Omega^2 + \text{sgn}(\beta_2)\Omega_c^2]^{1/2}, \tag{5.1.6}$$

where $\text{sgn}(\beta_2) = \pm 1$ depending on the sign of β_2,

$$\Omega_c^2 = \frac{4\gamma P_0}{|\beta_2|} = \frac{4}{|\beta_2|L_{\text{NL}}}, \tag{5.1.7}$$

and the nonlinear length L_{NL} is defined by Eq. (3.1.5). Because of the factor $\exp[i(\beta_0 z - \omega_0 t)]$ that has been factored out in Eq. (2.3.21), the actual wave number and the frequency of perturbation are $\beta_0 \pm K$ and $\omega_0 \pm \Omega$, respectively. With this factor in mind, the two terms in Eq. (5.1.5) represent two different frequency components, $\omega_0 + \Omega$ and $\omega_0 - \Omega$, that are present simultaneously. It will be seen later that these frequency components correspond to the two spectral sidebands that are generated when modulation instability occurs.

The dispersion relation (5.1.6) shows that steady-state stability depends critically on whether light experiences normal or anomalous GVD inside the fiber. In the case of normal GVD ($\beta_2 > 0$), the wave number K is real for all Ω, and the steady state is stable against small perturbations. By contrast, in the case of anomalous GVD ($\beta_2 < 0$), K becomes imaginary for $|\Omega| < \Omega_c$, and the perturbation $a(z, T)$ grows exponentially with z as seen from Eq. (5.1.5). As a result, the CW solution (5.1.2) is inherently unstable for $\beta_2 < 0$. This instability is referred to as modulation instability because it leads to a spontaneous temporal modulation of the CW beam and transforms it into a pulse train. Similar instabilities occur in many other nonlinear systems and are often called self-pulsing instabilities [31]–[34].

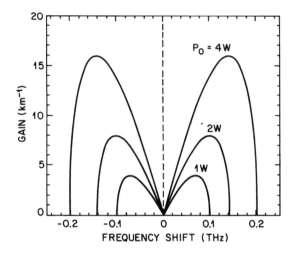

Figure 5.1 Gain spectra of modulation instability at three power levels for an optical fiber with $\beta_2 = -20 \text{ ps}^2/\text{km}$ and $\gamma = 2 \text{ W}^{-1}/\text{km}$.

5.1.2 Gain Spectrum

The gain spectrum of modulation instability is obtained from Eq. (5.1.6) by setting $\text{sgn}(\beta_2) = -1$ and $g(\Omega) = 2\,\text{Im}(K)$, where the factor of 2 converts g to power gain. The gain exists only if $|\Omega| < \Omega_c$ and is given by

$$g(\Omega) = |\beta_2 \Omega|(\Omega_c^2 - \Omega^2)^{1/2}. \tag{5.1.8}$$

Figure 5.1 shows the gain spectra at three power levels using parameter values appropriate for standard silica fibers in the wavelength region near 1.55 μm. The gain spectrum is symmetric with respect to $\Omega = 0$ such that $g(\Omega)$ vanishes at $\Omega = 0$. The gain becomes maximum at two frequencies given by

$$\Omega_{\text{max}} = \pm\frac{\Omega_c}{\sqrt{2}} = \pm\left(\frac{2\gamma P_0}{|\beta_2|}\right)^{1/2}, \tag{5.1.9}$$

with a peak value

$$g_{\text{max}} \equiv g(\Omega_{\text{max}}) = \tfrac{1}{2}|\beta_2|\Omega_c^2 = 2\gamma P_0, \tag{5.1.10}$$

where Eq. (5.1.7) was used to relate Ω_c to P_0. The peak gain is independent of the GVD parameter β_2 and increases linearly with the incident power.

The modulation-instability gain is affected by the loss parameter α that has been neglected in the derivation of Eq. (5.1.8). The main effect of fiber losses is to decrease the gain along fiber length because of reduced power [15]–[17]. In effect, Ω_c in Eq. (5.1.8) is replaced by $\Omega_c \exp(-\alpha z/2)$. Modulation instability still occurs as long as $\alpha L_{NL} < 1$. The effect of higher-order dispersive and nonlinear effects such as self-steepening and intrapulse Raman scattering can also be included using Eq. (2.3.39) in place of Eq. (5.1.1) as the starting point [20]–[22]. The third-order dispersion β_3 does not affect the gain spectrum of modulation instability. The main effect of self-steepening is to reduce the growth rate and the frequency range over which gain occurs from the values seen in Fig. 5.1. Equation (5.1.8) provides a simple estimate of the modulation-instability gain in most cases of practical interest.

As discussed in Chapter 10, modulation instability can be interpreted in terms of a four-wave-mixing process that is phase-matched by SPM. If a probe wave at a frequency $\omega_1 = \omega_0 + \Omega$ were to copropagate with the CW beam at ω_0, it would experience a net power gain given by Eq. (5.1.8) as long as $|\Omega| < \Omega_c$. Physically, the energy of two photons from the intense pump beam is used to create two different photons, one at the probe frequency ω_1 and the other at the idler frequency $2\omega_0 - \omega_1$. The case in which a probe is launched together with the intense pump wave is referred to as *induced* modulation instability.

Even when the pump wave propagates by itself, modulation instability can lead to spontaneous breakup of the CW beam into a periodic pulse train. Noise photons (vacuum fluctuations) act as a probe in this situation and are amplified by the gain provided by modulation instability. As the largest gain occurs for frequencies $\omega_0 \pm \Omega_{max}$, where Ω_{max} is given by Eq. (5.1.9), these frequency components are amplified most. Thus, a clear-cut evidence of *spontaneous* modulation instability at the fiber output is provided by two spectral sidebands located symmetrically at $\pm\Omega_{max}$ on each side of the central line at ω_0. In the time domain, the CW beam is converted into a periodic pulse train with a period $T_m = 2\pi/\Omega_{max}$.

One may wonder whether modulation instability can occur in the normal-dispersion region of optical fibers under certain conditions. It turns out that cross-phase modulation, occurring when two optical beams at different wavelengths or with orthogonal polarizations propagate simultaneously, can lead to modulation instability even in normally dispersive fibers. This case is discussed in Chapters 6 and 7. Even a single CW beam can become unstable in normally dispersive media if the medium response is sluggish. The gain

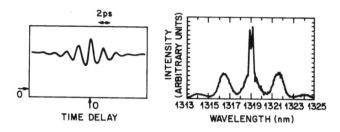

Figure 5.2 Autocorrelation trace and optical spectrum of 100-ps input pulses showing evidence of modulation instability at a peak power of 7.1 W. (After Ref. [18].)

peak occurs at a frequency $\Omega_{\max} = T_{\mathrm{NL}}^{-1}$, where T_{NL} is the nonlinear response time [30]. For silica fibers T_{NL} is so short (a few femtoseconds) and $\Omega_{\max}$ is so large that even the use of the NLS equation becomes questionable. However, when such fibers are doped with other materials (rare-earth ions, dyes, or semiconductors), it may be possible to observe the effects of a finite nonlinear response time.

5.1.3   Experimental Observation

Modulation instability in the anomalous-dispersion regime of optical fibers was first observed in an experiment in which 100-ps (FWHM) pulses from a Nd:YAG laser operating at 1.319 μm were transmitted through a 1-km-long fiber having $\beta_2 \approx -3$ ps^2/km [18]. Figure 5.2 shows the autocorrelation trace and the optical spectrum measured at the fiber output for a peak power $P_0 = 7.1$ W. The location of spectral sidebands is in agreement with the prediction of Eq. (5.1.9). Furthermore, the interval between the oscillation peaks in the autocorrelation trace is inversely related to $\Omega_{\max}$ as predicted by theory. The secondary sidebands seen in Fig. 5.2 are also expected when pump depletion is included. In this experiment, it was necessary to use 100-ps pulses rather than CW radiation to avoid stimulated Brillouin scattering (see Chapter 9). However, as the modulation period is $\sim$1 ps, the relatively broad 100-ps input pulses provide a quasi-CW environment for the observation of modulation instability.

In a related experiment, modulation instability was induced by sending a weak CW probe wave together with the intense pump pulses [19]. The probe was obtained from a single-mode semiconductor laser whose wavelength could be tuned over a few nanometers in the vicinity of the pump wavelength. The

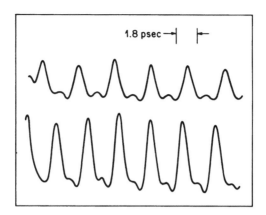

1.8 psec

Figure 5.3 Autocorrelation traces showing induced modulation instability at two different probe wavelengths. The modulation period can be adjusted by tuning the semiconductor laser acting as a probe (after Ref. [18].)

CW probe power of 0.5 mW was much smaller compared with the pump-pulse peak power of $P_0 = 3$ W. However, its presence led to breakup of each pump pulse into a periodic pulse train whose period was inversely related to the frequency difference between the pump and probe waves. Moreover, the period could be adjusted by tuning the wavelength of the probe laser. Figure 5.3 shows the autocorrelation traces for two different probe wavelengths. As the observed pulse width is < 1 ps, this technique is useful for generating subpicosecond pulses whose repetition rate can be controlled by tuning the probe wavelength.

When optical pulses with widths <100 ps are used, modulation instability can be initiated by SPM. If spectral broadening induced by SPM is large enough to exceed Ω_{max}, the SPM-generated frequency components near Ω_{max} can act as a probe and get amplified by modulation instability. This phenomenon is called SPM-induced modulation instability. One can estimate the fiber length L at which the spectral width approaches Ω_{max} by using $\delta\omega_{max}$ from Eq. (4.1.9) and requiring that $\Omega_{max} \approx \delta\omega_{max}$. In the case of a Gaussian pulse this condition is satisfied when

$$L \approx (2L_D L_{NL})^{1/2}, \tag{5.1.11}$$

where $L_D = T_0^2/|\beta_2|$ is the dispersion length introduced in Section 3.1. Numerical solutions of Eq. (5.1.1) confirm the occurrence of SPM-induced modulation instability [23]. In particular, the input pulse develops deep modulations at the frequency $\Omega_{max}/2\pi$ and the spectrum exhibits sidelobes at that frequency. The SPM-induced modulation instability has also been observed experimentally [23].

5.1.4 Ultrashort Pulse Generation

The linear stability analysis of the steady-state solution of Eq. (5.1.1) provides only the initial exponential growth of weak perturbations with the power gain given by Eq. (5.1.8). Clearly, an exponential growth cannot be sustained indefinitely because the frequency components at $\omega_0 \pm \Omega$ grow at the expense of the pump wave at ω_0, and pump depletion slows down the growth rate. Moreover, the sidebands at $\omega_0 \pm \Omega$ eventually become strong enough, and the perturbation becomes large enough, that the linear stability analysis breaks down. Evolution of the modulated state is then governed by the NLS equation (5.1.1). A simple approach solves this equation in the frequency domain as a four-wave mixing problem [29]; it is discussed in detail in Chapter 10. The main disadvantage of this approach is that it cannot treat generation of higher-order sidebands located at $\omega_0 \pm m\Omega$ ($m = 2, 3, \ldots$) that are invariably created when the first-order sidebands ($m = 1$) become strong.

The time-domain approach solves the NLS equation directly. Numerical solutions of Eq. (5.1.1) obtained with the input corresponding to a CW beam with weak sinusoidal modulation imposed on it show that the nearly CW beam evolves into a train of narrow pulses, separated by the period of initial modulation [14]. The fiber length required to realize such a train of narrow pulses depends on the initial modulation depth and is typically $\sim 5L_D$. With further propagation, the multipeak structure deforms and eventually returns to the initial input form. This behavior is found to be generic when Eq. (5.1.1) is solved by considering arbitrary periodic modulation of the steady state [35]. The foregoing scenario suggests that the NLS equation should have periodic solutions whose form changes with propagation. Indeed, it turns out that the NLS equation has a multiparameter family of periodic solutions [35]–[43]. In their most general form, these solutions are expressed in the form of Jacobian elliptic functions. In some specific cases, the solution can be written in terms of trigonometric and hyperbolic functions [39].

From a practical standpoint, with a proper choice of fiber length, modulation instability can be used for generating a train of short optical pulses whose repetition rate can be externally controlled. As early as 1989, 130-fs pulses at a 2-THz repetition rate were generated through induced modulation instability [44]. Since then, this technique has been used to create optical sources capable of producing periodic trains of ultrashort pulses at high but controllable repetition rates. Several experiments have used dispersion-decreasing fibers for this purpose [45]–[49]. Initial sinusoidal modulation in these experiments was imposed by beating two optical signals. In a 1992 experiment [46], the outputs of two distributed feedback semiconductor lasers, operating continuously at slightly different wavelengths near 1.55 μm, were combined in a fiber coupler to produce a sinusoidally modulated signal at a beat frequency that could be varied in the 70–90 GHz range by controlling the laser temperature. The beat signal was amplified to power levels ∼0.3 W by using a fiber amplifier and then propagated through a 1-km dispersion-shifted fiber, followed by a dispersion-decreasing fiber whose GVD decreased from 10 to 0.5 ps/(km-nm) over a length of 1.6 km. The output consisted of a high-quality pulse train at 70 GHz with individual pulses of 1.3-ps width. By 1993, this technique led to generation of a 250-fs pulse train at repetition rates 80–120 GHz [49].

The use of a dispersion-decreasing fiber is not essential for producing pulse trains through modulation instability. In an interesting experiment, a comb-like dispersion profile was produced by splicing pieces of low- and high-dispersion fibers [50]. A dual-frequency fiber laser was used to generate the high-power signal modulated at a frequency equal to the longitudinal-mode spacing (59 GHz). When such a modulated signal was launched into the fiber, the output consisted of a 2.2-ps pulse train at the 59-GHz repetition rate. In another experiment [51], a periodic train of 1.3-ps pulses at the 123-GHz repetition rate was generated by launching the high-power beat signal into a 5-km-long dispersion-shifted fiber. The experimental results were in good agreement with the numerical simulations based on the NLS equation.

The main problem with the preceding technique is that its use requires a relatively long fiber (∼5 km) and relatively high input powers (∼100 mW) for the pulse train to build up. This problem can be solved by enclosing the fiber within a cavity. The gain provided by modulation instability converts such a device into a self-pulsing laser. As early as 1988, a ring-cavity configuration was used to generate a pulse train through modulation instability [52]. Sine then, modulation instability occurring inside an optical resonator has attracted considerable attention [53]–[57]. Mathematical treatment in the case of a Fabry–Perot resonator is quite cumbersome because one must use a set of two coupled NLS equations for the counterpropagating optical fields. It turns out that modulation instability can occur even in the normal-dispersion regime of the fiber because of the feedback provided by cavity mirrors [54]. Moreover, the relatively small feedback occurring at the fiber–air interface (about 4%) is enough for this fundamental change to occur [55]. As a result, a self-pulsing fiber laser can be made without actually using any mirrors. Numerical and analytical results show that such a laser can generate ultrashort pulse trains with repetition rates in the THz range by using CW pump beams with power levels ~ 10 mW [56].

5.1.5 Impact on Lightwave Systems

Modulation instability affects the performance of long-haul optical communication systems in which fiber loss is compensated periodically using optical amplifiers [58]–[65]. Computer simulations showed as early as 1990 that it can be a limiting factor for systems employing the nonreturn-to-zero (NRZ) format for data transmission [58]. Since then, the impact of modulation instability has been studied both numerically and experimentally [63]. Physically, spontaneous emission of amplifiers can provide a seed for the growth of sidebands through induced modulation instability. As a result, signal spectrum broadens substantially. Since GVD-induced broadening of optical pulses depends on their bandwidth, this effect degrades system performance. Experimental results on a lightwave system operating at 10 Gb/s showed considerable degradation for a transmission distance of only 455 km [62]. As expected, system performance improved when GVD was compensated partially using a dispersion-compensating fiber.

The use of optical amplifiers can induce modulation instability through another mechanism and generate additional sidebands in which noise can be amplified in both the normal and anomalous GVD regime of optical fibers [59]. The new mechanism has its origin in the periodic sawtoothlike variation of the average power P_0 occurring along the link length. To understand the physics more clearly, note that a periodic variation of P_0 in z is equivalent to the creation of a nonlinear index grating because the term γP_0 in Eq. (5.1.4) becomes a periodic function of z. The period of this grating is equal to the amplifier spacing and is typically in the range 40–50 km. Such a long-period grating provides a new coupling mechanism between the spectral sidebands located at $\omega_0 + \Omega$ and $\omega_0 - \Omega$ and allows them to grow when the perturbation frequency Ω satisfies the Bragg condition.

The analysis of Section 5.1.1 can be extended to include periodic variations of P_0. If we replace P_0 in Eq. (5.1.4) by $P_0 f(z)$, where $f(z)$ is a periodic function, expand $f(z)$ in a Fourier series as $f(z) = \sum c_m \exp(2\pi i m z / L_A)$, the frequencies at which the gain peaks are found to be [59]

$$\Omega_m = \pm \left(\frac{2\pi m}{\beta_2 L_A} - \frac{2\gamma P_0 c_0}{\beta_2} \right)^{1/2}, \tag{5.1.12}$$

where the integer m represents the order of Bragg diffraction, L_A is the spacing between amplifiers (grating period), and the Fourier coefficient c_m is related to the fiber loss α as

$$c_m = \frac{1 - \exp(-\alpha L_A)}{\alpha L_A + 2im\pi}. \tag{5.1.13}$$

In the absence of grating, or when $m = 0$, Ω_0 exists only for anomalous dispersion, in agreement with Eq. (5.1.9). However, when $m \neq 0$, modulation-instability sidebands can occur even for normal dispersion ($\beta_2 > 0$). Physically, this behavior can be understood by noting that the nonlinear index grating helps to satisfy the phase-matching condition necessary for four-wave mixing when $m \neq 0$. Fortunately, this phenomenon is not likely to affect system performance significantly because neither the amplifier spacing nor the fiber parameters are uniform in practice.

With the advent of wavelength-division multiplexing (WDM), it has become common to employ the technique of dispersion management to reduce the GVD globally while keeping it high locally by using a periodic dispersion map. The periodic variation of β_2 creates another grating that affects modulation instability considerably. Mathematically, the situation is similar to the case already discussed except that β_2 rather than P_0 in Eq. (5.1.4) is a periodic function of z. The gain spectrum of modulation instability is obtained following a similar technique [61]. The β_2 grating not only generates new sidebands but also affects the gain spectrum seen in Fig. 5.1. In the case of strong dispersion management (relatively large GVD variations), both the peak value and the bandwidth of the modulation-instability gain are reduced, indicating that such systems should not suffer much from amplification of noise induced by modulation instability. This does not mean dispersion-managed WDM systems are immune to modulation instability. Indeed, it has been shown that WDM systems suffer from a resonant enhancement of four-wave mixing that degrades the system performance considerably when channel spacing is close to the frequency at which the modulation-instability gain is strongest [65]. On the positive side, this enhancement can be used for low-power, high-efficiency, wavelength conversion [66]. Modulation instability has also been used for measuring the distribution of zero-dispersion wavelength along a fiber by noting that the instability gain becomes quite small in the vicinity of $|\beta_2| = 0$ [67].

5.2   Fiber Solitons

The occurrence of modulation instability in the anomalous-GVD regime of optical fibers is an indication of a fundamentally different character of Eq. (5.1.1) when $\beta_2 < 0$. It turns out that this equation has specific pulselike solutions that either do not change along fiber length or follow a periodic evolution pattern—such solutions are known as optical solitons. The history of solitons, in fact, dates back to 1834, the year in which Scott Russell observed a heap of water in a canal that propagated undistorted over several kilometers. Here is a quote from his report published in 1844 [68]:

> I was observing the motion of a boat which was rapidly drawn along a narrow channel by a pair of horses, when the boat suddenly stopped—not so the mass of water in the channel which it had put in motion; it accumulated round the prow of the vessel in a state of violent agitation, then suddenly leaving it behind, rolled forward with great velocity, assuming the form of a large solitary elevation, a rounded, smooth and well-defined heap of water, which continued its course along the channel apparently without change of form or diminution of speed. I followed it on horseback, and overtook it still rolling on at a rate of some eight or nine miles an hour, preserving its original figure some thirty feet long and a foot to a foot and a half in height. Its height gradually diminished, and after a chase of one or two miles I lost it in the windings of the channel. Such, in the month of August 1834, was my first chance interview with that singular and beautiful phenomenon which I have called the Wave of Translation.

Such waves were later called solitary waves. However, their properties were not understood completely until the inverse scattering method was developed [69]. The term *soliton* was coined in 1965 to reflect the particlelike nature of those solitary waves that remained intact even after mutual collisions [70]. Since then, solitons have been discovered and studied in many branches of physics including optics [71]–[79]. In the context of optical fibers, the use of solitons for optical communications was first suggested in 1973 [80]. By the year 1999, several field trials making use of fiber solitons have been completed [81]. The word "soliton" has become so popular in recent years that a search on the Internet returns thousands of hits. Similarly, scientific databases reveal that hundreds of research papers are published every year with the word "soliton" in their title. It should be stressed that the distinction between a soliton and a solitary wave is not always made in modern optics literature, and it is quite common to refer to all solitary waves as solitons.

5.2.1 Inverse Scattering Method

Only certain nonlinear wave equations can be solved by the inverse scattering method [71]. The NLS equation (5.1.1) belongs to this special class of equations. Zakharov and Shabat used the inverse scattering method in 1971 to solve the NLS equation [82]. This method is similar in spirit to the Fourier-transform method used commonly for solving linear partial differential equations. The approach consists of identifying a suitable scattering problem whose potential is the solution sought. The incident field at $z = 0$ is used to find the initial scattering data whose evolution along z is easily determined by solving the linear scattering problem. The propagated field is reconstructed from the evolved scattering data. Since details of the inverse scattering method are available in many texts [71]–[79], only a brief description is given here.

Similar to Chapter 4, it is useful to normalize Eq. (5.1.1) by introducing three dimensionless variables

$$U = \frac{A}{\sqrt{P_0}}, \qquad \xi = \frac{z}{L_D}, \qquad \tau = \frac{T}{T_0}, \qquad (5.2.1)$$

and write it in the form

$$i\frac{\partial U}{\partial \xi} = \text{sgn}(\beta_2)\frac{1}{2}\frac{\partial^2 U}{\partial \tau^2} - N^2|U|^2U, \qquad (5.2.2)$$

where P_0 is the peak power, T_0 is the width of the incident pulse, and the parameter N is introduced as

$$N^2 = \frac{L_D}{L_{\text{NL}}} = \frac{\gamma P_0 T_0^2}{|\beta_2|}. \qquad (5.2.3)$$

The dispersion length L_D and the nonlinear length L_{NL} are defined as in Eq. (3.1.5). Fiber losses are neglected in this section but will be included later.

The parameter N can be eliminated from Eq. (5.2.2) by introducing

$$u = NU = \sqrt{\gamma L_D} A.$$ (5.2.4)

Equation (5.2.2) then takes the standard form of the NLS equation:

$$i\frac{\partial u}{\partial \xi} + \frac{1}{2}\frac{\partial^2 u}{\partial \tau^2} + |u|^2 u = 0,$$ (5.2.5)

where the choice $\text{sgn}(\beta_2) = -1$ has been made to focus on the case of anomalous GVD; the other case is considered in the next section. Note that an important scaling relation holds for Eq. (5.2.5). If $u(\xi, \tau)$ is a solution of this equation, then $\varepsilon u(\varepsilon^2 \xi, \varepsilon \tau)$ is also a solution, where ε is an arbitrary scaling factor. The importance of this scaling will become clear later.

In the inverse scattering method, the scattering problem associated with Eq. (5.2.5) is [77]

$$i\frac{\partial v_1}{\partial \tau} + u v_2 = \zeta v_1,$$ (5.2.6)

$$i\frac{\partial v_2}{\partial \tau} + u^* v_1 = -\zeta v_2,$$ (5.2.7)

where v_1 and v_2 are the amplitudes of the two waves scattered by the potential $u(\xi, \tau)$. The eigenvalue ζ plays a role similar to that played by the frequency in the standard Fourier analysis except that ζ can take complex values when $u \neq 0$. This feature can be identified by noting that, in the absence of potential ($u = 0$), v_1 and v_2 vary as $\exp(\pm i\zeta\tau)$.

Equations (5.2.6) and (5.2.7) apply for all values of ξ. In the inverse scattering method, they are first solved at $\xi = 0$. For a given initial form of $u(0, \tau)$, Eqs. (5.2.6) and (5.2.7) are solved to obtain the initial scattering data. The direct scattering problem is characterized by a reflection coefficient $r(\zeta)$ that plays a role analogous to the Fourier coefficient. Formation of the bound states (solitons) corresponds to the poles of $r(\zeta)$ in the complex ζ plane. Thus, the initial scattering data consist of the reflection coefficient $r(\zeta)$, the complex poles ζ_j, and their residues c_j, where $j = 1$ to N if N such poles exist. Although the parameter N of Eq. (5.2.3) is not necessarily an integer, the same notation is used for the number of poles to stress that its integer values determine the number of poles.

Evolution of the scattering data along the fiber length is determined by using well-known techniques [71]. The desired solution $u(\xi, \tau)$ is reconstructed from the evolved scattering data using the inverse scattering method. This step is quite cumbersome mathematically because it requires the solution of a complicated linear integral equation. However, in the specific case in which $r(\zeta)$ vanishes for the initial potential $u(0, \tau)$, the solution $u(\xi, \tau)$ can be determined by solving a set of algebraic equations. This case corresponds to solitons. The soliton order is characterized by the number N of poles, or eigenvalues ζ_j ($j = 1$–N). The general solution can be written as [82]

$$u(\xi, \tau) = -2 \sum_{j=1}^{N} \lambda_j^* \psi_{2j}^*, \qquad (5.2.8)$$

where

$$\lambda_j = \sqrt{c_j} \exp(i\zeta_j \tau + i\zeta_j^2 \xi), \qquad (5.2.9)$$

and ψ_{2j}^* is obtained by solving the following set of algebraic linear equations:

$$\psi_{1j} + \sum_{k=1}^{N} \frac{\lambda_j \lambda_k^*}{\zeta_j - \zeta_k^*} \psi_{2k}^* = 0, \qquad (5.2.10)$$

$$\psi_{2j}^* - \sum_{k=1}^{N} \frac{\lambda_j^* \lambda_k}{\zeta_j^* - \zeta_k} \psi_{1k} = \lambda_j^*. \qquad (5.2.11)$$

The eigenvalues ζ_j are generally complex ($2\zeta_j = \delta_j + i\eta_j$). Physically, the real part δ_j produces a change in the group velocity associated with the jth component of the soliton. For the Nth-order soliton to remain bound, it is necessary that all of its components travel at the same speed. Thus, all eigenvalues ζ_j should lie on a line parallel to the imaginary axis, i.e., $\delta_j = \delta$ for all j. This feature simplifies the general solution in Eq. (5.2.9) considerably. It will be seen later that the parameter δ represents a frequency shift of the soliton from the carrier frequency ω_0.

5.2.2 Fundamental Soliton

The first-order soliton ($N = 1$) corresponds to the case of a single eigenvalue. It is referred to as the fundamental soliton because its shape does not change on propagation. Its field distribution is obtained from Eqs. (5.2.8)–(5.2.11) after setting $j = k = 1$. Noting that $\psi_{21} = \lambda_1(1 + |\lambda_1|^4/\eta^2)^{-1}$ and substituting it in Eq. (5.2.8), we obtain

$$u(\xi, \tau) = -2(\lambda_1^*)^2(1 + |\lambda_1|^4/\eta^2)^{-1}. \qquad (5.2.12)$$

After using Eq. (5.2.9) for λ_1 together with $\zeta_1 = (\delta + i\eta)/2$ and introducing the parameters τ_s and ϕ_s through $-c_1/\eta = \exp(\eta\tau_s - i\phi_s)$, we obtain the following general form of the fundamental soliton:

$$u(\xi, \tau) = \eta \operatorname{sech}[\eta(\tau - \tau_s + \delta\xi)] \exp[i(\eta^2 - \delta^2)\xi/2 - i\delta\tau + i\phi_s], \quad (5.2.13)$$

where η, δ, τ_s, and ϕ_s are four arbitrary parameters that characterize the soliton. Thus, an optical fiber supports a four-parameter family of fundamental solitons, all sharing the condition $N = 1$.

Physically, the four parameters η, δ, τ_s, and ϕ_s represent amplitude, frequency, position, and phase of the soliton, respectively. The phase ϕ_s can be dropped from the discussion because a constant absolute phase has no physical significance. It will become relevant later when nonlinear interaction between a pair of solitons is considered. The parameter τ_s can also be dropped because it denotes the position of the soliton peak: If the origin of time is chosen such that the peak occurs at $\tau = 0$ at $\xi = 0$, one can set $\tau_s = 0$. It is clear from the phase factor in Eq. (5.2.13) that the parameter δ represents a frequency shift of the soliton from the carrier frequency ω_0. Using the carrier part, $\exp(-i\omega_0 t)$, the new frequency becomes $\omega_0' = \omega_0 + \delta/T_0$. Note that a frequency shift also changes the soliton speed from its original value v_g. This can be seen more clearly by using $\tau = (t - \beta_1 z)/T_0$ in Eq. (5.2.13) and writing it as

$$|u(\xi, \tau)| = \eta \operatorname{sech}[\eta(t - \beta_1' z)/T_0], \quad (5.2.14)$$

where $\beta_1' = \beta_1 + \delta|\beta_2|/T_0$. As expected on physical grounds, the change in group velocity ($v_g = 1/\beta_1$) is a consequence of fiber dispersion.

The frequency shift δ can also be removed from Eq. (5.2.13) by choosing the carrier frequency appropriately. Fundamental solitons then form a single-parameter family described by

$$u(\xi, \tau) = \eta \operatorname{sech}(\eta\tau) \exp(i\eta^2\xi/2). \quad (5.2.15)$$

The parameter η determines not only the soliton amplitude but also its width. In real units, the soliton width changes with η as T_0/η, i.e., it scales inversely with the soliton amplitude. This inverse relationship between the amplitude and the width of a soliton is the most crucial property of solitons. Its relevance will become clear later. The canonical form of the fundamental soliton is obtained by choosing $u(0,0) = 1$ so that $\eta = 1$. With this choice, Eq. (5.2.15) becomes

$$u(\xi, \tau) = \operatorname{sech}(\tau) \exp(i\xi/2). \quad (5.2.16)$$

One can verify by direct substitution in Eq. (5.2.5) that this solution is indeed a solution of the NLS equation.

The solution in Eq. (5.2.16) can also be obtained by solving the NLS equation directly, without using the inverse scattering method. The approach consists of assuming that a shape-preserving solution of the NLS equation exists and has the form

$$u(\xi, \tau) = V(\tau) \exp[i\phi(\xi, \tau)], \qquad (5.2.17)$$

where V is independent of ξ for Eq. (5.2.17) to represent a fundamental soliton that maintains its shape during propagation. The phase ϕ can depend on both ξ and τ. If Eq. (5.2.17) is substituted in Eq. (5.2.5) and the real and imaginary parts are separated, one obtains two equations for V and ϕ. The phase equation shows that ϕ should be of the form $\phi(\xi, \tau) = K\xi - \delta\tau$, where K and δ are constants. Choosing $\delta = 0$ (no frequency shift), $V(\tau)$ is found to satisfy

$$\frac{d^2V}{d\tau^2} = 2V(K - V^2). \qquad (5.2.18)$$

This nonlinear equation can be solved by multiplying it by $2(dV/d\tau)$ and integrating over τ. The result is

$$(dV/d\tau)^2 = 2KV^2 - V^4 + C, \qquad (5.2.19)$$

where C is a constant of integration. Using the boundary condition that both V and $dV/d\tau$ vanish as $|\tau| \to \infty$, C is found to be 0. The constant K is determined from the condition that $V = 1$ and $dV/d\tau = 0$ at the soliton peak, assumed to occur at $\tau = 0$. Its use provides $K = \frac{1}{2}$, and hence $\phi = \xi/2$. Equation (5.2.19) is easily integrated to obtain $V(\tau) = \text{sech}(\tau)$. We have thus recovered the solution in Eq. (5.2.16) using a simple technique.

In the context of optical fibers, the solution (5.2.16) indicates that if a hyperbolic-secant pulse, whose width T_0 and the peak power P_0 are chosen such that $N = 1$ in Eq. (5.2.3), is launched inside an ideal lossless fiber, the pulse will propagate undistorted without change in shape for arbitrarily long distances. It is this feature of the fundamental solitons that makes them attractive for optical communication systems [80]. The peak power P_0 required to support the fundamental soliton is obtained from Eq. (5.2.3) by setting $N = 1$ and is given by

$$P_0 = \frac{|\beta_2|}{\gamma T_0^2} \approx \frac{3.11 \, |\beta_2|}{\gamma T_{\text{FWHM}}^2}, \qquad (5.2.20)$$

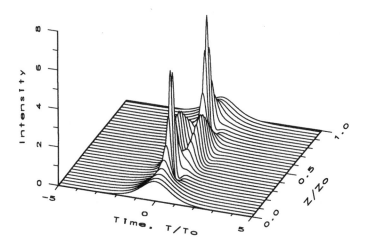

Figure 5.4 Temporal evolution over one soliton period for the third-order soliton. Note pulse splitting near $z_0 = 0.5$ and soliton recovery beyond that.

where the FWHM of the soliton is defined using $T_{\text{FWHM}} \approx 1.76\,T_0$ from Eq. (3.2.22). Using typical parameter values, $\beta_2 = -1$ ps^2/km and $\gamma = 3$ W^{-1}/km for dispersion-shifted fibers near the 1.55-μm wavelength, P_0 is ~ 1 W for $T_0 = 1$ ps but reduces to only 10 mW when $T_0 = 10$ ps because of its T_0^{-2} dependence. Thus, fundamental solitons can form in optical fibers at power levels available from semiconductor lasers even at a relatively high bit rate of 20 Gb/s.

5.2.3 Higher-Order Solitons

Higher-order solitons are also described by the general solution of Eq. (5.2.8). Various combinations of the eigenvalues η_j and the residues c_j generally lead to an infinite variety of soliton forms. If the soliton is assumed to be symmetric about $\tau = 0$, the residues are related to the eigenvalues by the relation [83]

$$c_j = \frac{\prod_{k=1}^{N} (\eta_j + \eta_k)}{\prod_{k \neq j}^{N} |\eta_j - \eta_k|}. \tag{5.2.21}$$

This condition selects a subset of all possible solitons. Among this subset, a special role is played by solitons whose initial shape at $\xi = 0$ is given by

$$u(0, \tau) = N \text{sech}(\tau), \tag{5.2.22}$$

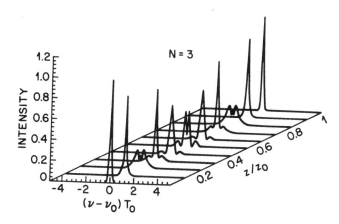

Figure 5.5 Spectral evolution over one soliton period for the third-order soliton.

where the soliton order N is an integer. The peak power necessary to launch the Nth-order soliton is obtained from Eq. (5.2.3) and is N^2 times of that required for the fundamental soliton. For the second-order soliton ($N = 2$), the field distribution is obtained from Eqs. (5.2.8)–(5.2.11). Using $\zeta_1 = i/2$ and $\zeta_2 = 3i/2$ for the two eigenvalues, the second-order soliton is given by [84]

$$u(\xi, \tau) = \frac{4[\cosh(3\tau) + 3\exp(4i\xi)\cosh(\tau)]\exp(i\xi/2)}{[\cosh(4\tau) + 4\cosh(2\tau) + 3\cos(4\xi)]}. \qquad (5.2.23)$$

An interesting property of the forementioned solution is that $|u(\xi, \tau)|^2$ is periodic in ξ with the period $\xi_0 = \pi/2$. In fact, this periodicity occurs for all higher-order solitons. Using the definition $\xi = z/L_D$ from Eq. (5.2.1), the soliton period z_0 in real units becomes

$$z_0 = \frac{\pi}{2}L_D = \frac{\pi}{2}\frac{T_0^2}{|\beta_2|} \approx \frac{T_{\text{FWHM}}^2}{2|\beta_2|}. \qquad (5.2.24)$$

Periodic evolution of a third-order soliton over one soliton period is shown in Fig. 5.4. As the pulse propagates along the fiber, it first contracts to a fraction of its initial width, splits into two distinct pulses at $z_0/2$, and then merges again to recover the original shape at the end of the soliton period at $z = z_0$. This pattern is repeated over each section of length z_0.

To understand the origin of periodic evolution for higher-order solitons, it is helpful to look at changes in the pulse spectra shown in Fig. 5.5 for the $N = 3$ soliton. The temporal and spectral changes result from an interplay between

SPM and GVD. The SPM generates a frequency chirp such that the leading edge of soliton is red-shifted while its trailing-edge is blue-shifted from the central frequency. The SPM-induced spectral broadening is clearly seen in Fig. 5.5 for $z/z_0 = 0.2$ with its typical oscillatory structure. In the absence of GVD, the pulse shape would have remained unchanged. However, anomalous GVD contracts the pulse as the pulse is positively chirped (see Section 3.2). Only the central portion of the pulse contracts because the chirp is nearly linear only over that part. However, as a result of a substantial increase in the pulse intensity near the central part of the pulse, the spectrum changes significantly as seen in Fig. 5.5 for $z/z_0 = 0.3$. It is this mutual interaction between the GVD and SPM effects that is responsible for the evolution pattern seen in Fig. 5.4.

In the case of a fundamental soliton $(N = 1)$, GVD and SPM balance each other in such a way that neither the pulse shape nor the pulse spectrum changes along the fiber length. In the case of higher-order solitons, SPM dominates initially but GVD soon catches up and leads to pulse contraction seen in Fig. 5.4. Soliton theory shows that for pulses with a hyperbolic-secant shape and with peak powers determined from Eq. (5.2.3), the two effects can cooperate in such a way that the pulse follows a periodic evolution pattern with original shape recurring at multiples of the soliton period z_0 given by Eq. (5.2.24). Near the 1.55-μm wavelength, typically $\beta_2 = -20$ ps^2/km for standard silica fibers. The soliton period is ~ 80 m for $T_0 = 1$ ps and scales as T_0^2, becoming 8 km when $T_0 = 10$ ps. For dispersion-shifted fibers with $\beta_2 \approx -2$ ps^2/km, z_0 increases by one order of magnitude for the same value of T_0.

5.2.4 Experimental Confirmation

The possibility of soliton formation in optical fibers was suggested as early as 1973 [80]. However, the lack of a suitable source of picosecond optical pulses at wavelengths >1.3 μm delayed their experimental observation until 1980. Solitons in optical fibers were first observed in an experiment [85] that used a mode-locked color-center laser capable of emitting short optical pulses $(T_{FWHM} \approx 7$ ps) near 1.55 μm, a wavelength near which optical fibers exhibit anomalous GVD together with minimum losses. The pulses were propagated inside a 700-m-long single-mode fiber with a core diameter of 9.3 μm. The fiber parameters for this experiment were estimated to be $\beta_2 \approx -20$ ps^2/ km and $\gamma \approx 1.3$ W^{-1}/km. Using $T_0 = 4$ ps in Eq. (5.2.20), the peak power for exciting a fundamental soliton is ~ 1 W.

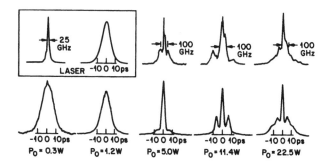

Figure 5.6 Autocorrelation traces (lower row) and pulse spectra (upper row) for several values of input peak power P_0. The corresponding traces for the input pulse are shown inside the rectangular box. (After Ref. [85].)

In the experiment, the peak power of optical pulses was varied over a range 0.3–25 W, and their pulse shape and spectrum were monitored at the fiber output. Figure 5.6 shows autocorrelation traces and pulse spectra at several power levels and compares them with those of the input pulse. The measured spectral width of 25 GHz of the input pulse is nearly transform limited, indicating that mode-locked pulses used in the experiment were unchirped. At a low power level of 0.3 W, optical pulses experienced dispersion-induced broadening inside the fiber, as expected from Section 3.2. However, as the power was increased, output pulses steadily narrowed, and their width became the same as the input width at $P_0 = 1.2$ W. This power level corresponds to the formation of a fundamental soliton and should be compared with the theoretical value of 1 W obtained from Eq. (5.2.20). The agreement is quite good in spite of many uncertainties inherent in the experiment.

At higher power levels, output pulses exhibited dramatic changes in their shape and developed a multipeak structure. For example, the autocorrelation trace for 11.4 W exhibits three peaks. Such a three-peak structure corresponds to two-fold splitting of the pulse, similar to that seen in Fig. 5.4 near $z/z_0 = 0.5$ for the third-order soliton. The observed spectrum also shows characteristic features seen in Fig. 5.5 near $z/z_0 = 0.5$. The estimated soliton period for this experiment is 1.26 km. Thus, at the fiber output $z/z_0 = 0.55$ for the 700-m-long fiber used in the experiment. As the power level of 11.4 W is also nearly nine times the fundamental soliton power, the data of Fig. 5.6 indeed correspond to the $N = 3$ soliton. This conclusion is further corroborated by the autocorrelation trace for $P_0 = 22.5$ W. The observed five-peak structure corresponds to three-fold splitting of the laser pulse, in agreement with the prediction of soliton theory for the fourth-order soliton $(N = 4)$.

The periodic nature of higher-order solitons implies that the pulse should restore its original shape and spectrum at distances that are multiples of the soliton period. This feature was observed for second- and third-order solitons in a 1983 experiment in which the fiber length of 1.3 km corresponded to nearly one soliton period [86]. In a different experiment, initial narrowing of higher-order solitons, seen in Fig. 5.4 for $N = 3$, was observed for values of N up to 13 [87]. Higher-order solitons also formed inside the cavity of a mode-locked dye laser operating in the visible region near 620 nm by incorporating an optical element with negative GVD inside the laser cavity [88]. Such a laser emitted asymmetric second-order solitons, under certain operating conditions, as predicted by inverse scattering theory.

5.2.5   Soliton Stability

A natural question is what happens if the initial pulse shape or the peak power is not matched to that required by Eq. (5.2.22) so that the input pulse does not correspond to an optical soliton. Similarly, one may ask how the soliton is affected if it is perturbed during its propagation inside the fiber. Such questions are answered by using perturbation methods developed for solitons and are discussed later in Section 5.4. This section focus on formation of solitons when the parameters of an input pulse do not correspond to a soliton.

Consider first the case when the peak power is not exactly matched and the value of N obtained from Eq. (5.2.3) is not an integer. Soliton perturbation theory has been used to study this case [84]. Because details are cumbersome, only results are summarized here. In physical terms, the pulse adjusts its shape and width as it propagates along the fiber and evolves into a soliton. A part of the pulse energy is dispersed away in the process. This part is known as the continuum radiation. It separates from the soliton as ξ increases and its contribution to soliton decays as $\xi^{-1/2}$. For $\xi \gg 1$, the pulse evolves asymptotically into a soliton whose order is an integer $\tilde{N}$ closest to the launched value of N. Mathematically, if

$$N = \tilde{N} + \varepsilon, \qquad |\varepsilon| < 1/2, \qquad (5.2.25)$$

the soliton part corresponds to an initial pulse shape of the form

$$u(0, \tau) = (\tilde{N} + 2\varepsilon) \operatorname{sech}[(1 + 2\varepsilon/\tilde{N})\tau]. \qquad (5.2.26)$$

The pulse broadens if $\varepsilon < 0$ and narrows if $\varepsilon > 0$. No soliton is formed when $N \leq 1/2$.

The effect of pulse shape on soliton formation can be investigated solving Eq. (5.2.5) numerically. Figure 4.7 (Chapter 4) shows evolution of a Gaussian pulse using the initial field $u(0, \tau) = \exp(-\tau^2/2)$. Even though $N = 1$, pulse shape changes along the fiber because of deviations from the 'sech' shape required for a fundamental soliton. The interesting feature of Fig. 4.7 is that the pulse adjusts its width and evolves asymptotically into a fundamental soliton. In fact, the evolution appears to be complete by $z/L_D = 5$, a distance that corresponds to about three soliton periods. An essentially similar evolution pattern occurs for other pulse shapes such as a super-Gaussian shape. The final width of the soliton and the distance needed to evolve into a fundamental soliton depend on the exact shape but the qualitative behavior remains the same.

As pulses emitted from laser sources are often chirped, we should also consider the effect of initial frequency chirp on soliton formation [89]–[97]. The chirp can be detrimental simply because it superimposes on the SPM-induced chirp and disturbs the exact balance between the GVD and SPM effects necessary for solitons. Its effect on soliton formation can be studied by solving Eq. (5.2.5) numerically with an input amplitude

$$u(0, \tau) = N \operatorname{sech}(\tau) \exp(-iC\tau^2/2), \qquad (5.2.27)$$

where C is the chirp parameter introduced in Section 3.2. The quadratic form of phase variation corresponds to a linear chirp such that the optical frequency increases with time (up-chirp) for positive values of C.

Figure 5.7 shows evolution of a fundamental soliton $(N = 1)$ in the case of a relatively low chirp $(C = 0.5)$. The pulse compresses initially mainly because of the positive chirp; initial compression occurs even in the absence of nonlinear effects. The pulse then broadens but is eventually compressed a second time with the tail separating from the main peak gradually. The main peak evolves into a soliton over a propagation distance $\xi > 15$. A similar behavior occurs for negative values of C. Formation of a soliton is expected for small values of $|C|$ because solitons are generally stable under weak perturbations. However, a soliton is destroyed if $|C|$ exceeds a critical value C_{cr}. For $N = 1$, a soliton does not form if C is increased from 0.5 to 2.

The critical value of the chirp parameter can be obtained using the inverse scattering method [93]–[95]. More specifically, Eqs. (5.2.6) and (5.2.7) are solved to obtain the eigenvalue ζ using u from Eq. (5.2.27). Solitons exist as

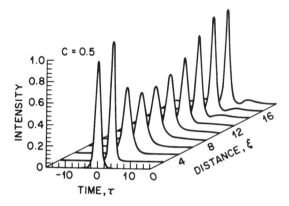

Figure 5.7 Soliton formation in the presence of an initial linear chirp for the case $N = 1$ and $C = 0.5$.

long as the imaginary part of ζ is positive. The critical value depends on N and is found to be about 1.64 for $N = 1$. It also depends on the form of the phase factor in Eq. (5.2.27). From a practical standpoint, initial chirp should be minimized as much as possible. This is necessary because, even if the chirp is not detrimental for $|C| < C_{\rm cr}$, a part of the pulse energy is shed as dispersive waves (continuum radiation) during the process of soliton formation [93]. For example, only 83% of the input energy is converted into a soliton in the case of $C = 0.5$ shown in Fig. 5.7, and this fraction reduces to 62% for $C = 0.8$.

It is clear from the preceding discussion that the exact shape of the input pulse used to launch a fundamental ($N = 1$) soliton is not critical. Moreover, as solitons can form for values of N in the range $0.5 < N < 1.5$, even the width and peak power of the input pulse can vary over a wide range [see Eq. (5.2.3)] without hindering soliton formation. It is this relative insensitivity to the exact values of input parameters that makes the use of solitons feasible in practical applications. However, it is important to realize that, when input parameters deviate substantially from their ideal values, a part of the pulse energy is invariably shed away in the form of dispersive waves as the pulse evolves to form a fundamental soliton [98]. Such dispersive waves are undesirable because they not only represent an energy loss but can also affect the performance of soliton communication systems [99]. Moreover, they can interfere with the soliton itself and modify its characteristics. In the case of an input pulse with N close to 1, such an interference introduces modulations on the pulse spectrum that have also been observed experimentally [100].

Starting in 1988, most of the experimental work on fiber solitons was devoted to their applications in fiber-optic communication systems [101]. Such systems make use of fundamental solitons for representing "1" bits in a digital bit stream and are covered in Part B (devoted to applications of nonlinear fiber optics). In a practical situation, solitons can be subjected to many types of perturbations as they propagate inside an optical fiber. Examples of perturbations include fiber losses, amplifier noise (if amplifiers are used to compensate fiber losses), third-order dispersion, and intrapulse Raman scattering. These effects are discussed later in this chapter.

5.3 Other Types of Solitons

The soliton solution given in Eq. (5.2.8) is not the only possible solution of the NLS equation. Many other kinds of solitons have been discovered depending on the dispersive and nonlinear properties of fibers. This section describes several of them, focusing mainly on dark and bistable solitons.

5.3.1 Dark Solitons

Dark solitons correspond to the solutions of Eq. (5.2.2) with $\mathrm{sgn}(\beta_2) = 1$ and occur in the normal-GVD region of fibers. They were discovered in 1973 and have attracted considerable attention since then [102]–[131]. The intensity profile associated with such solitons exhibits a dip in a uniform background, hence the name *dark* soliton. Pulse-like solitons discussed in Section 5.2 are called *bright* to make the distinction clear. The NLS equation describing dark solitons is obtained from Eq. (5.2.5) by changing the sign of the time-derivative term and is given by

$$i\frac{\partial u}{\partial \xi} - \frac{1}{2}\frac{\partial^2 u}{\partial \tau^2} + |u|^2 u = 0. \tag{5.3.1}$$

Similar to the case of bright solitons, the inverse scattering method has been used [103] to find dark-soliton solutions of Eq. (5.3.1) by imposing the boundary condition that $|u(\xi, \tau)|$ tends toward a constant nonzero value for large values of $|\tau|$. Dark solitons can also be obtained by assuming a solution of the form $u(\xi, \tau) = V(\tau) \exp[i\phi(\xi, \tau)]$, and then solving the ordinary differential equations satisfied by V and ϕ. The main difference compared with the case of bright solitons is that $V(\tau)$ becomes a constant (rather than being zero)

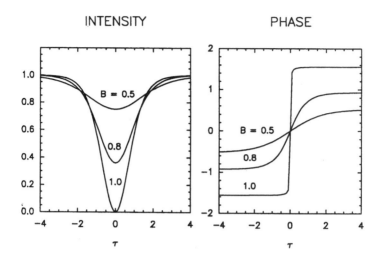

Figure 5.8 Intensity and phase profiles of dark solitons for several values of the blackness parameter B.

as $|\tau| \to \infty$. The general solution can be written as [77]

$$|u(\xi,\tau)| \equiv V(\tau) = \eta\{1 - B^2\text{sech}^2[\eta B(\tau - \tau_s)]\}^{1/2}, \qquad (5.3.2)$$

with the phase given by

$$\phi(\xi,\tau) = \tfrac{1}{2}\eta^2(3 - B^2)\xi + \eta\sqrt{1 - B^2}\,\tau + \tan^{-1}\left(\frac{B\tanh(\eta B\tau)}{\sqrt{1 - B^2}}\right). \qquad (5.3.3)$$

The parameters η and τ_s represent the soliton amplitude and the dip location, respectively. Similar to the bright-soliton case, τ_s can be chosen to be zero without loss of generality. In contrast with the bright-soliton case, the dark soliton has a new parameter B. Physically, B governs the depth of the dip ($|B| \leq 1$). For $|B| = 1$, the intensity at the dip center falls to zero. For other values of B, dip does not go to zero. Dark solitons for which $|B| < 1$ are called *gray* solitons to emphasize this feature; the parameter B governs the blackness of such gray solitons. The $|B| = 1$ case corresponds to a *black* soliton.

For a given value of η, Eq. (5.3.2) describes a family of dark solitons whose width increases inversely with B. Figure 5.8 shows the intensity and phase profiles of such dark solitons for several values of B. Whereas the phase of bright solitons [Eq. (5.2.15)] remains constant across the entire pulse, the phase of dark soliton changes with a total phase shift of $2\sin^{-1}B$, i.e., dark solitons are chirped. For the black soliton ($|B| = 1$), the chirp is such that the phase changes abruptly by π in the center. The phase change becomes more gradual and smaller for smaller values of $|B|$. The time-dependent phase or frequency chirp of dark solitons represents a major difference between bright and dark solitons. One consequence of this difference is that higher-order dark solitons neither form a bound state nor follow a periodic evolution pattern discussed in Section 5.2.3 in the case of bright solitons.

Dark solitons exhibit several interesting features [131]. Consider a black soliton whose canonical form is obtained from Eq. (5.3.2), choosing $\eta = 1$ and $B = 1$, and given by

$$u(\xi, \tau) = \tanh(\tau)\exp(i\xi), \tag{5.3.4}$$

where the phase jump of π at $\tau = 0$ is included in the amplitude part. Thus, an input pulse with "tanh" amplitude, exhibiting an intensity "hole" at the center, would propagate unchanged in the normal-dispersion region of optical fibers. One may ask, in analogy with the case of bright solitons, what happens when the input power exceeds the $N = 1$ limit. This question can be answered by solving Eq. (5.3.1) numerically with an input of the form $u(0, \tau) = N\tanh(\tau)$. Figure 5.9 shows the evolution pattern for $N = 3$; it should be compared with Fig. 5.4 where the evolution of a third-order bright soliton is shown. Two pairs of gray solitons appear and move away from the central black soliton as the propagation distance increases. At the same time, the width of the black soliton decreases [110]. This behavior can be understood by noting that an input pulse of the form $N\tanh(\tau)$ can form a fundamental black soliton of amplitude $N\tanh(N\tau)$ provided its width decreases by a factor of N. It sheds part of its energy in the process that appears in the form of gray solitons. These gray solitons move away from the central black soliton because of their different group velocities. The number of pairs of gray solitons is $N' - 1$, where $N' = N$ for integer values of N or the next integer close to it when N is not an integer. The important feature is that a fundamental dark soliton is always formed for $N > 1$.

Experimental realization of dark solitons is possible only with a finite background instead of the infinite background associated with ideal dark solitons. In practice, a pulse with a narrow dip at its center is used to excite a dark soliton. Numerical calculations show that dark solitons with a finite background pulse exhibit propagation properties nearly identical to those with infinite background if the background pulse is wider by a factor of 10 or more

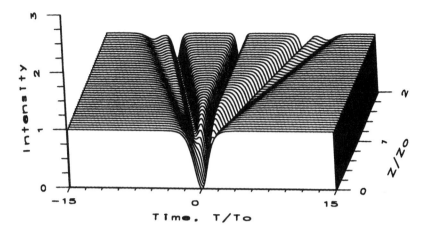

Figure 5.9 Evolution of a third-order dark soliton showing narrowing of the central dip and creation of two pairs of gray solitons.

compared with the soliton width [109]. Several techniques have been used to generate optical pulses with a narrow dip in the center [106]–[108]. Such pulses have been used for observing dark solitons in optical fibers. In one experiment [106], 26-ps input pulses (at 595 nm) with a 5-ps-wide central hole were launched along a 52-m fiber. In another experiment [107], the input to a 10-m fiber was a relatively wide 100-ps pulse (at 532 nm) with a 0.3-ps-wide hole that served as a dark pulse. However, as the phase was relatively constant over the hole width, such even-symmetry input pulses did not have the chirp appropriate for a dark soliton. Nonetheless, output pulses exhibited features that were in agreement with the predictions of Eq. (5.3.1).

The odd-symmetry input pulses appropriate for launching a dark soliton were used in a 1988 experiment [108]. A spatial mask, in combination with a grating pair, was used to modify the pulse spectrum such that the input pulse had a phase profile appropriate for forming the dark soliton represented by Eq. (5.3.4). The input pulses obtained from a 620-nm dye laser were ≈ 2-ps wide with a 185-fs hole in their center. The central hole widened at low powers but narrowed down to its original width when the peak power was high enough to sustain a dark soliton of that width. The experimental results agreed with the theoretical predictions of Eq. (5.3.1) quite well. In this experiment, the optical fiber was only 1.2 m long. In a 1993 experiment [121], 5.3-ps dark solitons, formed on a 36-ps wide pulse obtained from a 850-nm Ti:sapphire laser, were propagated over 1 km of fiber. The same technique was later extended to transmit dark-soliton pulse trains at a repetition rate of up to 60 GHz over 2 km of fiber. These results show that dark solitons can be generated and maintained over considerable fiber lengths.

During the 1990s, several practical techniques were introduced for generating dark solitons. In one method, a Mach–Zehnder modulator, driven by nearly rectangular electrical pulses, modulates the CW output of a semiconductor laser [118]. In an extension of this method, electric modulation is performed in one of the two arms of a Mach–Zehnder interferometer. A simple all-optical technique consists of propagating two optical pulses, with a relative time delay between them, in the normal-GVD region of the fiber [119]. The two pulse broaden, become chirped, and acquire a nearly rectangular shape as they propagate inside the fiber. As these chirped pulses merge into each other, they interfere. The result at the fiber output is a train of isolated dark solitons. In another all-optical technique, nonlinear conversion of a beat signal in a dispersion-decreasing fiber is used to generate a train of dark solitons [124]. The technique is similar to that discussed in Section 5.1 for generating a regular pulse train except that fiber GVD is chosen to be in the normal-dispersion regime everywhere along the fiber length. A 100-GHz train of 1.6-ps dark solitons was generated by using this technique and propagated over 2.2 km of (two soliton periods) of a dispersion-shifted fiber. Optical switching using a fiber-loop mirror, in which a phase modulator is placed asymmetrically, can also be used to generate dark solitons [125]. In another variation, a fiber with comblike dispersion profile was used to generate dark soliton pulses with a width of 3.8 ps at the 48-GHz repetition rate [128].

An interesting scheme uses electronic circuitry to generate a coded train of dark solitons directly from the NRZ data in electric form [126]. First, the NRZ data and its clock at the bit rate are passed through an AND gate. The resulting signal is then sent to a flip-flop circuit in which all rising slopes flip the signal. The resulting electrical signal drives a Mach–Zehnder LiNbO$_3$ modulator and converts the CW output from a semiconductor laser into a coded train of dark solitons. This technique was used for data transmission, and a 10-Gb/s signal was transmitted over 1200 km by using dark solitons [127]. Another relatively simple method uses spectral filtering of a mode-locked pulse train by using a fiber grating [129]. This scheme has also been used to generate a 6.1-GHz train and propagate it over a 7-km-long fiber [130].

Dark solitons remain a subject of continuing interest. Numerical simulations show that they are more stable in the presence of noise and spread more slowly in the presence of fiber loss compared with bright solitons. They are also relatively less affected by many other factors that have an impact on the use of bright solitons (amplifier-induced timing jitter, intrapulse Raman scattering, etc.). These properties point to potential application of dark solitons for optical communication systems. The reader is referred to Reference [131] for further details.

5.3.2   Dispersion-Managed Solitons

The NLS equation (5.2.5) and its soliton solutions assume that the GVD parameter β_2 is constant along the fiber. As discussed in Section 3.5, the technique of dispersion management is often used in the design of modern fiber-optic communication systems. This technique consists of using a periodic dispersion map by combining fibers with different characteristics such that the average GVD in each period is quite low while the local GVD at every point along the fiber link is relatively large. The period of the dispersion map is typically 50–60 km. In practice, just two kinds of fibers with opposite signs of β_2 are combined to reduce the average dispersion to a small value. Mathematically, Eq. (5.2.5) is replaced with

$$i\frac{\partial u}{\partial \xi} + \frac{d(\xi)}{2}\frac{\partial^2 u}{\partial \tau^2} + |u|^2 u = 0, \qquad (5.3.5)$$

where $d(\xi)$ is a periodic function of ξ with the period $\xi_{\mathrm{map}} = L_{\mathrm{map}}/L_D$. Here L_{map} is the length associated with the dispersion map.

Equation (5.3.5) does not appear to be integrable by the inverse scattering method. However, it has been found to have pulselike, periodic solutions. These solutions are referred to as *dispersion-managed* solitons [132]–[135]. It should be stressed that the term soliton is used loosely in this context because the properties of dispersion-managed solitons are quite different from those of the bright solitons discussed in Section 5.2. Not only the amplitude and the width of dispersion-managed solitons oscillate in a periodic manner, but their frequency also varies across the pulse, i.e., such solitons are chirped. Pulse shape is also close to being Gaussian rather than the "sech" shape of bright solitons found for constant-dispersion fibers. Even more surprisingly, such solitons can exist even when average dispersion along the fiber link is normal. Dispersion-managed solitons are covered in Chapter 8 of Part B.

5.3.3   Bistable Solitons

The discussion in this chapter is based on a specific form of the nonlinear polarization in Eq. (2.3.6), resulting in a refractive index that increases linearly with the mode intensity I, i.e.,

$$\tilde{n}(I) = n + n_2 I. \tag{5.3.6}$$

Such a form of refractive index is referred to as the Kerr nonlinearity. At very high intensity levels, the nonlinear response of any material begins to saturate, and it may become necessary to modify Eq. (5.3.6). For silica fibers, saturation of the Kerr nonlinearity occurs at quite high intensity levels. However, if fibers are made with other materials (such as chalcogenide glasses) or if a silica fiber is doped with other nonlinear materials (such as an organic dye), the nonlinear response can saturate at practical intensity levels. In that case, Eq. (5.3.6) should be replaced with

$$\tilde{n}(I) = n + n_2 f(I), \tag{5.3.7}$$

where $f(I)$ is some known function of the mode intensity I.

The NLS equation (5.2.5) can be generalized to accommodate Eq. (5.3.7) and takes the form [136]

$$i\frac{\partial u}{\partial \xi} + \frac{1}{2}\frac{\partial^2 u}{\partial \tau^2} + f(|u|^2)u = 0. \tag{5.3.8}$$

Equation (5.3.8) is not generally integrable by the inverse scattering method. However, it can be solved to find shape-preserving solutions by the method outlined in Section 5.2. The approach consists of assuming a solution of the form

$$u(\xi, \tau) = V(\tau)\exp(iK\xi), \tag{5.3.9}$$

where K is a constant and V is independent of ξ. If Eq. (5.3.9) is substituted in Eq. (5.3.8), $V(\tau)$ is found to satisfy

$$\frac{d^2 V}{d\tau^2} = 2V[K - f(V^2)]. \tag{5.3.10}$$

This equation can be solved by multiplying it by $2\,(dV/d\tau)$ and integrating over τ. Using the boundary condition $V = 0$ as $|\tau| \to \infty$, we obtain

$$(dV/d\tau)^2 = 4\int_0^V [K - f(V^2)]V\,dV. \tag{5.3.11}$$

This equation can be integrated to yield

$$2\tau = \int_0^V \left(\int_0^{V^2} [K - f(P)]\,dP \right)^{-1/2} dV. \qquad (5.3.12)$$

where $P = V^2$. For a given functional form of $f(P)$, we can determine the soliton shape $V(\tau)$ from Eq. (5.3.12) if K is known.

The parameter K can be related to the soliton energy defined as $E_s = \int_{-\infty}^{\infty} V^2\,d\tau$. Using Eq. (5.3.11), E_s depends on the wave number K as [136]

$$E_s(K) = \frac{1}{2} \int_0^{P_m} [K - F(P)]^{-1/2}\,dP, \qquad (5.3.13)$$

where

$$F(P) = \frac{1}{P} \int_0^P f(P)\,dP, \qquad F(0) = 0, \qquad (5.3.14)$$

and P_m is defined as the smallest positive root of $F(P) = K$; it corresponds to the peak power of the soliton.

Depending on the function $f(P)$, Eq. (5.3.13) can have more than one solution, each having the same energy E_s but different values of K and P_m. Typically, only two solutions correspond to stable solitons. Such solitons are called bistable solitons and have been studied extensively since their discovery in 1985 [136]–[148]. For a given amount of pulse energy, bistable solitons propagate in two different stable states and can be made to switch from one state to another [137]. An analytic form of the bistable soliton has also been found for a specific form of the saturable nonlinearity [142]. Bistable behavior has not yet been observed in optical fibers as the peak-power requirements are extremely high. Other nonlinear media with easily saturable nonlinearity may be more suitable for this purpose.

5.4   Perturbation of Solitons

Fiber-optic communication systems operating at bit rates of 10 Gb/s or more are generally limited by the GVD that tends to disperse optical pulses outside their assigned bit slot. Fundamental solitons are useful for such systems because they can maintain their width over long distances by balancing the effects of GVD and SPM, both of which are detrimental to system performance when solitons are not used. The use of solitons for optical communications was proposed as early as 1973 [80], and their use had reached the commercial stage by 2000 [81]. This success was possible only after the effects of fiber losses on solitons were understood and techniques for compensating them were developed [149]–[156]. The advent of erbium-doped fiber amplifiers fueled the development of soliton-based systems. However, with their use came the limitations imposed by the amplifier noise. In this section, we first discuss the method used commonly to analyze the effect of small perturbations on solitons and then apply it to study the impact of fiber losses, periodic amplification, amplifier noise, and soliton interaction.

5.4.1 Perturbation Methods

Consider the perturbed NLS equation written as

$$i\frac{\partial u}{\partial \xi} + \frac{1}{2}\frac{\partial^2 u}{\partial \tau^2} + |u|^2 u = i\varepsilon(u), \tag{5.4.1}$$

where $\varepsilon(u)$ is a small perturbation that can depend on u, u^*, and their derivatives. In the absence of perturbation ($\varepsilon = 0$), the soliton solution of the NLS equation is known and is given by Eq. (5.2.13). The question then becomes what happens to the soliton when $\varepsilon \neq 0$. Several perturbation techniques have been developed for answering this question [157]–[164]. They all assume that the functional form of the soliton remains intact in the presence of a small perturbation but the four soliton parameters change with ξ as the soliton propagates down the fiber. Thus, the solution of the perturbed NLS equation can be written as

$$u(\xi, \tau) = \eta(\xi)\operatorname{sech}[\eta(\xi)(\tau - q(\xi))]\exp[i\phi(\xi) - i\delta(\xi)\tau]. \tag{5.4.2}$$

The ξ dependence of η, δ, q, and ϕ is yet to be determined. In the absence of perturbation ($\varepsilon = 0$), η and δ are constants but $q(\xi)$ and $\phi(\xi)$ are obtained by solving the simple ordinary differential equations

$$\frac{dq}{d\xi} = -\delta, \qquad \frac{d\phi}{d\xi} = \frac{1}{2}(\eta^2 - \delta^2). \tag{5.4.3}$$

The perturbation techniques developed for solitons include the adiabatic perturbation method, the perturbed inverse scattering method, the Lie-transform method, and the variational method [77]. All of them attempt to obtain a set of four ordinary differential equations for the four soliton parameters. As an example, consider the variational method that makes use of the Lagrangian formalism developed for classical mechanics and was applied to solitons as early as 1979 [165]. It treats the soliton field u and its complex conjugate u^* as conjugate variables (similar to the position and the momentum of a particle in classical mechanics). The perturbed NLS equation can be restated as a variational problem by casting it in the form of the Euler–Lagrange equation

$$\frac{\partial}{\partial \xi}\left(\frac{\partial L_g}{\partial X_\xi}\right) + \frac{\partial}{\partial \tau}\left(\frac{\partial L_g}{\partial X_\tau}\right) - \frac{\partial L_g}{\partial X} = 0, \qquad (5.4.4)$$

where X represents either u or u^*, the subscripts τ and ξ denote differentiation with respect to that variable, and the Lagrangian density is given by [166]

$$L_g = \frac{i}{2}(uu_\xi^* - u^*u_\xi) - \frac{1}{2}(|u|^4 - |u_\tau|^2) + i(\varepsilon u^* - \varepsilon^* u). \qquad (5.4.5)$$

It is easy to verify that Eq. (5.4.4) reproduces the NLS equation (5.4.1) with the choice $X = u^*$.

The main step in the variational analysis consists of integrating the Lagrangian density over τ as

$$\bar{L}_g(\eta, \delta, q, \phi) = \int_{-\infty}^{\infty} L_g(u, u^*, \tau)\, d\tau, \qquad (5.4.6)$$

and then using the reduced Euler–Lagrange equation to determine how the four soliton parameters evolve with ξ. Using Eqs. (5.4.4)–(5.4.6), this procedure leads to the following set of four ordinary differential equations [77]:

$$\frac{d\eta}{d\xi} = \mathrm{Re}\int_{-\infty}^{\infty} \varepsilon(u)u^*(\tau)\, d\tau, \qquad (5.4.7)$$

$$\frac{d\delta}{d\xi} = -\mathrm{Im}\int_{-\infty}^{\infty} \varepsilon(u)\tanh[\eta(\tau-q)]u^*(\tau)\, d\tau, \qquad (5.4.8)$$

$$\frac{dq}{d\xi} = -\delta + \frac{1}{\eta^2}\,\mathrm{Re}\int_{-\infty}^{\infty} \varepsilon(u)(\tau-q)u^*(\tau)\, d\tau, \qquad (5.4.9)$$

$$\frac{d\phi}{d\xi} = \mathrm{Im}\int_{-\infty}^{\infty} \varepsilon(u)\{1/\eta - (\tau-q)\tanh[\eta(\tau-q)]\}u^*(\tau)\, d\tau$$
$$+ \tfrac{1}{2}(\eta^2 - \delta^2) + q\frac{d\delta}{d\xi}, \qquad (5.4.10)$$

where Re and Im stand for the real and imaginary parts, respectively. This set of four equations can also be obtained by using adiabatic perturbation theory or perturbation theory based on the inverse scattering method [157]–[164]. The reader is referred to Reference [77] for a comparison of various perturbation methods.

5.4.2 Fiber Losses

Because solitons result from a balance between the nonlinear and dispersive effects, the pulse must maintain its peak power if it has to preserve its soliton character. Fiber losses are detrimental simply because they reduce the peak power of solitons along the fiber length [see Eq. (1.2.3)]. As a result, the width of a fundamental soliton also increases with propagation because of power loss. Mathematically, fiber losses are accounted for by adding a loss term to Eq. (5.1.1) so that it takes the form of Eq. (2.3.41). In terms of the soliton units used in Section 5.2, the NLS equation becomes

$$i\frac{\partial u}{\partial \xi} + \frac{1}{2}\frac{\partial^2 u}{\partial \tau^2} + |u|^2 u = -\frac{i}{2}\Gamma u, \tag{5.4.11}$$

where

$$\Gamma = \alpha L_D = \alpha T_0^2 / |\beta_2|. \tag{5.4.12}$$

Equation (5.4.11) can be solved by using the variational method if $\Gamma \ll 1$ so that the loss term can be treated as a weak perturbation. Using $\varepsilon(u) = -\Gamma u/2$ in Eqs. (5.4.7)–(5.4.10) and performing the integrations, we find that only soliton amplitude η and phase ϕ are affected by fiber losses and vary along the fiber length as [149]

$$\eta(\xi) = \exp(-\Gamma\xi), \qquad \phi(\xi) = \phi(0) + [1 - \exp(-2\Gamma\xi)]/(4\Gamma), \tag{5.4.13}$$

where we assumed that $\eta(0) = 1$, $\delta(0) = 0$, and $q(0) = 0$. Both δ and q remain zero along the fiber.

Recalling that the amplitude and width of a soliton are related inversely, a decrease in soliton amplitude leads to broadening of the soliton. Indeed, if we write $\eta(\tau - q)$ in Eq. (5.4.2) as T/T_1 and use $\tau = T/T_0$, T_1 increases along the fiber exponentially as

$$T_1(z) = T_0 \exp(\Gamma\xi) \equiv T_0 \exp(\alpha z). \tag{5.4.14}$$

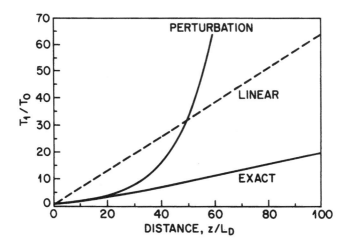

Figure 5.10 Variation of pulse width with distance in a lossy fiber for the fundamental soliton. The prediction of perturbation theory is also shown. Dashed curve shows the behavior expected in the absence of nonlinear effects. (After Ref. [167].)

An exponential increase in the soliton width with z cannot be expected to continue for arbitrarily large distances. This can be seen from Eq. (3.3.12), which predicts a linear increase with z when the nonlinear effects become negligible. Numerical solutions of Eq. (5.4.11) show that the perturbative solution is accurate only for values of z such that $\alpha z \ll 1$ [167]. Figure 5.10 shows the broadening factor T_1/T_0 as a function of ξ when a fundamental soliton is launched into a fiber with $\Gamma = 0.07$. The perturbative result is acceptable for up to $\Gamma\xi \approx 1$. In the regime ($\xi \gg 1$), pulse width increases linearly with a rate slower than that of a linear medium [168]. Higher-order solitons show a qualitatively similar asymptotic behavior. However, their pulse width oscillates a few times before increasing monotonically [167]. The origin of such oscillations lies in the periodic evolution of higher-order solitons.

How can a soliton survive inside lossy optical fibers? An interesting scheme restores the balance between GVD and SPM in a lossy fiber by changing dispersive properties of the fiber [169]. Such fibers are called *dispersion-decreasing* fibers (DDFs) because their GVD must decrease in such a way that it compensates for the reduced SPM experienced by the soliton as its energy is reduced by fiber loss. To see which GVD profile is needed, we modify Eq. (5.4.11) to allow for GVD variations along the fiber length and eliminate the loss term using $u = v\exp(-\Gamma\xi/2)$, resulting in the following equation:

$$i\frac{\partial v}{\partial \xi} + \frac{d(\xi)}{2}\frac{\partial^2 v}{\partial \tau^2} + e^{-\Gamma\xi}|v|^2 v = -\frac{i}{2}\Gamma u, \tag{5.4.15}$$

where $d(\xi) = |\beta_2(\xi)/\beta_2(0)|$ is the normalized local GVD. The distance ξ is normalized to the dispersion length $L_D = T_0^2/|\beta_2(0)|$, defined using the GVD value at the input end of the fiber.

Rescaling ξ using the transformation $\xi' = \int_0^\xi p(\xi)\,d\xi$, Eq. (5.4.15) becomes

$$i\frac{\partial v}{\partial \xi'} + \frac{1}{2}\frac{\partial^2 v}{\partial \tau^2} + \frac{e^{-\Gamma\xi}}{d(\xi)}|v|^2 v = 0, \tag{5.4.16}$$

If the GVD profile is chosen such that $d(\xi) = \exp(-\Gamma\xi)$, Eq. (5.4.16) reduces to the standard NLS equation. Thus, fiber losses have no effect on soliton propagation if the GVD of a fiber decreases exponentially along its length as

$$|\beta_2(z)| = |\beta_2(0)|\exp(-\alpha z). \tag{5.4.17}$$

This result can be easily understood from Eq. (5.2.3). If the soliton peak power P_0 decreases exponentially with z, the requirement $N = 1$ can still be maintained at every point along the fiber if $|\beta_2|$ were also to reduce exponentially.

Fibers with a nearly exponential GVD profile have been fabricated [170]. A practical technique for making such DDFs consists of reducing the core diameter along fiber length in a controlled manner during the fiber-drawing process. Variations in the core diameter change the waveguide contribution to β_2 and reduce its magnitude. Typically, GVD can be changed by a factor of 10 over a length of 20–40 km. The accuracy realized by the use of this technique is estimated to be better than 0.1 ps^2/km [171]. Since DDFs are not available commercially, fiber loss is commonly compensated by amplifying solitons. This is the topic discussed next.

5.4.3 Soliton Amplification

As already discussed, fiber losses lead to broadening of solitons. Such loss-induced broadening is unacceptable for many applications, especially when solitons are used for optical communications. To overcome the effect of fiber losses, solitons need to be amplified periodically so that their energy is restored to its initial value. Two different approaches have been used for soliton

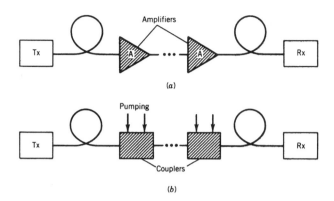

Figure 5.11 (a) Lumped and (b) distributed-amplification schemes used for compensation of fiber loss.

amplification [149]–[156]. These are known as lumped and distributed amplification schemes and are shown in Fig. 5.11 schematically. In the lumped scheme [150], an optical amplifier boosts the soliton energy to its input level after the soliton has propagated a certain distance. The soliton then readjusts its parameters to their input values. However, it also sheds a part of its energy as dispersive waves (continuum radiation) during this adjustment phase. The dispersive part is undesirable and can accumulate to significant levels over a large number of amplification stages.

This problem can be solved by reducing the spacing L_A between amplifiers such that $L_A \ll L_D$. The reason is that the dispersion length L_D sets the scale over which a soliton responds to external perturbations. If the amplifier spacing is much smaller than this length scale, soliton width is hardly affected over one amplifier spacing in spite of energy variations. In practice, the condition $L_A \ll L_D$ restricts L_A typically in the range 20–40 km even when the dispersion length exceeds 100 km [150]. Moreover, the lumped-amplification scheme becomes impractical at high bit rates requiring short solitons ($T_0 < 10$ ps) because dispersion length can then become quite short.

The distributed-amplification scheme uses either stimulated Raman scattering [151]–[154] (see Chapter 8) or erbium-doped fibers [172]–[176]. Both require periodic pumping along the fiber length. In the Raman case, a pump beam (up-shifted in frequency from the soliton carrier frequency by nearly 13 THz) is injected periodically into the fiber. For solitons propagating in the 1.55-μm wavelength region, one needs a high-power pump laser operating near 1.45 μm. In the case of erbium-doped fibers, the pump laser operates at 0.98 or 1.48 μm. The required pump power is relatively modest ($\sim$10 mW) in the case of erbium doping but exceeds 100 mW for Raman amplification. In both cases, optical gain is distributed over the entire fiber length. As a result, solitons can be amplified adiabatically while maintaining $N \approx 1$, a feature that reduces the dispersive part almost entirely [153].

Feasibility of the Raman-amplification scheme was first demonstrated in 1985 in an experiment in which soliton pulses of 10-ps width were propagated over a 10-km-long fiber [152]. In the absence of Raman gain, the width of solitons increased by $\approx$ 50% because of loss-induced broadening. This is in agreement with Eq. (5.4.14), which predicts $T_1/T_0 = 1.51$ for $z = 10$ km and $\alpha = 0.18$ dB/km, the values relevant for the experiment. The Raman gain was obtained by injecting a CW pump beam at 1.46 μm from a color-center laser in the direction opposite to that of soliton propagation. The pump power was adjusted close to 125 mW such that the total fiber loss of 1.8 dB was exactly balanced by the Raman gain. In a 1988 experiment [154], 55-ps solitons could be circulated up to 96 times through a 42-km fiber loop without significant increase in their width, resulting in an effective transmission distance of > 4000 km.

The lumped-amplification scheme was used starting in 1989 [155]. Since erbium-doped fiber amplifiers became available commercially after 1990, they have been used almost exclusively for loss compensation in spite of the lumped nature of amplification provided by them. To understand how solitons can survive in spite of large energy variations, we include the gain provided by lumped amplifiers in Eq. (5.4.11) by replacing Γ with a periodic function $\tilde{\Gamma}(\xi)$ such that $\tilde{\Gamma}(\xi) = \Gamma$ everywhere except at the location of amplifiers where it changes abruptly. If we make the transformation

$$u(\xi, \tau) = \exp\left(-\frac{1}{2}\int_0^\xi \tilde{\Gamma}(\xi)\,d\xi\right)v(\xi, \tau) \equiv a(\xi)v(\xi, \tau), \qquad (5.4.18)$$

where $a(\xi)$ contains rapid variations and $v(\xi, \tau)$ is a slowly varying function of ξ and use it in Eq. (5.4.11), $v(\xi, \tau)$ is found to satisfy

$$i\frac{\partial v}{\partial \xi} + \frac{1}{2}\frac{\partial^2 v}{\partial \tau^2} + a^2(\xi)|v|^2v = 0. \qquad (5.4.19)$$

Note that $a(\xi)$ is a periodic function of ξ with a period $\xi = L_A/L_D$, where L_A is the amplifier spacing. In each period, $a(\xi) \equiv a_0\exp(-\Gamma\xi/2)$ decreases exponentially and jumps to its initial value a_0 at the end of the period.

The concept of the guiding-center or path-averaged soliton [177] makes use of the fact that $a^2(\xi)$ in Eq. (5.4.19) varies rapidly in a periodic fashion. If the period $\xi_A \ll 1$, solitons evolve little over a short distance as compared with the dispersion length L_D. Over a soliton period, $a^2(\xi)$ varies so rapidly that its effects are averaged out, and we can replace $a^2(\xi)$ by its average value over one period. With this approximation, Eq. (5.4.19) reduces to the standard NLS equation:

$$i\frac{\partial v}{\partial \xi} + \frac{1}{2}\frac{\partial^2 v}{\partial \tau^2} + \langle a^2(\xi)\rangle |v|^2 v = 0. \tag{5.4.20}$$

The practical importance of the averaging concept stems from the fact that Eq. (5.4.20) describes soliton propagation quite accurately when $\xi_A \ll 1$ [77]. In practice, this approximation works reasonably well for values up to ξ_A as large as 0.25.

From a practical viewpoint, the input peak power P_s of the path-averaged soliton should be chosen such that $\langle a^2(\xi)\rangle = 1$ in Eq. (5.4.20). Introducing the amplifier gain $G = \exp(\Gamma\xi_A)$, the peak power is given by

$$P_s = \frac{\Gamma\xi_A P_0}{1 - \exp(-\Gamma\xi_A)} = \frac{G\ln G}{G-1}P_0, \tag{5.4.21}$$

where P_0 is the peak power in lossless fibers. Thus, soliton evolution in lossy fibers with periodic lumped amplification is identical to that in lossless fibers provided: (i) amplifiers are spaced such that $L_A \ll L_D$; and (ii) the launched peak power is larger by a factor $G\ln G/(G-1)$. As an example, $G = 10$ and $P_{in} \approx 2.56P_0$ for 50-km amplifier spacing and a fiber loss of 0.2 dB/km.

Figure 5.12 shows pulse evolution in the average-soliton regime over a distance of 10,000 km, assuming solitons are amplified every 50 km. When the soliton width corresponds to a dispersion length of 200 km, the soliton is well preserved even after 200 lumped amplifiers because the condition $\xi_A \ll 1$ is reasonably well satisfied. However, if the dispersion length reduces to 25 km, the soliton is destroyed because of relatively large loss-induced perturbations.

The condition $\xi_A \ll 1$ or $L_A \ll L_D$, required to operate within the average-soliton regime, can be related to the width T_0 by using $L_D = T_0^2/|\beta_2|$. The resulting condition is

$$T_0 \gg \sqrt{|\beta_2|L_A}. \tag{5.4.22}$$

The bit rate B of a soliton communication system is related to T_0 through $T_B = 1/B = 2q_0 T_0$, where T_B is the bit slot and q_0 represents the factor by

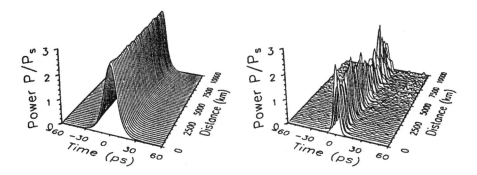

Figure 5.12 Evolution of loss-managed solitons over 10,000 km for $L_D = 200$ (left) and $L_D = 25$ km (right) when $L_A = 50$ km, $\alpha = 0.22$ dB/km, and $\beta_2 = 0.5$ ps²/km.

which it is larger than the soliton width. Thus, the condition (5.4.22) can be written in the form of a simple design criterion:

$$B^2 L_A \ll (4q_0^2 |\beta_2|)^{-1}. \qquad (5.4.23)$$

By choosing typical values $\beta_2 = -0.5$ ps²/km, $L_A = 50$ km, and $q_0 = 5$, we obtain $T_0 \gg 5$ ps and $B \ll 20$ GHz. Clearly, the use of amplifiers for soliton amplification imposes a severe limitation on both the bit rate and the amplifier spacing in practice.

Optical amplifiers, needed to restore the soliton energy, also add noise originating from spontaneous emission. The effect of spontaneous emission is to change randomly the four soliton parameters, η, δ, q, and ϕ in Eq. (5.4.2), at the output of each amplifier [161]. Amplitude fluctuations, as one might expect, degrade the signal-to-noise ratio (SNR). However, for applications of solitons in optical communications, frequency fluctuations are of much more concern. The reason can be understood from Eq. (5.4.2), and noting that a change in the soliton frequency by δ affects the speed at which the soliton propagates through the fiber. If δ fluctuates because of amplifier noise, soliton transit time through the fiber also becomes random. Fluctuations in the arrival time of a soliton are referred to as the Gordon–Haus timing jitter [178]. Practical implications of noise-induced timing jitter and the techniques developed for reducing it (optical filtering, synchronous modulation, etc.) are discussed in Part B in the context of soliton communication systems.

5.4.4 Soliton Interaction

The time interval T_B between two neighboring bits or pulses determines the bit rate of a communication system as $B = 1/T_B$. It is thus important to determine how close two solitons can come without affecting each other. Interaction between two solitons has been studied both analytically and numerically [179]–[191]. This section discusses the origin of mutual interaction and its affect on individual solitons.

It is clear on physical grounds that two solitons would begin to affect each other only when they are close enough that their tails overlap. Mathematically, the total field $u = u_1 + u_2$, where

$$u_j(\xi, \tau) = \eta_j \text{sech}[\eta_j(\tau - q_j)] \exp(i\phi_j - i\delta_j \tau), \qquad (5.4.24)$$

with $j = 1, 2$. It is u that satisfies the NLS equation, rather than u_1 and u_2 individually. In fact, by substituting $u = u_1 + u_2$ in Eq. (5.2.5), we can obtain the following perturbed NLS equation satisfied by the u_1 soliton:

$$i\frac{\partial u_1}{\partial \xi} + \frac{1}{2}\frac{\partial^2 u_1}{\partial \tau^2} + |u_1|^2 u_1 = -2|u_1|^2 u_2 - u_1^2 u_2^*. \qquad (5.4.25)$$

The NLS equation for u_2 is obtained by interchanging u_1 and u_2. The terms on the right-hand side act as a perturbation and are responsible for nonlinear interaction between two neighboring solitons.

Equations (5.4.7)–(5.4.10) can now be used to study how the four soliton parameters η_j, q_j, δ_j, and ϕ_j (with $j = 1, 2$) are affected by the perturbation. Introducing new variables as

$$\eta_\pm = \eta_1 \pm \eta_2, \qquad q_\pm = q_1 \pm q_2, \qquad (5.4.26)$$
$$\delta_\pm = \delta_1 \pm \delta_2, \qquad \phi_\pm = \phi_1 \pm \phi_2, \qquad (5.4.27)$$

one can obtain after some algebra the following set of equations [163]:

$$\frac{d\eta_+}{d\xi} = 0, \qquad \frac{d\eta_-}{d\xi} = \eta_+^3 \exp(-q_-)\sin\phi_-, \qquad (5.4.28)$$
$$\frac{d\delta_+}{d\xi} = 0, \qquad \frac{d\delta_-}{d\xi} = \eta_+^3 \exp(-q_-)\cos\phi_-, \qquad (5.4.29)$$
$$\frac{dq_-}{d\xi} = -\delta_-, \qquad \frac{d\phi_-}{d\xi} = \tfrac{1}{2}\eta_+\eta_-. \qquad (5.4.30)$$

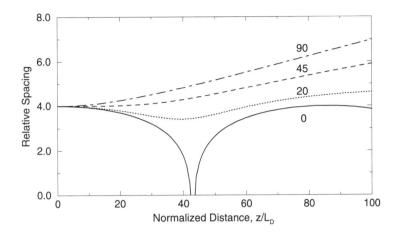

Figure 5.13 Relative spacing q between two interacting solitons as a function of fiber length for several values of initial phase difference ψ_0 (in degrees) when $q_0 = 4$.

Equations for q_+ and ϕ_+ are omitted since their dynamics do not affect soliton interaction. Further, η_+ and δ_+ remain constant during interaction. Using $\eta_+ = 2$ for two interacting fundamental solitons, the remaining four equations can be combined to yield:

$$\frac{d^2q}{d\xi^2} = -4e^{-2q}\cos(2\psi), \qquad \frac{d^2\psi}{d\xi^2} = 4e^{-2q}\sin(2\psi), \qquad (5.4.31)$$

where we introduced two new variables as $q = q_-/2$ and $\psi = \phi_-/2$. The same equations are obtained using the inverse scattering method [180]. They show that the relative separation q between two solitons depends only on their relative phase. Two solitons may attract (come closer) or repel (move apart) depending on the initial value of ψ.

Equations (5.4.31) can be solved analytically under quite general conditions [184]. In the case in which two solitons initially have the same amplitudes and frequencies, the solution becomes [77]

$$q(\xi) = q_0 + \tfrac{1}{2}\ln[\cosh^2(2\xi e^{-q_0}\sin\psi_0) + \cos^2(2\xi e^{-q_0}\cos\psi_0) - 1]. \quad (5.4.32)$$

where q_0 and ψ_0 are the initial values of q and ψ, respectively. Figure 5.13 shows how the relative separation $q(\xi)$ changes along the fiber length for two solitons with different phases. For ψ_0 below a certain value, q becomes zero periodically. This is referred to as a "collision" resulting from an attractive force between the two solitons. For values of ψ_0 larger than $\pi/8$, $q > q_0$ and increases monotonically with ξ. This is interpreted in terms of a nonlinearity-induced repulsive force between the two solitons. The specific cases $\psi_0 = 0$ and $\pi/2$ correspond to two solitons that are initially in phase or out of phase, respectively.

In the case of two in-phase solitons ($\psi_0 = 0$), the relative separation q changes with propagation periodically as

$$q(\xi) = q_0 + \ln|\cos(2\xi e^{-q_0})|. \tag{5.4.33}$$

Because $q(\xi) \leq q_0$ for all values of ξ, two in-phase solitons attract each other. In fact, q becomes zero after a distance

$$\xi = \tfrac{1}{2}e^{q_0}\cos^{-1}(e^{-q_0}) \approx \tfrac{\pi}{4}\exp(q_0), \tag{5.4.34}$$

where the approximate form is valid for $q_0 > 5$. At this distance, two solitons collide for the first time. Because of the periodic nature of the $q(\xi)$ in Eq. (5.4.33) the two solitons separate from each other and collide periodically. The oscillation period is called the collision length. In real units, the collision length is given by

$$L_{\text{col}} = \tfrac{\pi}{2}L_D\exp(q_0) \equiv z_0\exp(q_0), \tag{5.4.35}$$

where z_0 is the soliton period given in Eq. (5.2.24). This expression is quite accurate for $q_0 > 3$, as also found numerically [181]. A more accurate expression, valid for arbitrary values of q_0, is obtained using inverse scattering theory and is given by [187]

$$\frac{L_{\text{col}}}{L_D} = \frac{\pi\sinh(2q_0)\cosh(q_0)}{2q_0 + \sinh(2q_0)}. \tag{5.4.36}$$

In the case of two out-of-phase solitons ($\psi_0 = \pi/2$), the relative separation q changes with propagation as

$$q(\xi) = q_0 + \ln[\cosh(2\xi e^{-q_0})]. \tag{5.4.37}$$

As $\cosh(x) > 1$ for all values of x, it is clear that $q > q_0$ and increases monotonically with ξ.

Numerical solutions of the NLS equation are quite instructive and allow exploration of different amplitudes and different phases associated with a soliton pair by using the following form at the input end of the fiber:

$$u(0, \tau) = \text{sech}(\tau - q_0) + r\,\text{sech}[r(\tau + q_0)]e^{i\theta}, \tag{5.4.38}$$

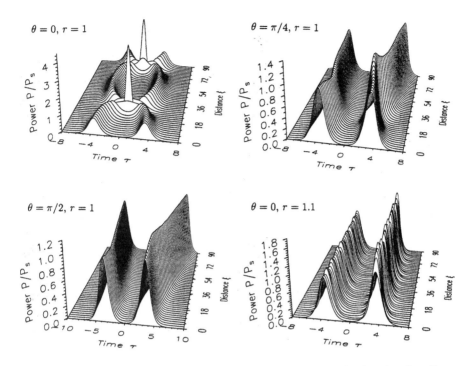

Figure 5.14 Evolution of a soliton pair over 90 dispersion lengths showing the effects of soliton interaction for four different choices of amplitude ratio r and relative phase θ. Initial separation $q_0 = 3.5$ in all four cases.

where r is relative amplitude, $\theta = 2\psi_0$ is the initial phase difference, and $2q_0$ is the initial separation between the two solitons. Figure 5.14 shows evolution of a soliton pair with an initial separation $q_0 = 3.5$ for several values of parameters r and θ. In the case of equal-amplitude solitons ($r = 1$), the two solitons attract each other in the in-phase case ($\theta = 0$) and collide periodically along the fiber length, just as predicted by perturbation theory. For $\theta = \pi/4$, the solitons separate from each other after an initial attraction stage in agreement with the results shown in Fig. 5.13. For $\theta = \pi/2$, the solitons repel each other even more strongly, and their spacing increases with distance monotonically. The last case shows the effect of slightly different soliton amplitudes by choosing $r = 1.1$. Two in-phase solitons oscillate periodically but never collide or move far away from each other.

The periodic collapse of neighboring solitons is undesirable from a practical standpoint. One way to avoid the collapse is to increase soliton separation such that $L_{col} \gg L_T$, where L_T is the transmission distance. Because $L_{col} \approx 3000 z_0$ for $q_0 = 8$, and $z_0 \sim 100$ km typically, a value of $q_0 = 8$ is large enough for any communication system. Several schemes can be used to reduce the soliton separation further without inducing the collapse. The interaction between two solitons is quite sensitive to their relative phase θ and the relative amplitude r. If the two solitons have the same phase ($\theta = 0$) but different amplitudes, the interaction is still periodic but without collapse [187]. Even for $r = 1.1$, the separation does not change by more than 10% during each period if $q_0 > 4$. Soliton interaction can also be modified by other factors such as higher-order effects [189], bandwidth-limited amplification [190], and timing jitter [191]. Higher-order effects are discussed in the following section.

5.5   Higher-Order Effects

The properties of optical solitons considered so far are based on the NLS equation (5.1.1). As discussed in Section 2.3, when input pulses are so short that $T_0 < 5$ ps, it is necessary to include higher-order nonlinear and dispersive effects through Eq. (2.3.39). In terms of the soliton units introduced in Section 5.2, Eq. (2.3.40) takes the form

$$i\frac{\partial u}{\partial \xi} + \frac{1}{2}\frac{\partial^2 u}{\partial \tau^2} + |u|^2 u = i\delta_3 \frac{\partial^3 u}{\partial \tau^3} - is\frac{\partial}{\partial \tau}(|u|^2 u) + \tau_R u\frac{\partial |u|^2}{\partial \tau}, \qquad (5.5.1)$$

where the pulse is assumed to propagate in the region of anomalous GVD ($\beta_2 < 0$) and fiber losses are neglected ($\alpha = 0$). The parameters δ_3, s, and τ_R govern, respectively, the effects of third-order dispersion (TOD), self-steepening, and intrapulse Raman scattering. Their explicit expressions are

$$\delta_3 = \frac{\beta_3}{6|\beta_2|T_0}, \qquad s = \frac{1}{\omega_0 T_0}, \qquad \tau_R = \frac{T_R}{T_0}. \qquad (5.5.2)$$

All three parameters vary inversely with pulse width and are negligible for $T_0 \gg 1$ ps. They become appreciable for femtosecond pulses. As an example, $\delta_3 \approx 0.03$, $s \approx 0.03$, and $\tau_R \approx 0.1$ for a 50-fs pulse ($T_0 \approx 30$ fs) propagating at 1.55 μm in a standard silica fiber if we take $T_R = 3$ fs.

5.5.1   Third-Order Dispersion

When optical pulses propagate relatively far from the zero-dispersion wavelength of an optical fiber, the TOD effects on solitons are small and can be treated perturbatively. To study such effects as simply as possible, let us set $s = 0$ and $\tau_R = 0$ in Eq. (5.5.1) and treat the δ_3 term as a small perturbation. Using Eqs. (5.4.7)–(5.4.10) with $\varepsilon(u) = \delta_3(\partial^3 u/\partial \tau^3)$, it is easy to show that amplitude η, frequency δ, and phase ϕ of the soliton are not affected by TOD. In contrast, the peak position q changes as [77]

$$\frac{dq}{d\xi} = -\delta + \delta_3 \eta^2. \qquad (5.5.3)$$

For a fundamental soliton with $\eta = 1$ and $\delta = 0$, the soliton peak shifts linearly with ξ as $q(\xi) = \delta_3 \xi$. Physically speaking, the TOD slows down the soliton and, as a result, the soliton peak is delayed by an amount that increases linearly with distance. This TOD-induced delay is negligible in most fibers for picosecond pulses for distances as large as $\xi = 100$ as long as β_2 is not nearly zero.

What happens if an optical pulse propagates at or near the zero-dispersion wavelength of an optical fiber such that β_2 is nearly zero. Considerable work has been done to understand propagation behavior in this regime [192]–[198]. The case $\beta_2 = 0$ has been discussed in Section 4.2 using Eq. (4.2.5). Equation (5.5.1) cannot be used in this case because the normalization scheme used for it becomes inappropriate. Normalizing the propagation distance to $L'_D = T_0^3/|\beta_3|$ through $\xi' = z/L'_D$, we obtain the following equation:

$$i\frac{\partial u}{\partial \xi'} - \text{sgn}(\beta_3)\frac{i}{6}\frac{\partial^3 u}{\partial \tau^3} + |u|^2 u = 0, \qquad (5.5.4)$$

where $u = \tilde{N}U$, with $\tilde{N}$ is defined by

$$\tilde{N}^2 = \frac{L'_D}{L_{\text{NL}}} = \frac{\gamma P_0 T_0^3}{|\beta_3|}. \qquad (5.5.5)$$

Figure 5.15 shows the pulse shape and the spectrum at $\xi' = 3$ for $\tilde{N} = 2$ and compares them with those of the input pulse at $\xi' = 0$. The most striking feature is splitting of the spectrum into two well-resolved spectral peaks [192]. These peaks correspond to the outermost peaks of the SPM-broadened spectrum (see Fig. 4.2). As the red-shifted peak lies in the anomalous-GVD regime,

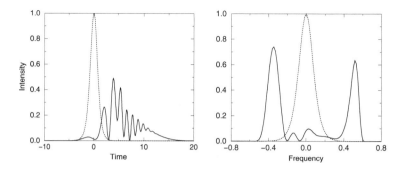

Figure 5.15 Pulse shape and spectrum at $z/L_D' = 3$ of a hyperbolic secant pulse propagating at the zero-dispersion wavelength with a peak power such that $\tilde{N} = 2$. Dotted curves show for comparison the initial profiles at the fiber input.

pulse energy in that spectral band can form a soliton. The energy in the other spectral band disperses away simply because that part of the pulse experiences normal GVD. It is the trailing part of the pulse that disperses away with propagation because SPM generates blue-shifted components near the trailing edge. The pulse shape in Fig. 5.15 shows a long trailing edge with oscillations that continues to separate away from the leading part with increasing ξ. The important point to note is that, because of SPM-induced spectral broadening, the input pulse does not really propagate at the zero-dispersion wavelength even if $\beta_2 = 0$ initially. In effect, the pulse creates its own $|\beta_2|$ through SPM. The effective value of $|\beta_2|$ is given by Eq. (4.2.7) and is larger for pulses with higher peak powers.

An interesting question is whether soliton-like solutions exist at the zero-dispersion wavelength of an optical fiber. Equation (5.5.4) does not appear to be integrable by the inverse scattering method. Numerical solutions show [194] that for $\tilde{N} > 1$, a "sech" pulse evolves over a length $\xi' \sim 10/\tilde{N}^2$ into a soliton that contains about half of the pulse energy. The remaining energy is carried by an oscillatory structure near the trailing edge that disperses away with propagation. These features of solitons have also been quantified by solving Eq. (5.5.4) approximately [194]–[198]. In general, solitons at the zero-dispersion wavelength require less power than those occurring in the anomalous-GVD regime. This can be seen by comparing Eqs. (5.2.3) and (5.5.5). To achieve the same values of N and $\tilde{N}$, the required power is smaller by a factor of $T_0|\beta_2/\beta_3|$ for pulses propagating at the zero-dispersion wavelength.

With the advent of wavelength-division multiplexing (WDM) and dispersion-management techniques, special fibers have been developed in which β_3 is nearly zero over a certain wavelength range while $|\beta_2|$ remains finite. Such fibers are called dispersion-flattened fibers. Their use requires consideration of the effects of fourth-order dispersion on solitons. The NLS equation then takes the following form:

$$i\frac{\partial u}{\partial \xi} + \frac{1}{2}\frac{\partial^2 u}{\partial \tau^2} + |u|^2 u = \delta_4 \frac{\partial^4 u}{\partial \tau^4}, \tag{5.5.6}$$

where $\delta_4 = \beta_4/(24|\beta_2|T_0^2)$.

The parameter δ_4 is relatively small for $T_0 > 1$ ps, and its effect can be treated perturbatively. However, δ_4 may become large enough for ultrashort pulses that a perturbative solution is not appropriate. A shape-preserving, solitary-wave solution of Eq. (5.5.6) can be found by assuming $u(\xi, \tau) = V(\tau)\exp(iK\xi)$ and solving the resulting ordinary differential equation for $V(\tau)$. This solution is given by [199]

$$u(\xi, \tau) = 3b^2\text{sech}^2(b\tau)\exp(8ib^2\xi/5), \tag{5.5.7}$$

where $b = (40\delta_4)^{-1/2}$. Note the sech^2-type form of the pulse amplitude rather than the usual "sech" form required for standard bright solitons. It should be stressed that both the amplitude and the width of the soliton are determined uniquely by the fiber parameters. Such fixed-parameter solitons are sometimes called autosolitons.

5.5.2 Self-Steepening

The phenomenon of self-steepening has been studied extensively [200]–[204]. Since it has already been covered in Section 4.3, its impact on solitons is discussed only briefly. To isolate the effects of self-steepening governed by the parameter s, it is useful to set $\delta_3 = 0$ and $\tau_R = 0$ in Eq. (5.5.1). Pulse evolution inside fibers is then governed by

$$i\frac{\partial u}{\partial \xi} + \frac{1}{2}\frac{\partial^2 u}{\partial \tau^2} + |u|^2 u + is\frac{\partial}{\partial \tau}(|u|^2 u) = 0. \tag{5.5.8}$$

As discussed in Section 4.3, self-steepening creates an optical shock on the trailing edge of the pulse in the absence of the GVD effects. This phenomenon is due to the intensity dependence of the group velocity that results in the peak of the pulse moving slower than the wings. The GVD dissipates the shock and

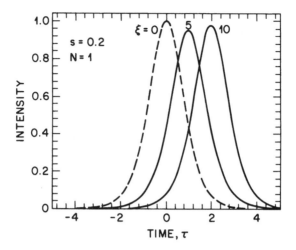

Figure 5.16 Pulse shapes at $\xi = 5$ and 10 for a fundamental soliton in the presence of self-steepening ($s = 0.2$). Dashed curve shows the initial shape for comparison. The solid curves coincide with the dashed curve when $s = 0$.

smoothes the trailing edge considerably. However, self-steepening would still manifest through a shift of the pulse center.

The self-steepening-induced shift is shown in Fig. 5.16 where pulse shapes at $\xi = 0$, 5, and 10 are plotted for $s = 0.2$ and $N = 1$ by solving Eq. (5.5.8) numerically with the input $u(0, \tau) = \mathrm{sech}(\tau)$. As the peak moves slower than the wings for $s \neq 0$, it is delayed and appears shifted toward the trailing side. The delay is well approximated by a simple expression $\tau_d = s\xi$ for $s < 0.3$. It can also be calculated by treating the self-steepening term in Eq. (5.5.8) as a small perturbation. Although the pulse broadens slightly with propagation (by $\sim 20\%$ at $\xi = 10$), it nonetheless maintains its soliton nature. This feature suggests that Eq. (5.5.8) has a soliton solution toward which the input pulse is evolving asymptotically. Such a solution indeed exists and has the form [166]

$$u(\xi, \tau) = V(\tau + M\xi) \exp[i(K\xi - M\tau)], \qquad (5.5.9)$$

where M is related to a shift of the carrier frequency. The group velocity changes as a result of the shift. The delay of the peak seen in Fig. 5.16 is due to this change in the group velocity. The explicit form of $V(\tau)$ depends on M and s [204]. In the limit $s = 0$, it reduces to the hyperbolic secant form of Eq. (5.2.16). Note also that Eq. (5.5.8) can be transformed into a so-called

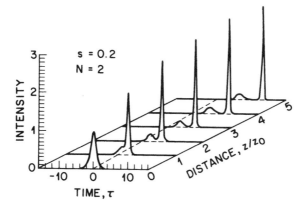

Figure 5.17 Decay of a second-order soliton ($N = 2$) induced by self-steepening ($s = 0.2$). Pulse evolution over five soliton periods is shown.

derivative NLS equation that is integrable by the inverse scattering method and whose solutions have been studied extensively in plasma physics [205]–[208].

The effect of self-steepening on higher-order solitons is remarkable in that it leads to breakup of such solitons into their constituents, a phenomenon referred to as soliton decay [201]. Figure 5.17 shows this behavior for a second-order soliton ($N = 2$) using $s = 0.2$. For this relatively large value of s, the two solitons have separated from each other within a distance of two soliton periods and continue to move apart with further propagation inside the fiber. A qualitatively similar behavior occurs for smaller values of s except that a longer distance is required for the breakup of solitons. The soliton decay can be understood using the inverse scattering method, with the self-steepening term acting as a perturbation. In the absence of self-steepening ($s = 0$), the two solitons form a bound state because both of them propagate at the same speed (the eigenvalues have the same real part). The effect of self-steepening is to break the degeneracy so that the two solitons propagate at different speeds. As a result, they separate from each other, and the separation increases almost linearly with the distance [202]. The ratio of the peak heights in Fig. 5.17 is about 9 and is in agreement with the expected ratio $(\eta_2/\eta_1)^2$, where η_1 and η_2 are the imaginary parts of the eigenvalues introduced in Section 5.2. The third- and higher-order solitons follow a similar decay pattern. In particular, the third-order soliton ($N = 3$) decays into three solitons whose peak heights are again in agreement with inverse scattering theory.

5.5.3   Intrapulse Raman Scattering

Intrapulse Raman scattering plays the most important role among the higher-order nonlinear effects. Its effects on solitons are governed by the last term in Eq. (5.5.1) and were observed experimentally in 1985 [209]. The need to include this term became apparent when a new phenomenon, called the soliton self-frequency shift, was observed in 1986 [210] and explained using the delayed nature of the Raman response [211]. Since then, this higher-order nonlinear effect has been studied extensively [212]–[230].

To isolate the effects of intrapulse Raman scattering, it is useful to set $\delta_3 = 0$ and $s = 0$ in Eq. (5.5.1). Pulse evolution inside fibers is then governed by

$$i\frac{\partial u}{\partial \xi} + \frac{1}{2}\frac{\partial^2 u}{\partial \tau^2} + |u|^2 u = \tau_R u \frac{\partial |u|^2}{\partial \tau}. \tag{5.5.10}$$

Using Eqs. (5.4.7)–(5.4.10) with $\varepsilon(u) = -i\tau_R u(\partial |u|^2/\partial \tau)$, it is easy to see that the amplitude η of the soliton is not affected by the Raman effect but its frequency δ changes as

$$\frac{d\delta}{d\xi} = -\frac{8}{15}\tau_R \eta^4. \tag{5.5.11}$$

Because η is a constant, this equation is easily integrated with the result $\delta(\xi) = (8\tau_R/15)\eta^4 \xi$. Using $\eta = 1$ and $\xi = z/L_D = |\beta_2|z/T_0^2$, the Raman-induced frequency shift can be written in real units as

$$\Delta\omega_R(z) = -8|\beta_2|T_R z/(15T_0^4). \tag{5.5.12}$$

The negative sign shows that the carrier frequency is reduced, i.e., the soliton spectrum shifts toward longer wavelengths or the "red" side.

Physically, the red shift can be understood in terms of stimulated Raman scattering (see Chapter 8). For pulse widths ~ 1 ps or shorter, the spectral width of the pulse is large enough that the Raman gain can amplify the low-frequency (red) spectral components of the pulse, with high-frequency (blue) components of the same pulse acting as a pump. The process continues along the fiber, and the energy from blue components is continuously transferred to red components. Such an energy transfer appears as a red shift of the soliton spectrum, with shift increasing with distance. As seen from Eq. (5.5.12) the frequency shift increases linearly along the fiber. More importantly, it scales with the pulse width as T_0^{-4}, indicating that it can become quite large for short pulses. As an example, soliton frequency changes at a rate of ~ 50 GHz/km

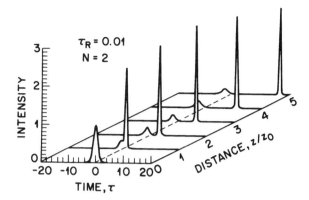

Figure 5.18 Decay of a second-order soliton ($N = 2$) induced by intrapulse Raman scattering ($\tau_R = 0.01$).

for 1-ps pulses ($T_0 = 0.57$ ps) in standard fibers with $\beta_2 = -20$ ps²/km and $T_R = 3$ fs. The spectrum of such pulses will shift by 1 THz after 20 km of propagation. This is a large shift if we note that the spectral width (FWHM) of such a soliton is < 0.5 THz. Typically, the Raman-induced frequency shift cannot be neglected for pulses shorter than 5 ps.

The Raman-induced red shift of solitons was observed in 1986 using 0.5-ps pulses obtained from a passively mode-locked color-center laser [210]. The pulse spectrum was found to shift as much as 8 THz for a fiber length under 0.4 km. The observed spectral shift was called the soliton self-frequency shift because it is induced by the soliton itself. In fact, it was in an attempt to explain the observed red shift that the importance of the delayed nature of the Raman response for transmission of ultrashort pulses was first realized [211].

The effect of intrapulse Raman scattering on higher-order solitons is similar to the case of self-steepening. In particular, even relatively small values of τ_R lead to the decay of higher-order solitons into its constituents [218]. Figure 5.18 shows such a decay for a second-order soliton ($N = 2$) by solving Eq. (5.5.10) numerically with $\tau_R = 0.01$. A comparison of Figs. 5.17 and 5.18 shows the similarity and the differences for two different higher-order nonlinear mechanisms. An important difference is that relatively smaller values of τ_R compared with s can induce soliton decay over a given distance. For example, if $s = 0.01$ is chosen in Fig. 5.17, the soliton does not split over the distance $z = 5z_0$. This feature indicates that the effects of τ_R are likely to dominate in practice over those of self-steepening.

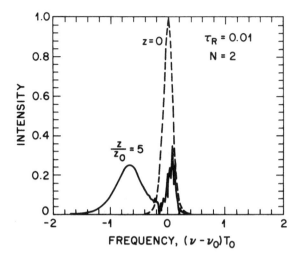

Figure 5.19 Pulse spectrum at $z/z_0 = 5$ for parameter values identical to those of Fig. 5.18. Dashed curve shows the spectrum of input pulses.

Another important difference seen in Figs. 5.17 and 5.18 is that both solitons are delayed in the case of self-steepening, while in the Raman case the low-intensity soliton is advanced and appears on the leading side of the incident pulse. This behavior can be understood qualitatively from Fig. 5.19 where the pulse spectrum at $z = 5z_0$ is compared with the input spectrum for the second-order soliton (whose evolution is shown in Fig. 5.18). The most noteworthy feature is the huge red shift of the soliton spectrum, about four times the input spectral width for $\tau_R = 0.01$ and $z/z_0 = 5$. The red-shifted broad spectral peak corresponds to the intense soliton shifting toward the right in Fig. 5.18, whereas the blue-shifted spectral feature corresponds to the other peak moving toward the left in that Figure. Because the blue-shifted components travel faster than the red-shifted ones, they are advanced while the others are delayed with respect to the input pulse. This is precisely what is seen in Fig. 5.18.

A question one may ask is whether Eq. (5.5.10) has soliton-like solutions. It turns out that pulselike solutions do not exist when the Raman term is included mainly because the resulting perturbation is of non-Hamiltonian type [162]. This feature of the Raman term can be understood by noting that the Raman-induced spectral red shift does not preserve pulse energy because a part of the energy is dissipated through the excitation of molecular vibrations.

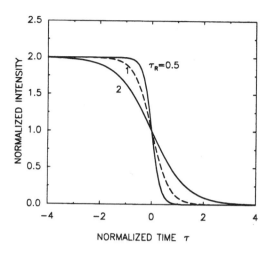

Figure 5.20 Temporal intensity profiles of kink solitons in the form of an optical shock for several values of τ_R. (After Ref. [227].)

However, a kink-type topological soliton (with infinite energy) has been found and is given by [227]

$$u(\xi, \tau) = \left(\frac{3\tau}{4\tau_R}\right)\left[\exp\left(\frac{3\tau}{\tau_R}\right) + 1\right]^{-1/2}\exp\left(\frac{9i\xi}{8\tau_R^2}\right). \qquad (5.5.13)$$

Kink solitons appear in many physical systems whose dynamics are governed by the sine–Gordon equation [71]. In the context of optical fibers, the kink soliton represents an optical shock front that preserves its shape when propagating through the fiber. Figure 5.20 shows the shock profiles by plotting $|u(\xi, \tau)|^2$ for several values of τ_R. Steepness of the shock depends on τ_R such that the shock front becomes increasingly steeper as τ_R is reduced. Even though the parameter N increases as τ_R is reduced, the power level P_0 (defined as the power at $\tau = 0$) remains the same. This can be seen by expressing P_0 in terms of the parameter T_R using Eqs. (5.2.3) and (5.5.2) so that $P_0 = 9|\beta_2|/(16\gamma T_R^2)$. Using typical values for fiber parameters, $P_0 \sim 10$ kW. It is difficult to observe such optical shocks experimentally because of large power requirements.

The kink soliton given in Eq. (5.5.13) is obtained assuming $u(\xi, \tau) = V(\tau)\exp(iK\xi)$, and solving the resulting ordinary differential equation for $V(\tau)$. The solution shows that kink solitons form a one-parameter family for various values of K and exist even in the normal-dispersion region of the

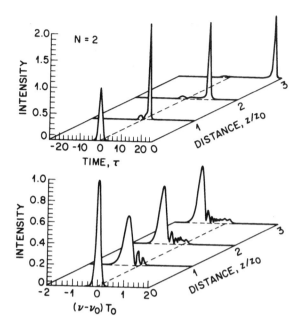

Figure 5.21 Evolution of pulse shapes and spectra for the case $N = 2$. The other parameter values are $\delta_3 = 0.03$, $s = 0.05$, and $\tau_R = 0.1$.

fiber [228]. They continue to exist even when the self-steepening term in Eq. (5.5.1) is included. The analytic form in Eq. (5.5.13) is obtained only for a specific value $K = 9/(8\tau_R^2)$. When $K < \tau_R^2$, the monotonically decaying tail seen in Fig. 5.20 develops an oscillatory structure.

5.5.4 Propagation of Femtosecond Pulses

For femtosecond pulses having widths $T_0 < 1$ ps, it becomes necessary to include all the higher-order terms in Eq. (5.5.1) because all three parameters δ_3, s, and τ_R become non-negligible. Evolution of such ultrashort pulses in optical fibers is studied by solving Eq. (5.5.1) numerically [231]–[234]. As an example, Fig. 5.21 shows the pulse shapes and spectra when a second-order soliton is launched at the input end of a fiber after choosing $\delta_3 = 0.03$, $s = 0.05$, and $\tau_R = 0.1$. These values are appropriate for a 50-fs pulse ($T_0 \approx 30$ fs) propagating in the 1.55-μm region of a standard silica fiber. Soliton decay occurs within a soliton period ($z_0 \approx 5$ cm), and the main peak shifts toward the trailing side at a rapid rate with increasing distance. This temporal shift is due to the decrease in the group velocity v_g occurring as a result of the red shift of the soliton spectrum. A shift in the carrier frequency of the soliton changes its speed because $v_g = (d\beta/d\omega)^{-1}$ is frequency dependent. If we use $T_0 = 30$ fs to convert the results of Fig. 5.21 into physical units, the 50-fs pulse has shifted by almost 40 THz or 20% of the carrier frequency after propagating a distance of only 15 cm.

When the input peak power is large enough to excite a higher-order soliton such that $N \gg 1$, the pulse spectrum evolves into several bands, each corresponding to splitting of a fundamental soliton from the original pulse. Such an evolution pattern was seen when 830-fs pulses with peak powers up to 530 W were propagated in fibers up to 1 km long [232]. The spectral peak at the extreme red end was associated with a soliton whose width was narrowest (≈ 55 fs) after 12 m and then increased with a further increase in the fiber length. The experimental results were in agreement with the predictions of Eq. (5.5.1).

The combined effect of TOD, self-steepening, and intrapulse Raman scattering on a higher-order soliton is to split it into its constituents. In fact, the TOD can itself lead to soliton decay even in the absence of higher-order nonlinear effects when the parameter δ_3 exceeds a threshold value [233]. For a second-order soliton ($N = 2$), the threshold value is $\delta_3 = 0.022$ but reduces to ≈ 0.006 for $N = 3$. For standard silica fibers δ_3 exceeds 0.022 at 1.55 μm for pulses shorter than 70 fs. However, the threshold can be reached for pulses wider by a factor of 10 when dispersion-shifted fibers are used.

An interesting question is whether Eq. (5.5.1) permits shape-preserving, solitary-wave solutions under certain conditions. Several such solutions have been found using a variety of techniques [235]–[250]. In most cases, the solution exists only for a specific choice of parameter combinations. For example, fundamental and higher-order solitons have been found when $\tau_R = 0$ with $s = -2\delta_3$ or $s = -6\delta_3$ [242]. From a practical standpoint, such solutions of Eq. (5.5.1) are rarely useful because it is hard to find fibers whose parameters satisfy the required constraints.

As successful as Eq. (5.5.1) is in modeling the propagation of femtosecond pulses in optical fibers, it is still approximate. As discussed in Section 2.3, a more accurate approach should use Eq. (2.3.33), where $R(t)$ takes into account the time-dependent response of the fiber nonlinearity. In a simple model, $R(t)$ is assumed to obey Eq. (2.3.34) so that both the electronic (the Kerr effect) and molecular (the Raman effect) contributions to the fiber nonlinearity are accounted for [219]–[222]. The delayed nature of the molecular response not only leads to the soliton self-frequency shift but also affects the interaction between neighboring solitons [223]. Equation (2.3.33) has been used to study numerically how intrapulse stimulated Raman scattering affects evolution of femtosecond optical pulses in optical fibers [224]–[226]. For pulses shorter than 20 fs even the use of this equation becomes questionable because of the slowly varying envelope approximation made in its derivation (see Section 2.3). Because such short pulses can be generated by modern mode-locked lasers, attempts have been made to improve upon this approximation while still working with the pulse envelope [251]–[253]. For supershort pulses containing only a few optical cycles, it eventually becomes necessary to abandon the concept of the pulse envelope and solve the Maxwell equations directly using an appropriate numerical scheme (see Section 2.4).

Problems

5.1 Solve Eq. (5.1.4) and derive an expression for the modulation-instability gain. What is the peak value of the gain and at what frequency does this gain occur?

5.2 A 1.55-μm soliton communication system is operating at 10 Gb/s using dispersion-shifted fibers with $D = 2$ ps/(km-nm). The effective core area of the fiber is 50 μm^2. Calculate the peak power and the pulse energy required for launching fundamental solitons of 30-ps width (FWHM) into the fiber.

5.3 What is the soliton period for the communication system described in Problem 5.2?

5.4 Verify by direct substitution that the soliton solution given in Eq. (5.2.16) satisfies Eq. (5.2.5).

5.5 Develop a computer program capable of solving Eq. (5.2.5) numerically using the split-step Fourier method of Section 2.4. Test it by comparing its output with the analytical solution in Eq. (5.2.16) when a fundamental soliton is launched into the fiber.

5.6 Use the computer program developed in Problem 5.5 to study the case of an input pulse of the form given in Eq. (5.2.22) for $N = 0.2$, 0.6, 1.0, and 1.4. Explain the different behavior occurring in each case.

5.7 Why should the amplifier spacing be a fraction of the soliton period when lumped amplifiers are used for compensating fiber losses?

5.8 A soliton communication system is designed with an amplifier spacing of 50 km. What should the input value of the soliton parameter N be to ensure that a fundamental soliton is maintained in spite of 0.2-dB/km fiber losses? What should the amplifier gain be? Is there any limit on the bit rate of such a system?

5.9 Study the soliton interaction numerically using an input pulse profile given in Eq. (5.4.38). Choose $r = 1$, $q_0 = 3$, and vary θ in the range 0 to π.

5.10 A soliton system is designed to transmit a signal over 5000 km at $B = 5$ Gb/s. What should the pulse width (FWHM) be to ensure that the neighboring solitons do not interact during transmission? The dispersion parameter $D = 2$ ps/(km-nm) at the operating wavelength.

5.11 What is intrapulse Raman scattering? Why does it lead to a shift in the carrier frequency of solitons? Derive an expression for the frequency shift using the Raman term as a perturbation. Calculate the shift for 1-ps (FWHM) solitons propagating in a 10-km-long fiber with the GVD $D = 2$ ps/(km-nm).

5.12 Verify by direct substitution that the solution given in Eq. (5.5.13) is indeed the solution of Eq. (5.5.1) when $\delta_3 = 0$, $s = 0$, and $N = 3/(4\tau_R)$.

References

[1] L. A. Ostrovskii, *Sov. Phys. Tech. Phys.* **8**, 679 (1964); *Sov. Phys. JETP* **24**, 797 (1967).

[2] G. B. Whitham, *Proc. Roy. Soc.* **283**, 238 (1965)

[3] G. B. Whitham, *J. Fluid Mech.* **27**, 399 (1967).

[4] T. B. Benjamin and J. E. Feir, *J. Fluid Mech.* **27**, 417 (1967).

[5] V. I. Bespalov and V. I. Talanov, *JETP Lett.* **3**, 307 (1966).

[6] V. I. Karpman, *JETP Lett.* **6**, 277 (1967).

[7] V. I. Karpman and E. M. Krushkal, *Sov. Phys. JETP* **28**, 277 (1969).

[8] T. Taniuti and H. Washimi, *Phys. Rev. Lett.* **21**, 209 (1968).

[9] C. K. W. Tam, *Phys. Fluids* **12**, 1028 (1969).

[10] A. Hasegawa, *Phys. Rev. Lett.* **24**, 1165 (1970); *Phys. Fluids* **15**, 870 (1971).

[11] A. Hasegawa, *Plasma Instabilities and Nonlinear Effects* (Springer-Verlag, Heidelberg, 1975).

[12] A. Hasegawa and W. F. Brinkman, *IEEE J. Quantum Electron.* **QE-16**, 694 (1980).

[13] D. R. Andersen, S. Datta, and R. L. Gunshor, *J. Appl. Phys.* **54**, 5608 (1983).

[14] A. Hasegawa, *Opt. Lett.* **9**, 288 (1984).

[15] D. Anderson and M. Lisak, *Opt. Lett.* **9**, 468 (1984);

[16] B. Hermansson and D. Yevick, *Opt. Commun.* **52**, 99 (1984);

[17] K. Tajima, *J. Lightwave Technol.* **4**, 900 (1986).

[18] K. Tai, A. Hasegawa, and A. Tomita, *Phys. Rev. Lett.* **56**, 135 (1986).

[19] K. Tai, A. Tomita, J. L. Jewell, and A. Hasegawa, *Appl. Phys. Lett.* **49**, 236 (1986).

[20] P. K. Shukla and J. J. Rasmussen, *Opt. Lett.* **11**, 171 (1986).

[21] M. J. Potasek, *Opt. Lett.* **12**, 921 (1987).

[22] I. M. Uzunov, *Opt. Quantum Electron.* **22**, 529 (1990).

[23] M. J. Potasek and G. P. Agrawal, *Phys. Rev. A* **36**, 3862 (1987).

[24] V. A. Vysloukh and N. A. Sukhotskova, *Sov. J. Quantum Electron.* **17**,1509 (1987).

[25] M. N. Islam, S. P. Dijaili, and J. P. Gordon, *Opt. Lett.* **13**, 518 (1988).

[26] F. Ito, K. Kitayama, and H. Yoshinaga, *Appl. Phys. Lett.* **54**, 2503 (1989).

[27] C. J. McKinstrie and G. G. Luther, *Physica Scripta* **30**, 31 (1990).

[28] G. Cappellini and S. Trillo, *J. Opt. Soc. Am. B* **8**, 824 (1991).

[29] S. Trillo and S. Wabnitz, *Opt. Lett.* **16**, 986 (1991).

[30] J. M. Soto-Crespo and E. M. Wright, *Appl. Phys. Lett.* **59**, 2489 (1991).

[31] R. W. Boyd, M. G. Raymer, and L. M. Narducci (eds.), *Optical Instabilities* (Cambridge University Press, London, 1986).

[32] F. T. Arecchi and R. G. Harrison (eds.), *Instabilities and Chaos in Quantum Optics* (Springer-Verlag, Berlin, 1987).

[33] C. O. Weiss and R. Vilaseca, *Dynamics of Lasers* (Weinheim, New York, 1991).

[34] G. H. M. van Tartwijk and G. P. Agrawal, *Prog. Quantum Electron.* **22**, 43 (1998).

[35] N. N. Akhmediev, V. M. Eleonskii, and N. E. Kulagin, *Sov. Phys. JETP* **62**, 894 (1985); *Theor. Math. Phys.* (USSR) **72**, 809 (1987).

[36] H. Hadachira, D. W. McLaughlin, J. V. Moloney, and A. C. Newell, *J. Math. Phys.* **29**, 63 (1988).

[37] L. Gagnon, *J. Opt. Soc. Am. B* **6**, 1477 (1989); L. Gagnon and P. Winternitz, *Phys. Rev. A* **39**, 296 (1989).

[38] D. Mihalache and N. C. Panoiu, *Phys. Rev. A* **45**, 673 (1992); *J. Math. Phys.* **33**, 2323 (1992).

[39] D. Mihalache, F. Lederer, and D. M. Baboiu, *Phys. Rev. A* **47**, 3285 (1993).

[40] L. Gagnon, *J. Opt. Soc. Am. B* **10**, 469 (1993).

[41] S. Kumar, G. V. Anand, and A. Selvarajan, *J. Opt. Soc. Am. B* **10** , 697 (1993).

[42] N. N. Akhmediev, *Phys. Rev. A* **47**, 3213 (1993).

[43] A. M. Kamchatnov, *Phys. Rep.* **286**, 200 (1997).

[44] E. J. Greer, D. M. Patrick, and P. G. J. Wigley, *Electron. Lett.* **25**, 1246 (1989).

[45] P. V. Mamyshev, S. V. Chernikov, and E. M. Dianov, *IEEE J. Quantum Electron.* **27**, 2347 (1991).

[46] S. V. Chernikov, J. R. Taylor, P. V. Mamyshev, and E. M. Dianov, *Electron. Lett.* **28**, 931 (1992).

[47] S. V. Chernikov, D. J. Richardson, R. I. Laming, and E. M. Dianov, *Electron. Lett.* **28**, 1210 (1992).

[48] S. V. Chernikov, P. V. Mamyshev, E. M. Dianov, D. J. Richardson, R. I. Laming, and D. N. Payne, *Sov. Lightwave Commun.* **2**, 161 (1992).

[49] S. V. Chernikov, D. J. Richardson, and R. I. Laming, *Appl. Phys. Lett.* **63**, 293 (1993).

[50] S. V. Chernikov, J. R. Taylor, and R. Kashyap, *Electron. Lett.* **29**, 1788 (1993); *Opt. Lett.* **19**, 539 (1994).

[51] E. A. Swanson and S. R. Chinn, *IEEE Photon. Technol. Lett.* **6**, 796 (1994).

[52] M. Nakazawa, K. Suzuki, and H. A. Haus, *Phys. Rev. A* **38**, 5193 (1988).

[53] M. Nakazawa, K. Suzuki, H. Kubota, and H. A. Haus, *Phys. Rev. A* **39**, 5768 (1989).

[54] S. Coen and M. Haelterman, *Phys. Rev. Lett.* **79**, 4139 (1197).

[55] M. Yu, C. J. McKinistrie, and G. P. Agrawal, *J. Opt. Soc. Am. B* **15**, 607 (1998); *J. Opt. Soc. Am. B* **15**, 617 (1998).

[56] S. Coen and M. Haelterman, *Opt. Commun.* **146**, 339 (1998); *Opt. Lett.* **24**, 80 (1999).

[57] S. Coen, M. Haelterman, P. Emplit, L. Delage, L. M. Simohamed, and F. Reynaud, *J. Opt. Soc. Am. B* **15**, 2283 (1998); *J. Opt. B* **1**, 36 (1999).

[58] J. P. Hamide, P. Emplit, and J. M. Gabriagues, *Electron. Lett.* **26**, 1452 (1990).

[59] F. Matera, A. Mecozzi, M. Romagnoli, and M. Settembre, *Opt. Lett.* **18**, 1499 (1993); *Microwave Opt. Tech. Lett.* **7**, 537 (1994).

[60] N. Kikuchi and S. Sasaki, *Electron. Lett.* **32**, 570 (1996).

[61] N. J. Smith and N. J. Doran, *Opt. Lett.* **21**, 570 (1996).

[62] R. A. Saunders, B. A. Patel, and D. Garthe, *IEEE Photon. Technol. Lett.* **9**, 699 (1997).

[63] R. Q. Hui, M. O'Sullivan, A. Robinson, and M. Taylor, *J. Lightwave Technol.* **15**, 1071 (1997).

[64] D. F. Grosz and H. L. Fragnito, *Microwave Opt. Tech. Lett.* **18**, 275 (1998); *Microwave Opt. Tech. Lett.* **19**, 149 (1998); *Microwave Opt. Tech. Lett.* **20**, 389 (1999).

[65] D. F. Grosz, C. Mazzali, S. Celaschi, A. Paradisi, and H. L. Fragnito, *IEEE Photon. Technol. Lett.* **11**, 379 (1999).

[66] G. A. Nowak, Y. H. Kao, T. J. Xia, M. N. Islam, and D. Nolan, *Opt. Lett.* **23**, 936 (1998).

[67] S. Nishi and M. Saruwatari, *Electron. Lett.* **31**, 225 (1995).

[68] J. Scott Russell, Report of 14th Meeting of the British Association for Advancement of Science, York, September 1844, pp. 311–390.

[69] C. S. Gardner, J. M. Green, M. D. Kruskal, and R. M. Miura, *Phys. Rev. Lett.* **19**, 1095 (1967); *Commun. Pure Appl. Math.* **27**, 97 (1974).

[70] N. J. Zabusky and M. D. Kruskal, *Phys. Rev. Lett.* **15**, 240 (1965).

[71] M. J. Ablowitz and P. A. Clarkson, *Solitons, Nonlinear Evolution Equations, and Inverse Scattering* (Cambridge University Press, New York, 1991).

[72] J. T. Taylor (ed.), *Optical Solitons—Theory and Experiment*, (Cambridge University Press, New York, 1992).

[73] F. K. Abdullaev, *Optical Solitons*, (Springer-Varlag, New York, 1993).

[74] P. G. Drazin, *Solitons: An Introduction* (Cambridge University Press, Cambridge, UK, 1993).

[75] G. L. Lamb, Jr., *Elements of Soliton Theory* (Dover, New York, 1994).

[76] C. H. Gu, *Soliton Theory and its Applications* (Springer-Verlag, New York, 1995).

[77] H. Hasegawa and Y. Kodama, *Solitons in Optical Communications* (Oxford University Press, New York, 1995).

[78] N. N. Akhmediev and A. A. Ankiewicz, *Solitons* (Chapman and Hall, New York, 1997).

[79] T. Miwa, *Mathematics of Solitons* (Cambridge University Press, New York, 1999).

[80] A. Hasegawa and F. Tappert, *Appl. Phys. Lett.* **23**, 142 (1973).

[81] P. Andrekson, *Laser Focus World* **35** (5), 145 (1999).

[82] V. E. Zakharov and A. B. Shabat, *Sov. Phys. JETP* **34**, 62 (1972).

[83] H. A. Haus and M. N. Islam, *IEEE J. Quantum Electron.* **QE-21**, 1172 (1985).

[84] J. Satsuma and N. Yajima, *Prog. Theor. Phys. Suppl.* **55**, 284 (1974).

[85] L. F. Mollenauer, R. H. Stolen, and J. P. Gordon, *Phys. Rev. Lett.* **45**, 1095 (1980).

[86] R. H. Stolen, L. F. Mollenauer, and W. J. Tomlinson, *Opt. Lett.* **8**, 186 (1983).

[87] L. F. Mollenauer, R. H. Stolen, J. P. Gordon, and W. J. Tomlinson, *Opt. Lett.* **8**, 289 (1983).

[88] F. Salin, P. Grangier, G. Roger, and A. Brun, *Phys. Rev. Lett.* **56**, 1132 (1986); *Phys. Rev. Lett.* **6**, 569 (1988).

[89] R. Meinel, *Opt. Commun.* **47**, 343 (1983).

[90] E. M. Dianov, A. M. Prokhorov, and V. N. Serkin, *Sov. Phys. Dokl.* **28**, 1036 (1983).

[91] A. M. Fattakhov and A. S. Chirkin, *Sov. J. Quantum Electron.* **14**, 1556 (1984).

[92] H. E. Lassen, F. Mengel, B. Tromborg, N. C. Albertson, and P. L. Christiansen, *Opt. Lett.* **10**, 34 (1985).

[93] C. Desem and P. L. Chu, *Opt. Lett.* **11**, 248 (1986).

[94] K. J. Blow and D. Wood, *Opt. Commun.* **58**, 349 (1986).

[95] A. I. Maimistov and Y. M. Sklyarov, *Sov. J. Quantum Electron.* **17**, 500 (1987).

[96] M. N. Belov, *Sov. J. Quantum Electron.* **17**, 1033 (1987).

[97] A. S. Gouveia-Neto, A. S. L. Gomes, and J. R. Taylor, *Opt. Commun.* **64**, 383 (1987).

[98] J. P. Gordon, *J. Opt. Soc. Am. B* **9**, 91 (1992).

[99] G. P. Agrawal, *Fiber-Optic Communication Systems*, 2nd ed. (Wiley, New York, 1997), Chap. 10.

[100] M. W. Chbat, P. R. Prucnal, M. N. Islam, C. E. Soccolich, and J. P. Gordon, *J. Opt. Soc. Am. B* **10**, 1386 (1993).

[101] L. F. Mollenauer, J. P. Gordon, and P. V. Mamyshev, *Optical Fiber Telecommunications III*, I. P. Kaminow and T. L. Koch, eds. (Academic Press, San Diego, CA, 1997), Chap. 12.

[102] A. Hasegawa and F. Tappert, *Appl. Phys. Lett.* **23**, 171 (1973).

[103] V. E. Zakharov and A. B. Shabat, *Sov. Phys. JETP* **37**, 823 (1973).

[104] B. Bendow, P. D. Gianino, N. Tzoar, and M. Jain, *J. Opt. Soc. Am.* **70**, 539 (1980).

[105] K. J. Blow and N. J. Doran, *Phys. Lett.* **107A**, 55 (1985).

[106] P. Emplit, J. P. Hamaide, F. Reynaud, C. Froehly, and A. Barthelemy, *Opt. Commun.* **62**, 374 (1987).

[107] D. Krökel, N. J. Halas, G. Giuliani, and D. Grischkowsky, *Phys. Rev. Lett.* **60**, 29 (1988).

[108] A. M. Weiner, J. P. Heritage, R. J. Hawkins, R. N. Thurston, E. M. Krischner, D. E. Leaird, and W. J. Tomlinson, *Phys. Rev. Lett.* **61**, 2445 (1988).

[109] W. J. Tomlinson, R. J. Hawkins, A. M. Weiner, J. P. Heritage, and R. N. Thurston, *J. Opt. Soc. Am. B* **6**, 329 (1989).

[110] W. Zhao and E. Bourkoff, *Opt. Lett.* **14**, 703 (1989); *Opt. Lett.* **14**, 808 (1989).

[111] S. A. Gredeskul, Y. S. Kivshar, and M. V. Yanovskaya, *Phys. Rev. A* **41**, 3994 (1990); Y. S. Kivshar, *Phys. Rev. A* **42**, 1757 (1990).

[112] W. Zhao and E. Bourkoff, *Opt. Lett.* **14**, 1372 (1989).

[113] J. A. Giannini and R. I. Joseph, *IEEE J. Quantum Electron.* **26**, 2109 (1990).

[114] J. E. Rothenberg, *Opt. Lett.* **15**, 443 (1990); *Opt. Commun.* **82**, 107 (1991).

[115] R. N. Thurston and A. M. Weiner, *J. Opt. Soc. Am. B* **8**, 471 (1991).

[116] Y. S. Kivshar and V. V. Afanasjev, *Opt. Lett.* **16**, 285 (1991); *Phys. Rev. A* **44**, R1446 (1991).

[117] J. P. Hamaide, P. Emplit, and M. Haelterman, *Opt. Lett.* **16**, 1578 (1991).

[118] W. Zhao and E. Bourkoff, *Opt. Lett.* **15** , 405 (1990); *J. Opt. Soc. Am. B* **9**, 1134 (1992).

[119] J. E. Rothenberg and H. K. Heinrich, *Opt. Lett.* **17**, 261 (1992).

[120] Y. S. Kivshar, *IEEE J. Quantum Electron.* **29**, 250 (1993).

[121] P. Emplit, M. Haelterman, and J. P. Hamaide, *Opt. Lett.* **18**, 1047 (1993).

[122] J. A. R. Williams, K. M. Allen, N. J. Doran, and P. Emplit, *Opt. Commun.* **112**, 333 (1994).

[123] Y. S. Kivshar and X. Yang, *Phys. Rev. E* **49**, 1657 (1994).

[124] D. J. Richardson, R. P. Chamberlain, L. Dong, and D. N. Payne, *Electron. Lett.* **30**, 1326 (1994).

[125] O. G. Okhotnikov and F. M. Araujo, *Electron. Lett.* **31**, 2187 (1995).

[126] M. Nakazawa and K. Suzuki, *Electron. Lett.* **31**, 1084 (1995).

[127] M. Nakazawa and K. Suzuki, *Electron. Lett.* **31**, 1076 (1995).

[128] A. K. Atieh, P. Myslinski, J. Chrostowski, and P. Galko, *Opt. Commun.* **133**, 541 (1997).

[129] P. Emplit, M. Haelterman, R. Kashyap, and M. DeLathouwer, *IEEE Photon. Technol. Lett.* **9**, 1122 (1997).

[130] R. Leners, P. Emplit, D. Foursa, M. Haelterman, and R. Kashyap, *J. Opt. Soc. Am. B* **14**, 2339 (1997).

[131] Y. S. Kivshar and B. Luther-Davies, *Phys. Rep.* **298**, 81 (1998).

[132] N. J. Smith, N. J. Doran, W. Forysiak, and F. M. Knox, *J. Lightwave Technol.* **15**, 1808 (1997).

[133] L. F. Mollenauer and P. V. Mamyshev, *IEEE J. Quantum Electron.* **34**, 2089 (1998).

[134] R. M. Mu, C. R. Menyuk, G. M. Carter, and J. M. Jacob, *IEEE J. Sel. Topics Quantum Electron.* **6**, 248 (2000).

[135] S. K. Turitsyn, M. P. Fedourk, E. G. Shapiro, V. K. Mezentsev, and E. G. Turitsyna, *IEEE J. Sel. Topics Quantum Electron.* **6**, 263 (2000).

[136] A. E. Kaplan, *Phys. Rev. Lett.* **55**, 1291 (1985); *IEEE J. Quantum Electron.* **QE-21**, 1538 (1985).

[137] R. H. Enns and S. S. Rangnekar, *Opt. Lett.* **12** , 108 (1987); *IEEE J. Quantum Electron.* **QE-23**, 1199 (1987).

[138] R. H. Enns, S. S. Rangnekar, and A. E. Kaplan, *Phys. Rev. A* **35**, 446 (1987); *Phys. Rev. A* **36**, 1270 (1987).

[139] R. H. Enns, R. Fung, and S. S. Rangnekar, *Opt. Lett.* **15**, 162, (1990); *IEEE J. Quantum Electron.* **27**, 252 (1991).

[140] R. H. Enns and S. S. Rangnekar, *Phys. Rev. A* **43**, 4047, (1991); *Phys. Rev. A* **44**, 3373 (1991).

[141] S. Gatz and J. Hermann, *J. Opt. Soc. Am. B* **8**, 2296 (1991); *Opt. Lett.* **17**, 484 (1992).

[142] W. Krolikowski and B. Luther-Davies, *Opt. Lett.* **17**, 1414 (1992).

[143] R. H. Enns, D. E. Edmundson, S. S. Rangnekar, and A. E. Kaplan, *Opt. Quantum Electron.* **24**, S2195 (1992).

[144] S. L. Eix, R. H. Enns, and S. S. Rangnekar, *Phys. Rev. A* **47**, 5009 1993).

[145] C. Deangelis, *IEEE J. Quantum Electron.* **30**, 818 (1994).

[146] G. H. Aicklen and L. S. Tamil, *J. Opt. Soc. Am. B* **13**, 1999 (1996).

[147] S. Tanev and D. I. Pushkarov, *Opt. Commun.* **141**, 322 (1997).

[148] A. Kumar, *Phys. Rev. E* **58**, 5021 (1998); A. Kumar and T. Kurz, *Opt. Lett.* **24**, 373 (1999).

[149] A. Hasegawa and Y. Kodama, *Proc. IEEE* **69**, 1145 (1981); *Opt. Lett.* **7**, 285 (1982).

[150] Y. Kodama and A. Hasegawa, *Opt. Lett.* **7**, 339 (1982); *Opt. Lett.* **8**, 342 (1983).

[151] A. Hasegawa, *Opt. Lett.* **8**, 650 (1983); *Appl. Opt.* **23**, 3302 (1984).

[152] L. F. Mollenauer, R. H. Stolen, and M. N. Islam, *Opt. Lett.* **10**, 229 (1985).

[153] L. F. Mollenauer, J. P. Gordon, and M. N. Islam, *IEEE J. Quantum Electron.* **QE-22**, 157 (1986).

[154] L. F. Mollenauer and K. Smith, *Opt. Lett.* **13**, 675 (1988).

[155] M. Nakazawa, Y. Kimura, and K. Suzuki, *Electron. Lett.* **25**, 199 (1989).

[156] M. Nakazawa, K. Suzuki, and Y. Kimura, *IEEE Photon. Technol. Lett.* **2**, 216 (1990).

[157] V. I. Karpman and E. M. Maslov, *Sov. Phys. JETP* **46**, 281 (1977).

[158] D. J. Kaup and A. C. Newell, *Proc. R. Soc. London*, Ser. A **361**, 413 (1978).

[159] V. I. Karpman, *Sov. Phys. JETP* **50**, 58 (1979); *Physica Scripta* **20**, 462 (1979).

[160] Y. S. Kivshar and B. A. Malomed, *Rev. Mod. Phys.* **61**, 761 (1989).

[161] H. Haus, *J. Opt. Soc. Am. B* **8**, 1122 (1991).

[162] C. R. Menyuk, *J. Opt. Soc. Am. B* **10**, 1585 (1993).

[163] T. Georges and F. Favre, *J. Opt. Soc. Am. B* **10**, 1880 (1993).

[164] T. Georges, *Opt. Fiber Technol.* **1**, 97 (1995).

[165] A. Bonderson, M. Lisak, and D. Anderson, *Physica Scripta* **20**, 479 (1979).

[166] D. Anderson and M. Lisak, *Phys. Rev. A* **27**, 1393 (1983).

[167] K. J. Blow and N. J. Doran, *Opt. Commun.* **52**, 367 (1985).

[168] D. Anderson and M. Lisak, *Opt. Lett.* **10**, 390 (1985).

[169] K. Tajima, *Opt. Lett.* **12**, 54 (1987).

[170] V. A. Bogatyrjov, M. M. Bubnov, E. M. Dianov, and A. A. Sysoliatin, *Pure Appl. Opt.* **4**, 345 (1995).

[171] D. J. Richardson, R. P. Chamberlin, L. Dong, and D. N. Payne, *Electron. Lett.* **31**, 1681 (1995).

[172] D. M. Spirit, I. W. Marshall, P. D. Constantine, D. L. Williams, S. T. Davey, and B. J. Ainslie, *Electron. Lett.* **27**, 222 (1991).

[173] M. Nakazawa, H. Kubota, K. Kurakawa, and E. Yamada, *J. Opt. Soc. Am. B* **8**, 1811 (1991).

[174] K. Kurokawa and M. Nakazawa, *IEEE J. Quantum Electron.* **28**, 1922 (1992).

[175] K. Rottwitt, J. H. Povlsen, and A. Bjarklev, *J. Lightwave Technol.* **11**, 2105 (1993).

[176] C. Lester, K. Rottwitt, J. H. Povlsen, P. Varming, M. A. Newhouse, and A. J. Antos, *Opt. Lett.* **20**, 1250 (1995).

[177] A. Hasegawa and Y. Kodama, *Phys. Rev. Lett.* **66**, 161 (1991).

[178] J. P. Gordon and H. A. Haus, *Opt. Lett.* **11**, 665 (1986).

[179] V. I. Karpman and V. V. Solov'ev, *Physica* **3D**, 487 (1981).

[180] J. P. Gordon, *Opt. Lett.* **8**, 596 (1983).

[181] K. J. Blow and N. J. Doran, *Electron. Lett.* **19**, 429 (1983).

[182] B. Hermansson and D. Yevick, *Electron. Lett.* **19**, 570 (1983).

[183] P. L. Chu and C. Desem, *Electron. Lett.* **19**, 956 (1983); *Electron. Lett.* **21**, 228 (1985).

[184] D. Anderson and M. Lisak, *Phys. Rev. A* **32**, 2270 (1985); *Opt. Lett.* **11**, 174 (1986).

[185] E. M. Dianov, Z. S. Nikonova, and V. N. Serkin, *Sov. J. Quantum Electron.* **16**, 1148 (1986).

[186] F. M. Mitschke and L. F. Mollenauer, *Opt. Lett.* **12**, 355 (1987).

[187] C. Desem and P. L. Chu, *Opt. Lett.* **12**, 349 (1987); *Electron. Lett.* **23**, 260 (1987).

[188] C. Desem and P. L. Chu, *IEE Proc.* **134**, Pt. J, 145 (1987).

[189] Y. Kodama and K. Nozaki, *Opt. Lett.* **12**, 1038 (1987).

[190] V. V. Afanasjev, *Opt. Lett.* **18**, 790 (193).

[191] A. N. Pinto, G. P. Agrawal, and J. F. da Rocha, *J. Lightwave Technol.* **18**, 515 (1998).

[192] G. P. Agrawal and M. J. Potasek, *Phys. Rev. A* **33**, 1765 (1986).

[193] G. R. Boyer and X. F. Carlotti, *Opt. Commun.* **60**, 18 (1986).

[194] P. K. Wai, C. R. Menyuk, H. H. Chen, and Y. C. Lee, *Opt. Lett.* **12**, 628 (1987); *IEEE J. Quantum Electron.* **24**, 373 (1988).

[195] M. Desaix, D. Anderson, and M. Lisak, *Opt. Lett.* **15**, 18 (1990).

[196] V. K. Mezentsev and S. K. Turitsyn, *Sov. Lightwave Commun.* **1**, 263 (1991).

[197] Y. S. Kivshar, *Phys. Rev. A* **43**, 1677 (1981); *Opt. Lett.* **16**, 892 (1991).

[198] V. I. Karpman, *Phys. Rev. E* **47**, 2073 (1993); *Phys. Lett. A* **181**, 211 (1993).

[199] M. Karlsson and A. Höök, *Opt. Commun.* **104**, 303 (1994).

[200] N. Tzoar and M. Jain, *Phys. Rev. A* **23**, 1266 (1981).

[201] E. A. Golovchenko, E. M. Dianov, A. M. Prokhorov, and V. N. Serkin, *JETP Lett.* **42**, 87 (1985); *Sov. Phys. Dokl.* **31**, 494 (1986).

[202] K. Ohkuma, Y. H. Ichikawa, and Y. Abe, *Opt. Lett.* **12**, 516 (1987).

[203] A. M. Kamchatnov, S. A. Darmanyan, and F. Lederer, *Phys. Lett. A* **245**, 259 (1998).

[204] W. P. Zhong and H. J. Luo, *Chin. Phys. Lett.* **17**, 577 (2000).

[205] E. Mjolhus, *J. Plasma Phys.* **16**, 321 (1976); **19** 437 (1978).

[206] K. Mio, T. Ogino, K. Minami, and S. Takeda, *J. Phys. Soc. Jpn.* **41**, 265 (1976).

[207] M. Wadati, K. Konno, and Y. H. Ichikawa, *J. Phys. Soc. Jpn.* **46**, 1965 (1979).

[208] Y. H. Ichikawa, K. Konno, M. Wadati, and H. Sanuki, *J. Phys. Soc. Jpn.* **48**, 279 (1980).

[209] E. M. Dianov, A. Y. Karasik, P. V. Mamyshev, A. M. Prokhorov, V. N. Serkin, M. F. Stel'makh, and A. A. Fomichev, *JETP Lett.* **41**, 294 (1985).

[210] F. M. Mitschke and L. F. Mollenauer, *Opt. Lett.* **11**, 659 (1986);

[211] J. P. Gordon, *Opt. Lett.* **11**, 662 (1986).

[212] Y. Kodama and A. Hasegawa, *IEEE J. Quantum Electron.* **QE-23**, 510 (1987).

[213] B. Zysset, P. Beaud, and W. Hodel, *Appl. Phys. Lett.* **50**, 1027 (1987).

[214] V. A. Vysloukh and T. A. Matveeva, *Sov. J. Quantum Electron.* **17**, 498 (1987).

[215] V. N. Serkin, *Sov. Tech. Phys. Lett.* **13**, 320 (1987); *Sov. Tech. Phys. Lett.* **13**, 366 (1987).

[216] A. B. Grudinin, E. M. Dianov, D. V. Korobkin, A. M. Prokhorov, V. N. Serkin, and D. V. Khaidarov, *JETP Lett.* **46**, 221 (1987).

[217] A. S. Gouveia-Neto, A. S. L. Gomes, and J. R. Taylor, *IEEE J. Quantum Electron.* **24**, 332 (1988).

[218] K. Tai, A. Hasegawa, and N. Bekki, *Opt. Lett.* **13**, 392 (1988).

[219] R. H. Stolen, J. P. Gordon, W. J. Tomlinson, and H. A. Haus, *J. Opt. Soc. Am. B* **6**, 1159 (1989).

[220] K. J. Blow and D. Wood, *IEEE J. Quantum Electron.* **25**, 2665 (1989).

[221] V. V. Afansasyev, V. A. Vysloukh, and V. N. Serkin, *Opt. Lett.* **15**, 489 (1990).

[222] P. V. Mamyshev and S. V. Chernikov, *Opt. Lett.* **15**, 1076 (1990).

[223] B. J. Hong and C. C. Yang, *Opt. Lett.* **15**, 1061 (1990); *J. Opt. Soc. Am. B* **8**, 1114 (1991).

[224] P. V. Mamyshev and S. V. Chernikov, *Sov. Lightwave Commun.* **2**, 97 (1992).

[225] R. H. Stolen and W. J. Tomlinson, *J. Opt. Soc. Am. B* **9**, 565 (1992).

[226] K. Kurokawa, H. Kubota, and M. Nakazawa, *Electron. Lett.* **28**, 2050 (1992).

[227] G. P. Agrawal and C. Headley III, *Phys. Rev. A* **46**, 1573 (1992).

[228] Y. S. Kivshar and B. A. Malomed, *Opt. Lett.* **18**, 485 (1993).

[229] V. N. Serkin, V. A. Vysloukh, and J. R. Taylor, *Electron. Lett.* **29**, 12 (1993).

[230] S. Liu and W. Wang, *Opt. Lett.* **18**, 1911 (1993).

[231] W. Hodel and H. P. Weber, *Opt. Lett.* **12**, 924 (1987).

[232] P. Beaud, W. Hodel, B. Zysset, and H. P. Weber, *IEEE J. Quantum Electron.* **QE-23**, 1938 (1987).

[233] P. K. A. Wai, C. R. Menyuk, Y. C. Lee, and H. H. Chen, *Opt. Lett.* **11**, 464 (1986).

[234] M. Trippenbach and Y. B. Band, *Phys. Rev. A* **57**, 4791 (1998).

[235] D. N. Christodoulides and R. I. Joseph, *Appl. Phys. Lett.* **47**, 76 (1985).

[236] L. Gagnon, *J. Opt. Soc. Am. B* **9**, 1477 (1989).

[237] A. B. Grudinin, V. N. Men'shov, and T. N. Fursa, *Sov. Phys. JETP* **70**, 249 (1990).

[238] L. Gagnon and P. A. Bélanger, *Opt. Lett.* **9**, 466 (1990).

[239] M. J. Potasek and M. Tabor, *Phys. Lett. A* **154**, 449 (1991).

[240] M. Florjanczyk and L. Gagnon, *Phys. Rev. A* **41**, 4478 (1990); *Phys. Rev. A* **45**, 6881 (1992).

[241] M. J. Potasek, *J. Appl. Phys.* **65**, 941 (1989); *IEEE J. Quantum Electron.* **29**, 281 (1993).

[242] S. Liu and W. Wang, *Phys. Rev. E* **49**, 5726 (1994).

[243] D. J. Frantzeskakis, K. Hizanidis, G. S. Tombrasand, and I. Belia, *IEEE J. Quantum Electron.* **31**, 183 (1995).

[244] K. Porsezian and K. Nakkeeran, *Phys. Rev. Lett.* **76**, 3955 (1996).

[245] G. J. Dong and Z. Z. Liu, *Opt. Commun.* **128**, 8 (1996).

[246] M. Gedalin, T. C. Scott, and Y. B. Band, *Phys. Rev. Lett.* **78**, 448 (1997).

[247] D. Mihalache, N. Truta, and L. C. Crasovan, *Phys. Rev. E* **56**, 1064 (1997).

[248] S. L. Palacios, A. Guinea, J. M. Fernandez-Diaz, and R. D. Crespo, *Phys. Rev. E* **60**, R45 (1999).

[249] C. E. Zaspel, *Phys. Rev. Lett.* **82**, 723 (1999).

[250] Z. Li, L. Li, H. Tian, and G. Zhou, *Phys. Rev. Lett.* **84**, 4096 (2000).

[251] T. Brabec and F. Krauszm, *Phys. Rev. Lett.* **78**, 3282 (1997).

[252] J. K. Ranka and A. L. Gaeta, *Opt. Lett.* **23**, 534 (1998).

[253] Q. Lin and E. Wintner, *Opt. Commun.* **150**, 185 (1998).

Chapter 6

Polarization Effects

As discussed in Section 2.3, a major simplification made in the derivation of the nonlinear Schrödinger (NLS) equation consists of assuming that the polarization state of the incident light is preserved during its propagating inside an optical fiber. This is not really the case in practice. In this chapter we focus on the polarization effects and consider the coupling between the two orthogonally polarized components of an optical field induced by the nonlinear phenomenon known as cross-phase modulation (XPM). The XPM is always accompanied by self-phase modulation (SPM) and can also occur between two optical fields of different wavelengths. The nondegenerate case involving different wavelengths is discussed in Chapter 7.

The chapter is organized as follows. The origin of nonlinear birefringence is discussed first in Section 6.1 and is followed by the derivation of a set of two coupled NLS equations that describes evolution of the two orthogonally polarized components of an optical field. The XPM-induced nonlinear birefringence has several practical applications discussed in Section 6.2. The next section considers nonlinear polarization changes with focus on polarization instability. Section 6.4 is devoted to the vector modulation instability occurring in birefringent fibers. In contrast with the scalar case discussed in Section 5.1, the vector modulation instability can occur even in the normal-dispersion regime of a birefringent fiber. Section 6.5 considers the effects of birefringence on solitons. The last section focuses on polarization-mode dispersion (PMD) occurring in fibers with randomly varying birefringence along their length and its implications for lightwave systems.

6.1   Nonlinear Birefringence

As mentioned in Section 2.2, even a single-mode fiber, in fact, supports two orthogonally polarized modes with the same spatial distribution. The two modes are degenerate in an ideal fiber (maintaining perfect cylindrical symmetry along its entire length) in the sense that their effective refractive indices, n_x and n_y, are identical. In practice, all fibers exhibit some modal birefringence ($n_x \neq n_y$) because of unintentional variations in the core shape and anisotropic stresses along the fiber length. Moreover, the degree of modal birefringence, $B_m = |n_x - n_y|$, and the orientation of x and y axes change randomly over a length scale ~ 10 m unless special precautions are taken.

In polarization-maintaining fibers, the built-in birefringence is made much larger than random changes occurring due to stress and core-shape variations. As a result, such fibers exhibit nearly constant birefringence along their entire length. This kind of birefringence is called *linear* birefringence. When the nonlinear effects in optical fibers become important, a sufficiently intense optical field can induce *nonlinear* birefringence whose magnitude is intensity dependent. Such self-induced polarization effects were observed as early as 1964 in bulk nonlinear media [1] and have been studied extensively since then [2]–[10]. In this section, we discuss the origin of nonlinear birefringence and develop mathematical tools that are needed for studying the polarization effects in optical fibers assuming a constant modal birefringence. Fibers in which linear birefringence changes randomly over their length are considered later in this chapter.

6.1.1   Origin of Nonlinear Birefringence

A fiber with constant modal birefringence has two principal axes along which the fiber is capable of maintaining the state of linear polarization of the incident light. These axes are called slow and fast axes based on the speed at which light polarized along them travels inside the fiber. Assuming $n_x > n_y$, n_x and n_y are the mode indices along the slow and fast axes, respectively. When low-power, continuous-wave (CW) light is launched with its polarization direction oriented at an angle with respect to the slow (or fast) axis, the polarization state of the CW light changes along the fiber from linear to elliptic, elliptic to circular, and then back to linear in a periodic manner (see Fig. 1.9) over a distance known as the beat length and defined as $L_B = \lambda / B_m$. The beat length can be as small as 1 cm in high-birefringence fibers with $B_m \sim 10^{-4}$. In low-birefringence fibers, typically $B_m \sim 10^{-6}$, and the beat length is ~ 1 m.

The electric field associated with an arbitrarily polarized optical wave can be written as

$$\mathbf{E}(\mathbf{r},t) = \tfrac{1}{2}(\hat{x}E_x + \hat{y}E_y)\exp(-i\omega_0 t) + \text{c.c.}, \qquad (6.1.1)$$

where E_x and E_y are the complex amplitudes of the polarization components of the field with the carrier frequency ω_0. The axial component E_z is assumed to remain small enough that it can be ignored.

The nonlinear part of the induced polarization[1] $\mathbf{P}_{\text{NL}}$ is obtained by substituting Eq. (6.1.1) in Eq. (2.3.6). In general, the third-order susceptibility is a fourth-rank tensor with 81 elements. In an isotropic medium, such as silica glass, only three elements are independent of one another, and the third-order susceptibility can be written in terms of them as [10]

$$\chi_{ijkl}^{(3)} = \chi_{xxyy}^{(3)}\delta_{ij}\delta_{kl} + \chi_{xyxy}^{(3)}\delta_{ik}\delta_{jl} + \chi_{xyyx}^{(3)}\delta_{il}\delta_{jk}, \qquad (6.1.2)$$

where δ_{ij} is the Kronecker delta function defined such that $\delta_{ij} = 1$ when $i = j$ and zero otherwise. Using this result in Eq. (2.3.6), $\mathbf{P}_{\text{NL}}$ can be written as

$$\mathbf{P}_{\text{NL}}(\mathbf{r},t) = \tfrac{1}{2}(\hat{x}P_x + \hat{y}P_y)\exp(-i\omega_0 t) + \text{c.c.}, \qquad (6.1.3)$$

with P_x and P_y given by

$$P_i = \frac{3\varepsilon_0}{4}\sum_j \left(\chi_{xxyy}^{(3)}E_iE_jE_j^* + \chi_{xyxy}^{(3)}E_jE_iE_j^* + \chi_{xyyx}^{(3)}E_jE_jE_i^* \right), \qquad (6.1.4)$$

where $i, j = x$ or y. From Eq. (6.1.2), we also obtain the relation

$$\chi_{xxxx}^{(3)} = \chi_{xxyy}^{(3)} + \chi_{xyxy}^{(3)} + \chi_{xyyx}^{(3)}, \qquad (6.1.5)$$

where $\chi_{xxxx}^{(3)}$ is the element appearing in the scalar theory of Section 2.3 and used in Eq. (2.3.13) to define the nonlinear parameter n_2.

The relative magnitudes of the three components in Eq. (6.1.5) depend on the physical mechanisms that contribute to $\chi^{(3)}$. In the case of silica fibers, the dominant contribution is of electronic origin [4], and the three components

---

[1]Polarization induced inside a dielectric medium by an electromagnetic field should not be confused with the state of polarization of that field. The terminology is certainly confusing but is accepted for historical reasons.

have nearly the same magnitude. If they are assumed to be identical, the polarization components P_x and P_y in Eq. (6.1.4) take the form

$$P_x = \frac{3\varepsilon_0}{4}\chi_{xxxx}^{(3)}\left[\left(|E_x|^2 + \frac{2}{3}|E_y|^2\right)E_x + \frac{1}{3}(E_x^*E_y)E_y\right], \qquad (6.1.6)$$

$$P_y = \frac{3\varepsilon_0}{4}\chi_{xxxx}^{(3)}\left[\left(|E_y|^2 + \frac{2}{3}|E_x|^2\right)E_y + \frac{1}{3}(E_y^*E_x)E_x\right]. \qquad (6.1.7)$$

The last term in Eqs. (6.1.6) and (6.1.7) leads to degenerate four-wave mixing. Its importance will be discussed later.

The nonlinear contribution Δn_x to the refractive index is governed by the term proportional to E_x in Eq. (6.1.6). Writing $P_j = \varepsilon_0\varepsilon_j^{\text{NL}}E_j$ and using

$$\varepsilon_j = \varepsilon_j^L + \varepsilon_j^{\text{NL}} = (n_j^L + \Delta n_j)^2, \qquad (6.1.8)$$

where n_j^L is the linear part of the refractive index ($j = x, y$), the nonlinear contributions Δn_x and Δn_y are given by

$$\Delta n_x = n_2\left(|E_x|^2 + \frac{2}{3}|E_y|^2\right), \qquad \Delta n_y = n_2\left(|E_y|^2 + \frac{2}{3}|E_x|^2\right), \qquad (6.1.9)$$

where n_2 is a nonlinear parameter as defined in Eq. (2.3.13). The physical meaning of the two terms on the right-hand side of these equations is quite clear. The first term is responsible for SPM. The second term results in XPM because the nonlinear phase shift acquired by one polarization component depends on the intensity of the other polarization component. The presence of this term induces a nonlinear coupling between the field components E_x and E_y. The nonlinear contributions Δn_x and Δn_y are in general unequal and thus create nonlinear birefringence whose magnitude depends on the intensity and the polarization state of the incident light. In the case of CW light propagating inside a fiber, nonlinear birefringence manifests as a rotation of the polarization ellipse [1].

6.1.2 Coupled-Mode Equations

The propagation equations governing evolution of the two polarization components along a fiber can be obtained following the method of Section 2.3. Assuming that the nonlinear effects do not affect the fiber mode significantly, the transverse dependence of E_x and E_y can be factored out using

$$E_j(\mathbf{r}, t) = F(x, y)A_j(z, t)\exp(i\beta_{0j}z), \qquad (6.1.10)$$

where $F(x, y)$ is the spatial distribution of the single mode supported by the fiber, $A_j(z, t)$ is the slowly varying amplitude, and β_{0j} is the corresponding propagation constant $(j = x, y)$. The dispersive effects are included by expanding the frequency-dependent propagation constant in a manner similar to Eq. (2.3.23). The slowly varying amplitudes, A_x and A_y, are found to satisfy the following set of two coupled-mode equations:

$$
\frac{\partial A_x}{\partial z} + \beta_{1x} \frac{\partial A_x}{\partial t} + \frac{i\beta_2}{2} \frac{\partial^2 A_x}{\partial t^2} + \frac{\alpha}{2} A_x
$$
$$
= i\gamma \left(|A_x|^2 + \frac{2}{3} |A_y|^2 \right) A_x + \frac{i\gamma}{3} A_x^* A_y^2 \exp(-2i\Delta\beta z), \quad (6.1.11)
$$

$$
\frac{\partial A_y}{\partial z} + \beta_{1y} \frac{\partial A_y}{\partial t} + \frac{i\beta_2}{2} \frac{\partial^2 A_y}{\partial t^2} + \frac{\alpha}{2} A_y
$$
$$
= i\gamma \left(|A_y|^2 + \frac{2}{3} |A_x|^2 \right) A_y + \frac{i\gamma}{3} A_y^* A_x^2 \exp(2i\Delta\beta z), \quad (6.1.12)
$$

where

$$
\Delta\beta = \beta_{0x} - \beta_{0y} = (2\pi/\lambda) B_m = 2\pi/L_B \quad (6.1.13)
$$

is related to the modal birefringence of the fiber. Note that modal birefringence also leads to different group velocities for the two polarization components because $\beta_{1x} \neq \beta_{1y}$ in general. In contrast, the parameters β_2 and γ are the same for both polarization components having the same wavelength λ.

The last term in Eqs. (6.1.11) and (6.1.12) is due to coherent coupling between the two polarization components and leads to degenerate four-wave mixing. Its importance to the process of polarization evolution depends on the extent to which the phase-matching condition is satisfied (see Chapter 10). If the fiber length $L \gg L_B$, the last term in Eqs. (6.1.11) and (6.1.12) changes sign often and its contribution averages out to zero. In highly birefringent fibers ($L_B \sim 1$ cm typically), the four-wave-mixing term can often be neglected for this reason. In contrast, this term should be included in weakly birefringent fibers, especially for short lengths. In that case, it is often convenient to rewrite Eqs. (6.1.11) and (6.1.12) using circularly polarized components defined as

$$
A_+ = (\bar{A}_x + i\bar{A}_y)/\sqrt{2}, \qquad A_- = (\bar{A}_x - i\bar{A}_y)/\sqrt{2}, \quad (6.1.14)
$$

where $\bar{A}_x = A_x \exp(i\Delta\beta z/2)$ and $\bar{A}_y = A_y \exp(-i\Delta\beta z/2)$. The A_+ and A_- represent right- and left-handed circularly polarized (σ_+ and σ_-) states, respectively, and satisfy somewhat simpler equations:

$$
\begin{aligned}
\frac{\partial A_+}{\partial z} + \beta_1 \frac{\partial A_+}{\partial t} + \frac{i\beta_2}{2} \frac{\partial^2 A_+}{\partial t^2} + \frac{\alpha}{2} A_+ \\
= \frac{i}{2}(\Delta\beta)A_- + \frac{2i\gamma}{3}\left(|A_+|^2 + 2|A_-|^2\right)A_+,
\end{aligned}
\tag{6.1.15}
$$

$$
\begin{aligned}
\frac{\partial A_-}{\partial z} + \beta_1 \frac{\partial A_-}{\partial t} + \frac{i\beta_2}{2} \frac{\partial^2 A_-}{\partial t^2} + \frac{\alpha}{2} A_- \\
= \frac{i}{2}(\Delta\beta)A_+ + \frac{2i\gamma}{3}\left(|A_-|^2 + 2|A_+|^2\right)A_-,
\end{aligned}
\tag{6.1.16}
$$

where we assumed that $\beta_{1x} \approx \beta_{1y} = \beta_1$ for fibers with relatively low birefringence. Notice that the four-wave-mixing terms appearing in Eqs. (6.1.11) and (6.1.12) are replaced by a linear-coupling term containing $\Delta\beta$. At the same time, the relative strength of XPM changes from $\frac{2}{3}$ to 2 when circularly polarized components are used to describe wave propagation.

6.1.3 Elliptically Birefringent Fibers

The derivation of Eqs. (6.1.11) and (6.1.12) assumes that the fiber is linearly birefringent, i.e., it has two principal axes along which linearly polarized light remains linearly polarized in the absence of nonlinear effects. Although this is ideally the case for polarization-maintaining fibers, *elliptically* birefringent fibers can be made by twisting a fiber preform during the draw stage [11].

The coupled-mode equations are modified considerably for elliptically birefringent fibers. This case can be treated by replacing Eq. (6.1.1) with

$$
\mathbf{E}(\mathbf{r},t) = \frac{1}{2}(\hat{e}_x E_x + \hat{e}_y E_y)\exp(-i\omega_0 t) + \text{c.c.},
\tag{6.1.17}
$$

where $\hat{e}_x$ and $\hat{e}_y$ are orthonormal polarization eigenvectors related to the unit vectors $\hat{x}$ and $\hat{y}$ used before as [12]

$$
\hat{e}_x = \frac{\hat{x} + ir\hat{y}}{\sqrt{1+r^2}}, \qquad \hat{e}_y = \frac{r\hat{x} - i\hat{y}}{\sqrt{1+r^2}}.
\tag{6.1.18}
$$

The parameter r represents the ellipticity introduced by twisting the preform. It is common to introduce the ellipticity angle θ as $r = \tan(\theta/2)$. The cases $\theta = 0$ and $\pi/2$ correspond to linearly and circularly birefringent fibers, respectively.

Following a procedure similar to that outlined here for linearly birefringent fibers, the slowly varying amplitudes A_x and A_y are found to satisfy the following set of coupled-mode equations [12]:

$$\frac{\partial A_x}{\partial z} + \beta_{1x}\frac{\partial A_x}{\partial t} + \frac{i\beta_2}{2}\frac{\partial^2 A_x}{\partial t^2} + \frac{\alpha}{2}A_x$$
$$= i\gamma[(|A_x|^2 + B|A_y|^2)A_x + CA_x^*A_y^2 e^{-2i\Delta\beta z}]$$
$$+ i\gamma D[A_y^*A_x^2 e^{i\Delta\beta z} + (|A_y|^2 + 2|A_x|^2)A_y e^{-i\Delta\beta z}], \qquad (6.1.19)$$

$$\frac{\partial A_y}{\partial z} + \beta_{1y}\frac{\partial A_y}{\partial t} + \frac{i\beta_2}{2}\frac{\partial^2 A_y}{\partial t^2} + \frac{\alpha}{2}A_y$$
$$= i\gamma[(|A_y|^2 + B|A_x|^2)A_y + CA_y^*A_x^2 e^{2i\Delta\beta z}]$$
$$+ i\gamma D[A_x^*A_y^2 e^{-i\Delta\beta z} + (|A_x|^2 + 2|A_y|^2)A_x e^{i\Delta\beta z}], \qquad (6.1.20)$$

where the parameters $B, C,$ and D are related to the ellipticity angle θ as

$$B = \frac{2 + 2\sin^2\theta}{2 + \cos^2\theta}, \quad C = \frac{\cos^2\theta}{2 + \cos^2\theta}, \quad D = \frac{\sin\theta\cos\theta}{2 + \cos^2\theta}. \qquad (6.1.21)$$

For a linearly birefringent fiber ($\theta = 0$), $B = \frac{2}{3}, C = \frac{1}{3}, D = 0$, and Eqs. (6.1.19) and (6.1.20) reduce to Eqs. (6.1.11) and (6.1.12), respectively.

Equations (6.1.19) and (6.1.20) can be simplified considerably for optical fibers with large birefringence. For such fibers, the beat length L_B is much smaller than typical propagation distances. As a result, the exponential factors in the last three terms of Eqs. (6.1.19) and (6.1.20) oscillate rapidly, contributing little to the pulse evolution process on average. If these terms are neglected, propagation of optical pulses in an elliptically birefringent fiber is governed by the following set of coupled-mode equations:

$$\frac{\partial A_x}{\partial z} + \beta_{1x}\frac{\partial A_x}{\partial t} + \frac{i\beta_2}{2}\frac{\partial^2 A_x}{\partial t^2} + \frac{\alpha}{2}A_x = i\gamma(|A_x|^2 + B|A_y|^2)A_x, \quad (6.1.22)$$

$$\frac{\partial A_y}{\partial z} + \beta_{1y}\frac{\partial A_y}{\partial t} + \frac{i\beta_2}{2}\frac{\partial^2 A_y}{\partial t^2} + \frac{\alpha}{2}A_y = i\gamma(|A_y|^2 + B|A_x|^2)A_y. \quad (6.1.23)$$

These equations represent an extension of the scalar NLS equation, derived in Section 2.3 without the polarization effects [see Eq. (2.3.27)], to the vector case and are referred to as the coupled NLS equations. The coupling parameter B depends on the ellipticity angle θ [see Eq. (6.1.21)] and can vary from $\frac{2}{3}$ to 2 for values of θ in the range 0 to $\pi/2$. For a linearly birefringent fiber, $\theta = 0$, and $B = \frac{2}{3}$. In contrast, $B = 2$ for a circularly birefringent fiber ($\theta = \pi/2$). Note also that $B = 1$ when $\theta \approx 35°$. As discussed later, this case is of particular interest because Eqs. (6.1.22) and (6.1.23) can be solved by the inverse scattering method only when $B = 1$ and $\alpha = 0$.

6.2 Nonlinear Phase Shift

As seen in Section 6.1, a nonlinear coupling between the two orthogonally polarized components of an optical wave changes the refractive index by different amounts for the two components. As a result, the nonlinear effects in birefringent fibers are polarization dependent. In this section we use the coupled NLS equations obtained in the case of high-birefringence fibers to study the XPM-induced nonlinear phase shift and its device applications.

6.2.1 Nondispersive XPM

Equations (6.1.22) and (6.1.23) need to be solved numerically when ultrashort optical pulses propagate inside birefringent fibers. However, they can be solved analytically in the case of CW radiation. The CW solution is also applicable for pulses whenever the fiber length L is much shorter than both the dispersion length $L_D = T_0^2/|\beta_2|$ and the walk-off length $L_W = T_0/|\Delta\beta|$, where T_0 is a measure of the pulse width. As this case can be applicable to pulses as short as 100 ps and sheds considerable physical insight, we discuss it first.

Neglecting the terms with time derivatives in Eqs. (6.1.22) and (6.1.23), we obtain the following two simpler equations:

$$\frac{dA_x}{dz} + \frac{\alpha}{2}A_x = i\gamma(|A_x|^2 + B|A_y|^2)A_x, \tag{6.2.1}$$

$$\frac{dA_y}{dz} + \frac{\alpha}{2}A_y = i\gamma(|A_y|^2 + B|A_x|^2)A_y. \tag{6.2.2}$$

These equations describe nondispersive XPM in birefringent fibers and extend the scalar theory of SPM in Section 4.1 to the vector case. They can be solved by using

$$A_x = \sqrt{P_x}\,e^{-\alpha z/2}e^{i\phi_x}, \qquad A_y = \sqrt{P_y}\,e^{-\alpha z/2}e^{i\phi_y}, \tag{6.2.3}$$

where P_x and P_y are the powers and ϕ_x and ϕ_y are the phases associated with the two polarization components. It is easy to deduce that P_x and P_y do not change with z. However, the phases ϕ_x and ϕ_y do change and evolve as

$$\frac{d\phi_x}{dz} = \gamma e^{-\alpha z}(P_x + BP_y), \qquad \frac{d\phi_y}{dz} = \gamma e^{-\alpha z}(P_y + BP_x). \tag{6.2.4}$$

Since P_x and P_y are constants, the phase equations can be solved easily with the result

$$\phi_x = \gamma(P_x + BP_y)L_{\text{eff}}, \qquad \phi_y = \gamma(P_y + BP_x)L_{\text{eff}}, \tag{6.2.5}$$

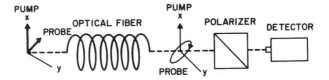

Figure 6.1 Schematic illustration of a Kerr shutter. Pump and probe beams are linearly polarized at 45° to each other at the input end. Polarizer blocks probe transmission in the absence of pump.

where the effective fiber length $L_{\text{eff}} = [1 - \exp(-\alpha L)]/\alpha$ is defined in the same way as in the SPM case [see Eq. (4.1.6)].

It is clear from Eq. (6.2.5) that both polarization components develop a nonlinear phase shift whose magnitude is the sum of the SPM and XPM contributions. In practice, the quantity of practical interest is the relative phase difference given by

$$\Delta\phi_{\text{NL}} \equiv \phi_x - \phi_y = \gamma L_{\text{eff}}(1 - B)(P_x - P_y). \tag{6.2.6}$$

No relative phase shift occurs when $B = 1$. However, when $B \neq 1$, a relative nonlinear phase shift between the two polarization components occurs if input light is launched such that $P_x \neq P_y$. As an example, consider a linearly birefringent fiber for which $B = \frac{2}{3}$. If CW light with power P_0 is launched such that it is linearly polarized at an angle θ from the slow axis, $P_x = P_0 \cos^2\theta$, $P_y = P_0 \sin^2\theta$, and the relative phase shift becomes

$$\Delta\phi_{\text{NL}} = (\gamma P_0 L_{\text{eff}}/3)\cos(2\theta). \tag{6.2.7}$$

This θ-dependent phase shift has several applications discussed next.

6.2.2 Optical Kerr Effect

In the optical Kerr effect, the nonlinear phase shift induced by an intense, high-power, pump beam is used to change the transmission of a weak probe through a nonlinear medium [4]. This effect can be used to make an optical shutter with picosecond response times [6]. It was first observed in optical fibers in 1973 [13] and has attracted considerable attention since then [14]–[24].

The operating principle of a Kerr shutter can be understood from Fig. 6.1. The pump and probe beams are linearly polarized at the fiber input with a 45° angle between their directions of polarization. A crossed polarizer at the fiber output blocks probe transmission in the absence of the pump beam. When the pump is turned on, the refractive indices for the parallel and perpendicular components of the probe (with respect to the direction of pump polarization) become slightly different because of pump-induced birefringence. The phase difference between the two components at the fiber output manifests as a change in the probe polarization, and a portion of the probe intensity is transmitted through the polarizer. The probe transmittivity depends on the pump intensity and can be controlled simply by changing it. In particular, a pulse at the pump wavelength opens the Kerr shutter only during its passage through the fiber. As the probe output at one wavelength can be modulated through a pump at a different wavelength, this device is also referred to as the Kerr modulator. It has potential applications in fiber-optical networks requiring all-optical switching.

Equation (6.2.6) cannot be used to calculate the phase difference between the x and y components of the probe because the pump and probe beams have different wavelengths in Kerr shutters. We follow a slightly different approach and neglect fiber losses for the moment; they can be included later by replacing L with L_{eff}. The relative phase difference for the probe at the output of a fiber of length L can always be written as

$$\Delta\phi = (2\pi/\lambda)(\tilde{n}_x - \tilde{n}_y)L, \tag{6.2.8}$$

where λ is the probe wavelength and

$$\tilde{n}_x = n_x + \Delta n_x, \qquad \tilde{n}_y = n_y + \Delta n_y. \tag{6.2.9}$$

As discussed earlier, the linear parts n_x and n_y of the refractive indices are different because of modal birefringence. The nonlinear parts Δn_x and Δn_y are different because of pump-induced birefringence.

Consider the case of a pump polarized linearly along the x axis. The x component of the probe is polarized parallel to the pump but its wavelength is different. For this reason, the corresponding index change Δn_x must be obtained by using the theory of Section 7.1. If the SPM contribution is neglected,

$$\Delta n_x = 2n_2|E_p|^2, \tag{6.2.10}$$

where $|E_p|^2$ is the pump intensity. When the pump and probe are orthogonally polarized, only the first term in Eq. (6.1.4) contributes to Δn_y because of different wavelengths of the pump and probe beams [9]. Again neglecting the

SPM term, Δn_y becomes

$$\Delta n_y = 2n_2 b|E_p|^2, \qquad b = \chi^{(3)}_{xxyy}/\chi^{(3)}_{xxxx}. \tag{6.2.11}$$

If the origin of $\chi^{(3)}$ is purely electronic, $b = \frac{1}{3}$. Combining Eqs. (6.2.8)–(6.2.11), the phase difference becomes

$$\Delta\phi \equiv \Delta\phi_L + \Delta\phi_{NL} = (2\pi L/\lambda)(\Delta n_L + n_{2B}|E_p|^2), \tag{6.2.12}$$

where $\Delta n_L = n_x - n_y$ accounts for linear birefringence, and the Kerr coefficient n_{2B} is given by

$$n_{2B} = 2n_2(1-b). \tag{6.2.13}$$

The probe transmittivity T_p can now be obtained noting that probe light is blocked by the polarizer when $\Delta\phi = 0$ (see Fig. 6.1). When $\Delta\phi \neq 0$, fiber acts as a birefringent phase plate, and some probe light passes through the polarizer. The probe transmittivity is related to $\Delta\phi$ by the simple relation

$$T_p = \frac{1}{4}|1 - \exp(i\Delta\phi)|^2 = \sin^2(\Delta\phi/2). \tag{6.2.14}$$

It becomes 100% when $\Delta\phi = \pi$ or an odd multiple of π. On the other hand, a phase shift by an even multiple of π blocks the probe completely.

To observe the optical Kerr effect, a polarization-maintaining fiber is generally used to ensure that the pump maintains its state of polarization. The constant phase shift $\Delta\phi_L$ resulting from linear birefringence can be compensated by inserting a quarter-wave plate before the polarizer in Fig. 6.1. However, in practice, $\Delta\phi_L$ fluctuates because of temperature and pressure variations, making it necessary to adjust the wave plate continuously. An alternative approach is to use two identical pieces of polarization-maintaining fibers, spliced together such that their fast (or slow) axes are at right angles to each other [18]. As Δn_L changes sign in the second fiber, the net phase shift resulting from linear birefringence is canceled.

Under ideal conditions, the response time of a Kerr shutter would be limited only by the response time of the Kerr nonlinearity (<10 fs for optical fibers). In practice, however, fiber dispersion limits the response time to values that can range from 1 ps to 1 ns depending on the operating parameters [14]. A major limiting factor is the group-velocity mismatch between the pump and the probe. The relative group delay is given by

$$\Delta t_g = |L/v_{g1} - L/v_{g2}|. \tag{6.2.15}$$

It can easily exceed 1 ns for a 100-m-long fiber unless special precautions are taken to reduce the group-velocity mismatch. One possibility is to choose the pump and probe wavelengths on opposite sides of the zero-dispersion wavelength.

Modal birefringence of the fiber sets another limit on the response time. Because of the index difference Δn_L, the orthogonally polarized components of the probe travel at different speeds and develop a relative delay $\Delta t_p = L\Delta n_L/c$. For a 100-m-long fiber with $\Delta n_L = 5 \times 10^{-5}$, $\Delta t_p \approx 17$ ps. It can be reduced by using fibers with smaller birefringence. The use of two fibers spliced together with their fast axes at right angles to each other can nearly eliminate Δt_p. The fundamental limit on the response time is then set by GVD that broadens the pump pulse during its propagation inside the fiber. It can be reduced to 1 ps or less either by reducing the fiber length or by bringing the pump wavelength closer to the zero-dispersion wavelength.

The minimum pump power required for 100% probe transmission can be estimated by setting $\Delta\phi_L = 0$ (complete compensation) and $\Delta\phi_{NL} = \pi$ in Eq. (6.2.12). It is given by

$$P_p = |E_p|^2 A_{\text{eff}} = \lambda A_{\text{eff}}/(2n_{2B}L), \qquad (6.2.16)$$

where A_{eff} is the effective core area. The effect of fiber loss can be included by replacing L with the effective length L_{eff} introduced earlier. Using $n_{2B} = 4.5 \times 10^{-16}$ cm^2/W, $A_{\text{eff}} = 10$ μm^2, and $\lambda = 1.06$ μm, the pump power $P_p \approx 1$ W for a 100-m-long fiber. The power can be reduced by increasing fiber length, but only at the expense of a slower response time limited by Eq. (6.2.15). In one experiment [15], $P_p = 0.39$ W was measured for $L = 580$ m and $A_{\text{eff}} = 22$ μm^2. In another experiment [21], the effective core area was reduced to 2 μm^2, and a semiconductor laser operating at 1.3 μm was used as a pump. A phase shift of 17° was realized at a pump power of only 27 mW. The estimated value of $P_pL = 11$ W-m for this experiment indicates that pump powers ~ 50 mW may be sufficient for 100% probe transmission if 200-m-long fibers were used in each arm of a Mach–Zehnder interferometer.

Equation (6.2.16) can be used to estimate the Kerr coefficient n_{2B}. Most measurements indicate $n_{2B} \approx 4 \times 10^{-16}$ cm^2/W with an experimental uncertainty of $\sim 20\%$ [13]–[21]. This value is in agreement with Eq. (6.2.13) if we use $n_2 \approx 3 \times 10^{-16}$ cm^2/W and $b = \frac{1}{3}$. The parameter b has been measured in an experiment [18] designed to allow an independent measurement of the susceptibility ratio indicated in Eq. (6.2.11). The measured value $b = 0.34$

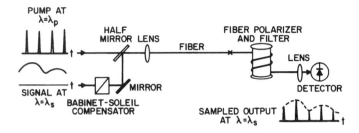

Figure 6.2 Schematic diagram of an all-optical Kerr shutter used for optical sampling. (After Ref. [16].)

suggests that the electronic contribution to $\chi^{(3)}$ dominates in silica fibers. This conclusion is in agreement with the measurements made using bulk glasses [5].

On the practical side, an all-fiber Kerr shutter has been used for optical sampling [16]. Figure 6.2 shows the experimental set up schematically. A Babinet–Soleil compensator was used to compensate for modal birefringence of the fiber. A highly birefringent piece of fiber was used as a polarizer with about 20-dB extinction ratio. It also served as a filter because fiber losses were quite high at the 1.06-μm pump wavelength. A laser diode at 0.84 μm served as the probe. The sampled probe output was in the form of a sequence of pulses whose separation and width were determined by pump pulses. In this experiment, pump pulses were fairly long ($\sim$ 300 ps). In a different experiment, 30-ps probe pulses at a repetition rate of 1.97 GHz (obtained from a 1.3-μm, gain-switched, distributed feedback semiconductor laser) were demultiplexed using 85-ps pump pulses from a mode-locked Nd:YAG laser [18].

In most experiments on Kerr shutters, it is generally necessary to use bulky high-power lasers to realize optical switching in silica fibers, making practical use of such devices difficult. As evident from Eq. (6.2.16), the product P_pL can be reduced considerably if optical fibers made with a high-nonlinearity material are used in place of silica fibers. Chalcogenide glasses offer such an opportunity because their nonlinear parameter n_2 is larger by a factor $\sim$100 compared with silica. Several experiments have shown that chalcogenide glass fibers offer a solution to making practical nonlinear Kerr shutters operating at high speeds [22]–[24]. In a 1992 experiment, a 1.319-μm mode-locked Nd:YAG laser in combination with a pulse compressor provided pump pulses of widths in the range 2.5–40 ps at the 100-MHz repetition rate [22]. The fiber length was kept < 1 m to avoid large losses associated with the As_2S_3-based chalcogenide fiber. In spite of such a small interaction length, the required pump power for optical switching was only ~ 5 W.

In a later experiment, all-optical switching was achieved using a semiconductor laser as a pump source [23]. Gain switching of a distributed feedback semiconductor laser, in combination with pulse compression, provided 8.2-ps pump pulses at the 100-MHz repetition rate. The peak power of pump pulses was increased to 13.9 W using an erbium-doped fiber amplifier. For a 1-m-long fiber the switched signal pulse had nearly the same width as the pump pulse, demonstrating ultrafast switching on a picosecond time scale. Signal pulses could be switched through the Kerr effect even when the signal was in the form of a 100-GHz pulse train, indicating the potential of a Kerr shutter for demultiplexing a 100-Gb/s communication channel.

6.2.3 Pulse Shaping

Nonlinear birefringence induced by an intense pulse can be used to modify the shape of the same pulse, even in the absence of a pump pulse, because its transmission through a combination of fiber and polarizer is generally intensity dependent. As a result, such a device can block low-intensity tails of a pulse while passing the central intense part of the same pulse. This phenomenon can be used to remove the low-intensity pedestal associated with some compressed pulses [25]–[27]. It can also be used to make fiber-optic logic gates [28].

The operating principle of an intensity discriminator is similar to that of the Kerr shutter shown in Fig. 6.1. The main difference is that instead of a pump, the signal pulse itself produces nonlinear birefringence and modifies its own state of polarization. To understand the physics behind such a device as simply as possible, let us neglect the GVD effects and use the nondispersive XPM theory of Section 6.2.1. Consider the case of an input beam linearly polarized at an angle θ with respect to one of the principal axes (x axis) of the fiber. The relative phase shift introduced between the two polarization components is then given by Eq. (6.2.7). This phase shift allows some power to be transmitted through the polarizer when $\theta \neq 0$. The transmittivity T_p is obtained by noting that

$$A_x = \sqrt{P_0}\cos\theta \exp(i\Delta\phi_{\mathrm{NL}}), \qquad A_y = \sqrt{P_0}\sin\theta, \qquad (6.2.17)$$

where $\Delta\phi_{\mathrm{NL}}$ is the nonlinear phase shift. Because the cross polarizer makes an angle $\pi/2 + \theta$ from the x axis, the total transmitted field becomes $A_t =$

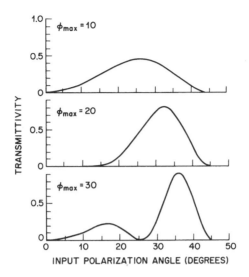

Figure 6.3 Transmittivity T_p as a function of input polarization angle θ for three different peak powers corresponding to $\phi_{max} = 10$, 20, and 30. (After Ref. [25].)

$\sqrt{P_0} \sin \theta \cos \theta [1 - \exp(i\Delta\phi_{NL})]$. As a result, T_p is given by [25]

$$T_p(\theta) = |A_t|^2/P_0 = \sin^2[(\gamma P_0 L/6) \cos(2\theta)] \sin^2(2\theta), \qquad (6.2.18)$$

where Eq. (6.2.7) was used. In the case of optical pulses propagating through the fiber, the product $\gamma P_0 L$ is related to the maximum phase shift ϕ_{max} induced by SPM [see Eq. (4.1.6)] and can also be related to the nonlinear length scale L_{NL} through the relation

$$\phi_{max} = \gamma P_0 L = L/L_{NL}. \qquad (6.2.19)$$

Pulse shaping occurs because T_p is power dependent at a given angle θ. If the angle θ is set to maximize the transmission at the pulse peak, the wings are removed because of their relatively low power levels. As a result, the output pulse becomes narrower than the input pulse. This behavior has been observed experimentally [26]. The optimum value of θ depends on the peak power P_0. Figure 6.3 shows T_p as a function of θ for three values of ϕ_{max}. The transmittivity can approach 90% at $\theta = 36.2°$ for $\phi_{max} = 30$.

Experimental results on pulse shaping indicate that the observed behavior does not always agree with Eq. (6.2.18). In particular, this equation predicts that $T_p = 0$ for $\theta = 45°$, i.e., the input light is blocked by the polarizer when the

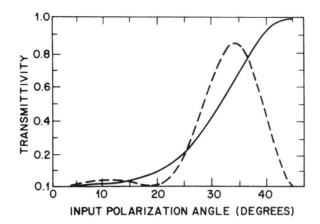

Figure 6.4 Transmittivity T_p as a function of input polarization angle θ when the effect of linear birefringence is included for $\Delta\beta L = 2\pi$ and $\phi_{max} = 6.5\pi$. Dashed line shows the behavior when $\Delta\beta = 0$. (After Ref. [29].)

two polarization components are excited with equal amplitudes. In practice, this is not the case. The reason for this discrepancy can be traced back to the neglect of the last term in Eqs. (6.1.11) and (6.1.12). A more accurate theory should include this term. In the CW or the quasi-CW case for which the dispersive effects are negligible, Eqs. (6.1.11) and (6.1.12) can be solved analytically by neglecting the time derivatives and the loss terms. The analytic solution is given in the next section. Its use shows that Eq. (6.2.18) is quite accurate in the case of highly birefringent fibers ($\Delta\beta L \gg 1$) except near $\theta = 45°$. In low-birefringence fibers, the transmittivity can be quite different than that given by Eq. (6.2.18). Figure 6.4 shows T_p as a function of θ for $\Delta\beta L = 2\pi$ and $\phi_{max} = 6.5\pi$. A comparison with the prediction of Eq. (6.2.18) reveals the importance of including linear birefringence. Physically, the linear and nonlinear birefringence contributions to the refractive index compete with each other, and both should be included.

6.3 Evolution of Polarization State

An accurate description of the nonlinear polarization effects in birefringent fibers requires simultaneous consideration of both the modal birefringence and self-induced nonlinear birefringence [29]–[46]. Evolution of the two polarization components along such fibers is governed by Eqs. (6.1.11) and (6.1.12) or their variants. However, before turning to the case of pulse propagation, it is useful to consider how the state of polarization evolves within the fiber when a CW (or quasi-CW) beam is launched at the input end.

6.3.1 Analytic Solution

In place of Eqs. (6.1.11) and (6.1.12), it is more convenient to use Eqs. (6.1.15) and (6.1.16) written in terms of the circularly polarized components. The terms containing time derivatives can be set to zero in the quasi-CW case. If we also neglect fiber losses, Eqs. (6.1.15) and (6.1.16) reduce to

$$\frac{dA_+}{dz} = \frac{i}{2}(\Delta\beta)A_- + \frac{2i\gamma}{3}(|A_+|^2 + 2|A_-|^2)A_+, \tag{6.3.1}$$

$$\frac{dA_-}{dz} = \frac{i}{2}(\Delta\beta)A_+ + \frac{2i\gamma}{3}(|A_-|^2 + 2|A_+|^2)A_-. \tag{6.3.2}$$

Consider first the low-power case and neglect the nonlinear effects ($\gamma = 0$). The resulting linear equations are easily solved. As an example, when the input beam with power P_0 is σ_+-polarized, the solution is given by

$$A_+(z) = \sqrt{P_0}\cos(\pi z/L_B), \qquad A_-(z) = i\sqrt{P_0}\sin(\pi z/L_B), \tag{6.3.3}$$

where the beat length $L_B = 2\pi/(\Delta\beta)$. The state of polarization is generally elliptical and evolves periodically with a period equal to the beat length. The ellipticity and the azimuth of the polarization ellipse at any point along the fiber can be obtained using

$$e_p = \frac{|A_+| - |A_-|}{|A_+| + |A_-|}, \qquad \theta = \frac{1}{2}\tan^{-1}\left(\frac{A_+}{A_-}\right). \tag{6.3.4}$$

Equations (6.3.1) and (6.3.2) can be solved analytically even when nonlinear effects become important. For this purpose, we use

$$A_\pm = \left(\frac{3\Delta\beta}{2\gamma}\right)^{1/2}\sqrt{p_\pm}\exp(i\phi_\pm), \tag{6.3.5}$$

and obtain the following three equations satisfied by the normalized powers p_+ and p_- and the phase difference $\psi \equiv \phi_+ - \phi_-$:

$$\frac{dp_+}{dZ} = \sqrt{2p_+p_-}\sin\psi, \tag{6.3.6}$$

$$\frac{dp_-}{dZ} = -\sqrt{2p_+p_-}\,\sin\psi, \tag{6.3.7}$$

$$\frac{d\psi}{dZ} = \frac{p_- - p_+}{\sqrt{p_+p_-}}\cos\psi + 2(p_- - p_+), \tag{6.3.8}$$

where $Z = (\Delta\beta)z/2$. These equations have the following two quantities that remain constant along the fiber [43]:

$$p = p_+ + p_-, \qquad \Gamma = \sqrt{p_+p_-}\cos\psi + p_+p_-. \tag{6.3.9}$$

Note that p is related to the total power P_0 launched into the fiber through $p = P_0/P_{cr}$, where P_{cr} is obtained from Eq. (6.3.5) and is given by

$$P_{cr} = 3|\Delta\beta|/(2\gamma). \tag{6.3.10}$$

Because of the two constants of motion, Eqs. (6.3.6)–(6.3.8) can be solved analytically in terms of the elliptic functions. The solution for p_+ is [32]

$$p_+(z) = \tfrac{1}{2}p - \sqrt{m|q|}\,\mathrm{cn}(x), \tag{6.3.11}$$

where $\mathrm{cn}(x)$ is a Jacobian elliptic function with the argument

$$x = \sqrt{|q|}(\Delta\beta)z + K(m), \tag{6.3.12}$$

$K(m)$ is the quarter period, and m and q are defined as

$$m = \tfrac{1}{2}[1 - \mathrm{Re}(q)/|q|], \qquad q = 1 + p\exp(i\psi_0). \tag{6.3.13}$$

Here ψ_0 is the value of ψ at $z = 0$. Both $p_-(z)$ and $\psi(z)$ can be obtained in terms of $p_+(z)$ using Eq. (6.3.9). The ellipticity and the azimuth of the polarization ellipse at any point along the fiber are then obtained from Eq. (6.3.4) after noting that $\theta = \psi/2$.

It is useful to display evolution of the polarization state as trajectories in the ellipticity–azimuth phase plane. Figure 6.5 shows such phase-space trajectories in the cases of (a) low input power ($p \ll 1$) and (b) high input power ($p = 3$). In the low-power case, all trajectories close, indicating oscillatory evolution of the polarization state [see Eq. (6.3.3)]. However, at power levels such that $p > 1$, a "seperatrix" divides the phase space into two distinct regions. In the region near $e_p = 0$ and $\theta = 0$ (light polarized close to the slow axis), trajectories form closed orbits, and polarization evolution is qualitatively

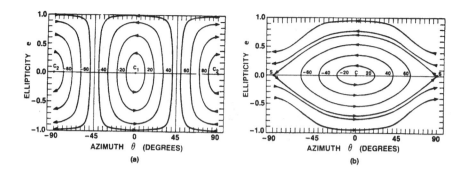

Figure 6.5 Phase-space trajectories representing evolution of the polarization state along fiber for (a) $p \ll 1$ and (b) $p = 3$. (After Ref. [32].)

similar to the low-power case. However, when light is polarized close to the fast axis, nonlinear rotation of the polarization ellipse leads to qualitatively different behavior because the fast axis corresponds to an unstable saddle point.

One can use the analytic solution to find the "fixed points" in the phase space. A fixed point represents a polarization state that does not change as light propagates inside the fiber. Below the critical power ($p < 1$), light polarized linearly ($e_p = 0$) along the slow and fast axes ($\theta = 0$ and $\pi/2$ represents two stable fixed points. At the critical power ($p = 1$),the fast-axis fixed point exhibits a pitchfork bifurcation. Beyond this power level, the linear-polarization state along the fast axis becomes unstable, but two new elliptically polarized states emerge as fixed points. These new polarization eigenstates are discussed next using the Poincaré-sphere representation.

6.3.2 Poincaré-Sphere Representation

An alternative approach to describe evolution of the polarization state in optical fibers is based on the rotation of the Stokes vector on the Poincaré sphere [31]. In this case, it is better to write Eqs. (6.3.1) and (6.3.2) in terms of linearly polarized components using Eq. (6.1.14). The resulting equations are

$$\frac{d\bar{A}_x}{dz} - \frac{i}{2}(\Delta\beta)\bar{A}_x = \frac{2i\gamma}{3}\left(|\bar{A}_x|^2 + \frac{2}{3}|\bar{A}_y|^2\right)\bar{A}_x + \frac{i\gamma}{3}\bar{A}_x^*\bar{A}_y^2, \quad (6.3.14)$$

$$\frac{d\bar{A}_y}{dz} + \frac{i}{2}(\Delta\beta)\bar{A}_y = \frac{2i\gamma}{3}\left(|\bar{A}_y|^2 + \frac{2}{3}|\bar{A}_x|^2\right)\bar{A}_y + \frac{i\gamma}{3}\bar{A}_y^*\bar{A}_x^2. \quad (6.3.15)$$

These equations can also be obtained from Eqs. (6.1.11) and (6.1.12).

At this point, we introduce the four real variables known as the Stokes parameters and defined as

$$S_0 = |\bar{A}_x|^2 + |\bar{A}_y|^2, \qquad S_1 = |\bar{A}_x|^2 - |\bar{A}_y|^2,$$
$$S_2 = 2\,\text{Re}(\bar{A}_x^* \bar{A}_y), \qquad S_3 = 2\,\text{Im}(\bar{A}_x^* \bar{A}_y), \qquad (6.3.16)$$

and rewrite Eqs. (6.3.14) and (6.3.15) in terms of them. After considerable algebra, we obtain

$$\frac{dS_0}{dz} = 0, \qquad\qquad \frac{dS_1}{dz} = \frac{2\gamma}{3} S_2 S_3, \qquad (6.3.17)$$

$$\frac{dS_2}{dz} = -(\Delta\beta) S_3 - \frac{2\gamma}{3} S_1 S_3, \qquad \frac{dS_3}{dz} = (\Delta\beta) S_2. \qquad (6.3.18)$$

It can be easily verified from Eq. (6.3.16) that $S_0^2 = S_1^2 + S_2^2 + S_3^2$. As S_0 is independent of z from Eq. (6.3.17), the Stokes vector $\mathbf{S}$ with components S_1, S_2, and S_3 moves on the surface of a sphere of radius S_0 as the CW light propagates inside the fiber. This sphere is known as the Poincaré sphere and provides a visual description of the polarization state. In fact, Eqs. (6.3.17) and (6.3.18) can be written in the form of a single vector equation as [31]

$$\frac{d\mathbf{S}}{dz} = \mathbf{W} \times \mathbf{S}, \qquad (6.3.19)$$

where the vector $\mathbf{W} = \mathbf{W}_L + \mathbf{W}_{NL}$ such that

$$\mathbf{W}_L = (\Delta\beta, 0, 0), \qquad \mathbf{W}_{NL} = (0, 0, -2\gamma S_3/3). \qquad (6.3.20)$$

Equation (6.3.19) includes both linear and nonlinear birefringence. It describes evolution of the polarization state of a CW optical field within the fiber under quite general conditions.

Figure 6.6 shows motion of the Stokes vector on the Poincaré sphere in several different cases. In the low-power case, nonlinear effects can be neglected by setting $\gamma = 0$. As $\mathbf{W}_{NL} = 0$ in that case, the Stokes vector rotates around the S_1 axis with an angular velocity $\Delta\beta$ (upper left sphere in Fig. 6.6). This rotation is equivalent to the periodic solution given in Eq. (6.3.3) obtained earlier. If the Stokes vector is initially oriented along the S_1 axis, it remains fixed. This can also be seen from the steady-state (z-invariant) solution of Eqs. (6.3.17) and (6.3.18) because $(S_0, 0, 0)$ and $(-S_0, 0, 0)$ represent their fixed points. These

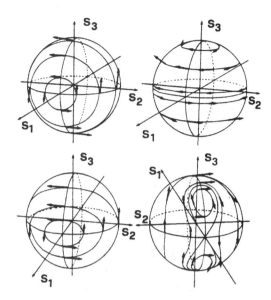

Figure 6.6 Trajectories showing motion of the Stokes vector on the Poincaré sphere. (a) Linear birefringence case (upper left); (b) nonlinear case with $\Delta\beta = 0$ (upper right); (c) mixed case with $\Delta\beta > 0$ and $P_0 > P_{cr}$ (lower row). Left and right spheres in the bottom row show the front and back of the Poincaré sphere. (After Ref. [31].)

two locations of the Stokes vector correspond to the linearly polarized incident light oriented along the slow and fast axes, respectively.

In the purely nonlinear case of isotropic fibers ($\Delta\beta = 0$), $\mathbf{W}_L = 0$. The Stokes vector now rotates around the the S_3 axis with an angular velocity $2\gamma S_3/3$ (upper right sphere in Fig. 6.6). This is referred to as self-induced ellipse rotation because it has its origin in the nonlinear birefringence. Two fixed points in this case correspond to the north and south poles of the Poincaré sphere and represent right and left circular polarizations, respectively.

In the mixed case, the behavior depends on the power level of the incident light. As long as $P_0 < P_{cr}$, nonlinear effects play a minor role, and the situation is similar to the linear case. At higher powers levels, the motion of the Stokes vector on the Poincaré sphere becomes quite complicated because $\mathbf{W}_L$ is oriented along the S_1 axis while $\mathbf{W}_{NL}$ is oriented along the S_3 axis. Moreover, the nonlinear rotation of the Stokes vector along the S_3 axis depends on the magnitude of S_3 itself. The bottom row in Fig. 6.6 shows motion of the Stokes vector on the front and back of the Poincaré sphere in the case $P_0 > P_{cr}$. When input light is polarized close to the slow axis (left sphere), the situation is similar to the linear case. However, the behavior is qualitatively different when input light is polarized close to the fast axis (right sphere).

To understand this asymmetry, let us find the fixed points of Eqs. (6.3.17) and (6.3.18) by setting the z derivatives to zero. The location and number of fixed points depend on the beam power P_0 launched inside the fiber. More specifically, the number of fixed points changes from two to four at a critical power level P_{cr} defined as in Eq. (6.3.10). For $P_0 < P_{cr}$, only two fixed points, $(S_0, 0, 0)$ and $(-S_0, 0, 0)$, occur; these are identical to the low-power case. In contrast, when $P_0 > P_{cr}$, two new fixed points emerge. The components of the Stokes vector at the location of the new fixed points on the Poincaré sphere are given by [45]

$$S_1 = -P_{cr}, \quad S_2 = 0, \quad S_3 = \pm\sqrt{P_0^2 - P_{cr}^2}. \tag{6.3.21}$$

These two fixed points correspond to elliptically polarized light and occur on the back of the Poincaré sphere in Fig. 6.6 (lower right). At the same time, the fixed point $(-S_0, 0, 0)$, corresponding to light polarized linearly along the fast axis, becomes unstable. This is equivalent to the pitchfork bifurcation discussed earlier. If the input beam is polarized elliptically with its Stokes vector oriented as indicated in Eq. (6.3.21), the polarization state will not change inside the fiber. When the polarization state is close to the new fixed points, the Stokes vector forms a close loop around the elliptically polarized fixed point. This behavior corresponds to the analytic solution discussed earlier. However, if the polarization state is close to the unstable fixed point $(-S_0, 0, 0)$, small changes in input polarization can induce large changes at the output. This issue is discussed next.

6.3.3 Polarization Instability

The polarization instability manifests as large changes in the output state of polarization when the input power or the polarization state of a CW beam is changed slightly [31]–[33]. The presence of polarization instability shows that slow and fast axes of a polarization-preserving fiber are not entirely equivalent.

The origin of polarization instability can be understood from the following qualitative argument [32]. When the input beam is polarized close to the slow axis (x axis if $n_x > n_y$), nonlinear birefringence adds to intrinsic linear birefringence, making the fiber more birefringent. By contrast, when the input beam is polarized close to the fast axis, nonlinear effects decrease total birefringence by an amount that depends on the input power. As a result, the fiber

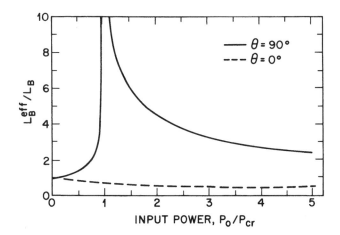

Figure 6.7 Effective beat length as a function of input power for beams polarized along the fast (solid line) and slow (dashed line) axes. (After Ref. [32].)

becomes less birefringent, and the effective beat length L_B^{eff} increases. At a critical value of the input power nonlinear birefringence can cancel intrinsic birefringence completely, and L_B^{eff} becomes infinite. With a further increase in the input power, the fiber again becomes birefringent but the roles of the slow and fast axes are reversed. Clearly large changes in the output polarization state can occur when the input power is close to the critical power necessary to balance the linear and nonlinear birefringences. Roughly speaking, the polarization instability occurs when the input peak power is large enough to make the nonlinear length L_{NL} comparable to the intrinsic beat length L_B.

The period of the elliptic function in Eq. (6.3.11) determines the effective beat length as [32]

$$L_B^{\text{eff}} = \frac{2K(m)}{\pi\sqrt{|q|}}L_B, \tag{6.3.22}$$

where L_B is the low-power beat length, $K(m)$ is the quarter-period of the elliptic function, and m and q are given by Eq. (6.3.13) in terms of the normalized input power defined as $p = P_0/P_{\text{cr}}$. In the absence of nonlinear effects, $p = 0$, $q = 1$, and we recover

$$L_B^{\text{eff}} = L_B = 2\pi/|\Delta\beta|. \tag{6.3.23}$$

Figure 6.7 shows how L_B^{eff} varies with p for $\theta = 0°$ and $\theta = 90°$. The effective beat length becomes infinite when $P_0 = P_{\text{cr}}$ and $\theta = 90°$ because of complete cancellation between the linear and nonlinear birefringences [33].

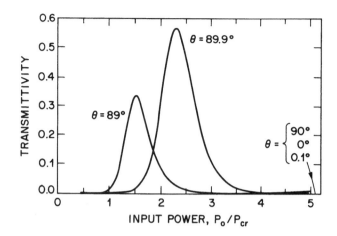

Figure 6.8 Transmittivity of a birefringent fiber of length $L = L_B$ as a function of input power for different input angles. (After Ref. [32].)

This is the origin of polarization instability. The critical power level P_{cr} at which L_B^{eff} becomes infinite is the same at which the number of fixed points on the Poincaré sphere changes from 2 to 4. Thus, polarization instability can be interpreted in terms of the emergence of elliptically polarized fixed points on the Poincaré sphere. The two viewpoints are identical.

As a result of large changes in L_B^{eff}, the output polarization state can change drastically when P_0 is close to P_{cr} and the input beam is polarized close to the fast axis. Figure 6.8 shows the transmittivity T_p as a function of the input power for several values of θ after assuming that a crossed polarizer at the fiber output blocks the light at low intensities (see Fig. 6.1). When $\theta = 0°$ or $90°$, T_p remains zero at all power levels. Small changes in θ near the slow axis still keep T_p near zero. However, T_p changes dramatically when θ is changed slightly near the fast axis. Note the extreme sensitivity of T_p to the input polarization angle as θ is varied from $89°$ to $90°$. Figure 6.8 is drawn for the case $(\Delta\beta)L = 2\pi$ or $L = L_B$. However, the qualitative behavior remains the same for other fiber lengths as well.

Polarization instability was first observed in 1986 by transmitting 80-ps pulses (at 532 nm) through a 53-cm-long fiber with a measured intrinsic beat length $L_B \approx 50$ cm [35]. The input pulses were right-circularly polarized and passed through a circular analyzer at the fiber output that transmitted only left-circularly polarized light. The shape of output pulses was found to change dramatically when the peak power exceeded a critical value. The measured critical power and the output pulse shapes were in agreement with the theoretical predictions. In a later experiment, polarization instability led to significant enhancement of weak intensity modulations when the input signal was polarized near the fast axis of a low-birefringence fiber [42]. The 200-ns input pulses were obtained from a Q-switched Nd:YAG laser operating at 1.06 μm. The intensity of these pulses exhibited 76-MHz modulation because of longitudinal-mode beating inside the laser. These small-amplitude modulations were unaffected when the signal was polarized near the slow axis of the fiber, but became amplified by as much as a factor of 6 when input pulses were polarized near the fast axis. The experimental results were in good qualitative agreement with theory, especially when the theory was generalized to include twisting of the fiber that resulted in elliptical birefringence [43].

The power-dependent transmittivity seen in Fig. 6.8 can be useful for optical switching. Self switching of linearly polarized beams induced by polarization instability has been demonstrated in silica fibers [44]. It can also be used to switch the polarization state of an intense beam through a weak pulse. Polarization switching can also occur for solitons [46]. In all case, the input power required for switching is quite large unless fibers with low modal birefringence are used. A fiber with a beat length $L_B = 1$ m requires $P_0 \sim 1$ kW if we use $\gamma = 10$ W^{-1}/km in Eq. (6.3.10). This value becomes larger by a factor of 100 or more when high-birefringence fibers are used. For this reason, polarization instability is not of concern when highly birefringent fibers are used because P_0 remains < 1 kW in most experiments.

6.3.4 Polarization Chaos

Polarization instability can lead to chaos in the output polarization state if linear birefringence of a fiber is modulated along its length. This can occur if the fiber is uniformly twisted while being wound onto a drum. Modulated birefringence can also be introduced during fiber fabrication through periodic rocking of the preform or by means of a periodic distribution of stress. The effects of modulated linear birefringence on evolution of the polarization state have been studied [37]–[40]. This section considers twisted fibers briefly.

Twisting of birefringent fibers produces two effects simultaneously. First, the principal axes are no longer fixed but rotate in a periodic manner along the fiber length. Second, shear strain induces circular birefringence in proportion to the twist rate. When both of these effects are included, Eqs. (6.3.1) and

(6.3.2) take the following form [43]:

$$\frac{dA_+}{dz} = ib_c A_+ + \frac{i}{2}(\Delta\beta)e^{2ir_t z}A_- + \frac{2i\gamma}{3}(|A_+|^2 + 2|A_-|^2)A_+, \quad (6.3.24)$$

$$\frac{dA_-}{dz} = ib_c A_- + \frac{i}{2}(\Delta\beta)e^{-2ir_t z}A_+ + \frac{2i\gamma}{3}(|A_-|^2 + 2|A_+|^2)A_-, \quad (6.3.25)$$

where $b_c = hr_t/2\bar{n}$ is related to circular birefringence, r_t is the twist rate per unit length, and $\bar{n}$ is the average mode index. The parameter h has a value of ~ 0.15 for silica fibers. The preceding equations can be used to find the fixed points, as done in Section 6.3.1 for an untwisted fiber. Above a critical power level, we again find four fixed points. As a result, polarization instability still occurs along the fast axis but the critical power becomes larger.

Birefringence modulation can also be included by making the parameter $\Delta\beta$ in Eqs. (6.3.1) and (6.3.2) a periodic function of z such that $\Delta\beta = \Delta\beta_0[1 - i\varepsilon\cos(b_m z)]$, where ε is the amplitude and b_m is the spatial frequency of modulation [40]. The resulting equations can not be solved analytically but one can use the phase-space or the Poincaré-sphere approach to study evolution of the polarization state approximately [37]–[40]. This approach shows that the motion of the Stokes vector on the Poincaré sphere becomes chaotic in the sense that polarization does not return to its original state after each successive period of modal birefringence $\Delta\beta$. Such studies are useful for estimating the range of parameter values that must be maintained to avoid chaotic switching if the fiber were to be used as an optical switch.

6.4 Vector Modulation Instability

This section extends the scalar analysis of Section 5.1 to the vector case in which a CW beam, when launched into a birefringent fiber, excites both polarization components simultaneously. Similar to the scalar case, modulation instability is expected to occur in the anomalous-GVD region of the fiber. The main issue is whether the XPM-induced coupling can destabilize the CW state even when the wavelength of the CW beam is in the normal-GVD regime of the fiber. Vector modulation instability in an isotropic nonlinear medium (no birefringence) was studied as early as 1970 using the coupled NLS equations [47]. In the context of birefringent fibers, it has been studied extensively since 1988, both theoretically and experimentally [48]–[63]. Since the qualitative behavior is different for weakly and strongly birefringent fibers, we consider the two cases separately.

6.4.1 Low-Birefringence Fibers

In the case of low-birefringence fibers, one must retain the coherent-coupling term in Eqs. (6.1.11) and (6.1.12) while considering modulation instability [48]. As before, it is easier to use Eqs. (6.1.15) and (6.1.16), written in terms of the circularly polarized components of the optical field. The steady-state or CW solution of these equations is given in Section 6.3 but is quite complicated to use for the analysis of modulation instability as it involves elliptic functions. The problem becomes tractable when the polarization state of the incident CW beam is oriented along a principal axis of the fiber.

Consider first the case in which the polarization state is oriented along the fast axis ($A_x = 0$). This case is especially interesting because the polarization instability discussed in Section 6.3 can also occur. If fiber losses are neglected by setting $\alpha = 0$, the steady-state solution becomes

$$\bar{A}_\pm(z) = \pm i\sqrt{P_0/2}\exp(i\gamma P_0 z), \tag{6.4.1}$$

where P_0 is the input power. Following the procedure of Section 5.1, stability of the steady state is examined by assuming a solution in the form

$$A_\pm(z,t) = \pm[i\sqrt{P_0/2} + a_\pm(z,t)]\exp(i\gamma P_0 z), \tag{6.4.2}$$

where $a_\pm(z,t)$ is a small perturbation. Using Eq. (6.4.2) in Eqs. (6.1.15) and (6.1.16) and linearizing in a_+ and a_-, we obtain a set of two coupled linear equations. These equations can be solved by assuming a solution of the form

$$a_\pm = u_\pm \exp[i(Kz - \Omega t)] + iv_\pm \exp[-i(Kz - \Omega t)], \tag{6.4.3}$$

where K is the wave number and Ω is the frequency of perturbation. We then obtain a set of four algebraic equations for $u_\pm$ and $v_\pm$. This set has a non-trivial solution only when the perturbation satisfies the following dispersion relation [48]

$$[(K - \beta_1\Omega)^2 - C_1][(K - \beta_1\Omega)^2 - C_2] = 0, \tag{6.4.4}$$

where

$$C_1 = \tfrac{1}{2}\beta_2\Omega^2(\tfrac{1}{2}\beta_2\Omega^2 + 2\gamma P_0), \tag{6.4.5}$$

$$C_2 = (\tfrac{1}{2}\beta_2\Omega^2 + \Delta\beta - 2\gamma P_0/3)(\tfrac{1}{2}\beta_2\Omega^2 + \Delta\beta). \tag{6.4.6}$$

As discussed in Section 5.1, the steady-state solution becomes unstable if the wave number K has an imaginary part for some values of Ω, indicating that a perturbation at that frequency would grow exponentially along the fiber with the power gain $g = 2\mathrm{Im}(K)$. The nature of modulation instability depends strongly on whether the input power P_0 is below or above the polarization-instability threshold P_{cr} given in Eq. (6.3.10). For $P_0 < P_{cr}$, modulation instability occurs only in the case of anomalous dispersion, and the results are similar to those of Section 5.1. The effect of XPM is to reduce the gain from that of Eq. (5.1.9) but the maximum gain occurs at the same value of Ω (see Fig. 5.1).

It is easy to deduce from Eq. (6.4.4) that modulation instability can occur even in the normal-dispersion regime of the fiber ($\beta_2 > 0$) provided $C_2 < 0$. This condition is satisfied for frequencies in the range $0 < |\Omega| < \Omega_{c1}$, where

$$\Omega_{c1} = (4\gamma/3\beta_2)^{1/2}\sqrt{P_0 - P_{cr}}. \tag{6.4.7}$$

Thus, $P_0 > P_{cr}$ is required for modulation instability to occur in the normal-dispersion regime of the fiber. When this condition is satisfied, the gain is given by

$$g(\Omega) = |\beta_2|\sqrt{(\Omega_{c2}^2 + \Omega^2)(\Omega_{c1}^2 - \Omega^2)}. \tag{6.4.8}$$

where

$$\Omega_{c2} = (2\Delta\beta/\beta_2)^{1/2}. \tag{6.4.9}$$

Consider now the case in which the CW beam is polarized along the slow axis ($A_y = 0$). We can follow essentially the same steps to find the dispersion relation $K(\Omega)$. In fact, Eqs. (6.4.4)–(6.4.6) remain applicable if we change the sign of $\Delta\beta$. Modulation instability can still occur in the normal-dispersion regime of the fiber but the gain exists only for frequencies in the range $\Omega_{c2} < |\Omega| < \Omega_{c3}$, where

$$\Omega_{c3} = (4\gamma/3\beta_2)^{1/2}\sqrt{P_0 + P_{cr}}. \tag{6.4.10}$$

The instability gain is now given by

$$g(\Omega) = |\beta_2|\sqrt{(\Omega_{c2}^2 - \Omega^2)(\Omega_{c3}^2 - \Omega^2)}. \tag{6.4.11}$$

Figure 6.9 compares the gain spectra for light polarized along the slow and fast axes using $\beta_2 = 60$ ps²/km and $\gamma = 23$ W⁻¹/km for a fiber with a beat length $L_B = 5.8$ m. For these parameter values, $p = 1$ at an input power of 70 W. At a power level of 112 W, $p = 1.6$ (left part) while $p > 2$ at 152 W (right part). The most noteworthy feature of Fig. 6.9 is that, in contrast with

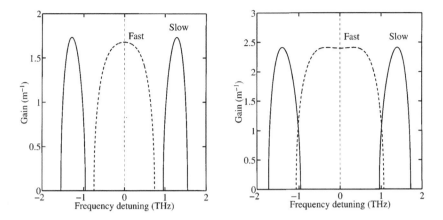

Figure 6.9 Gain spectra of modulation instability at power levels of 112 W (left) and 152 W (right) for a CW beam polarized along the slow or fast axis of a low-birefringence fiber. (After Ref. [60].)

the gain spectra of Fig. 5.1, the gain does not vanish near $\Omega = 0$ when light is polarized along the fast axis. When $p < 2$, the gain is in fact maximum at $\Omega = 0$, indicating that low-frequency or CW fluctuations would grow rapidly. This is a manifestation of the polarization instability discussed in Section 6.3, occurring only when the input beam is polarized along the fast axis. When $p > 2$, the gain peak occurs for a finite value of Ω. In that case, the CW beam would develop spectral sidebands irrespective of whether it is polarized along the slow or fast axis. This situation is similar to the scalar case of Section 5.1. The new feature is that such sidebands can develop even in the normal-GVD regime of a birefringent fiber.

6.4.2 High-Birefringence Fibers

For high-birefringence fibers, the last term representing coherent coupling (or four-wave mixing) can be neglected in Eqs. (6.1.11) and (6.1.12). These equations then reduce to Eqs. (6.1.22) and (6.1.23) with $B = \frac{2}{3}$ and exhibit a different kind of modulation instability [50]–[53]. This case is mathematically similar to the two-wavelength case discussed in Chapter 7.

To obtain the steady-state solution, the time derivatives in Eqs. (6.1.22) and (6.1.23) can be set to zero. If fiber losses are neglected by setting $\alpha = 0$, the steady-state solution is given by (see Section 6.2.1)

$$A_x(z) = \sqrt{P_x}\exp[i\phi_x(z)], \qquad A_y(z) = \sqrt{P_y}\exp[i\phi_y(z)], \qquad (6.4.12)$$

where P_x and P_y are the constant mode powers and

$$\phi_x(z) = \gamma(P_x + BP_y)z, \qquad \phi_y(z) = \gamma(P_y + BP_x)z. \qquad (6.4.13)$$

The phase shifts depend on the powers of both polarization components. In contrast with the case of weakly birefringent fibers, this solution is valid for a CW beam polarized at an arbitrary angle with respect to the slow axis.

Stability of the steady state is examined assuming a time-dependent solution of the form

$$A_j = \left(\sqrt{P_j} + a_j\right)\exp(i\phi_j), \qquad (6.4.14)$$

where $a_j(z,t)$ is a weak perturbation with $j = x, y$. We substitute Eq. (6.4.14) in Eqs. (6.1.22) and (6.1.23) and linearize them with respect to a_x and a_y. The resulting linear equations can again be solved in the form

$$a_j = u_j\exp[i(Kz - \Omega t)] + iv_j\exp[-i(Kz - \Omega t)], \qquad (6.4.15)$$

where $j = x, y$, K is the wave number, and Ω is the frequency of perturbation.

To simplify the algebra, let us focus on the case in which the input CW beam is polarized at $45°$ from the slow axis. As a result, both polarization modes have equal powers ($P_x = P_y = P$). The dispersion relation in this case can be written as [50]

$$[(K - b)^2 - H][(K + b)^2 - H] = C_X^2, \qquad (6.4.16)$$

where $b = (\beta_{1x} - \beta_{1y})\Omega/2$ takes into account the group-velocity mismatch,

$$H = \beta_2\Omega^2(\beta_2\Omega^2/4 + \gamma P), \qquad (6.4.17)$$

and the XPM coupling parameter C_X is defined as

$$C_X = B\beta_2\gamma P\Omega^2. \qquad (6.4.18)$$

As before, modulation instability occurs when K becomes complex for some values of Ω. Its gain is obtained from $g = 2\,\mathrm{Im}(K)$.

The most important conclusion drawn from Eq. (6.4.16) is that modulation instability can occur irrespective of the sign of the GVD parameter. In the case

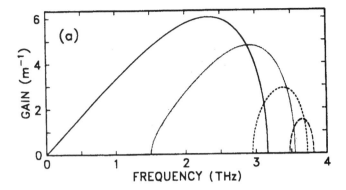

Figure 6.10 Gain spectra of modulation instability in a high-birefringence fiber when input beam is linearly polarized at 45° from the slow axis. Four curves correspond to (smallest to largest) for power levels of 60, 125, 250, and 500 W. (After Ref. [50].)

of normal GVD ($\beta_2 > 0$), the gain exists only if $C_X > |H - b|^2$. Figure 6.10 shows the gain spectra at several power levels using parameter values $\beta_2 = 65$ ps^2/km, $\gamma = 25.8$ W^{-1}/km, and a group-velocity mismatch of 1.6 ps/m. At low powers, the gain spectrum is quite narrow and its peak is located near $\Omega_m = |\beta_{1x} - \beta_{1y}|/\beta_2$. As input power increases, gain spectrum widens and its peak shifts to lower frequencies. In all four cases shown in Fig. 6.10, the CW beam develops temporal modulations at frequencies >2 THz as it propagates through the fiber. As Ω_m depends on birefringence of the fiber, it can be changed easily and provides a tuning mechanism for the modulation frequency. An unexpected feature is that modulation instability ceases to occur when the input power exceeds a critical value

$$P_c = 3(\beta_{1x} - \beta_{1y})^2/(4\beta_2\gamma). \tag{6.4.19}$$

Another surprising feature is that modulation instability ceases to occur when the input light is polarized close to a principal axis of the fiber [51].

Both of these features can be understood qualitatively if we interpret modulation instability in terms of a four-wave-mixing process that is phase matched by the modal birefringence of the fiber (see Section 10.3.3). In the case of normal GVD, the SPM- and XPM-induced phase shifts actually add to the GVD-induced phase mismatch. It is the fiber birefringence that cancels the phase mismatch. Thus, for a given value of birefringence, the phase-matching condition can only be satisfied if the nonlinear phase shifts remain below certain level. This is the origin of the critical power level in Eq. (6.4.19). An interesting feature of the four-wave-mixing process is that spectral sidebands are generated such that the low-frequency sideband at $\omega_0 - \Omega$ appears along the slow axis whereas the sideband at $\omega_0 + \Omega$ is polarized along the fast axis. This can also be understood from the phase-matching condition of Section 10.3.3.

6.4.3 Isotropic Fibers

As seen in the preceding, modal birefringence of fibers plays an important role for modulation instability to occur. A natural question is whether modulation instability can occur in isotropic fibers with no birefringence ($n_x = n_y$). Even though such fibers are hard to fabricate, fibers with extremely low birefringence ($|n_x - n_y| < 10^{-8}$) can be made by spinning the preform during the drawing stage. The question is also interesting from a fundamental standpoint and was discussed as early as 1970 [47].

The theory developed for high-birefringence fibers cannot be used in the limit $\Delta\beta = 0$ because the coherent-coupling term has been neglected. In contrast, theory developed for low-birefringence fibers remains valid in that limit. The main difference is that $P_{cr} = 0$ as polarization instability does not occur for isotropic fibers. As a result, $\Omega_{c2} = 0$ while $\Omega_{c1} = \Omega_{c3} \equiv \Omega_c$. The gain spectrum of modulation instability in Eq. (6.4.8) reduces to

$$g(\Omega) = |\beta_2\Omega|\sqrt{\Omega_c^2 - \Omega^2}, \tag{6.4.20}$$

irrespective of whether the input beam is polarized along the slow or fast axis. This is the same result obtained in Section 5.1 for the scalar case. It shows that the temporal and spectral features of modulation instability should not depend on the direction in which the input beam is linearly polarized. This is expected for any isotropic nonlinear medium on physical grounds.

The situation changes when the input beam is circularly or elliptically polarized. We can consider this case by setting $\Delta\beta = 0$ in Eqs. (6.1.15) and (6.1.16). Using $\alpha = 0$ for simplicity, these equations reduce to the following set of two coupled NLS equations [47]:

$$\frac{\partial A_+}{\partial z} + \frac{i\beta_2}{2}\frac{\partial^2 A_+}{\partial T^2} + i\gamma' \left(|A_+|^2 + 2|A_-|^2\right) A_+ = 0, \tag{6.4.21}$$

$$\frac{\partial A_-}{\partial z} + \frac{i\beta_2}{2}\frac{\partial^2 A_-}{\partial T^2} + i\gamma' \left(|A_-|^2 + 2|A_+|^2\right) A_- = 0, \tag{6.4.22}$$

where $T = t - \beta_1 z$ is the reduced time and $\gamma' = 2\gamma/3$. The steady-state solution of these equations is obtained easily and is given by

$$\bar{A}_\pm(z) = \sqrt{P_\pm}\exp(i\phi_\pm), \qquad (6.4.23)$$

where $P_\pm$ is the input power in the two circularly polarized components and $\phi_\pm(z) = \gamma'(P_\mp + 2P_\pm)z$ is the nonlinear phase shift.

As before, we perturb the steady-state solution using

$$A_\pm(z,t) = [\sqrt{P_\pm} + a_\pm(z,t)]\exp(i\phi_\pm), \qquad (6.4.24)$$

where $a_\pm(z,t)$ is a small perturbation. By using Eq. (6.4.24) in Eqs. (6.4.21) and (6.4.22) and linearizing in a_+ and a_-, we obtain a set of two coupled linear equations. These equations can be solved assuming a solution in the form of Eq. (6.4.3). We then obtain a set of four algebraic equations for $u_\pm$ and $v_\pm$. This set has a nontrivial solution only when the perturbation satisfies the following dispersion relation [47]

$$(K - H_+)(K - H_-) = C_X^2, \qquad (6.4.25)$$

where

$$H_\pm = \tfrac{1}{2}\beta_2\Omega^2(\tfrac{1}{2}\beta_2\Omega^2 + \gamma P_\pm), \qquad (6.4.26)$$

and the XPM coupling parameter C_X is now defined as

$$C_X = 2\beta_2\gamma\Omega^2\sqrt{P_+P_-}. \qquad (6.4.27)$$

A necessary condition for modulation instability to occur is $C_X^2 > H_+H_-$. As C_X depends on $\sqrt{P_+P_-}$ and vanishes for a circularly polarized beam, we can conclude that no instability occurs in that case. For an elliptically polarized beam, the instability gain depends on the ellipticity e_p defined as in Eq. (6.3.4).

6.4.4 Experimental Results

The vector modulation instability was first observed in the normal-dispersion region of a high-birefringence fiber [50]–[52]. In one experiment, 30-ps pulses at the 514-nm wavelength with 250-W peak power were launched into a 10-m fiber with a 45°-polarization angle [51]. At the fiber output, the pulse spectrum exhibited modulation sidebands with a 2.1-THz spacing, and the autocorrelation trace showed 480-fs intensity modulation. The observed sideband spacing was in good agreement with the value calculated theoretically. In another experiment, 600-nm input pulses were of only 9-ps duration [50]. As the 18-m-long fiber had a group-velocity mismatch of ≈ 1.6 ps/m, the two polarization components would separate from each other after only 6 m of fiber. The walk-off problem was solved by delaying the faster-moving polarization component by 25 ps at the fiber input. The temporal and spectral measurements indicated that both polarization components developed high-frequency (~ 3 THz) modulations, as expected from theory. Moreover, the modulation frequency decreased with an increase in the peak power. This experiment also revealed that each polarization component of the beam develops only one sideband, in agreement with theory. In a later experiment [52], modulation instability developed from temporal oscillations induced by optical wave breaking (see Section 4.2.3). This behavior can be understood from Fig. 4.12, noting that optical wave breaking manifests as spectral sidebands. If these sidebands fall within the bandwidth of the modulation-instability gain curve, their energy can seed the instability process.

Modulation instability in low-birefringence fibers was observed in 1995 using 60-ps pulses (with peak powers >1 kW) obtained from a krypton-ion laser operating at 647 nm [56]. Fibers used in the experiment were a few meters long, and their birefringence was controlled through stress induced by winding the fiber on a spool with a relatively small diameter. When input pulses were polarized along the slow axis, the two sidebands indicative of modulation instability had the same polarization and were polarized along the fast axis. Their spacing could be varied over a range of 20 nm or so by simply changing the spool size—a smaller spool diameter produced more stress-induced birefringence, resulting in larger sideband spacing. In a variation of this idea, fibers with periodically varying birefringence along their length were produced by wrapping the fiber around two spools [57]. Such a periodic variation can create new sidebands through quasi-phase matching, similar to the periodic variation of dispersion and nonlinearity discussed in Section 5.1.

A systematic study of induced modulation instability in low-birefringence fibers was performed in 1998 using a pump-probe configuration [60]. The probe beam was used to seed the process. In a series of experiments, the pump beam was obtained from a dye laser operating near 575 nm and consisted of 4-ns pulses that were wide enough to realize quasi-CW operation. The pump-probe wavelength separation was tunable; tuning allowed different regimes of modulation instability to be investigated. The fiber was drawn using a rapidly

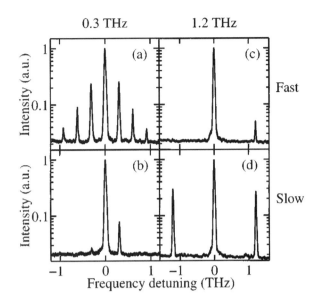

Figure 6.11 Modulation-instability sidebands observed in a low-birefringence fiber. Pump is polarized along the fast (top row) or slow axis (bottom row). Pump-probe detuning is 0.3 THz for left and 1.2 THz for right columns. (After Ref. [60].)

rotating preform so that its intrinsic birefringence averaged out to zero. A controlled amount of weak birefringence was introduced by winding the fiber onto a 14.5-cm-diameter spool. The measured beat length of 5.8 m for the fiber corresponded to a modal birefringence of only 10^{-7}. The critical power P_{cr} required for the onset of polarization instability [see Eq. (6.3.10)] was estimated to be 70 W for this fiber.

Figure 6.11 shows the pump spectra measured under several different experimental conditions. In all cases, the pump power was 112 W ($1.6P_{cr}$) while the probe power was kept low (~ 1 W). Consider first the case of a pump polarized along the fast axis (top row). For a pump-probe detuning of 0.3 THz, the probe frequency falls within the gain spectrum of modulation instability (see Fig. 6.9). As a result, the pump spectrum develops a series of sidebands spaced apart by 0.3 THz. In contrast, the probe frequency falls outside the gain spectrum for a detuning of 1.2 THz, and modulation instability does not occur. When the pump is polarized along the slow axis (bottom row), the situation is reversed. Now the 0.3-THz detuning falls outside the gain spectrum, and modulation-instability sidebands form only when the detuning is 1.2 THz.

These experimental results are in agreement with the theory given earlier. In the time domain, the pump pulse develops deep modulations that correspond to a train of dark solitons with repetition rates in the terahertz regime [59]. A dark-soliton is also formed when modulation instability occurs in high-birefringence fibers [61]. The formation of dark solitons is not surprising if we recall from Chapter 5 that optical fibers support only dark solitons in their normal-GVD regime.

In all of these experiments, fiber birefringence plays an important role. As discussed before, vector modulation instability can occur in isotropic fibers ($n_x = n_y$) such that the gain spectrum depends on the polarization state of the input CW beam. Unfortunately, it is difficult to make birefringence-free fibers. As an alternative, modulation instability was observed in a bimodal fiber in which the input beam excited two fiber modes (LP_{01} and LP_{11}) with nearly equal power levels, and the two modes had the same group velocity [58]. In a 1999 experiment [62], a nearly isotropic fiber was realized by winding 50 m of "spun" fiber with a large radius of curvature of 25 cm. The beat length for this fiber was ~ 1 km, indicating a birefringence level $< 10^{-8}$. Over the 50-m length used in the experiment, the fiber was nearly isotropic. Modulation-instability sidebands were observed when 230-ps pulses ($\lambda = 1.06$ μm) with a peak power of 120 W were launched into the fiber. The recorded spectra were almost identical when the polarization angle of linearly polarized light was changed over a 90° range. Sidebands disappeared for circularly polarized light. This behavior is expected since isotropic fibers have no preferred direction. When input light was elliptically polarized, the amplitude of spectral sidebands varied with the ellipticity, again in agreement with theory.

6.5 Birefringence and Solitons

The discussion of optical solitons in Chapter 5 neglected polarization effects and assumed implicitly that the fiber had no birefringence. The results presented there also apply for high-birefringence fibers when the input pulse is linearly polarized along one of the principal axes of a polarization-maintaining fiber. This section focuses on solitons forming when the input pulse is polarized at a finite angle from the slow axis [64]–[78]. There are two important issues. First, in a weakly birefringent fiber, the peak power of the soliton may exceed the critical power [see Eq. (6.3.10)] at which polarization instability occurs. This instability is likely to affect the solitons launched with their linear-polarization state aligned along the fast axis. Second, in a strongly bire-fringent fiber, the group-velocity mismatch between the two orthogonally po-larized components can lead to their physical separation within the fiber. Both of these issues are discussed in this section.

6.5.1 Low-Birefringence Fibers

Consider first the case of low-birefringence fibers. As the group-velocity mis-match is relatively small in such fibers, we can set $\beta_{1x} \approx \beta_{1y}$ in Eqs. (6.1.11) and (6.1.12) and use Eqs. (6.1.15) and (6.1.16) when circularly polarized com-ponents of the field are used in place of the linear ones. These equations can be scaled using the soliton units introduced in Section 5.2. The resulting coupled NLS equations take the form [64]

$$i\frac{\partial u_+}{\partial \xi} + \frac{i}{2}\frac{\partial^2 u_+}{\partial \tau^2} + bu_- + (|u_+|^2 + 2|u_-|^2)u_+ = 0, \qquad (6.5.1)$$

$$i\frac{\partial u_-}{\partial \xi} + \frac{i}{2}\frac{\partial^2 u_-}{\partial \tau^2} + bu_+ + (|u_-|^2 + 2|u_+|^2)u_- = 0, \qquad (6.5.2)$$

where $b = (\Delta\beta)L_D/2$ and fiber losses are neglected. The normalized variables ξ, τ, and $u_\pm$ are defined as

$$\xi = z/L_D, \quad \tau = (t - \beta_1 z)/T_0, \quad u_\pm = (2\gamma L_D/3)^{1/2}A_\pm, \qquad (6.5.3)$$

where $L_D = T_0^2/|\beta_2|$ is the dispersion length and T_0 is a measure of the pulse width. These equations generalize the scalar NLS equation of Section 5.2 to the vector case for low-birefringence fibers. They can be solved numerically using the split-step Fourier method of Section 2.4.

The numerical results show that the polarization instability affects soli-tons in a manner analogous to the CW case discussed in Section 6.2.3. If the nonlinear length L_{NL} is larger than the beat length $L_B = 2\pi/\Delta\beta$, solitons remain stable even if they are polarized close to the fast axis. By contrast, if $L_{NL} \ll L_B$, solitons polarized along the slow axis remain stable but become un-stable if polarized along the fast axis. A linearly polarized fundamental soliton ($N = 1$), launched with its polarization close to the fast axis with $L_{NL} \ll L_B$, follows the following evolution scenario [64]. Because of the onset of polar-ization instability, most of the pulse energy is transferred from the fast mode to the slow mode within a few soliton periods while a part of it is dispersed away. The pulse energy switches back and forth between the two modes a few times, a process similar to relaxation oscillations. Most of the input energy, however, appears eventually in a soliton-like pulse propagating along the slow axis. Higher-order solitons follow a somewhat different scenario. After going through an initial narrowing stage, they split into individual components, a behavior similar to that discussed in Section 5.5. A part of the energy is then transferred to the slow mode. A fundamental soliton eventually appears along the slow axis with a width narrower than the input width.

The CW instability condition can be used to obtain a condition on the soliton period. If we use Eq. (6.3.10), the condition $P_0 > P_{cr}$ becomes $(\Delta\beta)L_{NL} < \frac{2}{3}$, where $L_{NL} = (\gamma P_0)^{-1}$ is the nonlinear length. By using $\Delta\beta = 2\pi/L_B$, $N^2 = L_D/L_{NL}$ and $z_0 = (\pi/2)L_D$, this condition can be written as $z_0 < N^2 L_B/6$. The numerical results agree with it approximately [64]. Typically, $L_B \sim 1$ m for weakly birefringent fibers. Thus, polarization instability affects a fundamental soliton $(N = 1)$ only if $z_0 \ll 1$ m. Such values of z_0 are realized in practice only for femtosecond pulses $(T_0 < 100$ fs$)$.

6.5.2 High-Birefringence Fibers

In high-birefringence fibers, the group-velocity mismatch between the fast and slow components of the input pulse cannot be neglected. Such a mismatch would normally split a pulse into its two components polarized along the two principal axes if the input polarization angle θ deviates from 0 or 90°. The interesting question is whether such a splitting also occurs for solitons.

The effects of group-velocity mismatch are studied by solving Eqs. (6.1.22) and (6.1.23) numerically. If we assume anomalous dispersion $(\beta_2 < 0)$ and use the soliton units of Section 5.2, these equations become

$$i\left(\frac{\partial u}{\partial \xi} + \delta\frac{\partial u}{\partial \tau}\right) + \frac{1}{2}\frac{\partial^2 u}{\partial \tau^2} + (|u|^2 + B|v|^2)u = 0, \qquad (6.5.4)$$

$$i\left(\frac{\partial v}{\partial \xi} - \delta\frac{\partial v}{\partial \tau}\right) + \frac{1}{2}\frac{\partial^2 v}{\partial \tau^2} + (|v|^2 + B|u|^2)v = 0, \qquad (6.5.5)$$

where u and v are the normalized amplitudes of the field components polarized linearly along the x and y axes, respectively, and

$$\delta = (\beta_{1x} - \beta_{1y})T_0/2|\beta_2| \qquad (6.5.6)$$

governs the group-velocity mismatch between the two polarization components. The normalized time $\tau = (t - \bar{\beta}_1 z)/T_0$, where $\bar{\beta}_1 = \frac{1}{2}(\beta_{1x} + \beta_{1y})$ is

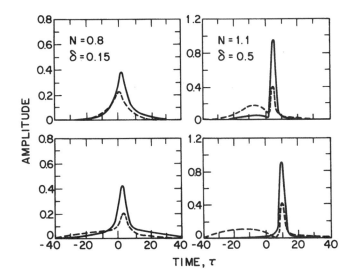

Figure 6.12 Pulse amplitudes $|u|$ (solid line) and $|v|$ (dashed line) at $\xi = 5\pi$ (upper row) and $\xi = 10\pi$ (lower row) for $\theta = 30°$. The parameters $N = 0.8$ and $\delta = 0.15$ for the left column and $N = 1.1$ and $\delta = 0.5$ for the right column. (After Ref. [67].)

inversely related to the average group velocity. Fiber losses are ignored for simplicity but can be included easily. The XPM coupling parameter $B = \frac{2}{3}$ for linearly birefringent fibers.

When an input pulse is launched with a polarization angle θ (measured from the slow axis), Eqs. (6.5.4) and (6.5.5) should be solved with the input

$$u(0, \tau) = N\cos\theta \operatorname{sech}(\tau), \qquad v(0, \tau) = N\sin\theta \operatorname{sech}(\tau), \qquad (6.5.7)$$

where N is the soliton order. In the absence of XPM-induced coupling, the two polarization components evolve independently and separate from each other because of their different group velocities. The central question is how this behavior is affected by the XPM. This question is answered by solving Eqs. (6.5.4) and (6.5.5) numerically with $B = 2/3$ for various values of N, θ, and δ [65]–[67]. The numerical results can be summarized as follows.

When the two modes are equally excited ($\theta = 45°$), the two components remain bound together if N exceeds a critical value N_{th} that depends on δ; $N_{th} \approx 0.7$ for $\delta = 0.15$, but $N_{th} \approx 1$ for $\delta = 0.5$. For values of $\delta \sim 1$, the threshold value exceeds 1.5. In this case, solitons can form even when $N < N_{th}$ but the two components travels at their own group velocities and become widely separated. When $N > N_{th}$, the two components remain close to each other but the distance between them changes in an oscillatory manner.

When $\theta \neq 45°$, the two modes have unequal amplitudes initially. In this case, if N exceeds N_{th}, a qualitatively different evolution scenario occurs depending on the value of δ. Figure 6.12 shows the pulse amplitudes of the two components for $\theta = 30°$ at $\xi = 5\pi$ and 10π [67]. The left-hand column corresponds to $\delta = 0.15$ and $N = 0.8$ while $\delta = 0.5$ and $N = 1.1$ for the right-hand column. For the case $\delta = 0.15$, the smaller pulse appears to have been captured by the larger one and the two move together. However, when $\delta = 0.5$, only a fraction of the energy in the smaller pulse is captured by the larger one; the remaining energy is dispersed away with propagation. Even more complex behavior occurs for larger values of δ and N.

The numerical results shown in Fig. 6.12 clearly indicate that under certain conditions the two orthogonally polarized solitons move with a common group velocity in spite of their different modal indices or polarization-mode dispersion (PMD). This phenomenon is called *soliton trapping* and, as discussed later, can be used for optical switching. It owes its existence solely to XPM. In the absence of the XPM term, Eqs. (6.5.4) and (6.5.5) become decoupled, indicating that each polarization component would propagate at a different group velocity dictated by the fiber birefringence. It is the XPM-induced nonlinear coupling between them that allows the two solitons to propagate at a common group velocity. Physically, the two solitons shift their carrier frequencies in the opposite directions to realize such a temporal synchronization. Specifically, the soliton along the fast axis slows down while the one along the slow axis speeds up. Indeed, the pulse spectra corresponding to the intensity profiles shown in Fig. 6.12 are found to be shifted exactly in such a way.

Because soliton trapping requires a balance between XPM and PMD, it can occur only when the peak power of the input pulse, or equivalently the soliton order N, exceeds a threshold value N_{th}. As N_{th} depends on both the polarization angle θ and δ, attempts have been made to estimate N_{th} analytically by solving Eqs. (6.5.4) and (6.5.5) approximately [69]–[77]. In a simple approach, the XPM term is treated as a perturbation within the Lagrangian formulation. In the case of equal amplitudes, realized by choosing $\theta = 45°$ in Eq. (6.5.7), the threshold value for soliton trapping is found to be [69]

$$N_{th} = [2(1+B)]^{-1/2} + (3/8B)^{1/2}\delta. \qquad (6.5.8)$$

For $B = \frac{2}{3}$, the predictions of Eq. (6.5.8) are in good agreement with the numerical results for small values of δ (up to 0.5). For large values of δ, the threshold value is well approximated by [77] $N_{th} = [(1 + 3\delta^2)/(1 + B)]^{1/2}$.

Soliton trapping was first observed [79] in 1989 by launching 0.3-ps pulses (obtained from a mode-locked color-center laser) into a 20-m single-mode fiber with modal birefringence $\Delta n \approx 2.4 \times 10^{-5}$, a value leading to polarization dispersion of 80 ps/km. For this experiment, the soliton period z_0 was 3.45 m while $\delta = 0.517$. The measured pulse spectra for the orthogonally polarized components were found to be separated by about 1 THz when the polarization angle was 45°. The autocorrelation trace indicated that the two pulses at the fiber output were synchronized temporally as expected for soliton trapping.

6.5.3 Soliton-Dragging Logic Gates

An important application of XPM interaction in birefringent fibers has led to the realization of all-optical, cascadable, ultrafast, logic gates, first demonstrated in 1989 [80]. Since then the performance of such logic gates has been studied extensively, both theoretically and experimentally [81]–[91].

The basic idea behind the operation of fiber-optic logic gates has its origin in the nonlinear phenomenon of soliton trapping discussed earlier. It can be understood as follows. In digital logic, each optical pulse is assigned a time slot whose duration is determined by the clock speed. If a signal pulse is launched together with an orthogonally polarized control pulse, and the control pulse is intense enough to trap the signal pulse during a collision, then both pulses can be dragged out of their assigned time slot because of the XPM-induced change in their group velocity. In other words, the absence or presence of a signal pulse at the fiber input dictates whether the control pulse ends up arriving within the assigned time slot or not. This temporal shift forms the basic logic element and can be used to perform more complex logic operations. Because the control pulse propagating as a soliton is dragged out of its time slot through the XPM interaction, such devices are referred to as soliton-dragging logic gates. In a network configuration, output signal pulse can be discarded while control pulse becomes the signal pulse for the next gate. This strategy makes the switching operation cascadable. In effect, each control pulse is used for switching only once irrespective of the number of gates in the network.

The experimental demonstration of various logic gates (such as exclusive OR, AND and NOR gates), based on the concept of soliton trapping, used femtosecond optical pulses (pulse width ~300 fs) from a mode-locked color-center laser operating at 1.685 μm [80]–[85]. In these experiments, orthogonally polarized signal and control pulses were launched into a highly birefringent fiber. In the implementation of a NOR gate, the experimental conditions were arranged such that the control pulse arrived in its assigned time slot of 1-ps duration in the absence of signal pulses (logical "1" state). In the presence of one or both signal pulses, the control pulse shifted by 2–4 ps because of soliton dragging and missed the assigned time slot (logical "0" state). The energy of each signal pulse was 5.8 pJ. The energy of the control pulse was 54 pJ at the fiber input but reduced to 35 pJ at the output, resulting in an energy gain by a factor of six. The experimental results can be explained quite well by solving Eqs. (6.5.4) and (6.5.5) numerically [90]. Since the first demonstration of such logic gates in 1989, considerable progress has been made. The use of soliton-dragging logic gates for soliton ring networks has also been proposed [91].

6.5.4 Vector Solitons

The phenomenon of soliton trapping suggests that the coupled NLS equations may possess exact solitary-wave solutions with the property that the orthogonally polarized components propagate in a birefringent fiber without change in shape. Such solitary waves are referred to as vector solitons to emphasize the fact that an input pulse maintains not only its intensity profile but also its state of polarization even when it is not launched along one of the principal axes of the fiber. A more general question one may ask is whether conditions exist under which two orthogonally polarized pulses of different widths and different peak powers propagate undistorted in spite of the XPM-induced nonlinear coupling between them.

Consider the case of high-birefringence fibers. To obtain soliton solutions of Eqs. (6.5.4) and (6.5.5), it is useful to simplify them using the transformation

$$u = \tilde{u}\exp(i\delta^2\xi/2 - i\delta\tau), \qquad v = \tilde{v}\exp(i\delta^2\xi/2 + i\delta\tau). \qquad (6.5.9)$$

The resulting equations are independent of δ and take the form

$$i\frac{\partial \tilde{u}}{\partial \xi} + \frac{1}{2}\frac{\partial^2 \tilde{u}}{\partial \tau^2} + (|\tilde{u}|^2 + B|\tilde{v}|^2)\tilde{u} = 0, \qquad (6.5.10)$$

$$i\frac{\partial \tilde{v}}{\partial \xi} + \frac{1}{2}\frac{\partial^2 \tilde{v}}{\partial \tau^2} + (|\tilde{v}|^2 + B|\tilde{u}|^2)\tilde{v} = 0. \qquad (6.5.11)$$

In the absence of XPM-induced coupling ($B = 0$), the two NLS equations become decoupled and have independent soliton solutions of the form discussed in Section 5.2. When $B \neq 0$, Eqs. (6.5.10) and (6.5.11) can be solved by the inverse scattering method only for a specific value of parameter B, namely $B = 1$. Manakov obtained such a solution in 1973 [92]. In its simplest form, the solution can be written as

$$\tilde{u}(\xi, \tau) = \cos\theta \operatorname{sech}(\tau) \exp(i\xi/2), \qquad (6.5.12)$$
$$\tilde{v}(\xi, \tau) = \sin\theta \operatorname{sech}(\tau) \exp(i\xi/2), \qquad (6.5.13)$$

where θ is an arbitrary angle. A comparison with Eq. (5.2.15) shows that this solution corresponds to a vector soliton that is identical with the fundamental soliton ($N = 1$) of Section 5.2 in all respects. The angle θ can be identified as the polarization angle.

The vector-soliton solution predicts that a 'sech' pulse with $N = 1$, linearly polarized at an arbitrary angle from a principal axis, can maintain both its shape and polarization provided the fiber is birefringent such that the XPM parameter $B = 1$. However, as discussed in Section 6.1, unless the fiber is especially designed, $B \neq 1$ in practice. In particular, $B = \frac{2}{3}$ for linearly birefringent fibers. For this reason, solitary-wave solutions of Eqs. (6.5.10) and (6.5.11) for $B \neq 1$ have been studied in many different contexts [93]–[117]. Such solutions are not solitons in a strict mathematical sense but, nevertheless, exhibit the shape-preserving property of solitons.

In the specific case of equal amplitudes ($\theta = 45°$), a solitary-wave solution of Eqs. (6.5.10) and (6.5.11) is given by [99]

$$\tilde{u} = \tilde{v} = \eta \operatorname{sech}[(1+B)^{1/2}\eta\tau] \exp[i(1+B)\eta^2\xi/2], \qquad (6.5.14)$$

where η represents the soliton amplitude. For $B = 0$, this solution reduces to the scalar soliton of Section 5.2. For $B \neq 0$, it represents a vector soliton polarized at 45° with respect to the principal axes of the fiber. Because of the XPM interaction, the vector soliton is narrower by a factor of $(1 + B)^{1/2}$ compared with the scalar soliton. For such a soliton, the combination of SPM and XPM compensates for the GVD. At the same time, the carrier frequencies of two polarization components must be different for compensating the PMD. This can be seen by substituting Eq. (6.5.14) in Eq. (6.5.9). The canonical form of the vector soliton, obtained by setting $\eta = 1$ is then given by

$$u(\xi, \tau) = \operatorname{sech}[(1+B)^{1/2}\tau] \exp[i(1+B+\delta^2)\xi/2 - i\delta\tau], \quad (6.5.15)$$
$$v(\xi, \tau) = \operatorname{sech}[(1+B)^{1/2}\tau] \exp[i(1+B+\delta^2)\xi/2 + i\delta\tau]. \quad (6.5.16)$$

The only difference between $u(\xi, \tau)$ and $v(\xi, \tau)$ is the sign of the last phase term involving the product $\delta\tau$. This sign change reflects the shift of the carrier frequency of the soliton components in the opposite directions.

The solution given by Eq. (6.5.14) represents one of the several solitary-wave solutions that have been discovered in birefringent fibers by solving Eqs. (6.1.19) and (6.1.20) under various approximations. In one case, the two components not only have an asymmetric shape but they can also have a double-peak structure [96]. In another interesting class of solutions, two solitary waves form bound states such that the state of polarization is not constant over the entire pulse but changes with time [97]. In some cases the state of polarization can even evolve periodically along the fiber length [111]. Several other solitary-wave solutions have been discovered during the 1990s [107]–[117].

Similar to the case of modulation instability, one may ask whether vector solitons exist in isotropic fibers with no birefringence. In this case, we should use Eqs. (6.5.1) and (6.5.2) with $b = 0$. These equations then become identical to Eqs. (6.1.17) and (6.1.18) with the choice $B = 2$. The only difference is that they are written in terms of the circularly polarized components as defined in Eq. (6.5.3). The vector soliton given in Eq. (6.5.14) thus exists even for isotropic fibers and can be written using $B = 2$ as

$$u_+ = u_- = \eta \, \text{sech}(\sqrt{3}\eta\tau) \exp(3i\eta^2\xi/2). \tag{6.5.17}$$

It corresponds to a linearly polarized pulse whose electric field vector may be oriented at any angle in the plane transverse to the fiber axis. Elliptically polarized solitons also exist for whom the the polarization ellipse rotates at a fixed rate [115]. The state of polarization is not uniform across the pulse for such solitons.

6.6 Random Birefringence

As mentioned in Section 6.1, modal birefringence in optical fibers changes randomly over a length scale ~ 10 m unless polarization-maintaining fibers are used. Because lightwave systems commonly use fibers with randomly varying birefringence, it is important to study how optical pulses are affected by random birefringence changes. Indeed, this issue has been investigated extensively [118]–[150]. In this section we consider the effects of random birefringence for both soliton and nonsoliton pulses.

6.6.1 Polarization-Mode Dispersion

It is intuitively clear that the polarization state of CW light propagating in fibers with randomly varying birefringence will generally be elliptical and would change randomly along the fiber during propagation. In the case of optical pulses, the polarization state can also be different for different parts of the pulse unless the pulse propagates as a soliton. Such random polarization changes typically are not of concern for lightwave systems because photodetectors used inside optical receivers are insensitive to the state of polarization of the incident light (unless a coherent-detection scheme is employed). What affects such systems is not the random polarization state but pulse broadening induced by random changes in the birefringence. This is referred to as PMD-induced pulse broadening.

The analytical treatment of PMD is quite complex in general because of its statistical nature. A simple model, first introduced in 1986 [118], divides the fiber into a large number of segments. Both the degree of birefringence and the orientation of the principal axes remain constant in each section but change randomly from section to section. In effect, each fiber section can be treated as a phase plate and a Jones matrix can be used for it [151]. Propagation of each frequency component associated with an optical pulse through the entire fiber length is then governed by a composite Jones matrix obtained by multiplying individual Jones matrices for each fiber section. The composite Jones matrix shows that two principal states of polarization exist for any fiber such that, when a pulse is polarized along them, the polarization state at fiber output is frequency independent to first order, in spite of random changes in fiber birefringence. These states are analogs of the slow and fast axes associated with polarization-maintaining fibers. Indeed, the differential group delay ΔT (relative time delay in the arrival time of the pulse) is largest for the principal states of polarization [135].

The principal states of polarization provide a convenient basis for calculating the moments of ΔT [119]. The PMD-induced pulse broadening is characterized by the root-mean-square (RMS) value of ΔT, obtained after averaging over random birefringence changes. Several approaches have been used to calculate this average using different models [119]–[123]. The variance $\sigma_T^2 \equiv \langle (\Delta T)^2 \rangle$ turns out to be the same in all cases and is given by [132]

$$\sigma_T^2(z) = 2\Delta'^2 l_c^2 [\exp(-z/l_c) + z/l_c - 1], \tag{6.6.1}$$

where the intrinsic modal dispersion $\Delta' = d(\Delta\beta)/d\omega$ is related to the difference in group velocities along the two principal states of polarization. The parameter l_c is the correlation length, defined as the length over which two polarization components remain correlated; its typical values are ~ 10 m.

For short distances such that $z \ll l_c$, $\sigma_T = \Delta'z$ from Eq. (6.6.1), as expected for a polarization-maintaining fiber. For distances $z > 1$ km, a good estimate of pulse broadening is obtained using $z \gg l_c$. For a fiber of length L, σ_T in this approximation becomes

$$\sigma_T \approx \Delta'\sqrt{2l_cL} \equiv D_p\sqrt{L}, \qquad (6.6.2)$$

where D_p is the PMD parameter. Measured values of D_p vary from fiber to fiber, typically in the range $D_p = 0.1$–2 ps/$\sqrt{\text{km}}$ [129]. Modern fibers are designed to have low PMD, with values of D_p as low as 0.05 ps/$\sqrt{\text{km}}$ [133]. Because of the $\sqrt{L}$ dependence, PMD-induced pulse broadening is relatively small compared with the GVD effects. Indeed, $\sigma_T \sim 1$ ps for fiber lengths ~ 100 km and can be ignored for pulse widths > 10 ps. However, PMD becomes a limiting factor for lightwave systems designed to operate over long distances at high bit rates near the zero-dispersion wavelength of the fiber [130].

Several schemes can be used for compensating the PMD effects occurring in lightwave systems [152]–[162]. In one scheme, the PMD-distorted signal is separated into its components along the two principal states of polarization, PMD is inferred from the measured relative phase, and the two components are synchronized after introducing appropriate delays [154]. The success of this technique depends on the ratio L/L_{PMD} for a fiber of length L, where $L_{\text{PMD}} = (T_0/D_p)^2$ is the PMD length for pulses of width T_0; considerable improvement is expected for fibers as long as $4L_{\text{PMD}}$ [163]. Because L_{PMD} can approach $10,000$ km for $D_p = 0.1$ ps/$\sqrt{\text{km}}$ and $T_0 = 10$ ps, first-order PMD effects can be compensated over transoceanic distances.

Several other factors need to be considered in practice. The derivation of Eq. (6.6.1) assumes that the fiber link has no elements exhibiting polarization-dependent loss or gain. The presence of polarization-dependent losses can induce additional broadening [134]. Similarly, the effects of second-order PMD should be considered for fibers with relatively low values of D_p. Such effects have been studied and lead to additional distortion of optical pulses [139]. Moreover, the effects of second-oder PMD depend on the chirp associated with an optical pulse and degrade the system performance when chirp is relatively large [137].

6.6.2 Polarization State of Solitons

As mentioned earlier, the polarization state of a pulse in general becomes nonuniform across the pulse because of random changes in fiber birefringence.

At the same time, PMD leads to pulse broadening. An interesting question is whether similar things happen to a soliton, or solitons are relatively immune to random birefringence changes because of their particle-like nature. This issue became important soon after solitons became a viable candidate for long-haul lightwave systems [164]–[176].

In the case of constant birefringence, it was seen in Section 6.5 that the orthogonally polarized components of a soliton can travel at the same speed, in spite of different group velocities associated with them at low powers. Solitons realize this synchronization by shifting their frequencies appropriately. It is thus not hard to imagine that solitons may avoid splitting and PMD-induced pulse broadening through the same mechanism. Indeed, numerical simulations based on Eqs. (6.1.11) and (6.1.12) indicate this to be the case [164] as long as the PMD parameter is small enough to satisfy the condition $D_p < 0.3\sqrt{D}$, where D is related to β_2 as indicated in Section 1.2.

To understand the relative insensitivity of solitons to PMD, let us discuss how Eqs. (6.1.11) and (6.1.12) can be adapted for fibers exhibiting random birefringence changes along their length. It is more convenient to write them in terms of the normalized amplitudes u and v defined as

$$u = A_x\sqrt{\gamma L_D}\,e^{i\Delta\beta z/2}, \qquad v = A_y\sqrt{\gamma L_D}\,e^{-i\Delta\beta z/2}. \tag{6.6.3}$$

If we also use soliton units and introduce normalized distance and time as

$$\xi = z/L_D, \qquad \tau = (t - \bar{\beta}_1 z)/T_0, \tag{6.6.4}$$

Eqs. (6.1.11) and (6.1.12) become

$$i\left(\frac{\partial u}{\partial \xi} + \delta\frac{\partial u}{\partial \tau}\right) + bu + \frac{1}{2}\frac{\partial^2 u}{\partial \tau^2} + \left(|u|^2 + \frac{2}{3}|v|^2\right)u + \frac{1}{3}v^2 u^* = 0, \tag{6.6.5}$$

$$i\left(\frac{\partial v}{\partial \xi} - \delta\frac{\partial v}{\partial \tau}\right) - bv + \frac{1}{2}\frac{\partial^2 v}{\partial \tau^2} + \left(|v|^2 + \frac{2}{3}|u|^2\right)v + \frac{1}{3}u^2 v^* = 0, \tag{6.6.6}$$

where

$$\delta = \frac{T_0}{2|\beta_2|}(\beta_{1x} - \beta_{1y}), \qquad b = \frac{T_0^2}{2|\beta_2|}(\beta_{0x} - \beta_{0y}). \tag{6.6.7}$$

Both δ and b vary randomly along the fiber because of random birefringence fluctuations.

Equations (6.6.5) and (6.6.6) can be written in a compact form by introducing a column vector U and the Pauli matrices as

$$U = \begin{pmatrix} u \\ v \end{pmatrix}, \quad \sigma_1 = \begin{pmatrix} 1 & 0 \\ 0 & -1 \end{pmatrix}, \quad \sigma_2 = \begin{pmatrix} 0 & 1 \\ 1 & 0 \end{pmatrix}, \quad \sigma_3 = \begin{pmatrix} 0 & -i \\ i & 0 \end{pmatrix}. \quad (6.6.8)$$

In terms of the column vector U, the coupled NLS equations become [165]

$$i\frac{\partial U}{\partial \xi} + \sigma_1 \left(bU + i\delta \frac{\partial U}{\partial \tau} \right) + \frac{1}{2}\frac{\partial^2 U}{\partial \tau^2} + s_0 U - \frac{1}{3}s_3 \sigma_3 U, \quad (6.6.9)$$

where the Stokes parameters are defined in terms of U as

$$s_0 = U^\dagger U = |u|^2 + |v|^2, \quad s_1 = U^\dagger \sigma_1 U = |u|^2 - |v|^2, \quad (6.6.10)$$
$$s_2 = U^\dagger \sigma_2 U = 2\,\mathrm{Re}(u^* v), \quad s_3 = U^\dagger \sigma_3 U = 2\,\mathrm{Im}(u^* v). \quad (6.6.11)$$

These Stokes parameters are analogous to those introduced in Section 6.3.2 for describing the polarization state of CW light on the Poincaré sphere. The main difference is that they are time dependent and describe the polarization state of a pulse. They can be reduced to those of Section 6.3.2 by integrating over time such that $S_j = \int_{-\infty}^{\infty} s_j(t)\,dt$ for $j = 0$ to 3.

As in the CW case, the Stokes vector with components s_1, s_2, and s_3 moves on the surface of the Poincaré sphere of radius s_0. When birefringence of the fiber varies randomly along the fiber, the tip of the Stokes vector moves randomly over the Poincaré sphere. The important question is the length scale over which such motion covers the entire surface of the Poincaré sphere and how this length compares with the dispersion length. To answer this question, one should consider random variations in b and δ as well as random changes in the orientation of the principal axes along fiber.

Random changes in b occur on a length scale ~ 1 m. As they only affect the phases of u and v, it is clear that such changes leave s_1 unchanged. As a result, the Stokes vector rotates rapidly around the s_1 axis. Changes in the orientation of the birefringence axes occur randomly over a length scale ~ 10 m. Such changes leave s_3 unchanged and thus rotate the Stokes vector around that axis. The combination of these two types of rotations forces the Stokes vector to fill the entire surface of the Poincaré sphere over a length scale ~ 1 km. As this distance is typically much shorter than the dispersion length, soliton parameters are not much affected by random changes in birefringence. The situation is similar to the case of energy variations occurring when fiber losses are compensated periodically using optical amplifiers (see Section 5.4). We can thus follow a similar approach and average Eq. (6.6.9) over random birefringence changes. When $\langle b \rangle = 0$ and $\langle \delta \rangle = 0$, the two terms containing σ_1 average out to zero. The last term in Eq. (6.6.9) requires the average $\langle s_3 \sigma_3 U \rangle$. This average turns out to be $s_0 U/3$ if we make use of the identity [165]

$$UU^\dagger = \sum_{j=1}^{3} s_j \sigma_j. \qquad (6.6.12)$$

After averaging over birefringence fluctuations, Eq. (6.6.9) reduces to

$$i\frac{\partial U}{\partial \xi} + \frac{1}{2}\frac{\partial^2 U}{\partial \tau^2} + \frac{8}{9}s_0 U = 0, \qquad (6.6.13)$$

The factor of $\frac{8}{9}$ can be absorbed in the normalization factor used for U and amounts to reducing the nonlinear parameter γ by this factor. In terms of the components u and v, Eq. (6.6.13) can be written as

$$i\frac{\partial u}{\partial \xi} + \frac{1}{2}\frac{\partial^2 u}{\partial \tau^2} + (|u|^2 + |v|^2)u = 0, \qquad (6.6.14)$$

$$i\frac{\partial v}{\partial \xi} + \frac{1}{2}\frac{\partial^2 v}{\partial \tau^2} + (|v|^2 + |u|^2)v = 0. \qquad (6.6.15)$$

As discussed in Section 6.5.3, this set of two coupled NLS equations is integrable by the inverse scattering method [92] and has the solution in the form of a fundamental vector soliton given in Eqs. (6.5.12) and (6.5.13). This solution shows that a fundamental soliton maintains the same polarization across the entire pulse "on average" in spite of random birefringence changes along the fiber. This is an extraordinary result and is indicative of the particle-like nature of solitons. In effect, solitons maintain uniform polarization across the entire pulse and resist small random changes in birefringence [164]. Extensive numerical simulations based on Eqs. (6.6.5) and (6.6.6) confirm that solitons can maintain a uniform polarization state approximately over long fiber lengths even when optical amplifiers are used for compensating fiber losses [165].

It is important to note that the vector soliton associated with Eqs. (6.6.14) and (6.6.15) represents the average behavior. The five parameters associated with this soliton (amplitude, frequency, position, phase, and polarization angle) will generally fluctuate along fiber length in response to random birefringence changes. Perturbation theory, similar to that used for scalar solitons in Section

5.4, can be used to study birefringence-induced changes in soliton parameters [166]–[170]. For example, the amplitude of the soliton decreases and its width increases because of perturbations produced by random birefringence. The reason behind soliton broadening is related to the generation of dispersive waves (continuum radiation) and resulting energy loss. The perturbation technique can also be used to study interaction between orthogonally polarized solitons [171] and timing jitter induced by amplifier-induced fluctuations in the polarization state of the soliton [176].

From a practical standpoint, uniformity of soliton polarization can be useful for polarization-division multiplexing. In this scheme, two orthogonally polarized bit streams are interleaved in the time domain. As a result, alternate pulses have orthogonal polarization states initially and are able to maintain their orthogonality if they propagate as solitons. This allows much tighter packing of solitons (resulting in a higher bit rate) because the interaction between two neighboring solitons is reduced when they are orthogonally polarized. However, extensive numerical simulations show that the technique of polarization-division multiplexing is useful in practice only when the PMD parameter D_p is relatively small [173]. When D_p is large, copolarized solitons provide an overall better system performance.

Problems

6.1 Derive an expression for the nonlinear part of the refractive index when an optical beams propagates inside a high-birefringence optical fiber.

6.2 Prove that Eqs. (6.1.15) and (6.1.16) indeed follow from Eqs. (6.1.11) and (6.1.12).

6.3 Prove that a high-birefringence fiber of length L introduces a relative phase shift of $\Delta\phi_{NL} = (\gamma P_0 L/3)\cos(2\theta)$ between the two linearly polarized components when a CW beam with peak power P_0 and polarization angle θ propagates through it. Neglect fiber losses.

6.4 Explain the operation of a Kerr shutter. What factors limit the response time of such a shutter when optical fibers are used as the Kerr medium?

6.5 How can fiber birefringence be used to remove the low-intensity pedestal associated with an optical pulse?

6.6 Solve Eqs. (6.3.1) and (6.3.2) in terms of the elliptic functions. You can consult Reference [43].

6.7 Prove that Eqs. (6.3.14) and (6.3.15) can be written in the form of Eq. (6.3.19) after introducing the Stokes parameters through Eq. (6.3.16).

6.8 What is meant by polarization instability in birefringent optical fibers? Explain the origin of this instability.

6.9 Derive the dispersion relation $K(\Omega)$ for modulation instability to occur in low-birefringence fibers starting from Eqs. (6.1.15) and (6.1.16). Discuss the frequency range over which the gain exists when $\beta_2 > 0$.

6.10 Derive the dispersion relation $K(\Omega)$ for modulation instability to occur in high-birefringence fibers starting from Eqs. (6.1.22) and (6.1.23). Discuss the frequency range over which the gain exists when $\beta_2 > 0$.

6.11 Solve Eqs. (6.5.4) and (6.5.5) numerically by using the split-step Fourier method. Reproduce the results shown in Fig. 6.12. Check the accuracy of Eq. (6.5.8) for $\delta = 0.2$ and $B = 2/3$.

6.12 Verify by direct substitution that the solution given by Eq. (6.5.14) satisfies Eqs. (6.5.4) and (6.5.5).

6.13 Explain the operation of soliton-dragging logic gates. How would you design a NOR gate by using such a technique?

6.14 Explain the origin of PMD in optical fibers. Why does PMD lead to pulse broadening. Do you expect PMD-induced broadening to occur for solitons?

References

[1] P. D. Maker, R. W. Terhune, and C. M. Savage, *Phys. Rev. Lett.* **12**, 507 (1964).

[2] G. Mayer and F. Gires, *Compt. Rend. Acad. Sci.* **258**, 2039 (1964).

[3] P. D. Maker and R. W. Terhune, *Phys. Rev. A* **137**, A801 (1965).

[4] M. A. Duguay and J. W. Hansen, *Appl. Phys. Lett.* **15**, 192 (1969).

[5] A. Owyoung, R. W. Hellwarth, and N. George, *Phys. Rev. B* **5**, 628 (1972).

[6] M. A. Duguay, in *Progress in Optics*, Vol. 14, E. Wolf, ed. (North-Holland, Amsterdam, 1976), p. 163.

[7] R. W. Hellwarth, *Prog. Quantum Electron.* **5**, 1 (1977).

[8] N. G. Phu-Xuan and G. Rivoire, *Opt. Acta* **25**, 233 (1978).

[9] Y. R. Shen, *The Principles of Nonlinear Optics* (Wiley, New York, 1984).

[10] R. W. Boyd, *Nonlinear Optics* (Academic Press, San Diego, 1992).

[11] R. Ulrich and A. Simon,, *Appl. Opt.* **18**, 2241 (1979).

[12] C. R. Menyuk, *IEEE J. Quantum Electron.* **25**, 2674 (1989).

[13] R. H. Stolen and A. Ashkin, *Appl. Phys. Lett.* **22**, 294 (1973).

[14] J. M. Dziedzic, R. H. Stolen, and A. Ashkin, *Appl. Opt.* **20**, 1403 (1981).

[15] J. L. Aryal, J. P. Pocholle, J. Raffy, and M. Papuchon, *Opt. Commun.* **49**, 405 (1984).

[16] K. Kitayama, Y. Kimura, and S. Sakai, *Appl. Phys. Lett.* **46**, 623 (1985).

[17] E. M. Dianov, E. A. Zakhidov, A. Y. Karasik, M. A. Kasymdzhanov, F. M. Mirtadzhiev, A. M. Prokhorov, and P. K. Khabibullaev, *Sov. J. Quantum Electron.* **17**, 517 (1987).

[18] T. Morioka, M. Saruwatari, and A. Takada, *Electron. Lett.* **23**, 453 (1987).

[19] K. C. Byron, *Electron. Lett.* **23**, 1324 (1987).

[20] T. Morioka and M. Saruwatari, *Electron. Lett.* **23**, 1330 (1987); *IEEE J. Sel. Areas Commun.* **6**, 1186 (1988).

[21] I. H. White, R. V. Penty, and R. E. Epworth, *Electron. Lett.* **24**, 340 (1988).

[22] M. Asobe, T. Kanamori, and K. Kubodera, *IEEE Photon. Technol. Lett.* **4**, 362 (1992); *IEEE J. Quantum Electron.* **29**, 2325 (1993).

[23] M. Asobe, H. Kobayashi, H. Itoh, and T. Kanamori, *Opt. Lett.* **18**, 1056 (1993).

[24] M. Asobe, *Opt. Fiber Technol.* **3**, 142 (1997).

[25] R. H. Stolen, J. Botineau, and A. Ashkin, *Opt. Lett.* **7**, 512 (1982).

[26] B. Nikolaus, D. Grischkowsky, and A. C. Balant, *Opt. Lett.* **8**, 189 (1983).

[27] N. J. Halas and D. Grischkowsky, *Appl. Phys. Lett.* **48**, 823 (1986).

[28] K. Kitayama, Y. Kimura, and S. Seikai, *Appl. Phys. Lett.* **46**, 317 (1985); *Appl. Phys. Lett.* **46**, 623 (1985).

[29] H. G. Winful, *Appl. Phys. Lett.* **47**, 213 (1985).

[30] B. Crosignani and P. Di Porto, *Opt. Acta* **32**, 1251 (1985).

[31] B. Daino, G. Gregori, and S. Wabnitz, *Opt. Lett.* **11**, 42 (1986).

[32] H. G. Winful, *Opt. Lett.* **11**, 33 (1986).

[33] G. Gregori and S. Wabnitz, *Phys. Rev. Lett.* **56**, 600 (1986).

[34] F. Matera and S. Wabnitz, *Opt. Lett.* **11**, 467 (1986).

[35] S. Trillo, S. Wabnitz, R. H. Stolen, G. Assanto, C. T. Seaton, and G. I. Stegeman, *Appl. Phys. Lett.* **49**, 1224 (1986).

[36] A. Vatarescu, *Appl. Phys. Lett.* **49**, 61 (1986).

[37] S. Wabnitz, *Phys. Rev. Lett.* **58**, 1415 (1987).

[38] A. Mecozzi, S. Trillo, S. Wabnitz, and B. Daino, *Opt. Lett.* **12**, 275 (1987).

[39] Y. Kimura and M. Nakazawa, *Jpn. J. Appl. Phys.* **2**, 1503 (1987).

[40] E. Caglioti, S. Trillo, and S. Wabnitz, *Opt. Lett.* **12**, 1044 (1987).

[41] S. Trillo, S. Wabnitz, E. M. Wright, and G. I. Stegeman, *Opt. Commun.* **70**, 166 (1989).

[42] S. F. Feldman, D. A. Weinberger, and H. G. Winful, *Opt. Lett.* **15**, 311 (1990).

[43] S. F. Feldman, D. A. Weinberger, and H. G. Winful, *J. Opt. Soc. Am. B* **10**, 1191 (1993).

[44] P. Ferro, S. Trillo, and S. Wabnitz, *Appl. Phys. Lett.* **64**, 2782 (1994); *Electron. Lett.* **30**, 1616 (1994).

[45] N. N. Akhmediev and J. M. Soto-Crespo, *Phys. Rev. E* **49**, 5742 (1994).

[46] Y. Barad and Y. Silberberg, *Phys. Rev. Lett.* **78**, 3290 (1997).

[47] A. L. Berkhoer and V. E. Zakharov, *Sov. Phys. JETP* **31**, 486 (1970).

[48] S. Wabnitz, *Phys. Rev. A* **38**, 2018 (1988).

[49] S. Trillo and S. Wabnitz, *J. Opt. Soc. Am. B* **6**, 238 (1989).

[50] J. E. Rothenberg, *Phys. Rev. A* **42**, 682 (1990).

[51] P. D. Drummond, T. A. B. Kennedy, J. M. Dudley, R. Leonhardt, and J. D. Harvey, *Opt. Commun.* **78**, 137 (1990).

[52] J. E. Rothenberg, *Opt. Lett.* **16**, 18 (1991).

[53] W. Huang and J. Hong, *J. Lightwave Technol.* **10**, 156 (1992).

[54] J. Hong and W. Huang, *IEEE J. Quantum Electron.* **28**, 1838 (1992).

[55] M. Haelterman and A. P. Sheppard, *Phys. Rev. E* **49**, 3389 (1994).

[56] S. G. Murdoch, R. Leonhardt, and J. D. Harvey, *Opt. Lett.* **20**, 866 (1995).

[57] S. G. Murdoch, M. D. Thomson, R. Leonhardt, and J. D. Harvey, *Opt. Lett.* **22**, 682 (1997).

[58] G. Millot, S. Pitois, P. Tchofo Dinda, M. Haelterman, *Opt. Lett.* **22**, 1686 (1997).

[59] G. Millot, E. Seve, S. Wabnitz, M. Haelterman, *Opt. Lett.* **23**, 511 (1998).

[60] G. Millot, E. Seve, S. Wabnitz, M. Haelterman, *J. Opt. Soc. Am. B* **15**, 1266 (1998).

[61] E. Seve, G. Millot, and S. Wabnitz, *Opt. Lett.* **23**, 1829 (1998).

[62] P. Kockaert, M. Haelterman, S. Pitois, and G. Millot, *Appl. Phys. Lett.* **75**, 2873 (1999).

[63] E. Seve, G. Millot, and S. Trillo, *Phys. Rev. E* **61**, 3139 (2000).

[64] K. J. Blow, N. J. Doran, and D. Wood, *Opt. Lett.* **12**, 202 (1987).

[65] C. R. Menyuk, *IEEE J. Quantum Electron.* **23**, 174 (1987).

[66] C. R. Menyuk, *Opt. Lett.* **12**, 614 (1987).

[67] C. R. Menyuk, *J. Opt. Soc. Am. B* **5**, 392 (1988).

[68] A. D. Boardman and G. S. Cooper, *J. Opt. Soc. Am. B* **5**, 403 (1988); *J. Mod. Opt.* **35**, 407 (1988).

[69] Y. S. Kivshar, *J. Opt. Soc. Am. B* **7**, 2204 (1990).

[70] R. J. Dowling, *Phys. Rev. A* **42**, 5553 (1990).

[71] B. A. Malomed, *Phys. Rev. A* **43**, 410 (1991).

[72] D. Anderson, Y. S. Kivshar, and M. Lisak, *Physica Scripta* **43**, 273 (1991).

[73] B. A. Malomed and S. Wabnitz, *Opt. Lett.* **16**, 1388 (1991).

[74] N. A. Kostov and I. M. Uzunov, *Opt. Commun.* **89**, 389 (1991).

[75] V. K. Mesentsev and S. K. Turitsyn, *Opt. Lett.* **17**, 1497 (1992).

[76] B. A. Malomed, *Phys. Rev. A* **43**, 410 (1991); *J. Opt. Soc. Am. B* **9**, 2075 (1992).

[77] X. D. Cao and C. J. McKinstrie, *J. Opt. Soc. Am. B* **10**, 1202 (1993).

[78] D. J. Kaup and B. A. Malomed, *Phys. Rev. A* **48**, 599 (1993).

[79] M. N. Islam, C. D. Poole, and J. P. Gordon, *Opt. Lett.* **14**, 1011 (1989).

[80] M. N. Islam, *Opt. Lett.* **14**, 1257 (1989); *Opt. Lett.* **15**, 417 (1990).

[81] M. N. Islam, C. E. Soccolich, and D. A. B. Miller, *Opt. Lett.* **15**, 909 (1990).

[82] M. N. Islam, C. E. Soccolich, C.-J. Chen, K. S. Kim. J. R. Simpson, and U. C. Paek, *Electron. Lett.* **27**, 130 (1991).

[83] M. N. Islam and J. R. Sauer, *IEEE J. Quantum Electron.* **27**, 843 (1991).

[84] M. N. Islam, C. R. Menyuk, C.-J. Chen, and C. E. Soccolich, *Opt. Lett.* **16**, 214 (1991).

[85] M. N. Islam, *Ultrafast Fiber Switching Devices and Systems* (Cambridge University Press, Cambridge, UK, 1992).

[86] C.-J. Chen, P. K. A. Wai, and C. R. Menyuk, *Opt. Lett.* **15**, 477 (1990).

[87] C. R. Menyuk, M. N. Islam, and J. P. Gordon, *Opt. Lett.* **16**, 566 (1991).

[88] C.-J. Chen, C. R. Menyuk, M. N. Islam, and R. H. Stolen, *Opt. Lett.* **16**, 1647 (1991).

[89] M. W. Chbat, B. Hong, M. N. Islam, C. E. Soccolich, and P. R. Prucnal, *J. Lightwave Technol.* **12**, 2011 (1992).

[90] Q. Wang, P. K. A. Wai, C.-J. Chen, and C. R. Menyuk, *Opt. Lett.* **17**, 1265 (1992); *J. Opt. Soc. Am. B* **10**, 2030 (1993).

[91] J. R. Sauer, M. N. Islam, and S. P. Dijali, *J. Lightwave Technol.* **11**, 2182 (1994).

[92] S. V. Manakov, *Sov. Phys. JETP* **38**, 248 (1974).

[93] Y. Inoue, *J. Plasma Phys.* **16**, 439 (1976); *J. Phys. Soc. Jpn.* **43**, 243 (1977).

[94] M. R. Gupta, B. K. Som, and B. Dasgupta, *J. Plasma Phys.* **25**, 499 (1981).

[95] V. E. Zakharov and E. I. Schulman, *Physica D* **4**, 270 (1982).

[96] D. N. Christodulides and R. I. Joseph, *Opt. Lett.* **13**, 53 (1988).

[97] M. V. Tratnik and J. E. Sipe, *Phys. Rev. A* **38**, 2011 (1988).

[98] N. N. Akhmediev, V. M. Elonskii, N. E. Kulagin, and L. P. Shilnikov, *Sov. Tech. Phys. Lett.* **15**, 587 (1989).

[99] T. Ueda and W. L. Kath, *Phys. Rev. A* **42**, 563 (1990).

[100] D. David and M. V. Tratnik, *Physica D* **51**, 308 (1991).

[101] S. Trillo and S. Wabnitz, *Phys. Lett.* **159**, 252 (1991).

[102] L. Gagnon, *J. Phys. A* **25**, 2649 (1992).

[103] B. A. Malomed, *Phys. Rev. A* **45**, R8821 (1992).

[104] M. V. Tratnik, *Opt. Lett.* **17**, 917 (1992).

[105] D. Kapor, M. Skrinjar, and S. Stojanovic, *J. Phys. A* **25**, 2419 (1992).

[106] R. S. Tasgal and M. J. Potasek, *J. Math. Phys.* **33**, 1280 (1992); M. J. Potasek, *J. Opt. Soc. Am. B* **10**, 941 (1993).

[107] M. Wadati, T. Iizuka, and M. Hisakado, *J. Phys. Soc. Jpn.* **61**, 2241 (1992).

[108] Y. S. Kivshar, *Opt. Lett.* **17**, 1322 (1992).

[109] Y. S. Kivshar and S. K. Turitsyn, *Opt. Lett.* **18**, 337 (1993).

[110] V. V. Afanasjev and A. B. Grudinin, *Sov. Lightwave Commun.* **3**, 77 (1993).

[111] M. Haelterman, A. P. Sheppard, and A. W. Snyder, *Opt. Lett.* **18**, 1406 (1993).

[112] D. J. Kaup, B. A. Malomed, and R. S. Tasgal, *Phys. Rev. E* **48**, 3049 (1993).

[113] J. C. Bhakta, *Phys. Rev. E* **49**, 5731 (1994).

[114] M. Haelterman and A. P. Sheppard, *Phys. Rev. E* **49**, 3376 (1994).

[115] Y. Silberberg and Y. Barad, *Opt. Lett.* **20**, 246 (1995).

[116] L. Slepyan, V. Krylov, and R. Parnes, *Phys. Rev. Lett.* **74**, 27256 (1995).

[117] Y. Chen and J. Atai, *Phys. Rev. E* **55**, 3652 (1997).

[118] C. D. Poole and R. E. Wagnar, *Electron. Lett.* **22**, 1029 (1986).

[119] C. D. Poole, *Opt. Lett.* **13**, 687 (1988).

[120] F. Curti, B. Diano, G. De Marchis, and F. Matera, *J. Lightwave Technol.* **8**, 1162 (1990).

[121] C. D. Poole, J. H. Winters, and J. A. Nagel, *Opt. Lett.* **16**, 372 (1991).

[122] N. Gisin, J. P. von der Weid, and J.-P. Pellaux *J. Lightwave Technol.* **9**, 821 (1991).

[123] G. J. Foschini and C. D. Poole, *J. Lightwave Technol.* **9**, 1439, (1991).

[124] P. K. A. Wai, C. R. Menyuk, H. H. Chen, *Opt. Lett.* **16**, 1231 (1991); *Opt. Lett.* **16**, 1735 (1991).

[125] N. Gisin and J. P. Pellaux *Opt. Commun.* **89**, 316 (1992).

[126] C. R. Menyuk and P. K. A. Wai, *J. Opt. Soc. Am. B* **11**, 1288 (1994); *Opt. Lett.* **19**, 1517 (1994).

[127] J. Zhou and M. J. O'Mahony, *IEEE Photon. Technol. Lett.* **6**, 1265 (1994).

[128] C. D. Poole and D. L. Favin, *J. Lightwave Technol.* **12**, 917 (1994).

[129] M. C. de Lignie, H. G. Nagel, and M. O. van Deventer, *J. Lightwave Technol.* **12**, 1325 (1994).

[130] E. Lichtman, *J. Lightwave Technol.* **13**, 898 (1995).

[131] Y. Suetsugu, T. Kato, and M. Nishimura, *IEEE Photon. Technol. Lett.* **7**, 887 (1995).

[132] P. K. A. Wai and C. R. Menyuk, *J. Lightwave Technol.* **14**, 148 (1996).

[133] F. Bruyère, *Opt. Fiber Technol.* **2**, 269 (1996).

[134] B. Huttner and N. Gisin, *Opt. Lett.* **22**, 504 (1997).

[135] C. D. Poole and J. Nagel, in *Optical Fiber Telecommunications IIIA* I. P. Kaminow and T. L. Koch, eds. (Academic Press, San Diego, CA, 1997), Chap. 6.

[136] M. Karlsson, *Opt. Lett.* **23**, 688 (1998).

[137] H. Bülow, *IEEE Photon. Technol. Lett.* **10**, 696 (1998).

[138] P. Ciprut, B. Gisin, N. Gisin, R. Passy, P. von der Weid, F. Prieto, and C. W. Zimmer, *J. Lightwave Technol.* **16**, 757 (1998).

[139] C. Francia, F. Bruyère, D. Penninckx, and M. Chbat, *IEEE Photon. Technol. Lett.* **10**, 1739 (1998).

[140] J. Cameron, X. Bao, and J. Stears, *Fiber Integ. Opt.* **18**, 49 (1999).

[141] M. Karlsson and J. Brentel, *Opt. Lett.* **24**, 939 (1999).

[142] G. J. Foschini, R. M. Jopson, L. E. Nelson, and H. Kogelnik, *J. Lightwave Technol.* **17**, 1560 (1999).

[143] M. Midrio, *J. Opt. Soc. Am. B* **17**, 169 (2000).

[144] H. Kogelnik, L. E. Nelson, J. P. Gordon, and R. M. Jopson, *Opt. Lett.* **25**, 19 (2000).

[145] A. O. Dal Forno, A. Paradisi, R. Passy, and J. P. von der Weid, *IEEE Photon. Technol. Lett.* **12**, 296 (2000).

[146] B. Huttner, C. Geiser, and N. Gisin, *IEEE J. Sel. Topics Quantum Electron.* **6**, 317 (2000).

[147] M. Shtaif and A. Mecozzi, *Opt. Lett.* **25**, 707 (2000).

[148] A. Eyal, Y. Li, W. K. Marshall, A. Yariv, and M. Tur, *Opt. Lett.* **25**, 875 (2000).

[149] M. Karlsson, J. Brentel, and P. A. Andrekson, *J. Lightwave Technol.* **18**, 941 (2000).

[150] Y. Li and A. Yariv, *J. Opt. Soc. Am. B* **17**, 1821 (2000).

[151] B. Saleh and M. C. Teich, *Fundamentals of Photonics*, (Wiley, New York, 1991), Chap. 6.

[152] Z. Hass, C. D. Poole, M. A. Santoro, and J. H. Winters, U. S. Patent 5 311 346 (1994).

[153] T. Takahashi, T. Imai, and M. Aiki, *Electron. Lett.* **30**, 348 (1994).

[154] B. W. Haaki, *IEEE Photon. Technol. Lett.* **9**, 121 (1997).

[155] D. A. Watley, K. S. Farley, B. J. Shaw, W. S. Lee, G. Bordogna, A. P. Hadjifotiou, and R. E. Epworth, *Electron. Lett.* **35**, 1094 (1999).

[156] S. Lee, R. Khosravani, J. Peng, V. Grubsky, D. S. Starodubov, A. E. Willner, and J. Feinberg, *IEEE Photon. Technol. Lett.* **11**, 1277 (1999).

[157] R. Noé, D. Sandel, M. Yoshida-Dierolf, S. Hinz *et al.*, *J. Lightwave Technol.* **17**, 1602 (1999).

[158] M. Shtaif, A. Mecozzi, M. Tur, and J. Nagel, *IEEE Photon. Technol. Lett.* **12**, 434 (2000).

[159] H. Y. Pua, K. Peddanarappagari, B. Zhu, C. Allen, K. Demarest, and R. Hui, *J. Lightwave Technol.* **18**, 832 (2000).

[160] C. K. Madsen, *Opt. Lett.* **25**, 878 (2000).

[161] T. Merker, N. Hahnenkamp, and P. Meissner, *Opt. Commun.* **182**, 135 (2000).

[162] L. Möller, *IEEE Photon. Technol. Lett.* **12**, 1258 (2000).

[163] D. Mahgerefteh and C. R. Menyuk, *IEEE Photon. Technol. Lett.* **11**, 340 (1999).

[164] L. F. Mollenauer, K. Smith, J. P. Gordon, and C. R. Menyuk, *Opt. Lett.* **9**, 1219 (1989).

[165] S. G. Evangelides, L. F. Mollenauer, J. P. Gordon, and N. S. Bergano, *J. Lightwave Technol.* **10**, 28 (1992).

[166] T. Ueda and W. L. Kath, *Physica D* **55**, 166 (1992).

[167] C. de Angelis, S. Wabnitz, and M. Haelterman, *Electron. Lett.* **29**, 1568 (1993).

[168] M. Matsumoto, Y. Akagi, and A. Hasegawa, *J. Lightwave Technol.* **15**, 584 (1997).

[169] D. Marcuse, C. R. Menyuk, and P. K. A. Wai, *J. Lightwave Technol.* **15**, 1735 (1997).

[170] T. L. Lakoba and D. J. Kaup, *Phys. Rev. E* **56**, 6147 (1997).

[171] C. de Angelis, P. Franco, M. Romagnoli, *Opt. Commun.* **157**, 161 (1998).

[172] X. Zhang, M. Karlsson, P. A. Andrekson, and K. Bertilsson, *IEEE Photon. Technol. Lett.* **10**, 376 (1998).

[173] X. Zhang, M. Karlsson, P. A. Andrekson, and E. Kolltveit, *IEEE Photon. Technol. Lett.* **10**, 1742 (1998).

[174] C. A. Eleftherianos, D. Syvridis, T. Sphicopoulos, and C. Caroubalos, *Electron. Lett.* **34**, 688 (1998); *Opt. Commun.* **154**, 14 (1998).

[175] E. Kolltveit, P. A. Andrekson, and X. Zhang, *Electron. Lett.* **35**, 319 (1999).

[176] S. M. Baker, J. N. Elgin, and H. J. Harvey, *Opt. Commun.* **165**, 27 (1999).

Chapter 7

Cross-Phase Modulation

So far in this book, only a single electromagnetic wave is assumed to propagate inside optical fibers. When two or more optical fields having different wavelengths propagate simultaneously inside a fiber, they interact with each other through the fiber nonlinearity. In general, such an interaction can generate new waves under appropriate conditions through a variety of nonlinear phenomena such as stimulated Raman or Brillouin scattering, harmonic generation, and four-wave mixing; these topics are covered in Chapters 8–10. The fiber non-linearity can also couple two fields through cross-phase modulation (XPM) without inducing any energy transfer between them. Cross-phase modulation is always accompanied by self-phase modulation (SPM) and occurs because the effective refractive index seen by an optical beam in a nonlinear medium depends not only on the intensity of that beam but also on the intensity of other copropagating beams [1].

The XPM-induced coupling among optical fields gives rise to a number of interesting nonlinear effects in optical fibers. This coupling between two fields of different wavelengths is considered in Section 7.1 where a set of two coupled nonlinear Schrödinger (NLS) equations is obtained. These equations are used in Section 7.2 to discuss how the XPM affects the phenomenon of modulation instability. Similar to the analysis in Section 6.4, this instability can occur even in the normal-dispersion regime of an optical fiber. Section 7.3 focuses on soliton pairs whose members support each other through XPM. The effects of XPM on the shape and the spectrum of copropagating ultrashort pulses are described in Section 7.4. Several applications of XPM-induced coupling in optical fibers are discussed in Section 7.5.

7.1 XPM-Induced Nonlinear Coupling

This section extends the theory of Section 2.3 to the case of two optical pulses at different wavelengths copropagating inside a single-mode fiber. In general, the two optical fields can differ not only in their wavelengths but also in their states of polarization. To simplify the presentation, we first focus on the case in which the two optical fields at different wavelengths are linearly polarized along one of the principal axes of a birefringent fiber. The case of arbitrarily polarized beams is discussed later in this section.

7.1.1 Nonlinear Refractive Index

In the quasi-monochromatic approximation, it is useful to separate the rapidly varying part of the electric field by writing it in the form

$$\mathbf{E}(\mathbf{r},t) = \tfrac{1}{2}\hat{x}[E_1 \exp(-i\omega_1 t) + E_2 \exp(-i\omega_2 t)] + \text{c.c.}, \qquad (7.1.1)$$

where $\hat{x}$ is the polarization unit vector, ω_1 and ω_2 are the carrier frequencies of the two pulses, and the corresponding amplitudes E_1 and E_2 are assumed to be slowly varying functions of time compared with an optical period. This assumption is equivalent to assuming that the spectral width of each pulse satisfies the condition $\Delta\omega_j \ll \omega_j$ ($j = 1, 2$), and holds quite well for pulse widths >0.1 ps. Evolution of the slowly varying amplitudes E_1 and E_2 is governed by the wave equation (2.3.1) with the linear and nonlinear parts of the induced polarization given by Eqs. (2.3.5) and (2.3.6).

To see the origin of XPM, we substitute Eq. (7.1.1) in Eq. (2.3.6) and find that the nonlinear polarization can be written as

$$\begin{aligned}
\mathbf{P}_{\text{NL}}(\mathbf{r},t) = \tfrac{1}{2}\hat{x}[&P_{\text{NL}}(\omega_1)\exp(-i\omega_1 t) + P_{\text{NL}}(\omega_2)\exp(-i\omega_2 t) \\
&+ P_{\text{NL}}(2\omega_1 - \omega_2)\exp[-i(2\omega_1 - \omega_2)t] \\
&+ P_{\text{NL}}(2\omega_2 - \omega_1)\exp[-i(2\omega_2 - \omega_1)t] + \text{c.c.},
\end{aligned} \qquad (7.1.2)$$

where the four terms depend on E_1 and E_2 as

$$P_{\text{NL}}(\omega_1) = \chi_{\text{eff}}(|E_1|^2 + 2|E_2|^2)E_1, \qquad (7.1.3)$$

$$P_{\text{NL}}(\omega_2) = \chi_{\text{eff}}(|E_2|^2 + 2|E_1|^2)E_2, \qquad (7.1.4)$$

$$P_{\text{NL}}(2\omega_1 - \omega_2) = \chi_{\text{eff}}E_1^2 E_2^*, \qquad (7.1.5)$$

$$P_{\text{NL}}(2\omega_2 - \omega_1) = \chi_{\text{eff}}E_2^2 E_1^*, \qquad (7.1.6)$$

with $\chi_{\text{eff}} = (3\varepsilon_0/4)\chi_{xxxx}^{(3)}$ acting as an effective nonlinear parameter.

The induced nonlinear polarization in Eq. (7.1.2) has terms oscillating at the new frequencies $2\omega_1 - \omega_2$ and $2\omega_2 - \omega_1$. These terms result from the phenomenon of four-wave mixing discussed in Chapter 10. It is necessary to satisfy the phase-matching condition if the new frequency components are to build up significantly, a condition not generally satisfied in practice unless special precautions are taken. The four-wave-mixing terms are neglected in this chapter after assuming that phase matching does not occur. The remaining two terms provide a nonlinear contribution to the refractive index. This can be seen writing $P_{\text{NL}}(\omega_j)$ in the form $(j = 1, 2)$

$$P_{\text{NL}}(\omega_j) = \varepsilon_0 \varepsilon_j^{\text{NL}} E_j, \tag{7.1.7}$$

and combining it with the linear part so that the total induced polarization is given by

$$P(\omega_j) = \varepsilon_0 \varepsilon_j E_j, \tag{7.1.8}$$

where

$$\varepsilon_j = \varepsilon_j^L + \varepsilon_j^{\text{NL}} = (n_j^L + \Delta n_j)^2, \tag{7.1.9}$$

n_j^L is the linear part of the refractive index and Δn_j is the change induced by the third-order nonlinear effects. Using the approximation $\Delta n_j \ll n_j^L$, the nonlinear part of the refractive index is given by $(j = 1, 2)$

$$\Delta n_j \approx \varepsilon_j^{\text{NL}}/2n_j \approx n_2(|E_j|^2 + 2|E_{3-j}|^2), \tag{7.1.10}$$

where $n_1^L \approx n_2^L = n$ has been assumed. The nonlinear parameter n_2 is defined as in Eq. (2.3.13).

Equation (7.1.10) shows that the refractive index seen by an optical field inside an optical fiber depends not only on the intensity of that field but also on the intensity of other copropagating fields [2]–[4]. As the optical field propagates inside the fiber, it acquires an intensity-dependent nonlinear phase shift

$$\phi_j^{\text{NL}}(z) = (\omega_j/c)\Delta n_j z = n_2(\omega_j/c)(|E_j|^2 + 2|E_{3-j}|^2)z, \tag{7.1.11}$$

where $j = 1$ or 2. The first term is responsible for SPM discussed in Chapter 4. The second term results from phase modulation of one wave by the copropagating wave and is responsible for XPM. The factor of 2 on the right-hand side of Eq. (7.1.11) shows that XPM is twice as effective as SPM for the same intensity [1]. Its origin can be traced back to the number of terms that contribute to the triple sum implied in Eq. (2.3.6). Qualitatively speaking, the number of terms doubles when the two optical frequencies are distinct compared with that when the frequencies are degenerate. The XPM-induced phase shift in optical fibers was measured as early as 1984 by injecting two continuous-wave (CW) beams into a 15-km-long fiber [3]. Soon after, picosecond pulses were used to observe the XPM-induced spectral changes [4]–[6].

7.1.2 Coupled NLS Equations

The pulse-propagation equations for the two optical fields can be obtained by following the procedure of Section 2.3. Assuming that the nonlinear effects do not affect significantly the fiber modes, the transverse dependence can be factored out writing $E_j(\mathbf{r},t)$ in the form

$$E_j(\mathbf{r},t) = F_j(x,y)A_j(z,t)\exp(i\beta_{0j}z), \qquad (7.1.12)$$

where $F_j(x,y)$ is the transverse distribution of the fiber mode for the jth field $(j = 1,2)$, $A_j(z,t)$ is the slowly varying amplitude, and β_{0j} is the corresponding propagation constant at the carrier frequency ω_j. The dispersive effects are included by expanding the frequency-dependent propagation constant $\beta(\omega)$ for each wave in a way similar to Eq. (2.3.23) and retaining only up to the quadratic term. The resulting propagation equation for $A_j(z,t)$ becomes

$$\frac{\partial A_j}{\partial z} + \beta_{1j}\frac{\partial A_j}{\partial t} + \frac{i\beta_{2j}}{2}\frac{\partial^2 A_j}{\partial t^2} + \frac{\alpha_j}{2}A_j$$
$$= in_2(\omega_j/c)(f_{jj}|A_j|^2 + 2f_{jk}|A_k|^2), \qquad (7.1.13)$$

where $k \neq j$, $\beta_{1j} = 1/v_{gj}$, v_{gj} is the group velocity, β_{2j} is the GVD coefficient, and α_j is the loss coefficient. The overlap integral f_{jk} is defined as

$$f_{jk} = \frac{\iint_{-\infty}^{\infty}|F_j(x,y)|^2|F_k(x,y)|^2 dxdy}{\left(\iint_{-\infty}^{\infty}|F_j(x,y)|^2 dxdy\right)\left(\iint_{-\infty}^{\infty}|F_k(x,y)|^2 dxdy\right)}. \qquad (7.1.14)$$

The differences among the overlap integrals can be significant in multimode fibers if the two waves propagate in different fiber modes. Even in single-mode fibers, f_{11}, f_{22}, and f_{12} differ from each other because of the frequency dependence of the modal distribution $F_j(x,y)$. The difference is small, however, and can be neglected in practice. In that case, Eq. (7.1.13) can be written as the following set of two coupled NLS equations [7]–[10]

$$\frac{\partial A_1}{\partial z} + \frac{1}{v_{g1}} \frac{\partial A_1}{\partial t} + \frac{i\beta_{21}}{2} \frac{\partial^2 A_1}{\partial t^2} + \frac{\alpha_1}{2} A_1$$
$$= i\gamma_1 (|A_1|^2 + 2|A_2|^2) A_1, \qquad (7.1.15)$$

$$\frac{\partial A_2}{\partial z} + \frac{1}{v_{g2}} \frac{\partial A_2}{\partial t} + \frac{i\beta_{22}}{2} \frac{\partial^2 A_2}{\partial t^2} + \frac{\alpha_2}{2} A_2$$
$$= i\gamma_2 (|A_2|^2 + 2|A_1|^2) A_2, \qquad (7.1.16)$$

where the nonlinear parameter γ_j is defined as in Eq. (2.3.28),

$$\gamma_j = n_2 \omega_j / (c A_{\text{eff}}), \qquad (j = 1, 2), \qquad (7.1.17)$$

and A_{eff} is the effective core area ($A_{\text{eff}} = 1/f_{11}$), assumed to be the same for both optical waves. Typically $A_{\text{eff}} = 50 \ \mu m^2$ in the 1.55-μm wavelength region. The corresponding values of γ_1 and γ_2 are $\sim 1 \ W^{-1}$/km depending on the frequencies ω_1 and ω_2. Generally, the two pulses not only have different GVD coefficients but also propagate at different speeds because of the difference in their group velocities. The group-velocity mismatch plays an important role as it limits the XPM interaction as pulses walk off from each other. One can define the walk-off length L_W using Eq. (1.2.14); it is a measure of the fiber length during which two overlapping pulses separate from each other as a result of the group-velocity mismatch.

7.1.3   Propagation in Birefringent Fibers

In a birefringent fiber, the state of polarization of both waves changes with propagation. The orthogonally polarized components of each wave are then mutually coupled through XPM. The total optical field can be written as

$$\mathbf{E}(\mathbf{r}, t) = \frac{1}{2} \left[(\hat{x} E_{1x} + \hat{y} E_{1y}) e^{-i\omega_1 t} + (\hat{x} E_{2x} + \hat{y} E_{2y}) e^{-i\omega_2 t} \right] + \text{c.c.} \qquad (7.1.18)$$

The slowly varying amplitudes A_{1x}, A_{1y}, A_{2x}, and A_{2y} can be introduced similarly to Eq. (7.1.12) and the coupled amplitude equations for them are obtained by following the same method. These equations are quite complicated in the general case that includes the coherent-couplings terms similar to those present in Eqs. (6.1.12) and (6.1.13). However, they are considerably simplified in the case of high-birefringence fibers because such terms can then be neglected. The resulting set of four coupled NLS equations becomes [11]

$$\frac{\partial A_{1p}}{\partial z} + \frac{1}{v_{g1p}}\frac{\partial A_{1p}}{\partial t} + \frac{i\beta_{21}}{2}\frac{\partial^2 A_{1p}}{\partial t^2} + \frac{\alpha_1}{2}A_{1p}$$
$$= i\gamma_1(|A_{1p}|^2 + 2|A_{2p}|^2 + B|A_{1q}|^2 + B|A_{2q}|^2)A_{1p}, \quad (7.1.19)$$

$$\frac{\partial A_{2p}}{\partial z} + \frac{1}{v_{g2p}}\frac{\partial A_{2p}}{\partial t} + \frac{i\beta_{22}}{2}\frac{\partial^2 A_{2p}}{\partial t^2} + \frac{\alpha_2}{2}A_{2p}$$
$$= i\gamma_2(|A_{2p}|^2 + 2|A_{1p}|^2 + B|A_{1q}|^2 + B|A_{2q}|^2)A_{2p}, \quad (7.1.20)$$

where $p = x, y$ and $q = x, y$ such that $p \neq q$. The parameter B is given in Eq. (6.1.21) and equals 2/3 for linearly birefringent fibers. These equations reduce to Eqs. (7.1.15) and (7.1.16) when both waves are polarized along a principal axis ($A_{1y} = A_{2y} = 0$).

7.2 XPM-Induced Modulation Instability

This section extends the analysis of Section 5.1 to the case in which two CW beams of different wavelengths propagate inside a fiber simultaneously. Similar to the single-beam case, modulation instability is expected to occur in the anomalous-GVD region of the fiber. The main issue is whether XPM-induced coupling can destabilize the CW state even when one or both beams experience normal GVD [12]–[19].

7.2.1 Linear Stability Analysis

The following analysis is similar to that of Section 6.4.2. The main difference is that XPM-induced coupling is stronger and the parameters β_2 and γ are different for the two beams because of their different wavelengths. As usual, the steady-state solution is obtained by setting the time derivatives in Eqs. (7.1.15) and (7.1.16) to zero. If fiber losses are neglected, the solution is of the form

$$\bar{A}_j = \sqrt{P_j}\exp(i\phi_j), \quad (7.2.1)$$

where $j = 1$ or 2, P_j is the incident optical power, and ϕ_j is the nonlinear phase shift acquired by the jth field and given by

$$\phi_j(z) = \gamma_j(P_j + 2P_{3-j})z. \quad (7.2.2)$$

Following the procedure of Section 5.1, stability of the steady state is examined assuming a time-dependent solution of the form

$$A_j = \left(\sqrt{P_j} + a_j\right) \exp(i\phi_j), \tag{7.2.3}$$

where $a_j(z,t)$ is a small perturbation. By using Eq. (7.2.3) in Eqs. (7.1.15) and (7.1.16) and linearizing in a_1 and a_2, the perturbations a_1 and a_2 satisfy the following set of two coupled linear equations:

$$\frac{\partial a_1}{\partial z} + \frac{1}{v_{g1}} \frac{\partial a_1}{\partial t} + \frac{i\beta_{21}}{2} \frac{\partial^2 a_1}{\partial t^2}$$
$$= i\gamma_1 P_1(a_1 + a_1^*) + 2i\gamma_1 (P_1 P_2)^{1/2}(a_2 + a_2^*), \tag{7.2.4}$$

$$\frac{\partial a_2}{\partial z} + \frac{1}{v_{g2}} \frac{\partial a_2}{\partial t} + \frac{i\beta_{22}}{2} \frac{\partial^2 a_2}{\partial t^2}$$
$$= i\gamma_2 P_2(a_2 + a_2^*) + 2i\gamma_2 (P_1 P_2)^{1/2}(a_1 + a_1^*), \tag{7.2.5}$$

where the last term is due to XPM.

The above set of linear equations has the following general solution:

$$a_j = u_j \exp[i(Kz - \Omega t)] + iv_j \exp[-i(Kz - \Omega t)], \tag{7.2.6}$$

where $j = 1, 2$, K is the wave number and Ω is the frequency of perturbation. Equations (7.2.4)–(7.2.6) provide a set of four homogeneous equations for u_1, u_2, v_1, and v_2. This set has a nontrivial solution only when the perturbation satisfies the following dispersion relation:

$$[(K - \Omega/v_{g1})^2 - f_1][(K - \Omega/v_{g2})^2 - f_2] = C_{\text{XPM}}, \tag{7.2.7}$$

where

$$f_j = \tfrac{1}{2}\beta_{2j}\Omega^2(\tfrac{1}{2}\beta_{2j}\Omega^2 + 2\gamma_j P_j) \tag{7.2.8}$$

for $j = 1, 2$. The coupling parameter C_{XPM} is defined as

$$C_{\text{XPM}} = 4\beta_{21}\beta_{22}\gamma_1\gamma_2 P_1 P_2 \Omega^4. \tag{7.2.9}$$

The steady-state solution becomes unstable if for some values of Ω the wave number K has an imaginary part. The perturbations a_1 and a_2 then experience an exponential growth along the fiber length. In the absence of XPM coupling

($C_{\mathrm{XPM}} = 0$), Eq. (7.2.7) shows that the analysis of Section 5.1 applies to each wave independently.

In the presence of XPM coupling, Eq. (7.2.7) provides a fourth-degree polynomial in K whose roots determine the conditions under which K becomes complex. In general, these roots are obtained numerically. If the wavelengths of the two optical beams are so close to each other that the group-velocity mismatch is negligible or are located on opposite sides of the zero-dispersion wavelength such that ($v_{g1} \approx v_{g2}$), the four roots are given by [12]

$$K = \Omega/v_{g1} \pm \{\tfrac{1}{2}(f_1 + f_2) \pm [(f_1 - f_2)^2/4 + C_{\mathrm{XPM}}]^{1/2}\}^{1/2}. \qquad (7.2.10)$$

It is easy to verify that K can become complex only if $C_{\mathrm{XPM}} > f_1 f_2$. Using Eqs. (7.2.8) and (7.2.9), the condition for modulation instability to occur can be written as

$$[\Omega^2/\Omega_{c1}^2 + \mathrm{sgn}(\beta_{21})][\Omega^2/\Omega_{c2}^2 + \mathrm{sgn}(\beta_{22})] < 4, \qquad (7.2.11)$$

where Ω_{c1} and Ω_{c2} are defined as

$$\Omega_{cj} = (4\gamma_j P_j/|\beta_{2j}|)^{1/2}, \qquad (7.2.12)$$

with $j = 1$ or 2. When the condition (7.2.11) is satisfied, the gain spectrum of modulation instability is obtained from $g(\Omega) = 2\,\mathrm{Im}(K)$.

The modulation-instability condition (7.2.11) shows that there is a range of Ω over which the gain $g(\Omega)$ exists. The steady-state solution (7.2.3) is unstable to perturbations at those frequencies. The most important conclusion drawn from Eq. (7.2.11) is that modulation instability can occur irrespective of the signs of the GVD coefficients. Thus, whereas modulation instability requires anomalous GVD in the case of a single beam (see Section 5.1), it can occur in the two-beam case even if both beams experience normal GVD. The frequency range over which $g(\Omega) > 0$ depends on whether β_{21} and β_{22} are both positive, both negative, or one positive and the other negative. The smallest frequency range corresponds to the case in which both beams are in the normal-dispersion regime of the fiber. Because modulation instability in that case is due solely to XPM, only this case is discussed further.

Figure 7.1 shows the gain spectra of XPM-induced modulation instability for silica fibers in the visible region near 0.53 μm choosing $\beta_{2j} = 60$ ps^2/km and $\gamma_j = 15$ W^{-1}/km in Eq. (7.2.8). The group-velocity mismatch is neglected in the left graph where different curves correspond to values of the power ratio

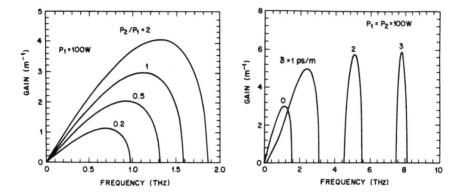

Figure 7.1 Gain spectra of XPM-induced modulation instability in the normal-dispersion regime of a fiber for (i) different power ratios with $\delta = 0$ and (ii) different values of δ with equal beam powers. (After Ref. [12].)

P_2/P_1 in the range 0–2. The right graph shows the effect of group-velocity mismatch for equal beam powers by varying the parameter $\delta = |v_{g1}^{-1} - v_{g2}^{-1}|$ in the range 0–3 ps/m. These results show that XPM-induced modulation instability can occur in the normal-GVD regime for relatively small values of δ. The peak gain of about 5 m^{-1} at the 100-W power level implies that the instability can develop in fibers a few meters long.

The set of four coupled equations obtained in Section 7.1.3 should be used to investigate XPM-induced modulation instability when two elliptically polarized CW beams propagate in a high-birefringence fiber. The new feature is that the dispersion relation, Eq. (7.2.7), becomes an eighth-degree polynomial in K. If one of the beams is polarized along a principal axis of the fiber, the dispersion relation reduces to a sixth-degree polynomial in K. The gain spectrum of XPM-induced modulation instability then depends on the polarization angle of the other beam [11]. In general, the gain bandwidth is reduced when the two beams are not polarized along the same axis.

7.2.2 Experimental Results

The experimental attempts to observe the XPM-induced modulation instability for normal GVD have focused mostly on the case of two polarization components of a single beam (see Section 6.4). It appears that this instability is difficult to observe in the case of two beams with different wavelengths. The reason is related to the fact that Eqs. (7.1.15) and (7.1.16) neglect four-wave mixing. The neglect of the four-wave-mixing terms can be justified when the wavelength difference is so large that phase matching cannot occur [17]–[19]. However, to observe modulation instability, the wavelength difference needs to be reduced to ~ 1 nm or less. Four-wave mixing then becomes nearly phase matched and cannot be ignored. Indeed, a careful analysis that includes GVD to all orders shows that XPM-induced modulation instability is not likely to occur in the normal-dispersion region of conventional silica fibers [19]. It can occur in especially designed dispersion-flattened fibers in which two normal-GVD regions are separated by an intermediate wavelength region of anomalous GVD. In such fibers, it is possible to match the group velocities even when the wavelengths of two beams differ by 100 nm or more.

The XPM-induced modulation instability has been observed when one of the beams propagates in the normal-GVD region while the other beam experiences anomalous dispersion. In one experiment [20], a pump-probe configuration was used such that the 1.06-μm pump pulses experienced normal GVD while 1.32-μm probe pulses propagated in the anomalous-GVD regime of the fiber. When the pump and probe pulses were launched simultaneously, the probe developed modulation sidebands, with a spacing of 260 GHz at the 0.4-W peak-power level of pump pulses, as a result of XPM-induced modulation instability. This configuration can be used to advantage if the pump beam is in the form of intense pulses whereas the other beam forms a weak CW signal. The weak CW beam can be converted into a train of ultrashort pulses because it is amplified through XPM-induced modulation instability only when the two waves are present simultaneously [8].

In an experimental realization of the preceding idea [21], 100-ps pump pulses were obtained from a 1.06-μm, mode-locked, Nd:YAG laser while an external-cavity semiconductor laser provided the weak CW signal (power <0.5 mW) whose wavelength was tunable over 1.43-1.56 μm. The zero-dispersion wavelength of the 1.2-km-long optical fiber was near 1.273 μm such that the group velocities were nearly equal at 1.06 and 1.51 μm. When 60-μW signal power was coupled into the fiber together with the pump pulses (peak power >500 W), the signal spectrum developed sidebands indicative of the XPM-induced modulation instability. The experimental results were in qualitative agreement with the numerical solutions of Eqs. (7.1.15) and (7.1.16) and indicated that the CW signal was converted into a train of picosecond pulses.

This technique has been used to generate a 10-GHz pulse train by launching the CW signal from a 1543-nm semiconductor laser into a fiber together with 13.7-ps pump pulses (10-GHz repetition rate) obtained from a 1558-nm, mode-locked semiconductor laser [22]. The 11-km-long dispersion-shifted fiber had its zero-dispersion wavelength at 1550 nm, resulting in nearly equal group velocities at the pump and signal wavelengths. The streak-camera measurements indicated that XPM-induced modulation instability converted the CW signal into a train of 7.4-ps pulses. If pump pulses were coded to carry digital information, signal pulses will reproduce the information faithfully because the XPM interaction requires the presence of a pump pulse. Such a device is useful for wavelength conversion of signals in optical communication systems.

7.3   XPM-Paired Solitons

Similar to the case of vector solitons discussed in Section 6.5.3, the XPM-induced modulation instability indicates that the coupled NLS equations may have solitary-wave solutions in the form of two paired solitons that preserve their shape through the XPM interaction. In fact, solitonic and periodic solutions of the coupled NLS equations have been studied since 1977 [23]–[43]. Because such solutions specify intensity profiles of both pulses and always occur in pairs, they are referred to as XPM-paired solitons (also called symbiotic solitons). Some of such paired solutions have been discussed in Section 6.5.3 in the context of vector solitons. However, in that case the two polarization components of a single beam experience the same GVD (either normal or anomalous). In the general case discussed here, the carrier frequencies of two solitons can be different enough that the two members of the soliton pair can have different signs for the GVD parameter.

7.3.1   Bright–Dark Soliton Pair

Solitons paired by XPM represent the specific solutions of Eqs. (7.1.15) and (7.1.16) for which the pulse shape does not change with z although the phase may vary along the fiber. Such solutions are not solitons in a strict mathematical sense and should be referred to more accurately as solitary waves. The group-velocity mismatch represents the biggest hurdle for the existence of XPM-paired solitons. It is possible to realize equal group velocities ($v_{g1} = v_{g2}$) if the wavelengths of two optical waves are chosen appropriately on opposite sides of the zero-dispersion wavelength such that one wave experiences normal GVD while the other wave lies in the anomalous-GVD region. Indeed, several examples of XPM-paired solitons were discovered exactly under such operating conditions [25]–[27].

An interesting example is provided by the bright–dark soliton pair formed when $\beta_{21} < 0$ and $\beta_{22} > 0$. If fiber losses are ignored ($\alpha_1 = \alpha_2 = 0$) and group velocities are assumed to be equal by setting $v_{g1} = v_{g2} = v_g$ in Eqs. (7.1.15) and (7.1.16), a bright–dark soliton pair is given by [25]

$$A_1(z,t) = B_1 \tanh[W(t - z/V)] \exp[i(K_1 z - \Omega_1 t)], \tag{7.3.1}$$
$$A_2(z,t) = B_2 \mathrm{sech}[W(t - z/V)] \exp[i(K_2 z - \Omega_2 t)], \tag{7.3.2}$$

where the soliton amplitudes are determined from

$$B_1^2 = (2\gamma_1 \beta_{22} + \gamma_2 |\beta_{21}|) W^2 / (3\gamma_1 \gamma_2), \tag{7.3.3}$$
$$B_2^2 = (2\gamma_2 |\beta_{21}| + \gamma_1 \beta_{22}) W^2 / (3\gamma_1 \gamma_2), \tag{7.3.4}$$

the wave numbers K_1 and K_2 are given by

$$K_1 = \gamma_1 B_1^2 - |\beta_{21}| \Omega_1^2 / 2, \qquad K_2 = \beta_{22}(\Omega_2^2 - W^2)/2, \tag{7.3.5}$$

and the effective group velocity of the soliton pair is obtained from

$$V^{-1} = v_g^{-1} - |\beta_{21}| \Omega_1 = v_g^{-1} + \beta_{22} \Omega_2. \tag{7.3.6}$$

As seen from Eq. (7.3.6), the frequency shifts Ω_1 and Ω_2 must have opposite signs and cannot be chosen independently. The parameter W governs the pulse width and determines the soliton amplitudes through Eqs. (7.3.3) and (7.3.4). Thus, two members of the soliton pair have the same width, the same group velocity, but different shapes and amplitudes such that they support each other through the XPM coupling. In fact, their shapes correspond to bright and dark solitons discussed in Chapter 5. The most striking feature of this soliton pair is that the dark soliton propagates in the anomalous-GVD regime whereas the bright soliton propagates in the normal-GVD regime, exactly opposite of the behavior expected in the absence of XPM. The physical mechanism behind such an unusual pairing can be understood as follows. Because XPM is twice as strong as SPM, it can counteract the temporal spreading of an optical pulse induced by the combination of SPM and normal GVD, provided the XPM-induced chirp is of the opposite kind than that produced by SPM. A dark soliton can generate this kind of chirp. At the same time, the XPM-induced chirp on the dark soliton is such that the pair of bright and dark solitons can support each other in a symbiotic manner.

7.3.2   Bright–Gray Soliton Pair

A more general form of an XPM-coupled soliton pair can be obtained by solving Eqs. (7.1.15) and (7.1.16) with the postulate

$$A_j(z,t) = Q_j(t - z/V)\exp[i(K_j z - \Omega_j t + \phi_j)], \qquad (7.3.7)$$

where V is the common velocity of the soliton pair, Q_j governs the soliton shape, K_j and Ω_j represent changes in the propagation constant and the frequency of two solitons, and ϕ_j is the phase ($j = 1, 2$). The resulting solution has the form [32]

$$Q_1(\tau) = B_1[1 - b^2\mathrm{sech}^2(W\tau)], \qquad Q_2(\tau) = B_2\mathrm{sech}(W\tau), \qquad (7.3.8)$$

where $\tau = t - z/V$. The parameters W and b depend on the soliton amplitudes, B_1 and B_2, and on fiber parameters through the relations

$$W = \left(\frac{3\gamma_1\gamma_2}{2\gamma_1\beta_{22} - 4\gamma_2\beta_{21}}\right)^{1/2} B_2, \qquad b = \left(\frac{2\gamma_1\beta_{22} - \gamma_2\beta_{21}}{\gamma_1\beta_{22} - 2\gamma_2\beta_{21}}\right)^{1/2}\frac{B_2}{B_1}. \quad (7.3.9)$$

The constants K_1 and K_2 are also fixed by various fiber parameters and soliton amplitudes. The phase of the bright soliton is constant but the dark-soliton phase ϕ_1 is time-dependent. The frequency shifts Ω_1 and Ω_2 are related to the soliton-pair speed as in Eq. (7.3.6).

The new feature of the XPM-coupled soliton pair in Eq. (7.3.8) is that the dark soliton is of "gray" type. The parameter b controls the depth of the intensity dip associated with a gray soliton. Both solitons have the same width W but different amplitudes. A new feature is that the two GVD parameters can be positive or negative. However, the soliton pair exists only under certain conditions. The solution is always possible if $\beta_{21} < 0$ and $\beta_{22} > 0$ and does not exist when $\beta_{21} > 0$ and $\beta_{22} < 0$. As discussed before, this behavior is opposite to what would normally be expected and is due solely to XPM. If both solitons experience normal GVD, the bright–gray soliton pair can exist if $\gamma_1\beta_{22} > 2\gamma_2\beta_{21}$. Similarly, if both solitons experience anomalous GVD, the soliton pair can exist if $2\gamma_1|\beta_{22}| < \gamma_2|\beta_{21}|$.

7.3.3   Other Soliton Pairs

The soliton-pair solutions given in the preceding are not the only possible solutions of Eqs. (7.1.15) and (7.1.16). These equations also support pairs with two bright or two dark solitons depending on various parameter values [27]. Moreover, the XPM-supported soliton pairs can exist even when group velocities are not equal because, similar to soliton trapping in birefringent fibers (see Section 6.5), two pulses can shift their carrier frequencies to equalize their group velocities. A simple way to find the conditions under which XPM-paired solitons can exist is to postulate an appropriate solution, substitute it in Eqs. (7.1.15) and (7.1.16), and then investigate whether soliton parameters (amplitude, width, group velocity, frequency shift, and wave number) can be determined with physically possible values [32]–[34]. As an example, consider the case when Eqs. (7.3.1) and (7.3.2) describe the postulated solution. Assume also that $K_1 = K_2$ and $\Omega_1 = \Omega_2$ so that the frequency shifts are equal. It turns out that the postulated solution is always possible if $\beta_{21} < 0$ and $\beta_{22} > 0$, but exists only under certain conditions if β_{21} and β_{22} have the same sign [34]. Further, the assumption $\Omega_1 = \Omega_2$ can be relaxed to obtain another set of solitary-wave solutions. Stability of the XPM-paired solitons is not always guaranteed and should be checked through numerical simulations.

The set of four coupled equations obtained in Section 7.1.3 should be used to study whether the XPM-paired vector solitons exist in birefringent fibers. By following the method discussed in the preceding, one finds that such soliton solutions indeed exist [11]. Depending on the parameter values, birefringent fibers can support a pair of bright vector solitons or a pair composed of one dark and one bright vector soliton. The XPM interaction of two elliptically polarized beams appears to have a rich variety of interesting features.

The coupled NLS equations (7.1.15) and (7.1.16) also have periodic solutions that represent two pulse trains that propagate undistorted through an optical fiber because of the XPM-induced coupling between them. One such periodic solution in terms of the elliptic functions was found in 1989 in the specific case in which both pulse trains have the same group velocity and experienced anomalous GVD inside the fiber [29]. By 1998, nine periodic solutions, written as different combinations of the elliptic functions, have been found [43]. All of these solutions assume equal group velocities and anomalous GVD for the two pulse trains. A further generalization considers XPM-induced coupling among more than two optical fields. In this case, one needs to solve a set of multiple coupled NLS equations of the form

$$\frac{\partial A_j}{\partial z} + \frac{1}{v_{gj}} \frac{\partial A_j}{\partial t} + \frac{i\beta_{2j}}{2} \frac{\partial^2 A_j}{\partial t^2} = i \left(\sum_{k=1}^{M} \gamma_{jk} |A_k|^2 \right) A_j, \tag{7.3.10}$$

where $j = 1–M$. These equations have periodic as well as soliton-pair solutions for certain combinations of parameter values [44].

7.4   Spectral and Temporal Effects

This section considers the spectral and temporal changes occurring as a result of XPM interaction between two copropagating pulses with nonoverlapping spectra [45]–[51]. For simplicity, the polarization effects are ignored assuming that the input beams preserve their polarization during propagation. Equations (7.1.15) and (7.1.16) then govern evolution of two pulses along the fiber length and include the effects of group-velocity mismatch, GVD, SPM, and XPM. If fiber losses are neglected for simplicity, these equations become

$$\frac{\partial A_1}{\partial z} + \frac{i\beta_{21}}{2}\frac{\partial^2 A_1}{\partial T^2} = i\gamma_1(|A_1|^2 + 2|A_2|^2)A_1, \qquad (7.4.1)$$

$$\frac{\partial A_2}{\partial z} + d\frac{\partial A_2}{\partial T} + \frac{i\beta_{22}}{2}\frac{\partial^2 A_2}{\partial T^2} = i\gamma_2(|A_2|^2 + 2|A_1|^2)A_2, \qquad (7.4.2)$$

where

$$T = t - \frac{z}{v_{g1}}, \qquad d = \frac{v_{g1} - v_{g2}}{v_{g1}v_{g2}}. \qquad (7.4.3)$$

Time T is measured in a reference frame moving with one pulse traveling at speed v_{g1}. The parameter d is a measure of group-velocity mismatch between the two pulses.

In general, two pulses can have different widths. Using the width T_0 of the first pulse at the wavelength λ_1 as a reference, we introduce the walk-off length L_W and the dispersion length L_D as

$$L_W = T_0/|d|, \qquad L_D = T_0^2/|\beta_{21}|. \qquad (7.4.4)$$

Depending on the relative magnitudes of L_W, L_D, and the fiber length L, the two pulses can evolve very differently. If L is small compared to both L_W and L_D, the dispersive effects do not play a significant role and can be neglected. This can occur for $T_0 > 1$ ns and $L < 10$ m if the center wavelengths of the two pulses are within 10 nm of each other ($|d| \sim 10$ ps/m). In this quasi-CW situation, the steady-state solution of Section 7.3 is applicable. If $L_W < L$ but $L_D \gg L$, the second derivatives in Eqs. (7.4.1) and (7.4.2) can be neglected but the first derivatives must be retained. Even though the pulse shape does not change, the combination of group-velocity mismatch and the nonlinearity-induced frequency chirp can affect the spectrum drastically. This is generally the case for $T_0 \sim 100$ ps, $L \sim 10$ m, and $d < 10$ ps/m. Finally, for ultrashort pulses ($T_0 < 10$ ps), the GVD terms should also be included; XPM then affects both the pulse shape and the spectrum. Both of these cases are discussed in what follows.

7.4.1 Asymmetric Spectral Broadening

Consider first the simple case $L \ll L_D$ for which the second-derivative terms in Eqs. (7.4.1) and (7.4.2) can be neglected. The group-velocity mismatch is included through the parameter d assuming $L_W < L$. As the pulse shapes do not change in the absence of GVD, Eqs. (7.4.1) and (7.4.2) can be solved analytically. The general solution at $z = L$ is given by [47]

$$A_1(L,T) = A_1(0,T)e^{i\phi_1}, \qquad A_2(L,T) = A_2(0,T-dL)e^{i\phi_2}, \qquad (7.4.5)$$

where the time-dependent nonlinear phase shifts are obtained from

$$\phi_1(T) = \gamma_1 \left(L|A_1(0,T)|^2 + 2\int_0^L |A_2(0,T-zd)|^2 \, dz \right), \qquad (7.4.6)$$

$$\phi_2(T) = \gamma_2 \left(L|A_2(0,T)|^2 + 2\int_0^L |A_1(0,T+zd)|^2 \, dz \right). \qquad (7.4.7)$$

The physical interpretation of Eqs. (7.4.5)–(7.4.7) is clear. As the pulse propagates through the fiber, its phase is modulated because of the intensity dependence of the refractive index. The modulated phase has two contributions. The first term in Eqs. (7.4.6) and (7.4.7) is due to SPM (see Section 4.1). The second term has its origin in XPM. Its contribution changes along the fiber length because of the group-velocity mismatch. The total XPM contribution to the phase is obtained by integrating over the fiber length.

The integration in Eqs. (7.4.6) and (7.4.7) can be carried out for specific pulse shapes. As an illustration, consider the case of two unchirped Gaussian pulses of the same width T_0 with the initial amplitudes

$$A_1(0,T) = \sqrt{P_1}\exp\left(-\frac{T^2}{2T_0^2}\right), \qquad A_2(0,T) = \sqrt{P_2}\exp\left(-\frac{(T-T_d)^2}{2T_0^2}\right),$$
$$(7.4.8)$$

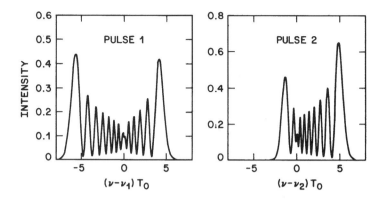

Figure 7.2 Optical spectra of two copropagating pulses exhibiting XPM-induced asymmetric spectral broadening. The parameters are $\gamma_1 P_1 L = 40$, $P_2/P_1 = 0.5$, $\gamma_2/\gamma_1 = 1.2$, $\tau_d = 0$, and $L/L_W = 5$.

where P_1 and P_2 are the peak powers and T_d is the initial time delay between the two pulses. Substituting Eq. (7.4.8) in Eq. (7.4.6), ϕ_1 is given by

$$\phi_1(\tau) = \gamma_1 L \left(P_1 e^{-\tau^2} + P_2 \frac{\sqrt{\pi}}{\delta} [\mathrm{erf}(\tau - \tau_d) - \mathrm{erf}(\tau - \tau_d - \delta)] \right), \qquad (7.4.9)$$

where $\mathrm{erf}(x)$ stands for the error function and

$$\tau = \frac{T}{T_0}, \qquad \tau_d = \frac{T_d}{T_0}, \qquad \delta = \frac{dL}{T_0}. \qquad (7.4.10)$$

A similar expression can be obtained for $\phi_2(T)$ using Eq. (7.4.7).

As discussed in Section 4.1, the time dependence of the phase manifests as spectral broadening. Similar to the case of pure SPM, the spectrum of each pulse is expected to broaden and develop a multipeak structure. However, the spectral shape is now governed by the combined contributions of SPM and XPM to the pulse phase. Figure 7.2 shows the spectra of two pulses using $\gamma_1 P_1 L = 40$, $P_2/P_1 = 0.5$, $\gamma_2/\gamma_1 = 1.2$, $\tau_d = 0$, and $\delta = 5$. These parameters correspond to an experimental situation in which a pulse at 630 nm, with 100-W peak power, was launched inside a fiber together with another pulse at 530 nm with 50-W peak power such that $T_d = 0$, $T_0 = 10$ ps, and $L = 5$ m. The most noteworthy feature of Fig. 7.2 is spectral asymmetry that is due solely to XPM. In the absence of XPM interaction the two spectra would be symmetric and would exhibit less broadening. The spectrum of pulse 2 is more asymmetric because the XPM contribution is larger for this pulse ($P_1 = 2P_2$).

A qualitative understanding of the spectral features seen in Fig. 7.2 can be developed from the XPM-induced frequency chirp using

$$\Delta v_1(\tau) = \frac{-1}{2\pi}\frac{\partial \phi_1}{\partial T} = \frac{\gamma_1 L}{\pi T_0}\left[P_1 \tau e^{-\tau^2} - \frac{P_2}{\delta}\left(e^{-(\tau - \tau_d)^2} - e^{-(\tau - \tau_d - \delta)^2}\right)\right],$$

(7.4.11)

where Eq. (7.4.9) was used. For $\tau_d = 0$ and $|\delta| \ll 1$ ($L \ll L_W$), the chirp is given by the simple relation

$$\Delta v_1(\tau) \approx \frac{\gamma_1 L}{\pi T_0}e^{-\tau^2}[P_1 \tau + P_2(2\tau - \delta)].$$

(7.4.12)

The chirp for pulse 2 is obtained following the same procedure and is given by

$$\Delta v_2(\tau) \approx \frac{\gamma_2 L}{\pi T_0}e^{-\tau^2}[P_2 \tau + P_1(2\tau + \delta)].$$

(7.4.13)

For positive values of δ, the chirp is larger near the leading edge for pulse 1 while the opposite occurs for pulse 2. Because the leading and trailing edges carry red- and blue-shifted components, respectively, the spectrum of pulse 1 is shifted toward red while that of pulse 2 is shifted toward blue. This is precisely what occurs in Fig. 7.2. The spectrum of pulse 2 shifts more because the XPM contribution is larger for it when $P_1 > P_2$. When $P_1 = P_2$ and $\gamma_1 \approx \gamma_2$, the spectra of two pulses would be the mirror images of each other.

The qualitative features of spectral broadening can be quite different if the two pulses do not overlap initially but have a relative time delay [47]. To isolate the effects of XPM, it is useful to consider the pump-probe configuration assuming $P_1 \ll P_2$. The pump-induced chirp imposed on the probe pulse is obtained from Eq. (7.4.11) by neglecting the SPM contribution and is of the form

$$\Delta v_1(\tau) = \text{sgn}(\delta)\Delta v_{\max} \exp[-(\tau - \tau_d)^2] - \exp[-(\tau - \tau_d - \delta)^2], \quad (7.4.14)$$

where $\Delta v_{\max}$ is the maximum XPM-induced chirp given by

$$\Delta v_{\max} = \frac{\gamma_1 P_2 L}{\pi T_0 |\delta|} = \frac{\gamma_1 P_2 L_W}{\pi T_0}.$$

(7.4.15)

Note that $\Delta v_{\max}$ is determined by the walk-off length L_W rather than the actual fiber length L. This is expected because the XPM interaction occurs as long as the two pulses overlap.

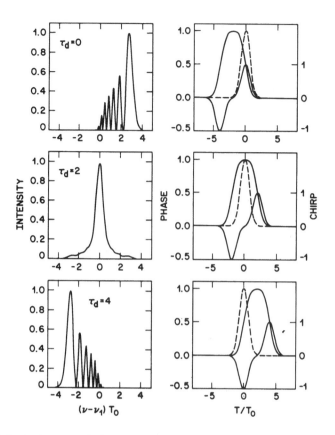

Figure 7.3 Optical spectra (left column) and XPM-induced phase and chirp (right column) for a probe pulse copropagating with a faster-moving pump pulse. Probe shape is shown by a dashed line. Three rows correspond to $\tau_d = 0$, 2, and 4, respectively. (After Ref. [47].)

Equation (7.4.14) shows that the XPM-induced chirp can vary significantly along the probe pulse if τ_d and δ are of opposite signs. As a result, the probe spectrum can have qualitatively different features depending on the relative values of τ_d and δ. Consider, for example, the case in which the pump pulse travels faster than the probe pulse ($\delta < 0$) and is delayed initially ($\tau_d \geq 0$). Figure 7.3 shows the probe spectrum together with the phase ϕ_1 and the chirp Δv_1 for $\delta = -4$ and $\tau_d = 0$, 2, and 4. The fiber length L and the pump peak power P_2 are chosen such that $\gamma_1 P_2 L = 40$ and $L/L_W = 4$. For reference, a 10-ps pump pulse with a group-velocity mismatch $d = 10$ ps/m has $L_W = 1$ m. The probe spectrum in Fig. 7.3 is shifted toward red with strong asymmetry for $\tau_d = 0$. For $\tau_d = 2$, it becomes symmetric while for $\tau_d = 4$ it is again asymmetric with a shift toward blue. In fact, the spectra for $\tau_d = 0$ and $\tau_d = 4$ are mirror images of each other about the central frequency $v_1 = \omega_1/2\pi$.

The probe spectra can be understood physically by considering the XPM-induced chirp shown in the right column of Fig. 7.3. For $\tau_d = 0$, the chirp is positive across the entire probe pulse, and the maximum chirp occurs at the pulse center. This is in contrast to the SPM case (shown in Fig. 4.1) where the chirp is negative near the leading edge, zero at the pulse center, and positive near the trailing edge. The differences in the SPM and XPM cases are due to group-velocity mismatch. When $\tau_d = 0$, the slow-moving probe pulse interacts mainly with the trailing edge of the pump pulse. As a result, the XPM-induced chirp is positive and the probe spectrum has only blue-shifted components. When $\tau_d = 4$, the pump pulse just catches up with the probe pulse at the fiber output. Its leading edge interacts with the probe; the chirp is therefore negative and the spectrum is shifted toward red. When $\tau_d = 2$, the pump pulse has time not only to the catch up but pass through the probe pulse in a symmetric manner. The chirp is zero at the pulse center similar to the case of SPM. However, its magnitude is considerably small across the entire pulse. As a result, the probe spectrum is symmetrically broadened but its tails carry a relatively small amount of pulse energy. The probe spectrum in this symmetric case depends quite strongly on the ratio L/L_W. If $L/L_W = 2$ with $\tau_d = 1$, the spectrum is broader with considerably more structure. By contrast, if $L \gg L_W$, the probe spectrum remains virtually unchanged.

The XPM-induced spectral broadening has been observed experimentally in the pump-probe configuration. In one experiment [5], the 10-ps pump pulses were obtained from a color-center laser operating at 1.51 μm while the probe pulses at 1.61 μm were generated using a fiber-Raman laser (see Section 8.2). The walk-off length was about 80 m while the dispersion length exceeded 10 km. Both the symmetric and asymmetric probe spectra were observed as the fiber length was increased from 50 to 400 m and the effective delay between the pulses was varied using time-dispersion tuning.

In a different experiment, a Nd:YAG laser was used to provide 33-ps pump pulses at 1.06 μm and 25-ps probe pulses at 0.53 μm [46]. The delay between two pulses was adjusted using a Mach–Zehnder interferometer. Because of a relatively large group-velocity mismatch ($d \approx 80$ ps/m), the walk-off length was only about 25 cm. For a 1-m-long fiber used in the experiment, $L/L_W = 4$. The probe spectra were recorded by varying the delay T_d and the peak power

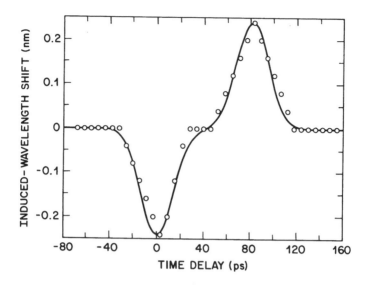

Figure 7.4 XPM-induced wavelength shift of a 0.53-μm probe pulse as a function of the initial time delay of the 1.06-μm pump pulse. Open circles show the experimental data while the solid line shows the theoretical prediction. (After Ref. [46].)

of the pump pulse. The spectra exhibited a shift toward the red or the blue side with some broadening as the multiple peaks could not be resolved. Such a XPM-induced shift is referred to as the induced frequency shift [46]. Figure 7.4 shows the induced shift as a function of the time delay T_d. The solid line is the theoretical prediction of Eq. (7.4.14). The frequency shift for a given time delay is obtained by maximizing $\Delta v_1(\tau)$. The maximum occurs near $\tau = 0$, and the frequency shift is given by

$$\Delta v_1 = \Delta v_{\max}\{\exp(-\tau_d^2) - \exp[-(\tau_d + \delta)^2]\}, \qquad (7.4.16)$$

where $\delta \approx -4$ for the experimental values of the parameters and $\tau_d = T_d/T_0$ with $T_0 \approx 20$ ps. Equation (7.4.16) shows that the maximum shift $\Delta v_{\max}$ occurs for $\tau_d = 0$ and $\tau_d = 4$, while the shift vanishes for $\tau_d = 2$. These features are in agreement with the experiment. According to Eq. (7.4.15) the maximum shift should increase linearly with the peak power of the pump pulse. This behavior is indeed observed experimentally as seen in Fig. 7.5. The XPM-induced shift of the probe wavelength is about 0.1 nm/kW. It is limited by the walk-off length and can be increased by an order of magnitude or more if the wavelength difference between the pump and probe is reduced to a few

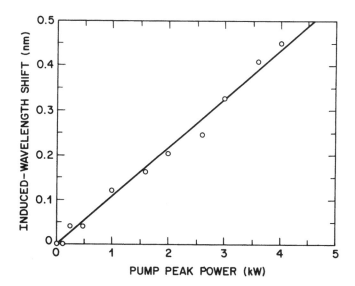

Figure 7.5 XPM-induced wavelength shift of a 0.53-μm probe pulse as a function of the peak power of copropagating a 1.06-μm pump pulse in the case of no initial time delay ($T_d = 0$) between the two pulses. (After Ref. [46].)

nanometers. The XPM-induced frequency shifts may be useful for optical communication applications.

7.4.2   Asymmetric Temporal Changes

In the preceding discussion the dispersion length L_D was assumed to be much larger than the fiber length L. As a result, both pulses maintained their shape during propagation through the fiber. If L_D becomes comparable to L or the walk-off length L_W, the combined effects of XPM, SPM, and GVD can lead to qualitatively new temporal changes that accompany the spectral changes discussed earlier. These temporal changes can be studied by solving Eqs. (7.1.15) and (7.1.16) numerically. It is useful to introduce the normalization scheme of Section 4.2 by defining

$$\xi = \frac{z}{L_D}, \qquad \tau = \frac{t - z/v_{g1}}{T_0}, \qquad U_j = \frac{A_j}{\sqrt{P_1}}, \qquad (7.4.17)$$

and write the coupled amplitude equations in the form [47]

$$\frac{\partial U_1}{\partial \xi} + \text{sgn}(\beta_{21}) \frac{i}{2} \frac{\partial^2 U_1}{\partial \tau^2} = iN^2(|U_1|^2 + 2|U_2|^2)U_1, \qquad (7.4.18)$$

$$\frac{\partial U_2}{\partial \xi} \pm \frac{L_D}{L_W} \frac{\partial U_2}{\partial \tau} + \frac{i}{2} \frac{\beta_{22}}{\beta_{21}} \frac{\partial^2 U_2}{\partial \tau^2} = iN^2 \frac{\omega_2}{\omega_1}(|U_2|^2 + 2|U_1|^2)U_2, \qquad (7.4.19)$$

where the parameter N is introduced as

$$N^2 = \frac{L_D}{L_{NL}} = \frac{\gamma_1 P_1 T_0^2}{|\beta_{21}|}. \qquad (7.4.20)$$

Fiber losses have been neglected assuming that $\alpha_j L \ll 1$ for $j = 1, 2$. The second term in Eq. (7.4.19) accounts for the group-velocity mismatch between the two pulses. The choice of plus or minus depends on the sign of the parameter d defined in Eq. (7.4.3).

To isolate the XPM effects, it is useful to consider a pump-probe configuration. Assuming $|U_2|^2 \ll |U_1|^2$, one can neglect the term containing $|U_2|^2$ in Eqs. (7.4.18) and (7.4.19). Evolution of the pump pulse, governed by Eq. (7.4.18), is then unaffected by the probe pulse. Evolution of the probe pulse is, however, affected considerably by the presence of the pump pulse because of XPM. Equation (7.4.19) governs the combined effects of XPM and GVD on the shape and the spectrum of the probe pulse. These equations can be solved numerically using the split-step Fourier method discussed in Section 2.4.

Figure 7.6 shows the shapes and the spectra of the pump and probe pulses at $\xi = 0.4$ for the case $N = 10$, $L_D/L_W = 10$, $\omega_2/\omega_1 = 1.2$, and $\beta_{22} \approx \beta_{21} > 0$. Both pulses at the fiber input are taken to be Gaussian of the same width with no initial time delay between them. The pump pulse is assumed to travel faster than the probe pulse $(d > 0)$. The shape and the spectrum of the pump pulse have features resulting from the combined effects of SPM and GVD (see Section 4.2). In contrast, the shape and the spectrum of the probe pulse are governed by the combined effects of XPM and GVD. For comparison, Fig. 7.7 shows the probe and pump spectra in the absence of GVD; asymmetric spectral broadening of the probe spectrum toward the blue side in the absence of GVD is discussed in Section 7.4.1. The effect of GVD is to reduce the extent of asymmetry; a part of the pulse energy is now carried by the red-shifted spectral components (see Fig. 7.6).

The most notable effect of GVD is seen in the shape of the probe pulse in Fig. 7.6. In the absence of GVD, the pulse shape remains unchanged as XPM

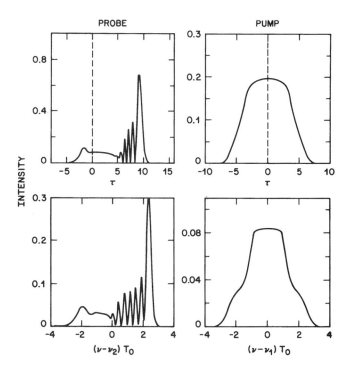

Figure 7.6 Shapes (upper row) and spectra (lower row) of probe and pump pulses at $\xi = 0.4$. Dashed line shows location of input pulses. Both pulses are Gaussian with the same width and overlap entirely at $\xi = 0$. (After Ref. [47].)

only affects the optical phase. However, when GVD is present, different parts of the probe pulse propagate at different speeds because of the XPM-induced chirp imposed on the probe pulse. This results in an asymmetric shape with considerable structure [47]. The probe pulse develops rapid oscillations near the trailing edge while the leading edge is largely unaffected. These oscillations are due to the phenomenon of optical wave breaking discussed in Section 4.2. There, the combination of SPM and GVD led to oscillations in the pulse wings (see Fig. 4.10). Here, it is the combination of XPM and GVD that results in oscillations over the entire trailing half of the probe pulse.

The features seen in Fig. 7.6 can be understood qualitatively noting that the XPM-induced chirp is maximum at the pulse center (as seen in the top row of Fig. 7.3). The combined effect of frequency chirp and positive GVD is to slow down the peak of the probe pulse with respect to its tails. The XPM-induced optical wave breaking occurs because the peak lags behind and interferes with

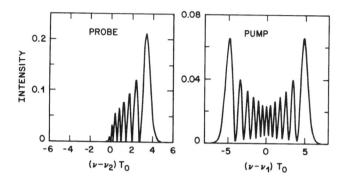

Figure 7.7 Spectra of probe and pump pulses under conditions identical to those of Fig. 7.6 except that the GVD effects are ignored. Pulse shapes are not shown as they remain unchanged.

the trailing edge. This can also be understood by noting that the faster moving pump pulse interacts mainly with the trailing edge of the probe pulse. In fact, if the probe and pump wavelengths were reversed so that the slower moving pump pulse interacts mainly with the leading edge, oscillations would develop near the leading edge because the XPM-induced chirp would speed up the peak of the probe pulse with respect to its tails. The effect of initial delay between the pump and probe pulses can lead to qualitative features quite different for the dispersive XPM compared with those shown in Fig. 7.3. For example, even if the pump pulse walks through the probe pulse in an asymmetric manner, the probe spectrum is no longer symmetric when the GVD effects are included.

The experimental observation of the XPM-induced asymmetric temporal effects requires the use of femtosecond pulses. This is so because $L_D > 1$ km for $T_0 > 5$ ps while $L_W \sim 1$ m for typical values of $|d| \sim 10$ ps/m. Because XPM occurs only during a few walk-off lengths, the interplay between the XPM and GVD effects can occur only if L_D and L_W become comparable. For example, if $T_0 = 100$ fs, L_D and L_W both become ~ 10 cm, and the temporal effects discussed in the preceding can occur in a fiber less than one meter long. For such short pulses, however, it becomes necessary to include the higher-order nonlinear effects.

7.4.3 Higher-Order Nonlinear Effects

As discussed in Section 2.3, several higher-order nonlinear effects should be considered for femtosecond optical pulses. The most important among them in practice is the Raman effect involving molecular vibrations. In the case of a single pulse propagating in the anomalous-GVD region, its inclusion leads to intrapulse Raman scattering that manifests through the Raman-induced frequency shift (see Section 5.5). The issue is how intrapulse Raman scattering affects the XPM interaction between two ultrashort optical pulses [52]–[54].

When the Raman contribution to the nonlinear polarization P_{NL} is included, one must use Eq. (2.3.31) in place of Eq. (2.3.6). The coupled NLS equations can still be obtained by following the procedure of Section 7.1 but mathematical details become quite cumbersome. By using Eq. (2.3.34) for the functional form of the Raman response function, the resulting equations can be written as [53]

$$
\frac{\partial A_j}{\partial z} + \frac{1}{v_{gj}} \frac{\partial A_j}{\partial t} + \frac{i\beta_{2j}}{2} \frac{\partial^2 A_j}{\partial t^2} + \frac{\alpha_j}{2} A_j = i\gamma_j (1 - f_R)(|A_j|^2 + 2|A_m|^2) A_j
$$
$$
+ i\gamma_j f_R \int_0^\infty ds\, h_R(s) \{ [|A_j(z, t - s)|^2 + |A_m(z, t - s)|^2] A_j(z, t)
$$
$$
+ A_j(z, t - s) A_m^*(z, t - s) \exp[i(\omega_j - \omega_m) s] A_m(z, t) \}, \tag{7.4.21}
$$

where $j = 1$ or 2 and $m = 3 - j$. The parameter f_R represents Raman contribution (about 18%) to the nonlinear polarization, and $h_R(t)$ is the Raman response function whose imaginary part is related to the Raman-gain spectrum through Eq. (2.3.35).

In spite of the complexity of Eq. (7.4.21), the physical meaning of various nonlinear terms is quite clear. On the right-hand side of Eq. (7.4.21), the first two terms represent the SPM and XPM contributions from the electronic response, the next two terms provide the SPM and XPM contributions from molecular vibrations, and the last term governs the energy transfer between the two pulses due to Raman amplification (see Chapter 8). When the Raman contribution is neglected by setting $f_R = 0$, Eq. (7.4.21) reduces to Eqs. (7.1.15) and (7.1.16). Similarly, if the two pulses are assumed to be much wider than the Raman response time (~ 70 fs) and $h_R(t)$ is replaced by a delta function, Eqs. (7.1.15) and (7.1.16) are again recovered provided the Raman amplification term is neglected.

Equation (7.4.21) shows that the XPM-induced coupling affects ultrashort optical pulses in several different ways when the Raman contribution is included. Energy transfer represented by the last term is discussed in Chapter 8 in the context of stimulated Raman scattering. The novel aspect of Eq. (7.4.21) is the SPM and XPM contributions from molecular vibrations. Similar to the single-pulse case, these contributions lead to shift of the carrier frequency. The most interesting feature is that such a shift results from both *intrapulse* and *interpulse* Raman scattering. In the context of solitons, the self-frequency shift is accompanied by a cross-frequency shift [53], occurring because of the simultaneous presence of a copropagating pulse. The self- and cross-frequency shifts may have the same or the opposite signs depending on whether the difference $\omega_1 - \omega_2$ in the carrier frequencies is smaller or larger than the frequency at which the Raman gain is maximum (see Chapter 8). As a result, the XPM interaction between two solitons of different carrier frequencies can either enhance or suppress the self-frequency shift expected when each soliton propagates alone [7].

7.5   Applications of XPM

The nonlinear phenomenon of XPM can be both beneficial and harmful. Perhaps its most direct impact is on the design of WDM lightwave systems where XPM often limits the system performance. This aspect of XPM is discussed in Chapter B.7. This section is devoted to other applications of XPM.

7.5.1   XPM-Induced Pulse Compression

SPM-induced chirp can be used to compress optical pulses (see Chapter B.6). Because XPM also imposes a frequency chirp on an optical pulse, it can be used for pulse compression as well [55]–[60]. An obvious advantage of XPM-induced pulse compression is that, in contrast to the SPM technique requiring the input pulse to be intense and energetic, XPM can compress weak input pulses because the frequency chirp is produced by a copropagating intense pump pulse. However, the XPM-induced chirp is affected by pulse walk-off and depends critically on the initial relative pump-signal delay. As a result, the practical use of XPM-induced pulse compression requires a careful control of the pump-pulse parameters such as its width, peak power, wavelength, and initial delay relative to the signal pulse.

Two cases must be distinguished depending on the relative magnitudes of the walk-off length L_W and the dispersion length L_D. If $L_D \gg L_W$ throughout the fiber, the GVD effects are negligible. In that case, an optical fiber generates the chirp through XPM, and a grating pair is needed to compress the chirped pulse. Equation (7.4.11) can be used to analyze the magnitude and the form of

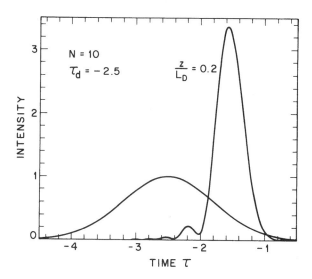

Figure 7.8 XPM-induced pulse compression in the normal-dispersion region of an optical fiber realized using a pump pulse of peak power such that $N = 10$. Shape of input pulses is also shown for comparison. (After Ref. [47].)

the chirp. A nearly linear chirp can be imposed across the signal pulse when the pump pulse is much wider compared with it [57]. The compression factor depends on the pump-pulse energy and can easily exceed 10.

Another pulse-compression mechanism can be used when L_D and L_W are comparable. In this case, the same piece of fiber that is used to impose the XPM-induced chirp also compresses the pulse through the GVD. Interestingly, in contrast to the SPM case where such compression can occur only in the anomalous-GVD region, the XPM offers the possibility of pulse compression even in the visible region (normal GVD) without the need of a grating pair. The performance of such a compressor can be studied by solving Eqs. (7.4.18) and (7.4.19) numerically for a given set of pump and signal pulses [47]. It is generally necessary to introduce a relative time delay τ_d between the pump and signal pulses such that the faster moving pulse overtakes the slower pulse and passes through it. The maximum compression occurs at a distance $\sim |\tau_d| L_W$ although the pulse quality is not necessarily the best at the point of maximum compression.

In general, a trade-off exists between the magnitude and the quality of compression. As an example, Fig. 7.8 compares the compressed pulse with the input signal pulse for the case in which at the fiber input both pulses are Gaussian pulses of the same width at visible wavelengths with the wavelength ratio $\lambda_1/\lambda_2 = 1.2$. The other parameters are $N = 1$, $L_D/L_W = 10$, and $\tau_d = -2.5$. The pulse is compressed by about a factor of 4 and, except for some ringing on the leading edge, is pedestal-free. Even this ringing can be suppressed if the pump pulse is initially wider than the signal pulse, although only at the expense of reduction in the amount of compression achievable at a given pump power. Of course, larger compression factors can be realized by increasing the pump power.

XPM-induced pulse compression in the normal-GVD region of a fiber can also occur when the XPM coupling is due to interaction between two orthogonally polarized components of a single beam [59]. An experiment in 1990 demonstrated pulse compression using just such a technique [58]. A polarizing Michelson interferometer was used to launch 2-ps pulses in a 1.4-m fiber (with a 2.1-mm beat length) such that the peak power and the relative delay of the two polarization components were adjustable. For a relative delay of 1.2 ps, the weak component was compressed by a factor of about 6.7 when the peak power of the other polarization component was 1.5 kW.

When both the pump and signal pulses propagate in the normal-GVD region of the fiber, the compressed pulse is necessarily asymmetric because of the group-velocity mismatch and the associated walk-off effects. The group velocities can be made nearly equal when wavelengths of the two pulses lie on opposite sides of the zero-dispersion wavelength (about 1.3 μm in conventional silica fibers). One possibility consists of compressing 1.55-μm pulses by using 1.06-μm pump pulses. The signal pulse by itself is too weak to form an optical soliton. However, the XPM-induced chirp imposed on it by a co-propagating pump pulse can be made strong enough that the signal pulse goes through an initial compression phase associated with higher-order solitons [8].

Figure 7.9 shows evolution of a signal pulse when the pump pulse has the same width as the signal pulse but is intense enough that $N = 30$ in Eq. (7.4.18). Because of the XPM-induced chirp, the signal pulse compresses by about a factor of 10 before its quality degrades. Both the compression factor and the pulse quality depend on the width and the energy of the pump pulse and can be controlled by optimizing pump-pulse parameters. This method of pulse compression is similar to that provided by higher-order solitons even though, strictly speaking, the signal pulse never forms a soliton. With the use of dispersion-shifted fibers, the technique can be used even when both pump and

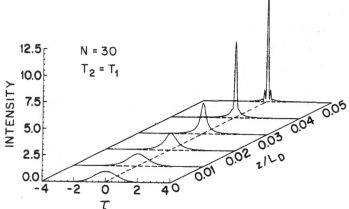

Figure 7.9 XPM-induced compression for a pulse experiencing anomalous GVD and copropagating with a pump pulse of peak power such that $N = 30$. Pump pulse propagates in the normal-GVD regime with the same group velocity. (After Ref. [47].)

signal wavelengths are in the 1.55-μm region as long as the zero-dispersion wavelength of the fiber lies in the middle. In a 1993 experiment, 10.6-ps signal pulses were compressed to 4.6 ps by using 12-ps pump pulses [60]. Pump and signal pulses were obtained from mode-locked semiconductor lasers operating at 1.56 and 1.54 μm, respectively, with a 5-GHz repetition rate. Pump pulses were amplified to an average power of 17 mW with the help of a fiber amplifier. This experiment demonstrates that XPM-induced pulse compression can occur at power levels achievable with semiconductor lasers.

7.5.2 XPM-Induced Optical Switching

The XPM-induced phase shift can also be used for optical switching. Several interferometric schemes have been used to take advantage of the XPM-induced phase shift for ultrafast optical switching [61]–[73]. The physics behind XPM-induced switching can be understood by considering a generic interferometer designed such that a weak signal pulse, divided equally between its two arms, experiences identical phase shifts in each arm and is transmitted through constructive interference. If a pump pulse at a different wavelength is injected into one of the arms of the interferometer, it would change the signal phase through XPM in that arm. If the XPM-induced phase shift is large enough (close to π), the signal pulse will not be transmitted because of the destructive interference occurring at the output. Thus, an intense pump pulse can switch the signal pulse through the XPM-induced phase shift.

XPM-induced optical switching was demonstrated in 1990 using a fiber-loop mirror acting as a Sagnac interferometer [63]. A dichroic fiber coupler, with 50:50 splitting ratio at 1.53 μm and 100:0 splitting ratio at 1.3 μm, was used to allow for dual-wavelength operation. A 1.53-μm color-center laser provided a low-power ($\sim$ 5 mW) CW signal. As expected, the counterpropagating signal beams experienced identical phase shifts, and the 500-m-long fiber loop acted as a perfect mirror, in the absence of a pump beam. When 130-ps pump pulses, obtained from a 1.3-μm Nd:YAG laser, were injected into the clockwise direction, the XPM interaction between the pump and the signal introduced a phase difference between the counterpropagating signal beams. Most of the signal power was transmitted when the peak power of the pump pulse was large enough to introduce a π phase shift.

The XPM-induced phase shift depends not only on the width and the shape of the pump pulse but also on the group-velocity mismatch. In the case in which both the pump and signal beams are pulsed, the phase shift also depends on the initial relative time delay between the pump and signal pulses. In fact, the magnitude and the duration of the XPM-induced phase shift can be controlled through the initial delay (see Fig. 7.3). The main point to note is that phase shift can be quite uniform over most of the signal pulse when the two pulses are allowed to completely pass through each other, resulting in complete switching of the signal pulse. The pump power required to produce π phase shift is generally quite large because of the group-velocity mismatch.

The group-velocity mismatch can be reduced significantly if the pump and signal pulses are orthogonally polarized but have the same wavelength. Moreover, even if the XPM-induced phase shift is less than π because of the birefringence-related pulse walk-off, the technique of cross-splicing can be used to accumulate it over long lengths [66]. In this technique, the fiber loop consists of multiple sections of polarization-maintaining fibers spliced together in such a way that the fast and slow axes are rotated by 90° in successive sections. As a result, the pump and signal pulses are forced to pass through each other in each section of the fiber loop, and the XPM-induced phase shift is enhanced by a factor equal to the number of sections.

7.5.3 XPM-Induced Nonreciprocity

XPM also occurs when two beams having the same (or different) wavelengths are propagated in opposite directions inside a fiber such that the counterprop-

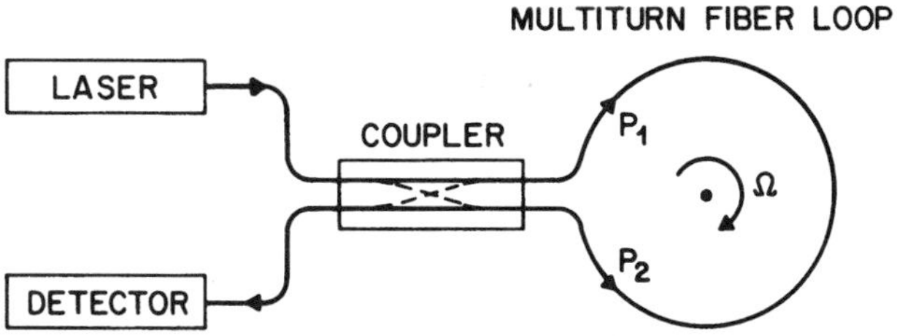

Figure 7.10 Schematic of a fiber gyroscope. Light from a laser is coupled through a 50% coupler to launch counterpropagating waves in a multiturn fiber loop. The rotation-induced phase difference is measured through a phase-sensitive detector.

agating waves interact with each other through XPM. Such an interaction can lead to new qualitative features, manifested through optical bistability and other instabilities when the fiber is used to construct a nonlinear ring resonator [74]–[85]. Of particular interest is the XPM-induced nonreciprocity that can affect the performance of fiber gyroscopes [86]–[91].

The origin of nonreciprocity between two counterpropagating waves can be understood by following the analysis of Section 7.1. If A_1 and A_2 are the amplitudes of the forward and backward propagating waves, they satisfy the coupled amplitude equations similar to Eqs. (7.1.15) and (7.1.16),

$$\pm\frac{\partial A_j}{\partial z} + \frac{1}{v_g}\frac{\partial A_j}{\partial t} + \frac{i\beta_2}{2}\frac{\partial^2 A_j}{\partial t^2} + \frac{\alpha}{2}A_j = i\gamma(|A_j|^2 + 2|A_{3-j}|^2)A_j, \qquad (7.5.1)$$

where the plus or minus sign corresponds to $j = 1$ or 2, respectively. In the case of CW beams, this set of two equations is readily solved. If fiber losses are neglected for simplicity, the solution is given as

$$A_j(z) = \sqrt{P_j}\exp(\pm i\phi_j), \qquad (7.5.2)$$

where P_j is the peak power and the nonlinear phase shift is given by

$$\phi_j = \gamma z(P_j + 2P_{3-j}), \qquad (7.5.3)$$

with $j = 1, 2$. If $P_1 \neq P_2$, the phase shifts ϕ_1 and ϕ_2 are not the same for the two counterpropagating waves. This nonreciprocity is due to the presence of the factor of two in the XPM term in Eq. (7.5.3).

XPM-induced nonreciprocity can be detrimental for high-precision fiber gyroscopes used to measure rotation rates as small as 0.01° per hour [92]. Figure 7.10 shows the design of a fiber gyroscope schematically. Its operation is based on the Sagnac effect, known to introduce a rotation-dependent phase difference between the counterpropagating waves [93]. The net phase difference is thus given by

$$\Delta\phi = \phi_1 - \phi_2 = \gamma L(P_2 - P_1) + S\Omega, \tag{7.5.4}$$

where L is the total fiber length, Ω is the rotation rate, and S is a scale factor that depends on the fiber length L as well as on the radius of the fiber loop [92]. If the powers P_1 and P_2 were constant, the XPM term in Eq. (7.5.4) would be of little concern. However, the power levels can fluctuate in practice. Even a power difference of 1 μW between the counterpropagating waves can change $\Delta\phi$ by $\sim 1 \times 10^{-6}$ rad if we use $\gamma \approx 10$ W^{-1}/km and $L \sim 100$ m. This value typically corresponds to an equivalent rotation rate of 0.1° per hour. For this reason, XPM severely limits the sensitivity of fiber gyroscopes unless the power levels are controlled to within 10 nW.

Several schemes can be used to mitigate the XPM problem and improve the gyroscope performance. In one scheme, the laser power is modulated before the counterpropagating waves are launched inside a fiber loop [87]. Because of the time dependence of optical fields, this case is analyzed by solving Eq. (7.5.1) with the appropriate boundary conditions [91]. The results show that the effect of nonreciprocity is reduced drastically if modulation frequency is chosen suitably. This can be understood physically by noting that XPM occurs only if the two pulses overlap temporally. On a more fundamental level, XPM-induced nonreciprocity results from interference between the counterpropagating waves. Modulation reduces the coherence between the counterpropagating waves, thereby reducing the effectiveness of such an interference. Indeed, the same result can also be obtained by using broadband sources with a limited coherence time [88]–[90]. Thermal sources or light-emitting diodes have been used for this purpose [92].

Let us consider briefly the effect of XPM on optical bistability. Any nonlinear medium placed inside a cavity can exhibit bistability [94], and optical fibers are no exception. If a fiber-ring cavity is used for this purpose, optical bistability can occur irrespective of whether the beam propagates in the clockwise or counterclockwise direction. An interesting situation occurs when the optical beams are launched in both directions. Because of the XPM-induced coupling between the counterpropagating beams, the device acts as two coupled bistable systems and can exhibit many new qualitative features [75]–[77]. Although optical bistability has been observed in the case of unidirectional propagation in a fiber-ring cavity [78], the bidirectional case has not attracted much attention.

The XPM interaction between two counterpropagating optical pulses is generally quite weak and can be neglected in the case of ultrashort pulses. The reason can be understood by noting that the XPM-induced phase shift decreases even for copropagating pulses as the relative group-velocity difference increases [see Eq. (7.4.9)]. For counterpropagating pulses the group-velocity mismatch is so large that the two pulses have little time to interact with each other. Nonetheless, measurable effects can occur for very intense pulses. For example, the spectral shift of a probe pulse observed in an experiment in which 0.7-ps pump pulses with peak intensities ~ 10 TW/cm^2 were propagated through a 1-mm-thick glass plate could be accounted for only when the XPM interaction between the counterpropagating pump and probe pulses was included [95]. In the case of optical fibers, XPM interaction between counterpropagating waves becomes important for fiber Bragg gratings (see Chapter B.1).

Problems

7.1 Explain what is meant by cross-phase modulation and why it occurs in optical fibers.

7.2 Derive an expression for the nonlinear part of the refractive index when two optical beams of different wavelengths but identical polarizations are propagating inside an optical fiber.

7.3 Derive the dispersion relation (7.2.7) for XPM-induced modulation instability starting from Eqs. (7.1.15) and (7.1.16). Under what conditions can modulation instability occur in the normal-GVD regime of a fiber?

7.4 Derive the dispersion relation for XPM-induced modulation instability starting from Eqs. (7.1.19) and (7.1.20). Assume that one of the CW beams is polarized along the slow axis while the other beam's polarization axis makes an angle θ from that axis.

7.5 Use the dispersion relation obtained in the previous problem to plot the gain spectra of modulation instability for several values of the polarization angle ($\theta = 0, 20, 45, 70$, and $90°$). Discuss your results physically.

7.6 Verify that the pair of bright and dark solitons given by Eqs. (7.3.1) and (7.3.2) indeed satisfies the coupled NLS equations.

7.7 Derive an expression for the XPM-induced phase shift imposed on a probe pulse by a copropagating pump pulse using Eq. (7.4.6). Assume that the two pulses have "sech" shape, the same width, and are launched simultaneously.

7.8 Use the result obtained in the previous problem to calculate the frequency chirp imposed on the probe pulse. Plot the chirp profile using suitable parameter values.

7.9 Make a figure similar to Fig. 7.3 using the same parameter values but assume that both pulses have "sech" shapes.

7.10 Explain why XPM produces a shift in the wavelength of a 0.53-μm probe pulse when a 1.06-μm pump pulse is launched with it simultaneously (no initial time delay). Can you predict the sign of wavelength shift for standard optical fibers?

7.11 Write a computer program using the split-step Fourier method (see Section 2.4) and solve Eqs. (7.4.18) and (7.4.19) numerically. Reproduce the results shown in Fig. 7.6.

References

[1] S. A. Akhmanov, R. V. Khokhlov, and A. P. Sukhorukov, in *Laser Handbook*, Vol. 2, F. T. Arecchi and E. O. Schulz-Dubois, Eds. (North-Holland, Amsterdam, 1972), Chap. E3.

[2] J. I. Gersten, R. R. Alfano, and M. Belic, *Phys. Rev. A* **21**, 1222 (1980).

[3] A. R. Chraplyvy and J. Stone, *Electron. Lett.* **20**, 996 (1984).

[4] R. R. Alfano, Q. X. Li, T. Jimbo, J. T. Manassah, and P. P. Ho, *Opt. Lett.* **14**, 626 (1986).

[5] M. N. Islam, L. F. Mollenauer, R. H. Stolen, J. R. Simpson, and H. T. Shang, *Opt. Lett.* **12**, 625 (1987).

[6] R. R. Alfano, P. L. Baldeck, F. Raccah, and P. P. Ho, *Appl. Opt.* **26**, 3491 (1987).

[7] D. Schadt and B. Jaskorzynska, *Electron. Lett.* **23**, 1090 (1987); *J. Opt. Soc. Am. B* **4**, 856 (1987); *J. Opt. Soc. Am. B* **5**, 2374 (1988).

[8] B. Jaskorzynska and D. Schadt, *IEEE J. Quantum Electron.* **24**, 2117 (1988).

[9] R. R. Alfano and P. P. Ho, *IEEE J. Quantum Electron.* **24**, 351 (1988).

[10] R. R. Alfano (ed.), *The Supercontinuum Laser Source*, (Springer-Verlag, New York, 1989).

[11] S. Kumar, A. Selvarajan, and G. V. Anand, *J. Opt. Soc. Am. B* **11**, 810 (1994).

[12] G. P. Agrawal, *Phys. Rev. Lett.* **59**, 880 (1987).

[13] G. P. Agrawal, P. L. Baldeck, and R. R. Alfano, *Phys. Rev. A* **39**, 3406 (1989).

[14] C. J. McKinstrie and R. Bingham, *Phys. Fluids B* **1**, 230 (1989).

[15] C. J. McKinstrie and G. G. Luther, *Physica Scripta* **30**, 31 (1990).

[16] W. Huang and J. Hong, *IEEE J. Quantum Electron.* **10**, 156 (1992).

[17] J. E. Rothenberg, *Phys. Rev. Lett.* **64**, 813 (1990).

[18] G. P. Agrawal, *Phys. Rev. Lett.* **64**, 814 (1990).

[19] M. Yu, C. J. McKinstrie, and G. P. Agrawal, *Phys. Rev. E* **48**, 2178 (1993).

[20] A. S. Gouveia-Neto, M. E. Faldon, A. S. B. Sombra, P. G. J. Wigley, and J. R. Taylor, *Opt. Lett.* **13**, 901 (1988).

[21] E. J. Greer, D. M. Patrick, P. G. J. Wigley, and J. R. Taylor, *Electron. Lett.* **18**, 1246 (1989); *Opt. Lett.* **15**, 851 (1990).

[22] D. M. Patrick and A. D. Ellis, *Electron. Lett.* **29**, 227 (1993).

[23] Y. Inoue, *J. Phys. Soc. Jpn.* **43**, 243 (1977).

[24] M. R. Gupta, B. K. Som, and B. Dasgupta, *J. Plasma Phys.* **25**, 499 (1981).

[25] S. Trillo, S. Wabnitz, E. M. Wright, and G. I. Stegeman, *Opt. Lett.* **13**, 871 (1988).

[26] V. V. Afanasjev, E. M. Dianov, A. M. Prokhorov, and V. N. Serkin, *JETP Lett.* **48**, 638 (1988).

[27] V. V. Afanasjev, Y. S. Kivshar, V. V. Konotop, and V. N. Serkin, *Opt. Lett.* **14**, 805 (1989).

[28] V. V. Afanasjev, E. M. Dianov, and V. N. Serkin, *IEEE J. Quantum Electron.* **25**, 2656 (1989).

[29] M. Florjanczyk and R. Tremblay, *Phys. Lett.* **141**, 34 (1989).

[30] L. Wang and C. C. Yang, *Opt. Lett.* **15**, 474 (1990).

[31] V. V. Afanasjev, L. M. Kovachev, and V. N. Serkin, *Sov. Tech. Phys. Lett.* **16**, 524 (1990).

[32] M. Lisak, A. Höök, and D. Anderson, *J. Opt. Soc. Am. B* **7**, 810 (1990).

[33] J. T. Manassah, *Opt. Lett.* **15**, 670 (1990).

[34] P. C. Subramaniam, *Opt. Lett.* **16**, 1560 (1991).

[35] Y. S. Kivshar, D. Anderson, M. Lisak, and V. V. Afanasjev, *Physica Scripta* **44**, 195 (1991).

[36] V. Y. Khasilev, *JETP Lett.* **56**, 194 (1992).

[37] M. Wadati, T. Iizuka, and M. Hisakado, *J. Phys. Soc. Jpn.* **61**, 2241 (1992).

[38] D. Anderson, A. Höök, M. Lisak, V. N. Serkin, and V. V. Afanasjev, *Electron. Lett.* **28**, 1797 (1992).

[39] S. G. Dinev, A. A. Dreischuh, and S. Balushev, *Physica Scripta* **47**, 792 (1993).

[40] A. Höök, D. Anderson, M. Lisak, V. N. Serkin, and V. V. Afanasjev, *J. Opt. Soc. Am. B* **10**, 2313 (1993).

[41] A. Höök and V. N. Serkin, *IEEE J. Quantum Electron.* **30**, 148 (1994).

[42] J. C. Bhakta, *Phys. Rev. E* **49**, 5731 (1994).

[43] F. T. Hioe, *Phys. Lett. A* **234**, 351 (1997); *Phys. Rev. E* **56**, 2373 (1997); *Phys. Rev. E* **58**, 6700 (1998).

[44] F. T. Hioe, *Phys. Rev. Lett.* **82**, 1152 (1999).

[45] J. T. Manassah, *Appl. Opt.* **26**, 3747 (1987).

[46] P. L. Baldeck, R. R. Alfano, and G. P. Agrawal, *Appl. Phys. Lett.* **52**, 1939 (1988).

[47] G. P. Agrawal, P. L. Baldeck, and R. R. Alfano, *Phys. Rev. A* **39**, 5063 (1989).

[48] T. Morioka and M. Saruwatari, *Electron. Lett.* **25**, 646 (1989).

[49] R. R. Alfano, P. L. Baldeck, P. P. Ho, and G. P. Agrawal, *J. Opt. Soc. Am. B* **6**, 824 (1989).

[50] M. Yamashita, K. Torizuka, T. Shiota, and T. Sato, *Jpn. J. Appl. Phys.* **29**, 294 (1990).

[51] D. M. Patrick and A. D. Ellis, *Electron. Lett.* **29**, 1391 (1993).

[52] A. Höök, *Opt. Lett.* **17**, 115 (1992).

[53] S. Kumar, A. Selvarajan, and G. V. Anand, *Opt. Commun.* **102**, 329 (1993).

[54] C. S. Aparna, S. Kumar, and A. Selvarajan, *Opt. Commun.* **131**, 267 (1996).

[55] E. M. Dianov, P. V. Mamyshev, A. M. Prokhorov, and S. V. Chernikov, *Sov. J. Quantum Electron.* **18**, 1211 (1988).

[56] J. T. Manassah, *Opt. Lett.* **13**, 755 (1988).

[57] G. P. Agrawal, P. L. Baldeck, and R. R. Alfano, *Opt. Lett.* **14**, 137 (1989).

[58] J. E. Rothenberg, *Opt. Lett.* **15**, 495 (1990).

[59] Q. Z. Wang, P. P. Ho, and R. R. Alfano, *Opt. Lett.* **15**, 1023 (1990); *Opt. Lett.* **16**, 496 (1991).

[60] A. D. Ellis and D. M. Patrick, *Electron. Lett.* **29**, 149 (1993).

[61] M. J. La Gasse, D. Liu-Wong, J. G. Fujimoto, and H. A. Haus, *Opt. Lett.* **14**, 311 (1989).

[62] T. Morioka and M. Saruwatari, *Opt. Eng.* **29**, 200 (1990).

[63] K. J. Blow, N. J. Doran, B. K. Nayar, and B. P. Nelson, *Opt. Lett.* **15**, 248 (1990).

[64] H. Vanherzeele and B. K. Nayar, *IEEE Photon. Technol. Lett.* **2**, 603 (1990).

[65] M. Jinno and T. Matsumoto, *IEEE Photon. Technol. Lett.* **2**, 349 (1990); *Opt. Lett.* **16**, 220 (1991).

[66] J. D. Moores, K. Bergman, H. A. Haus, and E. P. Ippen, *Opt. Lett.* **16**, 138 (1991); *J. Opt. Soc. Am. B* **8**, 594 (1991).

[67] H. Avrampoulos, P. M. W. French, M. C. Gabriel, H. H. Houh, N. A. Whitaker, and T. Morse, *IEEE Photon. Technol. Lett.* **3**, 235 (1991).

[68] H. Vanherzeele and B. K. Nayar, *Int. J. Nonlinear Opt. Phys.* **1**, 119 (1992).

[69] M. Jinno, *Opt. Lett.* **18**, 726 (1993).

[70] J. E. Rothenberg, *Opt. Lett.* **18**, 796 (1993).

[71] M. A. Franco, A. Alexandrou, and G. R. Boyer, *Pure Appl. Opt.* **4**, 451 (1995).

[72] P. M. Ramos and C. R. Pavia, *IEEE J. Sel. Topics Quantum Electron.* **3**, 1224 (1997).

[73] N. G. R. Broderick, D. Taverner, D. J. Richardson, and M. Ibsen, *J. Opt. Soc. Am. B* **17**, 345 (2000).

[74] G. P. Agrawal, *Appl. Phys. Lett.* **38**, 505 (1981).

[75] A. E. Kaplan, *Opt. Lett.* **6**, 360 (1981).

[76] A. E. Kaplan and P. Meystre, *Opt. Commun.* **40**, 229 (1982).

[77] G. P. Agrawal, *IEEE J. Quantum Electron.* **QE-18**, 214 (1982).

[78] H. Nakatsuka, S. Asaka, H. Itoh, K. Ikeda, and M. Matsuoka, *Phys. Rev. Lett.* **50**, 109 (1983).

[79] K. Ikeda, *J. Phys.* **44**, C2-183 (1983).

[80] K. Otsuka, *Opt. Lett.* **8**, 471 (1983).

[81] Y. Silberberg and I. Bar-Joseph, *J. Opt. Soc. Am. B* **1**, 662 (1984).

[82] W. J. Firth, A. Fitzgerald, and C. Paré, *J. Opt. Soc. Am. B* **7**, 1087 (1990).

[83] C. T. Law and A. E. Kaplan, *J. Opt. Soc. Am. B* **8**, 58 (1991).

[84] R. Vallé, *Opt. Commun.* **81**, 419 (1991); *Opt. Commun.* **93**, 389 (1992).

[85] M. Yu, C. J. McKinstrie, and G. P. Agrawal, *J. Opt. Soc. Am. B* **15**, 607 (1998).

[86] S. Ezekiel, J. L. Davis, and R. W. Hellwarth, *Opt. Lett.* **7**, 457 (1982).

[87] R. A. Bergh, H. C. Lefevre, and H. J. Shaw, *Opt. Lett.* **7**, 282 (1982).

[88] R. A. Bergh, B. Culshaw, C. C. Cutler, H. C. Lefevre, and H. J. Shaw, *Opt. Lett.* **7**, 563 (1982).

[89] K. Petermann, *Opt. Lett.* **7**, 623 (1982).

[90] N. J. Frigo, H. F. Taylor, L. Goldberg, J. F. Weller, and S. C. Rasleigh, *Opt. Lett.* **8**, 119 (1983).

[91] B. Crosignani and A. Yariv, *J. Lightwave Technol.* **LT-3**, 914 (1985).

[92] R. A. Bergh, H. C. Lefevre, and H. J. Shaw, *J. Lightwave Technol.* **LT-2**, 91 (1984).

[93] G. Sagnac, *Compt. Rend. Acad. Sci.* **95**, 708 (1913).

[94] H. M. Gibbs, *Optical Bistability: Controlling Light with Light* (Academic Press, Orlando, 1985).

[95] B. V. Vu, A. Szoke, and O. L. Landen, *Opt. Lett.* **18**, 723 (1993).

Chapter 8

Stimulated Raman Scattering

Stimulated Raman scattering (SRS) is an important nonlinear process that can turn optical fibers into broadband Raman amplifiers and tunable Raman lasers. It can also severely limit the performance of multichannel lightwave systems by transferring energy from one channel to the neighboring channels. This chapter describes both the useful and the harmful effects of SRS in optical fibers. Section 8.1 presents the basic theory behind SRS with emphasis on the pump power required to reach the Raman threshold; SRS under continuous-wave (CW) and quasi-CW conditions is considered in Section 8.2 where we also discuss the performance of fiber-based Raman lasers and amplifiers. Ultrafast SRS occurring for pulses of 100-ps width or less is considered in Sections 8.3 and 8.4 for normal and anomalous group-velocity dispersion (GVD), respectively. In both cases, attention is paid to the walk-off effects together with those resulting from self-phase modulation (SPM) and cross-phase modulation (XPM).

8.1 Basic Concepts

In any molecular medium, spontaneous Raman scattering can transfer a small fraction (typically $\sim 10^{-6}$) of power from one optical field to another field, whose frequency is downshifted by an amount determined by the vibrational modes of the medium. This process is called the Raman effect [1]. It is described quantum-mechanically as scattering of a photon by one of the molecules to a lower-frequency photon, while the molecule makes transition to a higher-energy vibrational state. Incident light acts as a pump for generating the frequency-shifted radiation called the Stokes wave. It was observed in 1962 that, for intense pump fields, the nonlinear phenomenon of SRS can occur in which the Stokes wave grows rapidly inside the medium such that most of the pump energy is transferred to it [2]. Since then, SRS has been studied extensively in a variety of molecular media [3]–[8]. This section introduces the basic concepts, such as the Raman gain and the Raman threshold, and provides a theoretical framework for describing SRS in optical fibers.

8.1.1 Raman-Gain Spectrum

In a simple approach valid under the CW and quasi-CW conditions, the initial growth of the Stokes wave is described by [8]

$$dI_s/dz = g_R I_p I_s, \tag{8.1.1}$$

where I_s is the Stokes intensity, I_p is the pump intensity, and the Raman-gain coefficient g_R is related to the cross section of spontaneous Raman scattering [7]. On a more fundamental level, g_R is related to the imaginary part of the third-order nonlinear susceptibility.

The Raman-gain spectrum $g_R(\Omega)$, where Ω represents the frequency difference between the pump and Stokes waves, is the most important quantity for describing SRS. It was measured for silica fibers in the early experiments on SRS, with refinements continuing in later years [9]–[14]. In general, g_R depends on composition of the fiber core and can vary significantly with the use of different dopants. Figure 8.1 shows g_R for fused silica as a function of the frequency shift at a pump wavelength $\lambda_p = 1 \ \mu$m. For other pump wavelengths g_R can be obtained by using the inverse dependence of g_R on λ_p. The most significant feature of the Raman gain in silica fibers is that $g_R(\Omega)$ extends over a large frequency range (up to 40 THz) with a broad peak located near 13 THz. This behavior is due to the noncrystalline nature of silica glass. In amorphous materials such as fused silica, molecular vibrational frequencies spread out into bands that overlap and create a continuum [15]. As a result, in contrast to most molecular media for which the Raman gain occurs at specific well-defined frequencies, it extends continuously over a broad range in silica fibers. As will be seen later, optical fibers can act as broadband amplifiers because of this feature.

To see how the process of SRS develops, consider a CW pump beam propagating inside a fiber at the optical frequency ω_p. If a probe beam at the

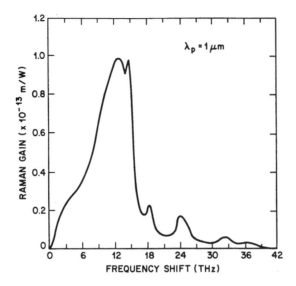

Figure 8.1 Raman-gain spectrum for fused silica at a pump wavelength $\lambda_p = 1$ μm. The Raman gain scales inversely with λ_p. (After Ref. [10].)

frequency ω_s is coincident with the pump at the fiber input, it will be amplified because of the Raman gain, as long as the frequency difference $\Omega = \omega_p - \omega_s$ lies within the bandwidth of the Raman-gain spectrum of Fig. 8.1. If only the pump beam is incident at the fiber input, spontaneous Raman scattering acts as a probe and is amplified with propagation. Because spontaneous Raman scattering generates photons within the entire bandwidth of the Raman-gain spectrum, all frequency components are amplified. However, the frequency component for which g_R is maximum builds up most rapidly. In the case of pure silica, g_R is maximum for the frequency component that is downshifted from the pump frequency by about 13.2 THz (440 cm^{-1}). It turns out that when the pump power exceeds a threshold value, this component builds up almost exponentially [16]. As a result, SRS leads to generation of the Stokes wave whose frequency is determined by the peak of the Raman gain. The corresponding frequency shift is called the Raman shift (or the Stokes shift).

8.1.2   Raman Threshold

To find the Raman threshold, we should consider nonlinear interaction between the pump and Stokes waves. In the CW case, this interaction is governed by the following set of two coupled equations:

$$\frac{dI_s}{dz} = g_R I_p I_s - \alpha_s I_s, \tag{8.1.2}$$

$$\frac{dI_p}{dz} = -\frac{\omega_p}{\omega_s} g_R I_p I_s - \alpha_p I_p, \tag{8.1.3}$$

where α_s and α_p account for fiber losses at the Stokes and pump frequencies, respectively. These equations can be derived rigorously using Maxwell's equations of Section 2.1. They can also be written phenomenologically by considering the processes through which photons appear in or disappear from each beam. One can readily verify that in the absence of losses,

$$\frac{d}{dz}\left(\frac{I_s}{\omega_s} + \frac{I_p}{\omega_p}\right) = 0. \tag{8.1.4}$$

This equation merely states that the total number of photons in the pump and Stokes beams remains constant during SRS.

Although pump depletion must be included for a complete description of SRS, it can be neglected for the purpose of estimating the Raman threshold [16]. Equation (8.1.3) is readily solved if we neglect the first term on its right-hand side that is responsible for pump depletion. If we substitute the solution in Eq. (8.1.2), we obtain

$$dI_s/dz = g_R I_0 \exp(-\alpha_p z)I_s - \alpha_s I_s, \tag{8.1.5}$$

where I_0 is the incident pump intensity at $z = 0$. Equation (8.1.5) can be easily solved, and the result is

$$I_s(L) = I_s(0)\exp(g_R I_0 L_{\text{eff}} - \alpha_s L), \tag{8.1.6}$$

where L is the fiber length and

$$L_{\text{eff}} = [1 - \exp(-\alpha_p L)]/\alpha_p. \tag{8.1.7}$$

The solution (8.1.6) shows that, because of pump absorption, the effective interaction length is reduced from L to L_{eff}.

The use of Eq. (8.1.6) requires an input intensity $I_s(0)$ at $z = 0$. In practice, SRS builds up from spontaneous Raman scattering occurring throughout the fiber length. It has been shown that this process is equivalent to injecting one fictitious photon per mode at the input end of the fiber [16]. Thus, we can calculate the Stokes power by considering amplification of each frequency component of energy $\hbar\omega$ according to Eq. (8.1.6) and then integrating over the whole range of the Raman-gain spectrum, that is,

$$P_s(L) = \int_{-\infty}^{\infty} \hbar\omega \, \exp[g_R(\omega_p - \omega)I_0 L_{\text{eff}} - \alpha_s L] \, d\omega, \tag{8.1.8}$$

where the fiber is assumed to support a single mode. The frequency dependence of g_R is shown in Fig. 8.1. Even though the functional form of $g_R(\Omega)$ is not known, the integral in Eq. (8.1.8) can be evaluated approximately using the method of steepest descent because the main contribution to the integral comes from a narrow region around the gain peak. Using $\omega = \omega_s$, we obtain

$$P_s(L) = P_{s0}^{\text{eff}} \exp[g_R(\Omega_R)I_0 L_{\text{eff}} - \alpha_s L], \tag{8.1.9}$$

where the effective input power at $z = 0$ is given by

$$P_{s0}^{\text{eff}} = \hbar\omega_s B_{\text{eff}}, \qquad B_{\text{eff}} = \left(\frac{2\pi}{I_0 L_{\text{eff}}}\right)^{1/2} \left|\frac{\partial^2 g_R}{\partial \omega^2}\right|_{\omega=\omega_s}^{-1/2}. \tag{8.1.10}$$

Physically, B_{eff} is the effective bandwidth of the Stokes radiation centered near the gain peak at $\Omega_R = \omega_p - \omega_s$. Although B_{eff} depends on the pump intensity and the fiber length, the spectral width of the dominant peak in Fig. 8.1 provides an order-of-magnitude estimate for it.

The Raman threshold is defined as the input pump power at which the Stokes power becomes equal to the pump power at the fiber output [16] or

$$P_s(L) = P_p(L) \equiv P_0 \exp(-\alpha_p L), \tag{8.1.11}$$

where $P_0 = I_0 A_{\text{eff}}$ is the input pump power and A_{eff} is the effective core area defined as in Section 2.3. In the case of multimode fibers A_{eff} is the inverse of the overlap integral in Eq. (7.1.14). Using Eq. (8.1.9) in Eq. (8.1.11) and assuming $\alpha_s \approx \alpha_p$, the threshold condition becomes

$$P_{s0}^{\text{eff}} \exp(g_R P_0 L_{\text{eff}}/A_{\text{eff}}) = P_0, \tag{8.1.12}$$

where P_{s0}^{eff} also depends on P_0 through Eqs. (8.1.10). The solution of Eq. (8.1.12) provides the critical pump power required to reach the Raman threshold. Assuming a Lorentzian shape for the Raman-gain spectrum, the critical pump power, to a good approximation, is given by [16]

$$\frac{g_R P_0^{cr} L_{\text{eff}}}{A_{\text{eff}}} \approx 16. \tag{8.1.13}$$

A similar analysis can be carried out for the backward SRS. The threshold condition in that case is still given by Eq. (8.1.13) but the numerical factor 16 is replaced with 20. As the threshold for forward SRS is reached first at a given pump power, backward SRS is generally not observed in optical fibers. Of course, the Raman gain can be used to amplify a backward propagating signal. Note also that the derivation of Eq. (8.1.13) assumes that the polarization of the pump and Stokes waves is maintained along the fiber. If polarization is not preserved, the Raman threshold is increased by a factor whose value lies between 1 and 2. In particular, if the polarization is completely scrambled, it increases by a factor of two.

In spite of various approximations made in the derivation of Eq. (8.1.13), it is able to predict the Raman threshold quite accurately. For long fibers such that $\alpha_p L \gg 1$, $L_{\text{eff}} \approx 1/\alpha_p$. At $\lambda_p = 1.55$ μm, a wavelength near which the fiber loss is minimum (about 0.2 dB/km), $L_{\text{eff}} \approx 20$ km. If we use a typical value $A_{\text{eff}} = 50$ μm^2, the predicted Raman threshold is $P_0^{cr} \approx 600$ mW. Because power levels inside a fiber are typically below 10 mW, SRS is not likely to occur in single-channel optical communication systems. With the advent of optical amplifiers, the input power can approach 100 mW for some applications, but still remains well below the critical value.

In practice, SRS is observed using high-power lasers. In the visible region, A_{eff} is typically 10 μm^2 in single-mode fibers. Equation (8.1.13) then yields $P_0^{cr} \sim 10$ W for a 10-m-long fiber. As such power levels are readily available (from Nd:YAG lasers, for example), SRS can be observed with fibers only a few meters long.

The simple theory of this section cannot explain the growth of the Stokes wave beyond the Raman threshold as it neglects pump depletion. Equations (8.1.2) and (8.1.3) should be solved to include the effect of pump depletion. These equations can be solved analytically in the specific case $\alpha_s = \alpha_p$ [17]. The results show that the threshold condition (8.1.13) remains fairly accurate. Once the Raman threshold is reached, the power is transferred from the pump to the Stokes rapidly. The theory predicts a complete transfer of pump power to the Stokes (except for fiber losses). In practice, however, the Stokes wave serves as a pump to generate a second-order Stokes wave if its power becomes large enough to satisfy Eq. (8.1.13). This process of cascade SRS can generate multiple Stokes waves whose number depends on the input pump power.

8.1.3 Coupled Amplitude Equations

The CW theory of SRS needs modification when optical pulses are used for pumping. This is almost always the case for optical fibers. In fact, in the case of CW pumping, stimulated Brillouin scattering (SBS) dominates and inhibits SRS because of its lower threshold (see Chapter 9). However, SBS can be nearly suppressed by using pump pulses of widths <1 ns. Each pump pulse then generates a Stokes (or Raman) pulse if the SRS threshold is reached. Similar to the CW case, the carrier frequency ω_s of the Raman pulse is downshifted from that of the pump by the Stokes shift of about 13 THz.

The dynamical description of SRS in optical fibers is considerably simplified if the medium is assumed to respond instantaneously. This assumption is often justified because the broad gain spectrum of Fig. 8.1 implies a response time of well below 100 fs. Except for ultrashort pump pulses (width ∼10 fs), the response time is generally much smaller than typical pulse widths. The mutual interaction between the Raman and pump pulses is then governed by a set of two coupled amplitude equations obtained such that they include the effects of Raman gain, pump depletion, SPM, XPM, and GVD. These equations can be derived following the analysis of Section 2.3.

A unified description should include nonlinear response function $R(t)$ given in Eq. (2.3.34) so that both the Kerr and Raman effects are included simultaneously [18]–[22]. The analysis is somewhat complicated because the optical field **E** in Eq. (2.3.31) is of the form

$$\mathbf{E}(\mathbf{r},t) = \tfrac{1}{2}\hat{x}\left\{A_p \exp[i(\beta_{0p} - \omega_p t)] + A_s \exp[i(\beta_{0s} - \omega_s t)]\right\} + \text{c.c.}, \quad (8.1.14)$$

where ω_p and ω_s are the carrier frequencies, β_{0p} and β_{0s} are the propagation constants, and A_p and A_s are the slowly varying envelopes associated with the pump and Raman pulses. After considerable algebra, one can obtain the following set of two equations [22]:

$$\frac{\partial A_p}{\partial z} + \frac{1}{v_{gp}}\frac{\partial A_p}{\partial t} + \frac{i\beta_{2p}}{2}\frac{\partial^2}{\partial t^2} + \frac{\alpha_p}{2}A_p$$
$$= i\gamma_p(1 - f_R)(|A_p|^2 + 2|A_s|^2)A_p + R_p(z,t), \quad (8.1.15)$$

$$\frac{\partial A_s}{\partial z} + \frac{1}{v_{gs}}\frac{\partial A_s}{\partial t} + \frac{i\beta_{2s}}{2}\frac{\partial^2 A_s}{\partial t^2} + \frac{\alpha_s}{2}A_s$$
$$= i\gamma_s(1 - f_R)(|A_s|^2 + 2|A_p|^2)A_s + R_s(z,t) \quad (8.1.16)$$

where v_{gj} is the group velocity, β_{2j} is the GVD coefficient, and γ_j is the non-linear parameter [defined in Eq. (7.1.19)] with $j = p$ or s. The Raman contributions R_p and R_s are obtained from

$$R_j(z,t) = i\gamma_j f_R A_j \int_{-\infty}^{t} h_R(t-t')[|A_j(z,t')|^2 + |A_k(z,t')|^2]\,dt' + i\gamma_j f_R A_k$$

$$\times \int_{-\infty}^{t} h_R(t-t')A_j(z,t')A_k^*(z,t')\exp[\pm i\Omega_R(t-t')]\,dt'. \qquad (8.1.17)$$

where $j,k = p$ or s such that $j \neq k$, $\Omega_R = \omega_p - \omega_s$ is the Stokes shift, and f_R represents the fractional Raman contribution (see Section 2.3). The Raman response function $h_R(t)$ leads to SRS. The effects of third-order dispersion can be included adding a third-derivative term proportional to β_3. Spontaneous Raman scattering can also be included adding a noise term in this set of two equations [21].

In the picosecond regime in which pulse widths exceed 1 ps, Eqs. (8.1.15) and (8.1.16) can be simplified considerably [22]. The reason is that u_p and u_s vary little over the time scale over which the Raman response function $h_R(t)$ changes. Treating u_p and u_s as constants, the integral in Eq. (8.1.17) can be performed analytically to obtain

$$R_j = i\gamma_j f_R[(|A_j|^2 + 2|A_k|^2)A_j + \tilde{h}_R(\pm\Omega_R)A_jA_k^*], \qquad (8.1.18)$$

where $\tilde{h}_R$ is the Fourier transform of the $h_R(t)$ and the negative sign is chosen for $j = s$. At the gain peak located at Ω_R, the real part of $\tilde{h}_R$ vanishes while the imaginary part is related to the Raman gain. Introducing the gain coefficients

$$g_p = 2\gamma_p f_R|\tilde{h}_R(\Omega_R)|, \qquad g_s = 2\gamma_s f_R|\tilde{h}_R(\Omega_R)|, \qquad (8.1.19)$$

the coupled amplitude equations become

$$\frac{\partial A_p}{\partial z} + \frac{1}{v_{gp}}\frac{\partial A_p}{\partial t} + \frac{i}{2}\beta_{2p}\frac{\partial^2}{\partial t^2} + \frac{\alpha_p}{2}A_p$$
$$= i\gamma_p[|A_p|^2 + (2 - f_R)|A_s|^2]A_p - \frac{g_p}{2}|A_s|^2A_p, \qquad (8.1.20)$$

$$\frac{\partial A_s}{\partial z} + \frac{1}{v_{gs}}\frac{\partial A_s}{\partial t} + \frac{i}{2}\beta_{2s}\frac{\partial^2 A_s}{\partial t^2} + \frac{\alpha_s}{2}A_s$$
$$= i\gamma_s[|A_s|^2 + (2 - f_R)|A_p|^2]A_s + \frac{g_s}{2}|A_p|^2A_s. \qquad (8.1.21)$$

Note that the XPM factor is $2 - f_R$, rather than being 2, when the Raman contribution is included [19]. The parameter f_R has a value of about 0.18 [12]. Under certain conditions, it is necessary to include the transient Raman dynamics by adding a third equation governing vibrations of molecules [23].

Equations (8.1.20) and (8.1.21) are solved in Section 8.3 where we consider SRS with picosecond pump pulses. The most important new feature is the group-velocity mismatch that limits the SRS process to a duration during which the pump and Raman pulses overlap. The walk-off length can be introduced as in Section 7.4 using

$$L_W = T_0 / |v_{gp}^{-1} - v_{gs}^{-1}|, \qquad (8.1.22)$$

where T_0 represents duration of the pump pulse. Typically $L_W \sim 1$ m in the visible region for $T_0 \sim 5$ ps. For pump pulses of width $T_0 > 1$ ns, L_W exceeds 200 m and is larger than fiber lengths commonly used to observe SRS. The GVD effects are negligible for such pulses and the CW theory is approximately valid in this quasi-CW regime. In fact, Eqs. (8.1.2) and (8.1.3) can be obtained from Eqs. (8.1.20) and (8.1.21) if the time derivatives are neglected and $I_j = |A_j|^2 / A_{\text{eff}}$ is used ($j = p$ or s). Note that Eqs. (8.1.20) and (8.1.21) are not valid for femtosecond pump pulses whose spectral width exceeds the Raman shift. This case is considered later in this chapter.

8.2   Quasi-Continuous SRS

Since the initial observation of SRS in optical fibers [10], SRS has been studied extensively using pump pulses of widths in the range 1–100 ns, a situation that corresponds to the quasi-CW regime [24]–[41]. In the single-pass geometry, each pump pulse launched at one end of the fiber generates a Stokes pulse at the other end. In the multipass geometry, the fiber is placed inside a cavity, resulting in a tunable Raman laser. Another application consists of using SRS for signal amplification. This section discusses all three aspects of SRS in optical fibers.

8.2.1   Single-Pass Raman Generation

The 1972 demonstration of SRS in silica fibers was carried out in the visible region using 532-nm pulses from a frequency-doubled Nd:YAG laser [9]. About 75 W of pump power was required to generate the Stokes radiation at

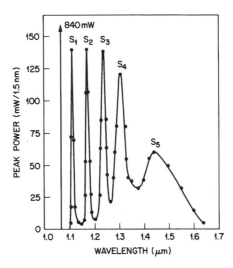

Figure 8.2 Five Stokes lines S_1 to S_5 generated simultaneously using 1.06-μm pump pulses. Vertical line corresponds to residual pump. Powers were measured through a monochromator with 1.5-nm resolution. (After Ref. [27] ©1978 IEEE.)

545 nm in a single-mode fiber of 9-m length and 4-μm core diameter. In later experiments, 150-ns infrared pump pulses, from a Nd:YAG laser operating at 1.06 μm, were used to initiate SRS [25]. In one experiment, the first-order Stokes line at 1.12 μm was observed at a pump power of 70 W [27]. Higher-order Stokes lines appeared at higher pump powers when the Stokes power became large enough to pump the next-order Stokes line. Figure 8.2 shows the optical spectrum at a pump power of about 1 kW with five Stokes lines clearly seen. Each successive Stokes line is broader than the preceding one. This broadening is due to several competing nonlinear processes and limits the total number of Stokes lines. Stokes line of up to 15th order have been generated in the visible region [31].

In these experiments no attempt was made to resolve spectral details of each Stokes line. In a subsequent experiment [36], the resolution was fine enough to resolve the line shape of the first-order Stokes line, generated by launching in a 100-m-long fiber pump pulses of about 1-ns duration, obtained from a mode-locked argon laser ($\lambda_p = 514.5$ nm). Figure 8.3 shows the observed spectra at three pump powers. The spectra exhibit a broad peak at 440 cm^{-1} (13.2 THz) and a narrow peak at 490 cm^{-1} (14.7 THz). As pump power is increased, the peak power of the broad peak saturates while that of

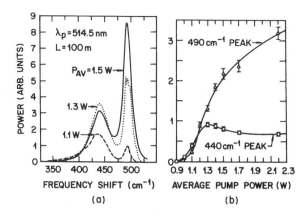

Figure 8.3 (a) Stokes spectra at three pump powers and (b) variations of peak powers with pump power. (After Ref. [36].)

the narrow peak keeps increasing.

The appearance of the double-peak Stokes spectrum can be understood by noting from Fig. 8.1 that the dominant peak in the Raman-gain spectrum actually consists of two peaks whose locations exactly coincide with the two peaks in the Stokes spectra of Fig. 8.3. A detailed numerical model, in which the shape of the Raman-gain spectrum is included and each spectral component of the Stokes line is propagated along the fiber including both the Raman gain and spontaneous Raman scattering, predicts line shapes in agreement with the experimentally observed spectra [36].

The features seen in Fig. 8.3 can be understood qualitatively as follows. Spontaneous Raman scattering generates Stokes light across the entire frequency range of the Raman-gain spectrum. After a short length of fiber, these weak signals are amplified with the appropriate gain coefficients while more spontaneous light is added. At low pump powers, the observed Stokes spectrum looks like $\exp[g_R(\Omega)]$ because of the exponential amplification process. As pump power is increased, the high-frequency peak at 440 cm^{-1} can pump the low-frequency peak at 490 cm^{-1} through the Raman-amplification process. This is precisely what is seen in Fig. 8.3. Eventually, the Stokes power becomes high enough to generate a second-order Stokes line. Even though this model is based on the CW theory of SRS, it is able to explain the qualitative features of Fig. 8.3 because the GVD effects are of minor importance for pulse widths ~ 1 ns. When the pump pulses become shorter than 1 ns, it becomes increasingly more important to include the GVD effects, especially the group-

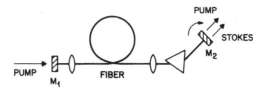

Figure 8.4 Schematic illustration of a tunable Raman laser. Mirrors M_1 and M_2 form a Fabry–Perot cavity. (After Ref. [46].)

velocity mismatch that leads to pulse walk-off. These effects are considered in Section 8.3.

The Stokes radiation generated through SRS is generally noisy because it builds up from spontaneous Raman scattering. As a result, both the width and the energy of Stokes pulses exhibit shot-to-shot fluctuations even when pump pulses have constant width and energy. The statistics associated with such fluctuations has been quantified by using Q-switched pulses emitted from a Nd:YAG laser at a repetition rate of 1 kHz [40]. The relative noise level of pulse energy decreases rapidly as the peak power of pump pulses is increased beyond the Raman threshold. Close to the threshold, the distribution of pulse energies is nearly exponential, as expected from quantum-noise theory [42]. However, just before the onset of the second-order Stokes line, the energy distribution of the first-order Stokes pulse becomes considerably narrower with a nearly Gaussian shape. The experimental results can be simulated solving Eqs. (8.1.2) and (8.1.3) with a random Stokes seed [41].

8.2.2 Raman Fiber Lasers

An important application of the SRS phenomenon in optical fibers has resulted in the development of fiber-based Raman lasers [43]–[65]. Such lasers can be tuned over a wide frequency range ($\sim$10 THz). Figure 8.4 shows a schematic of a Raman laser. A piece of single-mode fiber is placed inside a Fabry–Perot cavity formed by two partially reflecting mirrors M_1 and M_2. The cavity provides wavelength-selective feedback for the Stokes light generated inside the fiber through SRS. An intracavity prism allows tuning of the laser wavelength by dispersing various Stokes wavelengths spatially, each of which can be selected by turning the mirror M_2. The laser threshold corresponds to the pump power at which Raman amplification during a round trip is large enough to balance cavity losses consisting mainly of the transmission loss at the mirrors and coupling losses at the two ends of the fiber. If we assume a typical value of 10 dB for the round-trip loss, the threshold condition is

$$G = \exp(2g_R P_0 L_{eff}/A_{eff}) = 10, \qquad (8.2.1)$$

where L_{eff} is given by Eq. (8.1.7) for a fiber of length L. If optical fiber does not preserve polarization, g_R in Eq. (8.2.1) is reduced by a factor of two because of scrambling of the relative polarization between the pump and the Stokes waves [29]. A comparison of Eqs. (8.1.13) and (8.2.1) shows that the threshold pump power for a Raman laser is lower by at least one order of magnitude than that of single-pass SRS.

In the 1972 demonstration of a Raman laser [9], the threshold power was relatively large (about 500 W) because of a short fiber length ($L = 1.9$ m) used in the experiment. In subsequent experiments [43]–[45], the threshold was reduced to a level ~ 1 W by using longer fibers ($L \sim 10$ m). This feature permitted CW operation of a Raman laser at wavelengths in the range 0.50–0.53 μm using an argon-ion laser as the pump. Stimulated Brillouin scattering was suppressed by ensuring that the spectral width of the multimode pump was much larger than the Brillouin-gain bandwidth (see Section 9.1). The use of an intracavity prism allowed tuning of the laser wavelength over a range of about 10 nm.

At high pump powers, higher-order Stokes wavelengths are generated inside the fiber. These wavelengths are dispersed spatially by the intracavity prism. By adding separate mirrors for each Stokes beam, such a Raman laser can be operated at several wavelengths simultaneously, each of which can be independently tuned by turning a cavity mirror [44]. In one experiment, a ring-cavity configuration was used to generate five orders of tunable Stokes bands [50]. Raman lasers have been operated in the infrared region extending from 1.1–1.6 μm, a region useful for optical fiber communications, using a Nd:YAG laser as a pump [51].

When a Raman laser is pumped by a train of pump pulses, each Raman pulse after one round trip should be properly synchronized with one of the succeeding pump pulses. It is relatively easy to achieve synchronization in Raman lasers. The reason is that the laser can select a particular wavelength that fulfills the synchronous-pumping requirement among the wide range of possible wavelengths near the peak of the Raman-gain spectrum (see Fig. 8.1). Moreover, the laser wavelength can be tuned by simply changing the cavity length. This technique is referred to as time-dispersion tuning [46] to distin-

guish it from prism tuning (see Fig. 8.4) that works because of spatial dispersion provided by the prism. The technique is very effective in tuning pulsed
Raman lasers over a wide wavelength range. The tuning rate can be obtained
as follows. If the cavity length is changed by ΔL, the time delay Δt should be
exactly compensated by a wavelength change $\Delta\lambda$ such that

$$\Delta t \equiv \Delta L/c = |D(\lambda)|L\Delta\lambda, \qquad (8.2.2)$$

where L is the fiber length and D is the dispersion parameter introduced in
Section 1.2.3. The tuning rate is therefore given by

$$\frac{\Delta\lambda}{\Delta L} = \frac{1}{cL|D(\lambda)|} = \frac{\lambda^2}{2\pi c^2 L|\beta_2|}, \qquad (8.2.3)$$

where Eq. (1.2.11) was used to relate D to the GVD coefficient β_2. The tuning
rate depends on the fiber length L and the wavelength λ, and is typically $\sim$
1 nm/cm. In one experiment, a tuning rate of 1.8 nm/cm with a tuning range
of 24 nm was obtained for $L = 600$ m and $\lambda = 1.12$ μm [47].

Synchronously pumped Raman lasers have attracted attention for generating ultrashort optical pulses [59]. In general, it becomes necessary to take into
account the effects of GVD, group-velocity mismatch, SPM, and XPM when
such lasers are pulsed using pump pulses of widths below 100 ps. These effects
are discussed in Section 8.3. If the Raman pulse falls in the anomalous GVD
regime of the fiber, the soliton effects can create pulses of widths $\sim$100 fs or
less. Such lasers are called Raman soliton lasers and are covered in Section
8.4.

The development of Raman lasers advanced considerably during the 1990s.
A new feature was the integration of cavity mirrors within the fiber to make a
compact device. In an early approach, a ring-cavity configuration was used to
make a low-threshold, all-fiber Raman laser using a fiber loop and a fiber coupler [60]. With the advent of fiber-Bragg gratings, it has become possible to
replace cavity mirrors with such gratings [61]. Fused fiber couplers can also be
used for this purpose. In an interesting approach, three pairs of fiber gratings
or couplers are arranged such that they form three cavities for the three Raman
lasers operating at wavelengths 1.117, 1.175, and 1.24 μm, corresponding to
first, second, and third-order Stokes line of a 1.06-μm pump [63]. The resulting 1.24-μm Raman laser is useful for amplifying 1.31-μm signals [64]. The
same approach can be used for making a 1.48-μm Raman laser if a phospho-
silicate fiber is used [65]. Such a fiber provides a Stokes shift of nearly 40 THz

and can convert 1.06-μm pump to 1.48-μm radiation through the second-order Stokes line. Output powers of more than 1 W were generated by this technique using 4.5 W of pump power from a 976-nm diode-laser array that pumped a double-clad Yb-doped fiber to obtain 3.3 W of 1.06 μm radiation.

Raman lasers, operating in the visible and ultraviolet regions and tunable over a wide range, have also been made using a double-pass scheme with multimode fibers [62]. Tuning over a wide wavelength range (540–970 nm) with high peak-power levels (>12 kW) was realized when a 50-m-long multimode fiber (core diameter 200 μm) was pumped at 532 nm using second-harmonic pulses from a Q-switched Nd:YAG laser having peak powers of more than 400 kW. The same technique provided tuning over the wavelength range 360–527 nm when Q-switched pulses at at 335 nm were used using third-harmonic generation. As the broadband light generated by SRS passes through the fiber only twice, such cavityless lasers are not real lasers in the usual sense. They are nonetheless useful as a tunable source.

8.2.3 Raman Fiber Amplifiers

Optical fibers can be used to amplify a weak signal if that signal is launched together with a strong pump wave such that their frequency difference lies within the bandwidth of the Raman-gain spectrum. Because SRS is the physical mechanism behind amplification, such amplifiers are called Raman fiber amplifiers. They were made as early as 1976 and developed further during the 1980s for their potential applications in fiber-optic communication systems [66]–[91]. The experimental setup is similar to that of Fig. 8.4 except that mirrors are not needed. In the forward-pumping configuration, the pump propagates with the signal in the same direction whereas the two counterpropagate in the backward-pumping configuration.

The gain provided by Raman amplifiers under CW or quasi-CW operation can be obtained from Eqs. (8.1.2) and (8.1.3). If the signal intensity $I_s(z)$ remains much smaller than the pump intensity, pump depletion can be ignored. The signal intensity at the amplifier output at $z = L$ is then given by Eq. (8.1.6). Because $I_s(L) = I_s(0)\exp(-\alpha_s L)$ in the absence of pump, the amplification factor is given by

$$G_A = \exp(g_R P_0 L_{\text{eff}} / A_{\text{eff}}), \qquad (8.2.4)$$

where $P_0 = I_0 A_{\text{eff}}$ is the pump power at the amplifier input and L_{eff} is given by Eq. (8.1.7). If we use typical parameter values, $g_R = 1 \times 10^{-13}$ m/W, $L_{\text{eff}} =$

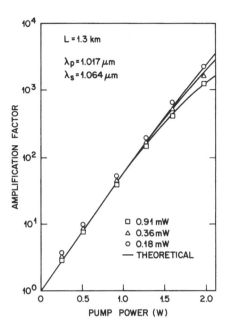

Figure 8.5 Variation of amplifier gain G_A with pump power P_0. Different symbols show the experimental data for three values of input signal power. Solid curves show the theoretical prediction using $g_R = 9.2 \times 10^{-14}$ m/W. (After Ref. [67].)

100 m, and $A_{\text{eff}} = 10 \ \mu\text{m}^2$, the signal is amplified considerably for $P_0 > 1$ W. Figure 8.5 shows the observed variation of G_A with P_0 when a 1.3-km-long fiber was used to amplify a 1.064-μm signal by using a 1.017-μm pump [67]. The amplification factor G_A increases exponentially with P_0 initially but starts to deviate for $P_0 > 1$ W. This is due to gain saturation occurring because of pump depletion. The solid lines in Fig. 8.5 are obtained by solving Eqs. (8.1.2) and (8.1.3) numerically to include pump depletion. The numerical results are in excellent agreement with the data.

An approximate expression for the saturated gain G_s in Raman amplifiers can be obtained by solving Eqs. (8.1.2) and (8.1.3) analytically [17] with the assumption $\alpha_s = \alpha_p \equiv \alpha$. Making the transformation $I_j = \omega_j F_j \exp(-\alpha z)$ with $j = s$ or p, we obtain two simple equations:

$$\frac{dF_s}{dz} = \omega_p g_R F_p F_s, \qquad \frac{dF_p}{dz} = -\omega_p g_R F_p F_s. \tag{8.2.5}$$

Noting that $F_p(z) + F_s(z) = C$, where C is a constant, the differential equation

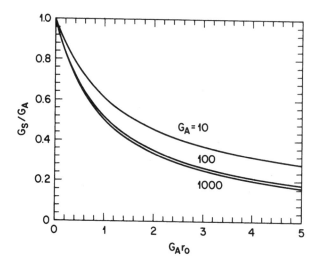

Figure 8.6 Gain-saturation characteristics of Raman amplifiers for several values of the unsaturated amplifier gain G_A.

for F_s can be integrated over the amplifier length, and the result is

$$G_s = \frac{F_s(L)}{F_s(0)} = \left(\frac{C - F_s(L)}{C - F_s(0)}\right) \exp(\omega_p g_R C L_{\text{eff}}). \tag{8.2.6}$$

Using $C = F_p(0) + F_s(0)$ in this equation the saturated gain of the amplifier is given by

$$G_s = \frac{1 + r_0}{r_0 + G_A^{-(1+r_0)}}, \tag{8.2.7}$$

where r_0 is related to the signal-to-pump power ratio at the fiber input as

$$r_0 = \frac{F_s(0)}{F_p(0)} = \frac{\omega_p}{\omega_s}\frac{P_s(0)}{P_0}, \tag{8.2.8}$$

and $G_A = \exp(g_R P_0 L_{\text{eff}}/A_{\text{eff}})$ is the small-signal (unsaturated) gain.

Figure 8.6 shows the saturation characteristics by plotting G_s/G_A as a function of $G_A r_0$ for several values of G_A. The saturated gain is reduced by a factor of two when $G_A r_0 \approx 1$. This condition is satisfied when the power in the amplified signal starts to approach the input pump power P_0. In fact, P_0 is a good measure of the saturation power of Raman amplifiers. As typically

$P_0 \sim 1$ W, the saturation power of Raman amplifiers is much larger compared with that of other optical amplifiers [87].

As seen in Fig. 8.5, Raman amplifiers can easily amplify an input signal by a factor of 1000 (30-dB gain) at a pump power of about 1 W [67]. In a 1983 experiment, a 1.24-μm signal from a semiconductor laser was amplified by 45 dB using a 2.4-km-long fiber [70]. This experiment used the forward-pumping configuration. In a different experiment [69], a 1.4-μm signal was amplified using both the forward- and backward-pumping configurations. The CW light from a Nd:YAG laser operating at 1.32 μm acted as the pump. Gain levels of up to 21 dB were obtained at a pump power of 1 W. The amplifier gain was nearly the same in both pumping configurations.

For optimum performance of Raman amplifiers, the frequency difference between the pump and signal beams should correspond to the peak of the Raman gain in Fig. 8.1. In the near-infrared region, the most practical pump source is the Nd:YAG laser operating at 1.06 or 1.32 μm. For this laser, the maximum gain occurs for signal wavelengths near 1.12 and 1.40 μm, respectively. However, the signal wavelengths of most interest from the standpoint of optical fiber communications are near 1.3 and 1.5 μm. A Nd:YAG laser can still be used if a higher-order Stokes line is used as a pump. For example, the third-order Stokes line at 1.24 μm from a 1.06-μm laser can act as a pump to amplify the signals at 1.3 μm. Similarly, the first-order Stokes line at 1.4 μm from a 1.32-μm laser can act as a pump to amplify the signals near 1.5 μm. As early as 1984, amplification factors of more than 20 dB were realized by using such schemes [72]–[74]. These experiments also indicated the importance of matching the polarization directions of the pump and probe waves as SRS nearly ceases to occur in the case of orthogonal polarizations. The use of a polarization-preserving fiber with a high-germania core has resulted in 20-dB gain at 1.52 μm by using only 3.7 W of input power from a Q-switched 1.34-μm laser [56].

A possible application of Raman amplifiers is as a preamplifier before the signal is detected at the receiver of an optical communication system [88]. Experimental measurements show that the signal-to-noise ratio (SNR) at the receiver is determined by spontaneous Raman scattering that invariably accompanies the amplification process [79]. The output consists not only of the desired signal but also of amplified spontaneous noise extending over a wide frequency range (~ 10 THz). It is possible to obtain an analytic expression for the noise power in the undepleted-pump approximation [76].

From a practical standpoint, the quantity of interest is the so-called on–off ratio, defined as the ratio of the signal power with the pump on to that with the pump off. This ratio can be measured experimentally. The experimental results for a 1.34-μm pump show that the on–off ratio is about 24 dB for the first-order Stokes line at 1.42 μm but degrades to 8 dB when the first-order Stokes line is used to amplify a 1.52-μm signal. The on–off ratio is also found to be smaller in the backward-pumping configuration [84]. It can be improved if the output is passed through an optical filter that passes the amplified signal but reduces the bandwidth of the spontaneous noise.

An attractive feature of Raman amplifiers is related to their broad bandwidth ($\sim$5 THz). It can be used to amplify several channels simultaneously in a wavelength-division-multiplexed (WDM) lightwave system. This feature was demonstrated in a 1987 experiment [90] in which signals from three distributed-feedback semiconductor lasers operating in the range 1.57–1.58 μm were amplified simultaneously using a 1.47-μm pump. The gain of 5 dB was obtained at a pump power of only 60 mW. A theoretical analysis shows that a trade-off exists between the on–off ratio and channel gains [91]. During the 1980s considerable attention focused on improving the performance of optical communication systems using Raman amplification [92]–[97]. This scheme is called distributed amplification as fiber losses are compensated over the entire fiber length in a distributed manner. It was used in 1988 to demonstrate soliton transmission over 4000 km using 55-ps pulses [97].

The main drawback of Raman amplifiers from the standpoint of lightwave system applications is that a high-power laser is required for pumping. The experiments performed near 1.55 μm often use tunable color-center lasers as a pump; such lasers are too bulky for communication applications. Indeed, with the advent of erbium-doped fiber amplifiers in 1989, Raman amplifiers were rarely used in the 1.55-μm wavelength region. The situation changed with the availability of compact high-power semiconductor and fiber lasers. Indeed, the development of Raman amplifiers has undergone a virtual renaissance during the 1990s [98]–[110]. As early as 1992, a Raman amplifier was pumped using a 1.55-μm semiconductor laser whose output was amplified though an erbium-doped fiber amplifier [99]. The 140-ns pump pulses had a 1.4-W peak-power level at the 1-kHz repetition rate and were capable of amplifying 1.66-μm signal pulses by more than 23 dB in a 20-km-long dispersion-shifted fiber. The resulting 200-mW peak power of 1.66-μm pulses was large enough for their use for optical time-domain reflection measurements, a technique commonly used for supervising and maintaining fiber-optic networks [100].

The use of Raman amplifiers in the 1.3-μm region has attracted considerable attention since 1995 [102]–[110]. In one approach, three pairs of fiber gratings are inserted within the fiber used for Raman amplification [102]. The Bragg wavelengths of these gratings are chosen such that they form three cavities for three Raman lasers operating at wavelengths 1.117, 1.175, and 1.24 μm that correspond to first-, second-, and third-order Stokes line of a 1.06-μm pump. All three lasers are pumped using a diode-pumped Nd-fiber laser through cascaded SRS. The 1.24-μm laser then pumps the Raman amplifier to provide signal amplification in the 1.3-μm region. The same idea of cascaded SRS was used to obtain 39-dB gain at 1.3 μm by using WDM couplers in place of fiber gratings [103]. In a different approach, the core of silica fiber is doped heavily with germania. Such a fiber can be pumped to provide 30-dB gain at a pump power of only 350 mW [104]. Such pump powers can be obtained by using two or more semiconductor lasers. A dual-stage configuration has also been used in which a 2-km-long germania-doped fiber is placed in series with a 6-km-long dispersion-shifted fiber in a ring geometry [110]. Such a Raman amplifier, when pumped by a 1.24-μm Raman laser, provided 22-dB gain in the 1.3-μm wavelength region with a noise figure of about 4 dB.

A second application of Raman amplifiers is to extend the bandwidth of WDM lightwave systems operating in the 1.55-μm region [111]–[114]. Erbium-doped fiber amplifiers, used commonly in this wavelength regime, have a bandwidth of under 35 nm. Moreover, a gain-flattening technique is needed to use the entire 35-nm bandwidth. Massive WDM systems (80 or more channels) typically require optical amplifiers capable of providing uniform gain over a 70–80-nm wavelength range. Hybrid amplifiers made by combining erbium doping with Raman gain have been developed for this purpose. In one implementation of this idea [114], a nearly 80-nm bandwidth was realized by combining an erbium-doped fiber amplifier with two Raman amplifiers, pumped simultaneously at three different wavelengths (1471, 1495, and 1503 nm) using four pump modules, each module launching more than 150 mW of power into the fiber. The pump laser for erbium dopants launched 62-mW of power at 1465 nm. The combined gain of about 30 dB was nearly uniform over the wavelength range 1.53–1.61 μm.

A third application of the Raman gain is being pursued for distributed amplification of signals [115]–[119]. In this case, relatively long spans ($\sim$50 km) of the transmission fiber are pumped bidirectionally for compensating fiber

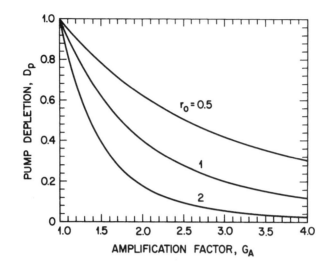

Figure 8.7 Pump-depletion characteristics showing variation of D_p with G_A for three values of r_0.

losses in a distributed manner. This technique is especially useful for solitons [117]. In a 2000 demonstration of this technique, 100 WDM channels with 25-GHz channel spacing, each operating at a bit rate of 10 Gb/s, were transmitted over 320 km [119]. All channels were amplified simultaneously by pumping each 80-km fiber span in the backward direction using four semiconductor lasers. Such a distributed Raman amplifier provided 15-dB gain at a pump power of 450 mW.

8.2.4   Raman-Induced Crosstalk

The same Raman gain that is beneficial for making fiber amplifiers and lasers is also detrimental for WDM systems. The reason is that a short-wavelength channel can act as a pump for longer-wavelength channels and thus transfer part of the pulse energy to neighboring channels. This leads to Raman-induced crosstalk among channels that can affect the system performance considerably [120]–[134].

Consider first a two-channel system with the short-wavelength channel acting as a pump. The power transfer between the two channels is governed by Eqs. (8.1.2) and (8.1.3). These equations can be solved analytically if the fiber loss is assumed to be the same for both channels ($\alpha_s = \alpha_p$), an assumption easily justified for typical channel spacings near 1.55 μm. The amplification factor G_s for the longer-wavelength channel is given by Eq. (8.2.7). The associated reduction in the short-wavelength channel power is obtained from the pump-depletion factor

$$D_p = \frac{I_p(L)}{I_p(0)\exp(-\alpha_p L)} = \frac{1+r_0}{1+r_0\,G_A^{1+r_0}}, \qquad (8.2.9)$$

where G_A and r_0 are defined by Eqs. (8.2.4) and (8.2.8), respectively. Figure 8.7 shows the pump-depletion characteristics by plotting D_p as a function of G_A for several values of r_0. These curves can be used to obtain the Raman-induced power penalty defined as the relative increase in the pump power necessary to maintain the same level of output power as that in the absence of Raman crosstalk. The power penalty can be written as (in decibels)

$$\Delta = 10\log(1/D_p). \qquad (8.2.10)$$

A 1-dB penalty corresponds to $D_p \approx 0.8$. If we assume equal channel powers at the fiber input ($r_0 \approx 1$), $D_p = 0.8$ corresponds to $G_A \approx 1.22$. The input channel powers corresponding to 1-dB penalty can be obtained from Eq. (8.2.4). If we use the typical values for 1.55-μm optical communication systems, $g_R = 7 \times 10^{-14}$ m/W, $A_{\text{eff}} = 50\ \mu\text{m}^2$, and $L_{\text{eff}} = 1/\alpha_p \approx 20$ km, $G_A = 1.22$ corresponds to $P_0 = 7$ mW. If the Raman gain is reduced by a factor of 2 to account for polarization scrambling [29], this value increases to $P_0 = 14$ mW. The experimental measurements of the power penalty are in agreement with the predictions of Eqs. (8.2.9) and (8.2.10).

The situation is more complicated for multichannel WDM systems. The intermediate channels not only transfer energy to the longer-wavelength channels but, at the same time, also receive energy from the shorter-wavelength channels. For an M-channel system one can obtain the output powers for each channel by solving a set of M coupled equations similar to Eqs. (8.1.2) and (8.1.3). The shortest-wavelength channel is most affected by Raman-induced crosstalk because it transfers a part of its energy to all channels lying within the Raman-gain bandwidth. The transferred amount, however, is different for each channel as it is determined by the amount of the Raman gain corresponding to the relative wavelength spacing. In one approach, the Raman-gain spectrum of Fig. 8.1 is approximated by a triangular profile [122]. The results show that for a 10-channel system with 10-nm separation, the input power of each channel should not exceed 3 mW to keep the power penalty below 0.5 dB. In a 10-channel experiment with 3-nm channel spacing, no power penalty related to Raman crosstalk was observed when the input channel power was kept below 1 mW [124]. See Section B.7.2 for further details.

8.3 SRS with Short Pump Pulses

The quasi-CW regime of SRS considered in Section 8.2 applies for pump pulses of widths >1 ns because the walk-off length L_W, defined by Eq. (8.1.22), generally exceeds the fiber length L for such pulses. However, for ultrashort pulses of widths below 100 ps, typically $L_W < L$. SRS is then limited by the group-velocity mismatch and occurs only over distances $z \sim L_W$ even if the actual fiber length L is considerable larger than L_W. At the same time, the nonlinear effects such as SPM and XPM become important because of relatively large peak powers and affect considerably evolution of both pump and Raman pulses [135]–[161]. This section discusses the experimental and theoretical aspects of SRS in the normal-GVD regime of optical fibers. Section 8.4 is devoted to the case of anomalous GVD where the role of soliton effects becomes important. In both cases, pulse widths are assumed to be larger than the Raman response time ($\sim$50 fs) so that transient effects are negligible.

8.3.1 Pulse-Propagation Equations

In the general case in which GVD, SPM, XPM, pulse walk-off, and pump depletion all play an important role, Eqs. (8.1.20) and (8.1.21) should be solved numerically. If fiber loss is neglected because of relatively small fiber lengths used in most experiments, and if time is measured in a frame of reference moving with the pump pulse, these equations take the form

$$\frac{\partial A_p}{\partial z} + \frac{i\beta_{2p}}{2}\frac{\partial^2 A_p}{\partial T^2}$$
$$= i\gamma_p[|A_p|^2 + (2 - f_R)|A_s|^2]A_p - \frac{g_p}{2}|A_s|^2 A_p, \qquad (8.3.1)$$

$$\frac{\partial A_s}{\partial z} - d\frac{\partial A_s}{\partial T} + \frac{i\beta_{2s}}{2}\frac{\partial^2 A_s}{\partial T^2}$$
$$= i\gamma_s[|A_s|^2 + (2 - f_R)|A_p|^2]A_p + \frac{g_s}{2}|A_p|^2 A_s, \qquad (8.3.2)$$

where

$$T = t - z/v_{gp}, \qquad d = v_{gp}^{-1} - v_{gs}^{-1}. \tag{8.3.3}$$

The walk-off parameter d accounts for the group-velocity mismatch between the pump and Raman pulses and is typically 2–6 ps/m. The GVD parameter β_{2j}, the nonlinearity parameter γ_j, and the Raman-gain coefficient g_j ($j = p$ or s) are slightly different for the pump and Raman pulses because of the Raman shift of about 13 THz between their carrier frequencies. In terms of the wavelength ratio λ_p/λ_s, these parameters for pump and Raman pulses are related as

$$\beta_{2s} = \frac{\lambda_p}{\lambda_s}\beta_{2p}, \qquad \gamma_s = \frac{\lambda_p}{\lambda_s}\gamma_p, \qquad g_s = \frac{\lambda_p}{\lambda_s}g_p. \tag{8.3.4}$$

Four length scales can be introduced to determine the relative importance of various terms in Eqs. (8.3.1) and (8.3.2). For pump pulses of duration T_0 and peak power P_0, these are defined as

$$L_D = \frac{T_0^2}{|\beta_{2p}|}, \quad L_W = \frac{T_0}{|d|}, \quad L_{NL} = \frac{1}{\gamma_p P_0}, \quad L_G = \frac{1}{g_p P_0}. \tag{8.3.5}$$

The dispersion length L_D, the walk-off length L_W, the nonlinear length L_{NL}, and the Raman-gain length L_G provide, respectively, the length scales over which the effects of GVD, pulse walk-off, nonlinearity (both SPM and XPM), and Raman gain become important. The shortest length among them plays the dominant role. Typically, $L_W \sim 1$ m (for $T_0 < 10$ ps) while L_{NL} and L_G become smaller or comparable to it for $P_0 > 100$ W. In contrast, $L_D \sim 1$ km for $T_0 = 10$ ps. Thus, the GVD effects are generally negligible for pulses as short as 10 ps. The situation changes for pulse widths ~ 1 ps or less because L_D decreases faster than L_W with a decrease in the pulse width. The GVD effects can then affect SRS evolution significantly, especially in the anomalous-dispersion regime.

8.3.2 Nondispersive Case

When the second-derivative term in Eqs. (8.3.1) and (8.3.2) is neglected, these equations can be solved analytically [158]–[161]. The analytic solution takes a simple form if pump depletion during SRS is neglected. As this assumption is justified for the initial stages of SRS and permits us to gain physical insight, let us consider it in some detail. The resulting analytic solution includes the effects of both XPM and pulse walk-off. The walk-off effects without

XPM [123] and the XPM effects without walk-off [136] were considered relatively early. Both of them can be included by solving Eqs. (8.3.1) and (8.3.2) with $\beta_{2p} = \beta_{2s} = 0$ and $g_p = 0$. Equation (8.3.1) for the pump pulse then yields the solution

$$A_p(z,T) = A_p(0,T)\exp[i\gamma_p|A_p(0,T)|^2z], \qquad (8.3.6)$$

where the XPM term has been neglected assuming $|A_s|^2 \ll |A_p|^2$. For the same reason, the SPM term in Eq. (8.3.2) can be neglected. The solution of Eq. (8.3.2) is then given by [147]

$$A_s(z,T) = A_s(0,T+zd)\exp\{[g_s/2 + i\gamma_s(2 - f_R)]\psi(z,T)\}, \qquad (8.3.7)$$

where

$$\psi(z,T) = \int_0^z |A_p(0,T+zd-z'd)|^2dz'. \qquad (8.3.8)$$

Equation (8.3.6) shows that the pump pulse of initial amplitude $A_p(0,T)$ propagates without change in its shape. The SPM-induced phase shift imposes a frequency chirp on the pump pulse that broadens its spectrum (see Section 4.1). The Raman pulse, by contrast, changes both its shape and spectrum as it propagates through the fiber—temporal changes occur owing to Raman gain while spectral changes have their origin in XPM. Because of pulse walk-off, both kinds of changes are governed by an overlap factor $\psi(z,T)$ that takes into account the relative separation between the two pulses along the fiber. This factor depends on the pulse shape. For a Gaussian pump pulse with the input amplitude

$$A_p(0,T) = \sqrt{P_0}\exp(-T^2/2T_0^2), \qquad (8.3.9)$$

the integral in Eq. (8.3.8) can be performed in terms of error functions with the result

$$\psi(z,\tau) = [\mathrm{erf}(\tau+\delta) - \mathrm{erf}(\tau)](\sqrt{\pi}P_0z/\delta), \qquad (8.3.10)$$

where $\tau = T/T_0$ and δ is the propagation distance in units of the walk-off length, that is,

$$\delta = zd/T_0 = z/L_W. \qquad (8.3.11)$$

An analytic expression for $\psi(z,\tau)$ can also be obtained for pump pulses having 'sech' shape [158]. In both cases, the Raman pulse compresses initially, reaches a minimum width, and then begins to rebroaden as it is amplified through SRS. It also acquires a frequency chirp through XPM. This qualitative behavior persists even when pump depletion is included [158]–[160].

Equation (8.3.7) describes Raman amplification when a weak signal pulse is injected together with the pump pulse. The case in which the Raman pulse builds from noise is much more involved from a theoretical viewpoint. A general approach should use a quantum-mechanical treatment similar to that employed for describing SRS in molecular gases [42]. It can be considerably simplified for optical fibers if transient effects are ignored by assuming that pump pulses are much wider than the Raman response time. In that case, Eqs. (8.3.1) and (8.3.2) can be used provided a noise term (often called the Langevin force) is added to the right-hand side of these equations. The noise term leads to pulse-to-pulse fluctuations in the amplitude, the width, and the energy of the Raman pulse similar to those observed for SRS in molecular gases [21]. Its inclusion is essential if the objective is to quantify such fluctuations.

The *average* features of noise-seeded Raman pulses can be described by using the theory of Section 8.1.2 where the effective Stokes power at the fiber input is obtained by using one photon per mode at all frequencies within the Raman-gain spectrum. Equation (8.1.10) then provides the input peak power of the Raman pulse, while its shape remains undetermined. Numerical solutions of Eqs. (8.3.1) and (8.3.2) show that the *average* pulse shapes and spectra at the fiber output are not dramatically affected by different choices of the shape of the seed pulse. A simple approximation consists of assuming

$$A_s(0, T) = (P_{s0}^{\text{eff}})^{1/2}, \tag{8.3.12}$$

where P_{s0}^{eff} is given by Eq. (8.1.10). Alternatively, one may take the seed pulse to be Gaussian with a peak power P_{s0}^{eff}.

As a simple application of the analytic solution (8.3.7), consider the Raman threshold for SRS induced by short pump pulses of width T_0 and peak power P_0 [141]. The peak power of the Raman pulse at the fiber output ($z = L$) is given by

$$P_s(L) = |A_s(L, 0)|^2 = P_{s0}^{\text{eff}} \exp(\sqrt{\pi} g_s P_0 L_W), \tag{8.3.13}$$

where Eq. (8.3.10) was used with $\tau = 0$ and $L/L_W \gg 1$. If we define the Raman threshold in the same way as for the CW case, the threshold is achieved when $P_s(L) = P_0$. The comparison of Eqs. (8.1.12) and (8.3.13) shows that one can use the CW criterion provided the effective length is taken to be

$$L_{\text{eff}} = \sqrt{\pi} L_W \approx T_{\text{FWHM}}/|d|. \tag{8.3.14}$$

In particular, Eq. (8.1.13) can be used to obtain the critical peak power of the pump pulse if L_{eff} is obtained from Eq. (8.3.14). This change is expected because the effective interaction length between the pump and Raman pulses is determined by the length L_W—SRS ceases to occur when the two pulses move apart enough that they stop overlapping significantly. Equations (8.1.13) and (8.3.14) show that the Raman threshold depends on the width of the pump pulse and increases inversely with T_{FWHM}. For pulse widths ~ 10 ps ($L_W \sim 1$ m), the threshold pump powers are ~ 100 W.

The analytic solution (8.3.7) can be used to obtain both the shape and the spectrum of the Raman pulse during the initial stages of SRS [147]. The spectral evolution is governed by the XPM-induced frequency chirp. The chirp behavior has been discussed in Section 7.4 in the context of XPM-induced asymmetric spectral broadening (see Fig. 7.3). The qualitative features of the XPM-induced chirp are identical to these shown there as long as the pump remains undepleted. Note, however, that the Raman pulse travels faster than the pump pulse in the normal-GVD regime. As a result, chirp is induced mainly near the trailing edge. It should be stressed that both pulse shapes and spectra are considerably modified when pump depletion is included [144]. The growing Raman pulse affects itself through SPM and the pump pulse through XPM.

8.3.3   Effects of GVD

When the fiber length is comparable to the dispersion length L_D, it is important to include the GVD effects. Such effects cannot be described analytically, and a numerical solution of Eqs. (8.3.1) and (8.3.2) is necessary to understand the SRS evolution. A generalization of the split-step Fourier method of Section 2.4 can be used for this purpose. The method requires specification of the Raman pulse at the fiber input by using Eq. (8.1.10).

For numerical purposes, it is useful to introduce the normalized variables. A relevant length scale along the fiber length is provided by the walk-off length L_W. By defining

$$z' = \frac{z}{L_W}, \qquad \tau = \frac{T}{T_0}, \qquad U_j = \frac{A_j}{\sqrt{P_0}}, \qquad (8.3.15)$$

and using Eq. (8.3.4), Eqs. (8.3.1) and (8.3.2) become

$$\frac{\partial U_p}{\partial z'} + \frac{i}{2}\frac{L_W}{L_D}\frac{\partial^2 U_p}{\partial \tau^2}$$

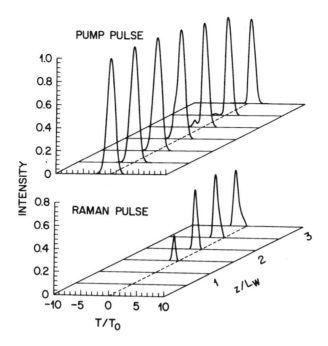

Figure 8.8 Evolution of pump and Raman pulses over three walk-off lengths for the case $L_D/L_W = 1000$, $L_W/L_{NL} = 24$, and $L_W/L_G = 12$.

$$= \frac{iL_W}{L_{NL}}[|U_p|^2 + (2 - f_R)|U_s|^2]U_p - \frac{L_W}{2L_G}|U_s|^2U_p, \qquad (8.3.16)$$

$$\frac{\partial U_s}{\partial z'} - \frac{\partial U_s}{\partial \tau} + \frac{ir}{2}\frac{L_W}{L_D}\frac{\partial^2 U_s}{\partial \tau^2}$$

$$= \frac{irL_W}{L_{NL}}[|U_s|^2 + (2 - f_R)|U_p|^2]U_s + \frac{rL_W}{2L_G}|U_p|^2U_s, \qquad (8.3.17)$$

where the lengths L_D, L_W, L_{NL}, and L_G are given by Eq. (8.3.5). The parameter $r = \lambda_p/\lambda_s$ and is about 0.95 at $\lambda_p = 1.06\ \mu$m. Figure 8.8 shows the evolution of the pump and Raman pulses over three walk-off lengths for the case $L_D/L_W = 1000$, $L_W/L_{NL} = 24$, and $L_W/L_G = 12$. The pump pulse is taken to be a Gaussian while the Raman seed is obtained from Eq. (8.3.12) with $P_{s0}^{eff} = 2 \times 10^{-7}$ W. The results shown in Fig. 8.8 are applicable to a wide variety of input pulse widths and pump wavelengths by using the scaling of Eqs. (8.3.5) and (8.3.15). The choice $L_W/L_G = 12$ implies that

$$\sqrt{\pi}g_sP_{0L}W \approx 21, \qquad (8.3.18)$$

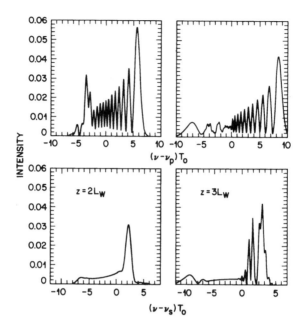

Figure 8.9 Spectra of pump (upper row) and Raman (lower row) pulses at twice (left column) and thrice (right column) the walk-off length for parameter values of Fig. 8.8.

and corresponds to a peak power 30% above the Raman threshold.

Several features of Fig. 8.8 are noteworthy. The Raman pulse starts to build up after one walk-off length. Energy transfer to the Raman pulse from the pump pulse is nearly complete by $z = 3L_W$ because the two pulses are then physically separated because of the group-velocity mismatch. As the Raman pulse moves faster than the pump pulse in the normal-GVD regime, the energy for SRS comes from the leading edge of the pump pulse. This is clearly apparent at $z = 2L_W$ where energy transfer has led to a two-peak structure in the pump pulse as a result of pump depletion—the hole near the leading edge corresponds exactly to the location of the Raman pulse. The small peak near the leading edge disappears with further propagation as the Raman pulse walks through it. The pump pulse at $z = 3L_W$ is asymmetric in shape and appears narrower than the input pulse as it consists of the trailing portion of the input pulse. The Raman pulse is also narrower than the input pulse and is asymmetric with a sharp leading edge.

The spectra of pump and Raman pulses display many interesting features resulting from the combination of SPM, XPM, group-velocity mismatch, and pump depletion. Figure 8.9 shows the pump and Raman spectra at $z/L_W = 2$ (left column) and $z/L_W = 3$ (right column). The asymmetric nature of these spectra is due to XPM (see Section 7.4.1). The high-frequency side of the pump spectra exhibits an oscillatory structure that is characteristic of SPM (see Section 4.1). In the absence of SRS, the spectrum would be symmetric with the same structure appearing on the low-frequency side. As the low-frequency components occur near the leading edge of the pump pulse and because the pump is depleted on the leading side, the energy is transferred mainly from the low-frequency components of the pump pulse. This is clearly seen in the pump spectra of Fig. 8.9. The long tail on the low-frequency side of the Raman spectra is also partly for the same reason. The Raman spectrum is nearly featureless at $z = 2L_W$ but develops considerable internal structure at $z = 3L_W$. This is due to the combination of XPM and pump depletion; the frequency chirp across the Raman pulse induced by these effects can vary rapidly both in magnitude and sign and leads to a complicated spectral shape [144].

The temporal and spectral features seen in Figs. 8.8 and 8.9 depend on the peak power of input pulses through the lengths L_G and L_{NL} in Eqs. (8.3.16) and (8.3.17). When peak power is increased, both L_G and L_{NL} decrease by the same factor. Numerical results show that because of a larger Raman gain, the Raman pulse grows faster and carries more energy than that shown in Fig. 8.8. More importantly, because of a decrease in L_{NL}, both the SPM and XPM contributions to the frequency chirp are enhanced, and the pulse spectra are wider than those shown in Fig. 8.9. An interesting feature is that the Raman-pulse spectrum becomes considerably wider than the pump spectrum. This is due to the stronger effect of XPM on the Raman pulse compared with that of the pump pulse. The XPM-enhanced frequency chirp for the Raman pulse was predicted as early as 1976 [135]. In a theoretical study that included XPM but neglected group-velocity mismatch and pump depletion, the spectrum of the Raman pulse was shown to be wider by a factor of two [136]. Numerical results that include all of these effects show an enhanced broadening by up to a factor of three, in agreement with experiments discussed later. Direct measurements of the frequency chirp also show an enhanced chirp for the Raman pulse compared with that of the pump pulse [155].

8.3.4 Experimental Results

The spectral and temporal features of ultrafast SRS have been studied in many experiments performed in the visible as well as in the near-infrared region. In

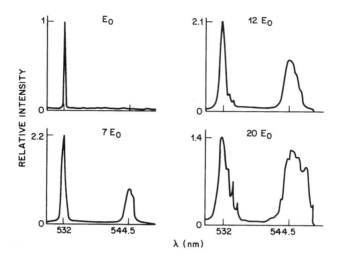

Figure 8.10 Experimental spectra observed when 25-ps pump pulses at 532 nm were propagated through a 10-m-long fiber. Four spectra correspond to different input pulse energies normalized to E_0. (After Ref. [146].)

one experiment, 60-ps pulses from a mode-locked Nd:YAG laser operating at 1.06 μm were propagated through a 10-m-long fiber [137]. When the pump power exceeded the Raman threshold ($\sim$1 kW), a Raman pulse was generated. Both the pump and Raman pulses were narrower than the input pulse as expected from the results shown in Fig. 8.8. The spectrum of the Raman pulse was much broader (spectral width $\approx$2 THz) than that of the pump pulse. The XPM-enhanced spectral broadening of the Raman pulse was quantified in an experiment in which 25-ps pulses at 532 nm were propagated through a 10-m-long fiber [146]. Figure 8.10 shows the observed spectra at four values of pump-pulse energies. The Raman spectral band located at 544.5 mm has a width about three times that of the pump. This is expected from theory and is due to the XPM-induced frequency chirp [136].

In the spectra of Fig. 8.10 the fine structure could not be resolved because of a limited spectrometer resolution. Details of the pump spectrum were resolved in another experiment in which 140-ps input pulses at 1.06 μm were propagated through a 150-m-long fiber [145]. Figure 8.11 shows the observed pump spectra at several values of input peak powers. The Raman threshold is about 100 W in this experiment. For $P_0 < 100$ W, the spectrum exhibits a multi-peak structure, typical of SPM (see Section 4.1). However, the pump spectrum

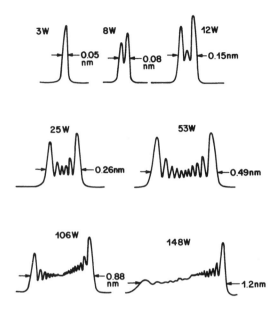

Figure 8.11 Experimental spectra of pump pulses at different input peak powers after 140-ps pump pulses have propagated through a 150-m-long fiber. The Raman threshold is reached near 100 W. (After Ref. [145].)

broadens and becomes highly asymmetric for $P_0 > 100$ W. In fact, the two spectra corresponding to peak powers beyond the Raman threshold show features that are qualitatively similar to those of Fig. 8.9 (upper row). The spectral asymmetry is due to the combined effects of XPM and pump depletion.

Another phenomenon that can lead to new qualitative features in the pump spectrum is the XPM-induced modulation instability. It has been discussed in Section 7.2 for the case in which two pulses at different wavelengths are launched at the fiber input. However, the same phenomenon should occur even if the second pulse is internally generated through SRS. Similar to the case of modulation instability occurring in the anomalous-dispersion regime (see Section 5.1), XPM-induced modulation instability manifests through the appearance of spectral sidelobes in the pulse spectra.

Figure 8.12 shows the observed spectra of the pump and Raman pulses in an experiment in which 25-ps pulses at 532 mm were propagated through a 3-m-long fiber [156]. The fiber-core diameter was only 3 μm to rule out the possibility of multimode four-wave mixing (see Chapter 10). The central peak in the pump spectrum contains considerable internal structure (Fig. 8.11) that

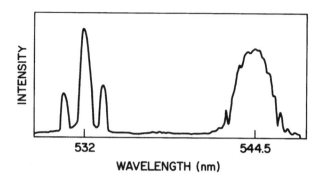

Figure 8.12 Spectra of pump and Raman pulses showing spectral sidelobes as a result of XPM-induced modulation instability. (After Ref. [156].)

remained unresolved in this experiment. The two sidebands provide a clear evidence of XPM-induced modulation instability. The position of sidebands changes with fiber length and with peak power of pump pulses. The Stokes spectrum also shows sidelobes, as expected from the theory of Section 7.2, although they are barely resolved because of XPM-induced broadening of the spectrum.

The temporal measurements of ultrafast SRS show features that are similar to those shown in Fig. 8.8 [139]–[142]. In one experiment, 5-ps pulses from a dye laser operating at 615 nm were propagated through a 12-m-long fiber with a core diameter of 3.3 μm [140]. Figure 8.13 shows the cross-correlation traces of the pump and Raman pulses at the fiber output. The Raman pulse arrives about 55 ps earlier than the pump pulse; this is consistent with the group-velocity mismatch at 620 nm. More importantly, the Raman pulse is asymmetric with a sharp leading edge and a long trailing edge, features qualitatively similar to those shown in Fig. 8.8. Similar results have been obtained in other experiments where the pulse shapes are directly recorded using a streak camera [145] or a high-speed photodetector [154].

The effects of pulse walk-off on SRS were studied by varying the peak power of 35-ps pump pulses (at 532 nm) over a range 140–210 W while fiber length was varied over a range 20–100 m [141]. Temporal measurements of the pump and Raman pulses were made with a high-speed CdTe photodetector and a sampling oscilloscope. The results show that the Raman pulse is produced within the first three to four walk-off lengths. The peak appears after about two walk-off lengths into the fiber for 20% energy conversion and moves

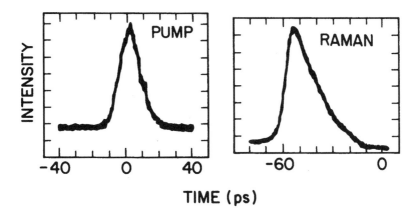

Figure 8.13 Cross-correlation traces of pump and Raman pulses at the output of a 12-m-long fiber. The intensity scale is arbitrary (after Ref. [140].)

closer to the input for higher peak powers. These conclusions are in agreement with the numerical results shown in Fig. 8.8.

So far, only first-order SRS has been considered. When the input peak power of pump pulses exceeds the Raman threshold considerably, the Raman pulse may become intense enough to act as a pump to generate the second-order Stokes pulse. Such a cascaded SRS was seen in a 615-nm experiment using 5-ps pump pulses with a peak power of 1.5 kW [140]. In the near-infrared region multiple orders of Stokes can be generated using 1.06-μm pump pulses. The efficiency of the SRS process can be improved considerably by using silica fibers whose core has been doped with P_2O_5 because of a relatively large Raman gain associated with P_2O_5 glasses [162]–[164].

From a practical standpoint ultrafast SRS limits the performance of fiber-grating compressors [165]; the peak power of input pulses must be kept below the Raman threshold to ensure optimum performance. Thus SRS not only acts as a loss mechanism but also limits the quality of pulse compression by distorting the linear nature of the frequency chirp resulting from the XPM interaction between pump and Raman pulses [166]. A spectral-filtering technique had been used to improve the quality of compressed pulses even in the presence of SRS [154]. In this approach a portion of the pulse spectrum is selected by using an asymmetric spectral window so that the filtered pulse has nearly linear chirp across its entire width. Good-quality compressed pulses can be obtained in the strong-SRS regime but only at the expense of substantial energy loss [158].

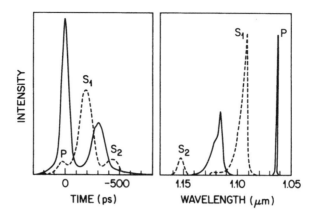

Figure 8.14 Temporal and spectral output of a Raman laser in the case of resonant (dashed curve) and single-pass (full curve) operation. (After Ref. [149].)

8.3.5   Synchronously Pumped Raman Lasers

The preceding section has focused on single-pass SRS. By placing the fiber inside a cavity (see Fig. 8.4), a single-pass SRS configuration can be turned into a Raman laser. Such lasers were discussed in Section 8.2.2 in the case of CW or quasi-CW operation ($T_0 \sim 1$ ns). This section considers Raman lasers pumped synchronously to emit short optical pulses with widths <100 ps. In a commonly used scheme, pump pulses are typically 100 ps wide and are obtained from a mode-locked Nd:YAG laser operating at 1.06 μm.

Figure 8.14 compares the temporal and spectral features at the output under single-pass (solid curve) and multipass (dashed curve) operations, the latter case corresponding to a Raman laser. In this experiment, the fiber was 150-m long and the pump-pulse width was about 120 ps [149]. In the single-pass case, the spectrum shows a SRS peak near 1.12 μm. The corresponding Raman pulse appears 300 ps earlier than the pump pulse, as expected from the walk-off effects. In the case of resonant operation as a Raman laser, the dominant spectral peak occurs at 1.093 μm, the wavelength corresponding to which the laser is synchronously pumped. Furthermore, this wavelength could be tuned over 50 nm through the time-dispersion technique by changing the fiber-cavity length by 10 cm [see Eq. (8.2.3)]. The second spectral peak in Fig. 8.14 corresponds to a nonresonant second-order Stokes line. In the time domain, the three-peak structure results from a superposition of the pump pulse and the two Raman pulses corresponding to the two spectral peaks. The first-Stokes

Raman pulse dominates because of its resonant nature.

Pulse widths generated from Raman lasers are about the same as those of pump pulses ($\sim$ 100 ps). However, because of the SPM and XPM effects, output pulses are chirped and, if the chirp is linear over a significant portion of the pulse, a fiber-grating compressor can be used to compress them (see Chapter B.6). In an important development, pulses as short as 0.8 ps were obtained by placing a fiber-grating compressor inside the cavity of a Raman laser [59]. The grating separation was adjusted to provide a slightly negative GVD over a complete round trip inside the ring cavity, that is, the grating pair not only compensated the positive GVD of the fiber but also provided a net negative-GVD environment to the pulses circulating inside the cavity. Pulses as short as 0.4 ps have been obtained by using this technique [148]. Furthermore, the Raman laser was tunable over 1.07–1.12 μm with pulse widths in the range 0.4–0.5 ps over the entire tuning interval. Such performance was achieved with the use of spectral filtering by placing an aperture inside the fiber-grating compressor.

A tunable Raman laser has been used to demonstrate amplification of femtosecond optical pulses in a Raman amplifier in both forward and backward-pumping configurations [148]. In the forward-pumping configuration, the 500-fs pulses are first passed through a 100-m-long fiber, where they broaden to about 23 ps as a result of SPM and GVD. The broadened pulses then enter a Raman amplifier, consisting of only a 1-m-long fiber and pumped by 50-ps pulses at 1.06 μm. The amplified pulses are compressed in a fiber-grating compressor. The compressed pulse was slightly broader (about 0.7 ps) than the input pulse but had its energy amplified by up to a factor of 15,000 when pumped by 150-kW pulses. The experiment demonstrated that the frequency chirp of the 23-ps input pulses was nearly unaffected by the process of Raman amplification. Such features indicate that ultrafast SRS in optical fibers not only is capable of generating femtosecond pulses but can also provide high peak powers.

8.4 Soliton Effects

When the wavelength of the pump pulse is close to or inside the anomalous-dispersion region of an optical fiber, the Raman pulse should experience the soliton effects (see Chapter 5). Such effects have attracted considerable attention both theoretically and experimentally [167]–[205]. Because the Raman pulse propagates as a soliton, it is common to refer to it as a Raman soliton [200]. These solitons must be distinguished from the pair of bright and dark solitons formed during transient SRS in molecular gases. In that case, it is necessary to include the dynamics of the vibrational mode participating in the process of transient SRS [206]–[210]. In contrast, Raman solitons discussed here occur in the stationary SRS regime.

8.4.1 Raman Solitons

In the anomalous-dispersion region of an optical fiber, under suitable conditions, almost all of the pump-pulse energy can be transferred to a Raman pulse that propagates undistorted as a fundamental soliton. Numerical results show that this is possible if the Raman pulse is formed at a distance at which the pump pulse, propagating as a higher-order soliton, achieves its minimum width [167]. By contrast, if energy transfer to the Raman pulse is delayed and occurs at a distance where the pump pulse has split into its components (see Fig 5.4 for the case $N = 3$), the Raman pulse does not form a fundamental soliton, and its energy rapidly disperses.

Equations (8.3.16) and (8.3.17) can be used to study ultrafast SRS in the anomalous-GVD regime by simply changing the sign of the second-derivative terms. As in the case of normal GVD shown in Fig. 8.8, energy transfer to the Raman pulse occurs near $z \approx L_W$. For the Raman pulse to form a soliton, $z_{opt} \approx L_W$, where z_{opt} is the optimum fiber length for the soliton-effect compressor (see Chapter B.6). This condition implies that L_W should not be too small in comparison with the dispersion length L_D. For example, $L_W = (\pi/8)L_D$ for the third-order soliton if we note that $z_{opt} = z_0/4$ for $N = 3$, where $z_0 = (\pi/2)L_D$ is the soliton period. Typically, L_W and L_D become comparable for femtosecond pulses of widths $T_0 \sim 100$ fs. For such ultrashort pump pulses, the distinction between pump and Raman pulses gets blurred as their spectra begin to overlap considerably. This can be seen by noting that the Raman-gain peak in Fig. 8.1 corresponds to a spectral separation of about 13 THz while the spectral width of a 100-fs pulse is ~ 10 THz. Equations (8.3.16) and (8.3.17) do not provide a realistic description of ultrafast SRS with femtosecond pump pulses, particularly in the case of anomalous GVD where the input pulse may shorten considerably during early stages of propagation.

An alternative approach is provided by the generalized propagation equation of Section 2.3. Equation (2.3.39) includes the effect of Raman gain through the last term proportional to the parameter T_R. As discussed there, T_R is related

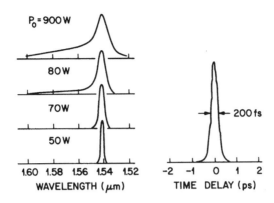

Figure 8.15 Pulse spectra of 30-ps input pulses at the output of a 250-m-long fiber for peak powers in the range 50–900 W. Autocorrelation trace of the Stokes tail in the topmost spectrum is shown on the right-hand side. (After Ref. [168].)

to the slope of the Raman gain near origin in Fig. 8.1. Some of the effects of this Raman-gain term on evolution of femtosecond pulses have already been discussed in Section 5.6. Figure 5.20 shows pulse shapes and spectra for a pump pulse whose peak power corresponds to a second-order soliton $(N = 2)$. The input pulse splits into two pulses within one soliton period.

The same behavior can also be interpreted in terms of intrapulse Raman scattering [168], a phenomenon that can occur even before the threshold of noise-induced SRS is reached. The basic idea is the following. The input pulse, propagating as a higher-order soliton, narrows its width and broadens its spectrum during the initial contraction phase. Spectral broadening on the red side provides a seed for Raman amplification, that is, the blue components of the pulse pump the red components through self-induced SRS. This is clearly seen in Fig. 5.20 where the dominant spectral peak moves continuously toward the red side. Such a shift is called the soliton self-frequency shift [170]. In the time domain, energy in the red-shifted components appears in the form of a Raman pulse that lags behind the input pulse because the red-shifted components travel slowly in the anomalous-GVD regime. The use of Eq. (2.3.40) becomes questionable for pulse widths 100 fs or less as it does not take into account the shape of the Raman-gain spectrum (see Fig. 8.1). Equation (2.3.33) should be used for such ultrashort pulses.

Intrapulse Raman scattering can occur even for picosecond input pulses as long as the soliton order N is large enough to broaden the input spectrum

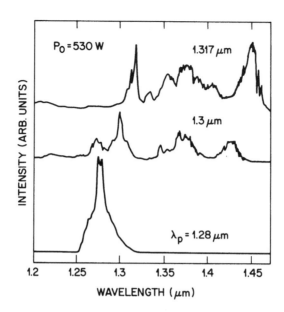

Figure 8.16 Pulse spectra at the output of a 150-m-long fiber when 0.83-ps input pulses are launched with the peak power $P_0 = 530$ W. The zero-dispersion wavelength of the fiber is 1.317 μm. (After Ref. [176] ©1987 IEEE.)

(through SPM) to widths ~ 1 THz. Indeed, in the first experimental demonstration of this phenomenon 30-ps input pulses at 1.54 μm were propagated through a 250-m-long fiber [168]. Figure 8.15 shows the observed pulse spectra for four input peak powers P_0 in the range 50–900 W. For $P_0 \sim 100$ W, a long tail toward the red side appears. This value corresponds to about $N = 30$ for the experimental values of parameters. Equations (6.4.2) and (6.4.3) predict compression of the input pulse by a factor of about 120 at a distance of 300 m if we use $z_0 = 27$ km for the soliton period ($T_0 \approx 17$ ps). The spectral width of such a compressed pulse is close to 2 THz. The autocorrelation trace of the energy in the Stokes tail of the topmost spectrum (downshifted 1.6 THz from the pump frequency) is also shown in Fig. 8.15. It corresponds to a pedestal-free Raman pulse of 200-fs width.

Intrapulse Raman scattering has attracted considerable attention as it provides a convenient way of generating Raman solitons whose carrier wavelength can be tuned by changing fiber length or input peak power [168]–[179]. In one experiment [176], the use of a dye laser permitted tuning of the input wavelength over a range 1.25–1.35 μm so that the input pulse could be launched in

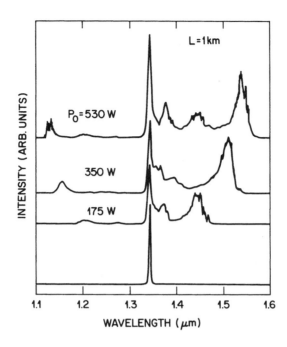

Figure 8.17 Pulse spectra at the output of a 1-km-long fiber for several peak powers of 0.83-ps input pulses. The input wavelength of 1.341 μm lies in the anomalous-GVD region of the fiber. (After Ref. [176] ©1987 IEEE.)

the normal or the anomalous GVD regime of the optical fiber (zero-dispersion wavelength $\approx$ 1.317 μm). Figure 8.16 shows the pulse spectra as the input wavelength λ_p is varied from 1.28 to 1.317 μm. The 0.83-ps input pulses with a peak power of 530 W were propagated over a 150-m-long single-mode fiber. No Stokes band was generated for $\lambda_p = 1.28$ μm because of the positive GVD of the fiber, although SPM-induced spectral broadening is clearly seen. Two Stokes bands formed for $\lambda_p = 1.3$ μm, even though the input wavelength is below the zero-dispersion wavelength, because a substantial part of the pulse energy appeared in the negative-GVD region after SPM-induced spectral broadening. For $\lambda_p = 1.317$, the Stokes bands are more intense because SRS is more effective in transferring pulse energy toward low frequencies.

Spectral changes for input pulses launched well into the anomalous-GVD regime are shown in Fig. 8.17 for $\lambda_p = 1.341$ μm and a fiber length of 1 km. Pulse spectra change considerably with an increase in the input peak power. Three Stokes bands appear for $P_0 = 530$ W (topmost curve). An anti-Stokes band carrying 5–10% of the input energy also appears. In the time domain, a separate Raman pulse is associated with each of the Stokes bands. The autocorrelation measurements show that the widths of these Raman pulses are ~300 fs [176]. The width depends considerably on the fiber length; the smallest width was about 55-fs for 150-ps wide input pulses. The input peak power of 5 kW was large enough to generate multiple Stokes in a 20-m-long fiber. The fourth-order Stokes band near 1.3 μm served as a pump to generate a continuum that extended up to 1.7 μm. Using a two-stage compression configuration, the pulse width could be reduced to 18 fs, a pulse consisting of only three optical cycles.

Raman solitons have also been generated using 100-ps pump pulses from a mode-locked Nd:YAG laser operating at 1.32 μm. In one experiment, the use of a conventional fiber with the zero-dispersion wavelength near 1.3 μm led to 100-fs Raman pulses near 1.4 μm [178]. In another experiment, a dispersion-shifted fiber, having its zero-dispersion wavelength at 1.46 μm, was used with the same laser. The first-order Stokes at 1.407 μm served as a pump to generate the second-order Stokes (at 1.516 μm) in the anomalous-dispersion regime of the fiber [177]. The Raman pulse associated with the second-order Stokes band was 130-fs wide. The pulse had a broad pedestal and carried only about 30% of its energy in the form of a soliton.

Modulation instability plays an important role in the formation of Raman solitons when pump wavelength lies in the anomalous-GVD regime of an optical fiber [200]. The role of modulation instability can be understood as follows. When the pump pulse experiences anomalous GVD, it develops sidebands indicative of modulation instability. The low-frequency sideband (typical spacing ~1 THz) falls within the bandwidth of the Raman-gain spectrum and seeds the formation of the Raman pulse. At high pump powers, the spectrum of the Raman pulse becomes so broad (~10 THz) that the fiber can support a Raman soliton of width ~100 fs even when injected pump pulses are more than 100-ps wide. In one experiment, Raman solitons of 60-fs duration were observed by pumping a 25-m-long, P_2O_5-doped, silica fiber with 150-ps pump pulses obtained from a 1.319-μm Nd:YAG laser [199]. The zero-dispersion wavelength of the fiber was chosen to be quite close to the pump wavelength to enhance the modulation frequency [see Eq. (5.1.8)].

Before closing this section, let us consider whether Eqs. (8.3.1) and (8.3.2) have solitary-wave solutions that can be interpreted as Raman solitons. As early as 1988, it was found that soliton-like Raman pulses with a 'sech' profile

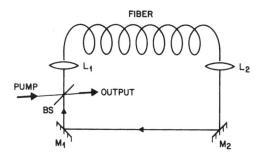

Figure 8.18 Schematic illustration of the ring-cavity geometry used for Raman soliton lasers. BS is a dichroic beamsplitter, M_1 and M_2 are mirrors of 100% reflectivity, and L_1 and L_2 are microscope objective lenses.

can form under certain conditions provided the pump remains largely undepleted [190]. When pump depletion is included, pulselike solutions of Eqs. (8.3.1) and (8.3.2), similar to the XPM-paired solitons discussed in Section 7.3.3, have not been found. However, as discussed in Section 5.5, these equations permit shocklike solutions in the form of a sharp front [211].

8.4.2   Raman Soliton Lasers

An interesting application of the soliton effects has led to the development of a new kind of laser known as the Raman soliton laser [181]–[187]. Such lasers provide their output in the form of solitons of widths ∼100 fs, but at a wavelength corresponding to the first-order Stokes wavelength. Furthermore, the wavelength can be tuned over a considerable range (∼10 nm) by using the time-dispersion technique discussed in Section 8.2.2. A ring-cavity configuration, shown schematically in Fig. 8.18, is commonly employed. A dichroic beamsplitter, highly reflective at the pump wavelength and partially reflective at the Stokes wavelength λ_s, is used to couple pump pulses into the ring cavity and to provide the laser output.

In a 1987 experimental demonstration of the Raman soliton laser [181], 10-ps pulses from a mode-locked color-center laser operating near 1.48 μm were used to pump the Raman laser synchronously. The ring cavity had a 500-m-long, polarization-preserving, dispersion-shifted fiber having its zero-dispersion wavelength λ_D near 1.536 μm. Such a fiber permitted the pump and Raman pulses to overlap over a considerable portion of the fiber as the pump and Raman wavelengths were on opposite sides of λ_D by nearly the same amount

($\lambda_s \approx 1.58~\mu$m). Output pulses with widths ~ 300 fs were produced with a low but broad pedestal. In an attempt to remove the pedestal, the ring cavity of Fig. 8.18 was modified by replacing the fiber with two fiber pieces with variable coupling between them. A 100-m-long section provided the Raman gain while another 500-m-long section was used for pulse shaping. The SRS did not occur in the second section because the coupler reduced pump power levels below the Raman threshold. It was possible to obtain pedestal-free pulses of 284-fs width when wavelength separation corresponded to 11.4 THz (about 90 nm). However, when the wavelengths were 13.2 THz apart (corresponding to maximum Raman gain), considerable pulse energy appeared in the form of a broad pedestal. This complex behavior is attributed to the XPM effects [182].

In later experiments, 100-ps pulses obtained from a mode-locked Nd:YAG laser operating at 1.32 μm were used to synchronously pump Raman soliton lasers [183]–[185]. This wavelength regime is of interest because conventional fibers with $\lambda_D \sim 1.3~\mu$m can be used. Furthermore, both the pump and Raman pulses are close to the zero-dispersion wavelength of the fiber so that they can overlap long enough to provide the required Raman gain (the walk-off length ≈ 300 m). Pulse widths as short as 160 fs were obtained in an experiment that employed a 1.1-km fiber that did not even preserve wave polarization [183]. Output pulses contained a broad pedestal with only 20% of the energy appearing in the form of a Raman soliton. In another experiment, a dispersion-shifted fiber with $\lambda_D = 1.46~\mu$m was used [184]. Raman solitons of about 200-fs width were then observed through the second- and third-order Stokes lines, generated near 1.5 and 1.6 μm, respectively. This process of cascaded SRS has also been used to generate Raman solitons near 1.5 μm by pumping the laser with 1.06-μm pump pulses [187]. The first three Stokes bands then lie in the normal-GVD regime of a conventional fiber ($\lambda_D > 1.3~\mu$m). The fourth and fifth Stokes bands form a broad spectral band encompassing the range 1.3–1.5 μm that contains about half of the input energy. Autocorrelation traces of output pulses in the spectral region near 1.35, 1.4, 1.45, and 1.5 μm showed that the energy in the pedestal decreased as the wavelength increased. In fact, output pulses near 1.5 μm were nearly pedestal free.

Even though Raman soliton lasers are capable of generating femtosecond solitons useful for many applications, they suffer from a noise problem that limits their usefulness. Measurements of intensity noise for a synchronously pumped Raman soliton laser indicated that the noise was more than 50 dB above the shot-noise level [194]. Pulse-to-pulse timing jitter was also found to be quite large, exceeding 5 ps at 1.6-W pump power. Such large noise levels can be understood if one considers the impact of the Raman-induced frequency shift (see Section 5.5) on the performance of such lasers. For synchronous pumping to be effective, the round-trip time of the Raman soliton inside the laser cavity must be an integer multiple of the spacing between pump pulses. However, the Raman-induced frequency shift changes the group velocity and slows down the pulse in such an unpredictable manner that synchronization is quite difficult to achieve in practice. As a result, Raman soliton lasers generate pulses in a way similar to single-pass Raman amplifiers and suffer from the noise problems associated with such amplifiers.

The performance of Raman soliton lasers can be significantly improved if the Raman-induced frequency shift can somehow be suppressed. It turns out that such frequency shifts can be suppressed with a proper choice of pump power and laser wavelength. In a 1992 experiment, a Raman soliton laser was synchronously pumped by using 200-ps pump pulses from a 1.32-μm Nd:YAG laser and was tunable over the 1.37–1.44 μm wavelength range [198]. The Raman-induced frequency shift was suppressed in the wavelength range 1.41–1.44 μm for which the Raman-gain spectrum in Fig. 8.1 had a positive slope. Measurements of noise indicated significant reduction in both intensity noise and timing jitter [203]. Physically, Raman-gain dispersion is used to cancel the effects of the last term in Eq. (5.5.1) that is responsible for the Raman-induced frequency shift [212].

8.4.3 Soliton-Effect Pulse Compression

In some sense, Raman solitons formed in Raman amplifiers or lasers take advantage of the soliton-effect pulse-compression technique (see Chapter B.6). Generally speaking, Raman amplification can be used for simultaneous amplification and compression of picosecond optical pulses by taking advantage of the anomalous GVD in optical fibers. In a 1991 experiment 5.8-ps pulses, obtained from a gain-switched semiconductor laser and amplified using an erbium-doped fiber amplifier to an energy level appropriate for a fundamental soliton, were compressed to 3.6-ps through Raman amplification in a 23-km-long dispersion-shifted fiber [197]. The physical mechanism behind simultaneous amplification and compression can be understood from Eq. (5.2.3) for the soliton order N. If a fundamental soliton is amplified adiabatically, the condition $N = 1$ can be maintained during amplification provided the soliton width changes as $P_0^{-1/2}$ with an increase in the peak power P_0.

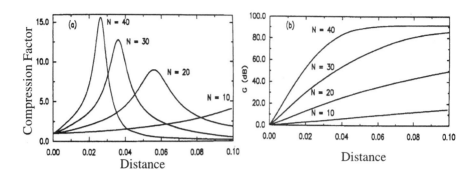

Figure 8.19 Compression factor (left) and amplification factor (right) as a function of normalized distance for several values of N in the case of Gaussian-shape pump and signal pulses. (After Ref. [201].)

In many practical situations, the input signal pulse is not intense enough to sustain a soliton inside the fiber during Raman amplification. It is possible to realize pulse compression even for such weak pulses through the XPM effects that invariably occur during SRS [201]. In essence, the pump and signal pulses are injected simultaneously into the fiber. The signal pulse extracts energy from the pump pulse through SRS and is amplified. At the same time, it interacts with the pump pulse through XPM that imposes a nearly linear frequency chirp on it and compresses it in the presence of anomalous GVD. As discussed in Section 7.5 in the context of XPM-induced pulse compression, the effectiveness of such a scheme depends critically on the relative group-velocity mismatch between the pump and signal pulses. A simple way to minimize the group-velocity mismatch and maximize the XPM-induced frequency chirp is to choose the zero-dispersion wavelength of the fiber in the middle of the pump and signal wavelengths.

Numerical simulations based on Eqs. (8.3.1) and (8.3.2) show that compression factors as large as 15 can be realized while the signal-pulse energy is amplified a millionfold [201]. Figure 8.19 shows (a) the compression factor and (b) the amplification factor as a function of the propagation distance ($\xi = z/L_D$) for several values of the parameter N, related to the peak power P_0 of pump pulses through $N = (\gamma_p P_0 L_D)^{1/2}$, where L_D is defined in Eq. (8.3.5). The pump and signal pulses are assumed to be Gaussian, have the same width, and propagate at the same speed. Pulse compression is maximum for an optimum fiber length, a feature similar to that of soliton-effect compressors (see

Chapter B.6). This behavior is easily understood by noting that GVD reduces the XPM-induced chirp to nearly zero at the point of maximum compression (see Section 3.2). The main point to note from Fig. 8.19 is that weak input pulses can be amplified by 50–60 dB while getting compressed simultaneously by a factor of 10 or more. The quality of compressed pulses is also quite good with no pedestal and little ringing. The qualitative features of pulse compression remain nearly the same when pulse widths or group velocities of the pump and signal pulses are not the same, making this technique quite attractive from a practical standpoint. Simultaneous amplification and compression of picosecond optical pulses were indeed observed in a 1996 experiment [202].

8.5 Effect of Four-Wave Mixing

Four-wave mixing, a nonlinear phenomenon discussed in Chapter 10, is known to affect the SRS in any medium [7]. Its impact on the SRS process in optical fibers has been studied extensively [213]–[224]. This section describes the relevant features qualitatively.

To understand how four-wave mixing can influence SRS, it is useful to reconsider the physics behind the SRS process. In quantum-mechanical terms, Raman scattering can be thought of as downconversion of a pump photon into a lower-frequency photon and a phonon associated with a vibrational mode of molecules. An up-conversion process in which a phonon combines with the pump photon to generate a higher-frequency photon is also possible, but occurs rarely because it requires the presence of a phonon of right energy and momentum. The optical wave associated with higher-frequency photons is called anti-Stokes and is generated at a frequency $\omega_a = \omega_p + \Omega$ for a Stokes wave of frequency $\omega_s = \omega_p - \Omega$, where ω_p is the pump frequency. Because $2\omega_p = \omega_a + \omega_s$, four-wave mixing, a process where two pump photons annihilate themselves to produce Stokes and anti-Stokes photons, can occur provided the total momentum is conserved. The momentum-conservation requirement leads to the phase-matching condition $\Delta k = 2k_p - k_a - k_s = 0$ that must be satisfied for four-wave mixing to take place (see Section 10.1).

The phase-matching condition in not easily satisfied in single-mode fibers for $\Omega \sim 10$ THz. For this reason, the anti-Stokes wave is rarely observed during SRS. As discussed in Section 10.3, the phase-matching condition may be nearly satisfied when GVD is not too large. In that case, Eqs. (8.1.20) and (8.1.21) should be supplemented with a third equation that describes propagation of the anti-Stokes wave and its coupling to the Stokes wave through four-wave mixing. The set of three equations can be solved approximately when pump depletion is neglected [214]. The results show that the Raman gain g_R depends on the mismatch Δk, and may increase or decrease from its value in Fig. 8.1 depending on the value of Δk. In particular g_R becomes small near $\Delta k = 0$, indicating that four-wave mixing can suppress SRS under appropriate conditions. Partial suppression of SRS was indeed observed in an experiment [214] in which the Raman gain was reduced by a factor of 2 when the pump power P_0 was large enough to make $|\Delta k| < 3g_R P_0$. A spectral component at the anti-Stokes frequency was also observed in the experiment.

The effects of four-wave mixing on SRS were also observed in another experiment [219] in which the spectrum of Raman pulses was observed to exhibit a double-peak structure corresponding to two peaks at 13.2 and 14.7 THz in Fig. 8.1. At low pump powers, the 13.2-THz peak dominated as the Raman gain is slightly larger (by about 1%) for this peak. However, as pump power was increased, the 14.7-THz peak began to dominate the Raman-pulse spectrum. These results can be understood by noting that the Raman-gain reduction induced by four-wave mixing is frequency dependent such that the effective Raman gain becomes larger for the 14.7-THz peak for pump intensities in excess of 1 GW/cm^2.

The effects of fiber birefringence have been ignored in this chapter. Its inclusion complicates the SRS analysis considerably [220]. For example, if a pump pulse is polarized at an angle with respect to a principal axis of the fiber so that it excites both the slow and fast polarization modes of the fiber, each of them can generate a Stokes pulse if its intensity exceeds the Raman threshold. These Stokes waves interact with the two pump-pulse components and with each other through XPM. To describe such an interaction, Eqs. (8.1.20) and (8.1.21) must be replaced by a set of four equations that includes all nonlinear terms appropriately as discussed in Section 7.1. The situation is even more complicated if the anti-Stokes wave is included: one must then solve a set of six coupled equations.

For pump pulses propagating in the anomalous-GVD regime of optical fibers, one should consider how modulation instability and SRS influence each other. As discussed in Section 5.1, modulation instability can be thought of as a four-wave mixing process phase-matched by SPM (see Section 10.3). It generates new waves at frequencies $\omega_p + \Omega_m$ and $\omega_p - \Omega_m$, where Ω_m depends on the pump power and is generally different than the Raman shift Ω. Thus, a unified analysis should consider five waves at frequencies ω_p, $\omega_p \pm \Omega_m$, and $\omega_p \pm \Omega$. Each of these five waves can have two orthogonally polarized components, resulting in a set of ten coupled-amplitude equations. The analysis becomes simpler for ultrashort pump pulses whose spectrum is so broad that the frequencies Ω_m and Ω fall within its bandwidth. The propagation of such pump pulses is described by a set of two coupled equations that include the effects of GVD, SPM, XPM, fiber birefringence, four-wave mixing, and intrapulse SRS. A linear stability analysis of such equations can be performed to obtain the effective gain spectrum when a CW pump beam enters an optical fiber [222]. Attempts have also been made to include the effects of intrapulse SRS and of fiber birefringence on propagation of two optical pulses in an optical fiber [20].

Problems

8.1 What is meant by Raman scattering? Explain its origin. What is the difference between spontaneous and stimulated Raman scattering?

8.2 Explain the meaning of Raman gain. Why does the Raman-gain spectrum extend over a wide range of 40 THz?

8.3 Solve Eqs. (8.1.2) and (8.1.3) neglecting pump deletion. Calculate the Stokes power at the output of a 1-km-long fiber when 1-μW power is injected together with an intense pump beam. Assume $g_R I_p(0) = 2 \text{ km}^{-1}$ and $\alpha_p = \alpha_s = 0.2$ dB/km.

8.4 Perform the integration in Eq. (8.1.9) using the method of steepest descent and derive Eq. (8.1.9).

8.5 Use Eq. (8.1.9) to derive the Raman-threshold condition given in Eq. (8.1.13).

8.6 Solve Eqs. (8.1.2) and (8.1.3) analytically after assuming $\alpha_p = \alpha_s$.

8.7 Calculate the threshold pump power of a 1.55-μm Raman laser whose cavity includes a 1-km-long fiber with 40-μm^2 effective core area. Use $\alpha_p = 0.3$ dB/km and total cavity losses of 6 dB. Use the Raman gain from Fig. 8.1.

8.8 Explain the technique of time-dispersion tuning used commonly for synchronously pumped Raman lasers. Estimate the tuning range for the laser of Problem 8.7.

8.9 Solve Eqs. (8.3.1) and (8.3.2) analytically after setting $\beta_{2p} = \beta_{2s} = 0$.

8.10 Use the results of Problem 8.9 to plot the output pulse shapes for Raman amplification in a 1-km-long fiber. Assume $\lambda_p = 1.06~\mu$m, $\lambda_s = 1.12~\mu$m, $\gamma_p = 10$ W^{-1}/km, $g_R = 1 \times 10^{-3}$ m/W, $d = 5$ ps/m, and $A_{\mathrm{eff}} = 40~\mu$m^2. Input pump and Stokes pulses are Gaussian with the same 100-ps width (FWHM) and with 1 kW and 10 mW peak powers, respectively.

8.11 Solve Eqs. (8.3.16) and (8.3.17) numerically using the split-step Fourier method and reproduce the results shown in Figs. 8.8 and 8.9.

8.12 Design an experiment for amplifying 50-ps (FWHM) Gaussian pulses through SRS by at least 30 dB such that they are also compressed by a factor of 10. Use numerical simulations to verify your design.

References

[1] C. V. Raman, *Indian J. Phys.* **2**, 387 (1928).

[2] E. J. Woodbury and W. K. Ng, *Proc. IRE* **50**, 2347 (1962).

[3] R. W. Hellwarth, *Phys. Rev.* **130**, 1850 (1963); *Appl. Opt.* **2**, 847 (1963).

[4] E. Garmire, E. Pandarese, and C. H. Townes, *Phys. Rev. Lett.* **11**, 160 (1963).

[5] Y. R. Shen and N. Bloembergen, *Phys. Rev.* **137**, A1786 (1965).

[6] W. Kaiser and M Maier, in *Laser Handbook*, Vol. 2, F. T. Arecchi and E. O. Schulz-Dubois, Eds. (North-Holland, Amsterdam, 1972), Chap. E2.

[7] Y. R. Shen, *The Principles of Nonlinear Optics* (Wiley, New York, 1984), Chap. 10.

[8] R. W. Boyd, *Nonlinear Optics* (Academic Press, San Diego, 1992).

[9] R. H. Stolen, E. P. Ippen, and A. R. Tynes, *Appl. Phys. Lett.* **20**, 62 (1972).

[10] R. H. Stolen and E. P. Ippen, *Appl. Phys. Lett.* **22**, 276 (1973).

[11] R. H. Stolen, *Proc. IEEE* **68**, 1232 (1980).

[12] R. H. Stolen, J. P. Gordon, W. J. Tomlinson, and H. A. Haus, *J. Opt. Soc. Am. B* **6**, 1159 (1989).

[13] D. J. Dougherty, F. X. Kärtner, H. A. Haus, and E. P. Ippen, *Opt. Lett.* **20**, 31 (1995).

[14] D. Mahgerefteh, D. L. Butler, J. Goldhar, B. Rosenberg, and G. L. Burdge, *Opt. Lett.* **21**, 2026 (1996).

[15] R. Shuker and R. W. Gammon, *Phys. Rev. Lett.* **25**, 222 (1970).

[16] R. G. Smith, *Appl. Opt.* **11**, 2489 (1972).

[17] J. AuYeung and A. Yariv, *IEEE J. Quantum Electron.* **QE-14**, 347 (1978).

[18] C. R. Menyuk, M. N. Islam, and J. P. Gordon, *Opt. Lett.* **16**, 566 (1991).

[19] A. Höök, *Opt. Lett.* **17**, 115 (1992).

[20] S. Kumar, A. Selvarajan, and G. V. Anand, *Opt. Commun.* **102**, 329 (1993); *J. Opt. Soc. Am. B* **11**, 810 (1994).

[21] C. Headley and G. P. Agrawal, *IEEE J. Quantum Electron.* **31**, 2058 (1995).

[22] C. Headley and G. P. Agrawal, *J. Opt. Soc. Am. B* **13**, 2170 (1996).

[23] A. Picozzi, C. Montes, J. Botineau, and E. Picholle, *J. Opt. Soc. Am. B* **15**, 1309 (1998).

[24] C. Lin and R. H. Stolen, *Appl. Phys. Lett.* **28**, 216 (1976).

[25] C. Lin, L. G. Cohen, R. H. Stolen, G. W. Tasker, and W. G. French, *Opt. Commun.* **20**, 426 (1977).

[26] V. V. Grigoryants, B. L. Davydov, M. E. Zhabotinsky, V. F. Zolin, G. A. Ivanov, V. I Smirnov, and Y. K. Chamorovski, *Opt. Quantum Electron.* **9**, 351 (1977).

[27] L. G. Cohen and C. Lin, *IEEE J. Quantum Electron.* **QE-14**, 855 (1978).

[28] V. S. Butylkin, V. V. Grigoryants, and V. I. Smirnov, *Opt. Quantum Electron.* **11**,141 (1979).

[29] R. H. Stolen, in *Optical Fiber Communications*, S. E. Miller and A. G. Chynoweth, Eds. (Academic Press, New York, 1979), Chap. 5.

[30] P.-J. Gao, C.-J. Nie, T.-L. Yang, and H.-Z. Su, *Appl. Phys.* **24**, 303 (1981).

[31] G. Rosman, *Opt. Quantum Electron.* **14**, 92 (1982).

[32] Y. Ohmori, Y. Sesaki, and T. Edahiro, *Trans. IECE Japan* **E66**, 146 (1983).

[33] R. Pini, M. Mazzoni, R. Salimbeni, M. Matera, and C. Lin, *Appl. Phys. Lett.* **43**, 6 (1983).

[34] M. Rothschild and H. Abad, *Opt. Lett.* **8**, 653 (1983).

[35] F. R. Barbosa, *Appl. Opt.* **22**, 3859 (1983).

[36] R. H. Stolen, C. Lee, and R. K. Jain, *J. Opt. Soc. Am. B* **1**, 652 (1984).

[37] C. Lin, *J. Lightwave Technol.* **LT-4**, 1103 (1986).

[38] A. S. L. Gomes, V. L. DaSilva, J. R. Taylor, B. J. Ainslie, and S. P. Craig, *Opt. Commun.* **64**, 373 (1987).

[39] K. X. Liu and E. Garmire, *IEEE J. Quantum Electron.* **27**, 1022 (1991).

[40] J. Chang,, D. Baiocchi, J. Vas, and J. R. Thompson, *Opt. Commun.* **150**, 339 (1998).

[41] E. Landahl, D. Baiocchi, and J. R. Thompson, *Opt. Commun.* **150**, 339 (1998).

[42] M. G. Raymer and I. A. Walmsley, in *Progress in Optics*, Vol. 30, E. Wolf, Ed. (Elsevier, Amsterdam, 1990), Chap. 3.

[43] K. O. Hill, B. S. Kawasaki, and D. C. Johnson, *Appl. Phys. Lett.* **28**, 608 (1976); *Appl. Phys. Lett.* **29**, 181 (1976).

[44] R. K. Jain, C. Lin, R. H. Stolen, W. Pleibel, and P. Kaiser, *Appl. Phys. Lett.* **30**, 162 (1977).

[45] D. C. Johnson, K. O. Hill, B. S. Kawasaki, and D. Kato, *Electron. Lett.* **13**, 53 (1977).

[46] R. H. Stolen, C. Lin, and R. K. Jain, *Appl. Phys. Lett.* **30**, 340 (1977).

[47] C. Lin, R. H. Stolen, and L. G. Cohen, *Appl. Phys. Lett.* **31**, 97 (1977).

[48] C. Lin, R. H. Stolen, W. G. French, and T. G. Malone, *Opt. Lett.* **1**, 96 (1977).

[49] R. K. Jain, C. Lin, R. H. Stolen, and A. Ashkin, *Appl. Phys. Lett.* **31**, 89 (1977).

[50] R. H. Stolen, C. Lin, J. Shah, and R. F. Leheny, *IEEE J. Quantum Electron.* **QE-14**, 860 (1978).

[51] C. Lin and W. G. French, *Appl. Phys. Lett.* **34**, 10 (1979).

[52] R. H. Stolen, *IEEE J. Quantum Electron.* **QE-15**, 1157 (1979).

[53] C. Lin and P. F. Glodis, *Electron. Lett.* **18**, 696 (1982).

[54] M. Kakazawa, T. Masamitsu, and N. Ichida, *J. Opt. Soc. Am. B* **1**, 86 (1984).

[55] A. R. Chraplyvy and J. Stone, *Opt. Lett.* **9**, 241 (1984).

[56] M. Nakazawa, *Opt. Lett.* **10**, 193 (1985).

[57] R. Pini, R. Salimbeni, M. Vannini, A. F. M. Y. Haider, and C. Lin, *Appl. Opt.* **25**, 1048 (1986).

[58] T. Mizunami, T. Miyazaki, and K. Takagi, *J. Opt. Soc. Am. B* **4**, 498 (1987).

[59] J. D. Kafka, D. F. Head, and T. Baer, in *Ultrafast Phenomena V*, G. R. Fleming and A. E. Siegman, Eds. (Springer-Verlag, Berlin, 1986), p. 51.

[60] E. Desurvire, A. Imamoglu, and H. J. Shaw, *J. Lightwave Technol.* **LT-5**, 89 (1987).

[61] P. N. Kean, K. Smith, B. D. Sinclair, and W. Sibbett, *J. Mod. Opt.* **35**, 397 (1988).

[62] I. K. Ilev, H. Kumagai, and H. Toyoda, *Appl. Phys. Lett.* **69**, 1846 (1996); *Appl. Phys. Lett.* **70**, 3200 (1997); *Opt. Commun.* **138**, 337 (1997).

[63] A. J. Stentz, *Proc. SPIE* **3263**, 91 (1998).

[64] S. V. Chernikov, N. S. Platonov, D. V. Gapontsev, D. I. Chang, M. J. Guy, and J. R. Taylor, *Electron. Lett.* **34**, 680 (1998).

[65] V. I. Karpov, E. M. Dianov, V. M. Paramonov, *et al.*, *Opt. Lett.* **24**, 887 (1999).

[66] C. Lin and R. H. Stolen, *Appl. Phys. Lett.* **29**, 428 (1976).

[67] M. Ikeda, *Opt. Commun.* **39**, 148 (1981).

[68] G. A. Koepf, D. M. Kalen, and K. H. Greene, *Electron. Lett.* **18**, 942 (1982).

[69] Y. Aoki, S. Kishida, H. Honmou, K. Washio, and M. Sugimoto, *Electron. Lett.* **19**, 620 (1983).

[70] E. Desurvire, M. Papuchon, J. P. Pocholle, J. Raffy, and D. B. Ostrowsky, *Electron. Lett.* **19**, 751 (1983).

[71] A. R. Chraplyvy, J. Stone, and C. A. Burrus, *Opt. Lett.* **8**, 415 (1983).

[72] M. Nakazawa, M. Tokuda, Y. Negishi, and N. Uchida, *J. Opt. Soc. Am. B* **1**, 80 (1984).

[73] K. Nakamura, M. Kimura, S. Yoshida, T. Hikada, and Y. Mitsuhashi, *J. Lightwave Technol.* **LT-2**, 379 (1984).

[74] M. Nakazawa, *Appl. Phys. Lett.* **46**, 628 (1985).

[75] M. Nakazawa, T. Nakashima, and S. Seikai, *J. Opt. Soc. Am. B* **2**, 215 (1985).

[76] M. L. Dakss and P. Melman, *J. Lightwave Technol.* **LT-3**, 806 (1985).

[77] J. P. Pocholle, J. Raffy, M. Papuchon, and E. Desurvire, *Opt. Exp.* **24**, 600 (1985).

[78] Y. Durteste, M. Monerie, and P. Lamouler, *Electron. Lett.* **21**, 723 (1985).

[79] N. A. Olsson and J. Hegarty, *J. Lightwave Technol.* **LT-4**, 391 (1986).

[80] K. Vilhelmsson, *J. Lightwave Technol.* **LT-4**, 400 (1986).

[81] E. Desurvire, M. J. F. Digonnet, and H. J. Shaw, *J. Lightwave Technol.* **LT-4**, 426 (1986).

[82] Y. Aoki, S. Kishida, and K. Washio, *Appl. Opt.* **25**, 1056 (1986).

[83] S. Seikai, T. Nakashima, and N. Shibata, *J. Lightwave Technol.* **LT-4**, 583 (1986).

[84] K. Mochizuki, N. Edagawa, and Y. Iwamoto, *J. Lightwave Technol.* **LT-4**, 1328 (1986).

[85] R. W. Davies, P. Melman, W. H. Nelson, M. L. Dakss, and B. M. Foley, *J. Lightwave Technol.* **LT-5**, 1068 (1987).

[86] A. S. Davison and I. H. White, *Electron. Lett.* **23**, 1344 (1987).

[87] M. J. O'Mahony, *J. Lightwave Technol.* **LT-6**, 531 (1988).

[88] J. Hegarty, N. A. Olsson, and L. Goldner, *Electron. Lett.* **21**, 290 (1985).

[89] K. Mochizuki, N. Edagawa, and Y. Iwamoto, *J. Lightwave Technol.* **LT-4**, 1328 (1986).

[90] N. Edagawa, K. Mochizuki, and Y. Iwamoto, *Electron. Lett.* **23**, 196 (1987).

[91] M. L. Dakss and P. Melman, *IEE Proc.* **135**, Pt. J., 96 (1988).

[92] K. Mochizuki, *J. Lightwave Technol.* **LT-3**, 688 (1985).

[93] Y. Aoki, *J. Lightwave Technol.* **LT-6**, 1225 (1988).

[94] A. Hasegawa, *Appl. Opt.* **23**, 3302 (1984).

[95] L. F. Mollenauer, R. H. Stolen, and M. N. Islam, *Opt. Lett.* **10**, 229 (1985).

[96] L. F. Mollenauer, J. P. Gordon, and M. N. Islam, *IEEE J. Quantum Electron.* **QE-22**, 157 (1986).

[97] L. F. Mollenauer and K. Smith, *Opt. Lett.* **13**, 675 (1988).

[98] E. J. Greer, J. M. Hickmann, and J. R. Taylor, *Electron. Lett.* **27**, 1171 (1991).

[99] T. Horiguchi, T. Sato, and Y. Koyamada, *IEEE Photon. Technol. Lett.* **4**, 64 (1992).

[100] T. Sato, T. Horiguchi, Y. Koyamada, and I. Sankawa, *IEEE Photon. Technol. Lett.* **4**, 923 (1992).

[101] T. Mizunami, H. Iwashita, and K. Takagi, *Opt. Commun.* **97**, 74 (1993).

[102] S. G. Grubb, *Proc. Conf. on Optical Amplifiers and Applications*, Optical Society of America, Washington, DC, 1995.

[103] S. V. Chernikov, Y. Zhu, R. Kashyap, and J. R. Taylor, *Electron. Lett.* **31**, 472 (1995).

[104] E. M. Dianov, *Laser Phys.* **6**, 579 (1996).

[105] D. I. Chang, S. V. Chernikov, M. J. Guy, J. R. Taylor, and H. J. Kong, *Opt. Commun.* **142**, 289 (1997).

[106] P. B. Hansen, L. Eskilden, S. G. Grubb, A. J. Stentz, T. A. Strasser, J. Judkins, J. J. DeMarco, J. R. Pedrazzani, and D. J. DiGiovanni, *IEEE Photon. Technol. Lett.* **9**, 262 (1997).

[107] A. Bertoni, *Opt. Quantum Electron.* **29**, 1047 (1997).

[108] A. Bertoni and G. C. Reali, *Appl. Phys. B* **67**, 5 (1998).

[109] E. M. Dianov, M. V. Grekov, I. A. Bufetov, V. M. Mashinsky, O. D. Sazhin, A. M. Prokhorov, G. G. Devyatykh, A. N. Guryanov, and V. F. Khopin, *Electron. Lett.* **34**, 669 (1998).

[110] D. V. Gapontsev, S. V. Chernikov, and J. R. Taylor, *Opt. Commun.* **166**, 85 (1999).

[111] H. Masuda, S. Kawai, K. Suzuki, and K. Aida, *IEEE Photon. Technol. Lett.* **10**, 516 (1998).

[112] J. Kani and M. Jinno, *Electron. Lett.* **35**, 1004 (1999).

[113] H. Kidorf, K. Rottwitt, M. Nissov, M. Ma, and E. Rabarijaona, *IEEE Photon. Technol. Lett.* **11**, 530 (1999).

[114] H. Masuda and S. Kawai, *IEEE Photon. Technol. Lett.* **11**, 647 (1999).

[115] K. Rottwitt, J. H. Povlsen, and A. Bajarklev, *J. Lightwave Technol.* **11**, 2105, (1993).

[116] A. Altuncu, L. Noel, W. A. Pender, A. S. Siddiqui, T. Widdowson, A. D. Ellis, M. A. Newhouse, A. J. Antos, G. Kar, and P. W. Chu, *Electron. Lett.* **32**, 233 (1996).

[117] Z. M. Liao and G. P. Agrawal, *IEEE Photon. Technol. Lett.* **11**, 818 (1999).

[118] H. Suzuki, N. Takachio, H. Masuda, and M. Koga, *Electron. Lett.* **35**, 1175 (1999).

[119] H. Suzuki, J. Kani, H. Masuda, N. Takachio, K. Iwatsuki, Y. Tada, and M. Sumida, *IEEE Photon. Technol. Lett.* **12**, 903 (2000).

[120] A. R. Chraplyvy and P. S. Henry, *Electron. Lett.* **19**, 641 (1983).

[121] A. Tomita, *Opt. Lett.* **8**, 412 (1983).

[122] A. R. Chraplyvy, *Electron. Lett.* **20**, 58 (1984).

[123] D. Cotter and A. M. Hill, *Electron. Lett.* **20**, 185 (1984).

[124] N. A. Olsson, J. Hegarty, R. A. Logan, L. F. Johnson, K. L. Walker, L. G. Cohen, B. L. Kasper, and J. C. Campbell, *Electron. Lett.* **21**, 105 (1985).

[125] A. R. Chraplyvy, *J. Lightwave Technol.* **8**, 1548 (1990).

[126] R. G. Waarts, A. A. Friesem, E. Lichtman, H. H. Yaffe, and R.-P. Braun, *IEE Proc.* **78**, 1344 (1990).

[127] S. Chi and S. C. Wang, *Electron. Lett.* **26** , 1509 (1992).

[128] S. Tariq and J. C. Palais, *J. Lightwave Technol.* **11**, 1914 (1993); *Fiber Integ. Opt.* **15**, 335 (1996).

[129] D. N. Christodoulides and R. B. Jander, *IEEE Photon. Technol. Lett.* **8**, 1722 (1996).

[130] J. Wang, X. Sun, and M. Zhang, *IEEE Photon. Technol. Lett.* **10**, 540 (1998).

[131] M. Zirngibl, *Electron. Lett.* **34**, 789 (1998).

[132] M. E. Marhic, F. S. Yang, and L. G. Kazovsky, *J. Opt. Soc. Am. B* **15**, 957 (1998).

[133] S. Bigo, S. Gauchard, A Bertaina, and J. P. Hamaide, *IEEE Photon. Technol. Lett.* **11**, 671 (1999).

[134] A. G. Grandpierre, D. N. Christodoulides, and J. Toulouse, *IEEE Photon. Technol. Lett.* **11**, 1271 (1999).

[135] V. N. Lugovoi, *Sov. Phys. JETP* **44**, 683 (1976).

[136] J. I. Gersten, R. R. Alfano, and M. Belic, *Phys. Rev. A* **21**, 1222 (1980).

[137] E. M. Dianov, A. Y. Karasik, P. V. Mamyshev, G. I. Onishchukov, A. M. Prokhorov, M. F. Stel'makh, and A. A. Fomichev, *JETP Lett.* **39**, 691 (1984).

[138] J.-H. Lu, Y.-L. Li, and J.-L. Jiang, *Opt. Quantum Electron.* **17**, 187 (1985).

[139] E. M. Dianov, A. Y. Karasik, P. V. Mamyshev, A. M. Prokhorov, and V. N. Serkin, *Sov. Phys. JETP* **62**, 448 (1985).

[140] B. Valk, W. Hodel, and H. P. Weber, *Opt. Commun.* **54**, 363 (1985).

[141] R. H. Stolen and A. M. Johnson, *IEEE J. Quantum Electron.* **QE-22**, 2154 (1986).

[142] P. M. W. French, A. S. L. Gomes, A. S. Gouveia-Neto, and J. R. Taylor, *IEEE J. Quantum Electron.* **QE-22**, 2230 (1986).

[143] D. Schadt, B. Jaskorzynska, and U. Österberg, *J. Opt. Soc. Am. B* **3**, 1257 (1986).

[144] D. Schadt and B. Jaskorzynska, *J. Opt. Soc. Am. B* **4**, 856 (1987).

[145] P. N. Kean, K. Smith, and W. Sibbett, *IEE Proc.* **134**, Pt. J, 163 (1987).

[146] R. R. Alfano, P. L. Baldeck, F. Raccah, and P. P. Ho, *Appl. Opt.* **26**, 3492 (1987).

[147] J. T. Manassah, *Appl. Opt.* **26**, 3747 (1987); J. T. Manassah and O. Cockings, *Appl. Opt.* **26**, 3749 (1987).

[148] E. M. Dianov, P. V. Mamyshev, A. M. Prokhorov, and D. G. Fursa, *JETP Lett.* **45**, 599 (1987).

[149] K. Smith, P. N. Kean, D. W. Crust, and W. Sibbett, *J. Mod. Opt.* **34**, 1227 (1987).

[150] E. M. Dianov, P. V. Mamyshev, A. M. Prokhorov, and D. G. Fursa, *JETP Lett.* **46**, 482 (1987).

[151] M. Nakazawa, M. S. Stix, E. P. Ippen, and H. A. Haus, *J. Opt. Soc. Am. B* **4**, 1412 (1987).

[152] P. L. Baldeck, P. P. Ho, and R. R. Alfano, *Rev. Phys. Appl.* **22**, 1677 (1987).

[153] R. R. Alfano and P. P. Ho, *IEEE J. Quantum Electron.* **QE-24**, 351 (1988).

[154] A. M. Weiner, J. P. Heritage, and R. H. Stolen, *J. Opt. Soc. Am. B* **5**, 364 (1988).

[155] A. S. L. Gomes, V. L. da Silva, and J. R. Taylor, *J. Opt. Soc. Am. B* **5**, 373 (1988).

[156] P. L. Baldeck, R. R. Alfano, and G. P. Agrawal, in *Ultrafast Phenomena VI*, T. Yajima *et al.*, eds. (Springer-Verlag, Berlin, 1988), p. 53.

[157] M. Kuckartz, R. Schultz, and A. Harde, *Opt. Quantum Electron.* **19**, 237 (1987); *J. Opt. Soc. Am. B* **5**, 1353 (1988).

[158] J. Hermann and J. Mondry, *J. Mod. Opt.* **35**, 1919 (1988).

[159] R. Osborne, *J. Opt. Soc. Am. B* **6**, 1726 (1989).

[160] D. N. Cristodoulides and R. I. Joseph, *IEEE J. Quantum Electron.* **25**, 273 (1989).

[161] Y. B. Band, J. R. Ackerhalt, and D. F. Heller, *IEEE J. Quantum Electron.* **26**, 1259 (1990).

[162] K. Suzuki, K. Noguchi, and N. Uesugi, *Opt. Lett.* **11**, 656 (1986); *J. Lightwave Technol.* **6**, 94 (1988).

[163] A. S. L. Gomes, V. L. da Silva, J. R. Taylor, B. J. Ainslie, and S. P. Craig, *Opt. Commun.* **64**, 373 (1987).

[164] E. M. Dianov, L. A. Bufetov, M. M. Bubnov, M. V. Grekov, S. A. Vasiliev, and O. I. Medvedkov, *Opt. Lett.* **25**, 402 (2000).

[165] T. Nakashima, M. Nakazawa, K. Nishi, and H. Kubota, *Opt. Lett.* **12**, 404 (1987).

[166] A. S. L. Gomes, A. S. Gouveia-Neto, and J. R. Taylor, *Opt. Quantum Electron.* **20**, 95 (1988).

[167] V. A. Vysloukh and V. N. Serkin, *JETP Lett.* **38**, 199 (1983).

[168] E. M. Dianov, A. Y. Karasik, P. V. Mamyshev, A. M. Prokhorov, V. N. Serkin, M. F. Stel'makh, and A. A. Fomichev, *JETP Lett.* **41**, 294 (1985).

[169] E. M. Dianov, A. M. Prokhorov, and V. N. Serkin, *Opt. Lett.* **11**, 168 (1986).

[170] F. M. Mitschke and L. F. Mollenauer, *Opt. Lett.* **11**, 659 (1986).

[171] A. B. Grudinin, E. M. Dianov, D. V. Korobkin, A. M. Prokhorov, V. N. Serkin, and D. V. Khaidarov, *JETP Lett.* **45**, 260 (1987).

[172] V. N. Serkin, *Sov. Tech. Phys. Lett.* **13**, 366 (1987).

[173] K. L. Vodop'yanov, A. B. Grudinin, E. M. Dianov, L. A. Kulevskii, A. M. Prokhorov, and D. V. Khaidarov, *Sov. J. Quantum Electron.* **17**, 1311 (1987).

[174] B. Zysset, P. Beaud, and W. Hodel, *Appl. Phys. Lett.* **50**, 1027 (1987).

[175] W. Hodel and H. P. Weber, *Opt. Lett.* **12**, 924 (1987).

[176] P. Beaud, W. Hodel, B. Zysset, and H. P. Weber, *IEEE J. Quantum Electron.* **QE-23**, 1938 (1987).

[177] A. S. Gouveia-Neto, A. S. L. Gomes, J. R. Taylor, B. J. Ainslie, and S. P. Craig, *Electron. Lett.* **23**, 1034 (1987).

[178] A. S. Gouveia-Neto, A. S. L. Gomes, and J. R. Taylor, *Opt. Lett.* **12**, 1035 (1987).

[179] K. J. Blow, N. J. Doran, and D. Wood, *J. Opt. Soc. Am. B* **5**, 381 (1988).

[180] A. S. Gouveia-Neto, A. S. L. Gomes, and J. R. Taylor, *IEEE J. Quantum Electron.* **QE-24**, 332 (1988).

[181] M. N. Islam, L. F. Mollenauer, and R. H. Stolen, in *Ultrafast Phenomena V*, G. R. Fleming and A. E. Siegman, eds. (Springer-Verlag, Berlin, 1986), p. 46.

[182] M. N. Islam, L. F. Mollenauer, R. H. Stolen, J. R. Simpson, and H. T. Shang, *Opt. Lett.* **12**, 814 (1987).

[183] J. D. Kafka and T. Baer, *Opt. Lett.* **12**, 181 (1987).

[184] A. S. Gouveia-Neto, A. S. L. Gomes, and J. R. Taylor, *Electron. Lett.* **23**, 537 (1987); *Opt. Quantum Electron.* **20**, 165 (1988).

[185] H. A. Haus and M. Nakazawa, *J. Opt. Soc. Am. B* **4**, 652 (1987).

[186] A. S. Gouveia-Neto, A. S. L. Gomes, J. R. Taylor, B. J. Ainslie, and S. P. Craig, *Opt. Lett.* **12**, 927 (1987); *Electron. Lett.* **23**, 1034 (1987).

[187] V. L. da Silva, A. S. L. Gomes, and J. R. Taylor, *Opt. Commun.* **66**, 231 (1988).

[188] A. S. Gouveia-Neto, A. S. L. Gomes, J. R. Taylor, and K. J. Blow, *J. Opt. Soc. Am. B* **5**, 799 (1988).

[189] A. S. Gouveia-Neto, M. E. Faldon, and J. R. Taylor, *Opt. Lett.* **13**, 1029 (1988); *Opt. Commun.* **69**, 325 (1989).

[190] A. Höök, D. Anderson, and M. Lisak, *Opt. Lett.* **13**, 1114 (1988).

[191] S. Chi and S. Wen, *Opt. Lett.* **14**, 84 (1989); L. K. Wang, and C. C. Yang, *Opt. Lett.* **14**, 84 (1989).

[192] A. S. Gouveia-Neto, P. G. J. Wigley, and J. R. Taylor, *Opt. Commun.* **72**, 119 (1989).

[193] A. Höök, D. Anderson, and M. Lisak, *J. Opt. Soc. Am. B* **6**, 1851 (1989).

[194] U. Keller, K. D. Li, M. Rodwell, and D. M. Bloom, *IEEE J. Quantum Electron.* **25**, 280 (1989).

[195] E. A. Golovchenko, E. M. Dianov, P. V. Mamyshev, A. M. Prokhorov, and D. G. Fursa, *J. Opt. Soc. Am. B* **7**, 172 (1990).

[196] E. J. Greer, D. M. Patrick, P. G. J. Wigley, J. I. Vukusic, and J. R. Taylor, *Opt. Lett.* **15**, 133 (1990).

[197] K. Iwaisuki, K. Suzuki, and S. Nishi, *IEEE Photon. Technol. Lett.* **3**, 1074 (1991).

[198] M. Ding and K. Kikuchi, *IEEE Photon. Technol. Lett.* **4**, 927 (1992); *IEEE Photon. Technol. Lett.* **4**, 1109 (1992).

[199] A. S. Gouveia-Neto, *J. Lightwave Technol.* **10**, 1536 (1992).

[200] E. M. Dianov, A. B. Grudinin, A. M. Prokhorov, and V. N. Serkin, in *Optical Solitons —Theory and Experiment*, J. R. Taylor, ed. (Cambridge University Press, Cambridge, UK, 1992), Chap. 7.

[201] C. Headley III and G. P. Agrawal, *J. Opt. Soc. Am. B* **10**, 2383 (1993).

[202] R. F. de Souza , E. J. S. Fonseca, M. J. Hickmann, and A. S. Gouveia-Neto, *Opt. Commun.* **124**, 79 (1996).

[203] M. Ding and K. Kikuchi, *Fiber Integ. Opt.* **13**, 337 (1994).

[204] E. L. Buckland, R. W. Boyd, and A. F. Evans, *Opt. Lett.* **22**, 454 (1997).

[205] K. Chan and W. Cao, *Opt. Commun.* **158**, 159 (1998).

[206] K. Drühl, R. G. Wenzel, and J. L. Carlsten, *Phys. Rev. Lett.* **51**, 1171 (1983); *J. Stat. Phys.* **39**, 615 (1985).

[207] C. R. Menyuk, *Phys. Rev. Lett.* **62**, 2937 (1989)

[208] J. C. Englund and C. M. Bowden, *Phys. Rev. Lett.* **57**, 2661 (1986); *Phys. Rev. A* **42**, 2870 (1990); *Phys. Rev. A* **46**, 578 (1992).

[209] M. Scalora, C. M. Bowden, and J. W. Haus, *Phys. Rev. Lett.* **69**, 3310 (1992).

[210] C. Montes, A. Picozzi, and D. Bahloul, *Phys. Rev. E* **55**, 1091 (1997).

[211] D. N. Christodoulides, *Opt. Commun.* **86**, 431 (1991).

[212] M. Ding and K. Kikuchi, *IEEE Photon. Technol. Lett.* **4**, 497 (1992).

[213] K. J. Blow and D. Wood, *IEEE J. Quantum Electron.* **25**, 2656 (1989).

[214] E. Golovchenko, P. V. Mamyshev, A. N. Pilipetskii, and E. M. Dianov, *IEEE J. Quantum Electron.* **26**, 1815 (1990).

[215] Y. Chen, *J. Opt. Soc. Am. B* **7**, 43 (1990).

[216] E. Golovchenko and A. N. Pilipetskii, *Sov. Lightwave Commun.* **1**, 271 (1991).

[217] P. N. Morgan and J. M. Liu, *IEEE J. Quantum Electron.* **27**, 1011 (1991).

[218] E. Golovchenko, P. V. Mamyshev, A. N. Pilipetskii, and E. M. Dianov, *J. Opt. Soc. Am. B* **8**, 1626 (1991).

[219] P. V. Mamyshev, A. M. Vertikov, and A. M. Prokhorov, *Sov. Lightwave Commun.* **2**, 73 (1992).

[220] A. M. Vertikov and P. V. Mamyshev, *Sov. Lightwave Commun.* **2**, 119 (1992).

[221] S. Trillo and S. Wabnitz, *J. Opt. Soc. Am. B* **9**, 1061 (1992).

[222] E. Golovchenko and A. N. Pilipetskii, *J. Opt. Soc. Am. B* **11**, 92 (1994).

[223] R. Schulz and H. Harde, *J. Opt. Soc. Am. B* **12**, 1279 (1995).

[224] P. T. Dinda, G. Millot, and S. Wabnitz, *J. Opt. Soc. Am. B* **15**, 1433 (1998).

Chapter 9

Stimulated Brillouin Scattering

Stimulated Brillouin scattering (SBS) is a nonlinear process that can occur in optical fibers at input power levels much lower than those needed for stimulated Raman scattering (SRS). It manifests through the generation of a backward-propagating Stokes wave that carries most of the input energy, once the Brillouin threshold is reached. Stimulated Brillouin scattering is typically harmful for optical communication systems. At the same time, it can be useful for making fiber-based Brillouin lasers and amplifiers. This chapter is devoted to the SBS phenomenon in optical fibers. Section 9.1 presents the basic concepts behind SBS with emphasis on the Brillouin gain. Section 9.2 discusses the theoretical aspects such as Brillouin threshold, pump depletion, and gain saturation. The dynamic aspects of SBS are the focus of Section 9.3 where we discuss nonlinear phenomena such as SBS-induced modulation instability and optical chaos. Section 9.4 is devoted to Brillouin lasers. Other practical aspects of SBS are covered in Section 9.5.

9.1 Basic Concepts

The nonlinear phenomenon of SBS, first observed in 1964, has been studied extensively [1]–[10]. It is similar to SRS inasmuch as it manifests through the generation of a Stokes wave whose frequency is downshifted from that of the incident light by an amount set by the nonlinear medium. However, major differences exist between SBS and SRS. For example, the Stokes wave propagates backward when SBS occurs in a single-mode optical fiber, in contrast to SRS that can occur in both directions. The Stokes shift ($\sim$10 GHz) is smaller

355

by three orders of magnitude for SBS compared with that of SRS. The threshold pump power for SBS depends on the spectral width associated with the pump wave. It can be as low as ~ 1 mW for a continuous-wave (CW) pump or when the pumping is in the form of relatively wide pump pulses (width > 1 μs). In contrast, SBS nearly ceases to occur for short pump pulses (width < 10 ns). All of these differences stem from a single fundamental change—acoustical phonons participate in SBS whereas optical phonons are involved in the case of SRS.

9.1.1 Physical Process

The process of SBS can be described classically as a nonlinear interaction between the pump and Stokes fields through an acoustic wave. The pump field generates an acoustic wave through the process of electrostriction [10]. The acoustic wave in turn modulates the refractive index of the medium. This pump-induced index grating scatters the pump light through Bragg diffraction. Scattered light is downshifted in frequency because of the Doppler shift associated with a grating moving at the acoustic velocity v_A. The same scattering process can be viewed quantum-mechanically as if annihilation of a pump photon creates a Stokes photon and an acoustic phonon simultaneously. As both the energy and the momentum must be conserved during each scattering event, the frequencies and the wave vectors of the three waves are related by

$$\Omega_B = \omega_p - \omega_s, \qquad \mathbf{k}_A = \mathbf{k}_p - \mathbf{k}_s, \qquad (9.1.1)$$

where ω_p and ω_s are the frequencies, and $\mathbf{k}_p$ and $\mathbf{k}_s$ are the wave vectors, of the pump and Stokes waves, respectively.

The frequency Ω_B and the wave vector $\mathbf{k}_A$ of the acoustic wave satisfy the standard dispersion relation

$$\Omega_B = v_A |\mathbf{k}_A| \approx 2 v_A |\mathbf{k}_p| \sin(\theta/2), \qquad (9.1.2)$$

where θ is the angle between the pump and Stokes fields, and we used $|\mathbf{k}_p| \approx |\mathbf{k}_s|$ in Eq. (9.1.1). Equation (9.1.2) shows that the frequency shift of the Stokes wave depends on the scattering angle. In particular, Ω_B is maximum in the backward direction ($\theta = \pi$) and vanishes in the forward direction ($\theta = 0$). In a single-mode optical fiber, only relevant directions are the forward and backward directions. For this reason, SBS occurs only in the backward direction with the Brillouin shift given by

$$v_B = \Omega_B / 2\pi = 2n v_A / \lambda_p, \qquad (9.1.3)$$

where Eq. (9.1.2) was used with $|\mathbf{k}_p| = 2\pi n/\lambda_p$ and n is the modal index at the pump wavelength λ_p. If we use $v_A = 5.96$ km/s and $n = 1.45$, the values appropriate for silica fibers, $v_B \approx 11.1$ GHz at $\lambda_p = 1.55$ μm.

Even though Eq. (9.1.2) predicts correctly that SBS should occur only in the backward direction in single-mode fibers, spontaneous Brillouin scattering can occur in the forward direction. This happens because the guided nature of acoustic waves leads to a relaxation of the wave-vector selection rule. As a result, a small amount of Stokes light is generated in the forward direction. This phenomenon is referred to as guided-acoustic-wave Brillouin scattering [11]. In practice, the Stokes spectrum shows multiple lines with frequency shifts ranging from 10–1000 MHz. Because of its extremely weak character, this phenomenon is not considered further in this chapter.

9.1.2 Brillouin-Gain Spectrum

Similar to the case of SRS, the growth of the Stokes wave is characterized by the Brillouin-gain spectrum $g_B(\Omega)$ peaking at $\Omega = \Omega_B$. However, in contrast with the SRS case, the spectral width of the gain spectrum is very small ($\sim$10 MHz in lieu of $\sim$10 THz) because it is related to the damping time of acoustic waves or the phonon lifetime. In fact, if acoustic waves are assumed to decay as $\exp(-\Gamma_B t)$, the Brillouin gain has a Lorentzian spectrum of the form [10]

$$g_B(\Omega) = g_p \frac{(\Gamma_B/2)^2}{(\Omega - \Omega_B)^2 + (\Gamma_B/2)^2}, \tag{9.1.4}$$

where the peak value of the Brillouin-gain coefficient occurring at $\Omega = \Omega_B$ is given by [4]

$$g_p \equiv g_B(\Omega_B) = \frac{2\pi^2 n^7 p_{12}^2}{c\lambda_p^2 \rho_0 v_A \Gamma_B}, \tag{9.1.5}$$

where p_{12} is the longitudinal elasto-optic coefficient and ρ_0 is the material density. The full width at half maximum (FWHM) of the gain spectrum is related to the Γ_B as $\Delta v_B = \Gamma_B/(2\pi)$. The phonon lifetime itself is related to Γ_B as $T_B = \Gamma_B^{-1} \sim 10$ ns.

Measurements of the Brillouin gain in bulk silica were performed as early as 1950 [12]. More recent measurements, performed using an argon-ion laser, show that $v_B = 34.7$ GHz and $\Delta v_B = 54$ MHz at $\lambda_p = 486$ nm [13]. These experiments also indicate that Δv_B depends on the Brillouin shift and varies slightly faster than v_B^2; a quadratic dependence is expected from theory. Noting

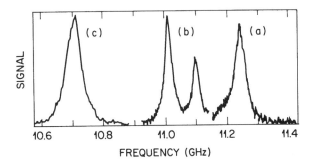

Figure 9.1 Brillouin-gain spectra of three fibers at $\lambda_p = 1.525$ μm: (a) silica-core fiber, (b) depressed-cladding fiber, and (c) dispersion-shifted fiber. (After Ref. [16].)

from Eq. (9.1.3) that v_B varies inversely with λ_p, Δv_B is expected to obey a λ_p^{-2} dependence on the pump wavelength. This narrowing of the Brillouin-gain profile with an increase in λ_p cancels the decrease in gain apparent from Eq. (9.1.5). As a result, the peak value g_p is nearly independent of the pump wavelength. If parameter values typical of fused silica are used in Eq. (9.1.5), $g_p \approx 5 \times 10^{-11}$ m/W. This value is larger by nearly three orders of magnitude compared with that of the Raman-gain coefficient (see Section 8.1).

The Brillouin-gain spectrum for silica fibers can differ significantly from that observed in bulk silica because of the guided nature of optical modes and the presence of dopants in the fiber core [14]–[20]. Figure 9.1 shows the gain spectra measured for three different fibers having different structures and different doping levels of germania in their cores. The measurements were made by using an external-cavity semiconductor laser operating at 1.525 μm and employed a heterodyne-detection technique with 3-MHz resolution [16]. The fiber (a) has a core of nearly pure silica (germania concentration of about 0.3% per mole). The measured Brillouin shift $v_B \approx 11.25$ GHz is in agreement with Eq. (9.1.3) for this fiber if we use the acoustic velocity of bulk silica. The Brillouin shift is reduced for fibers (b) and (c) with nearly inverse dependence on the germania concentration. The fiber (b) has a double-peak structure that results from an inhomogeneous distribution of germania within the core [16]. In a different experiment [18], a three-peak gain spectrum was observed and interpreted to result from different acoustic velocities in the core and cladding regions of the fiber.

The Brillouin-gain bandwidth in Fig. 9.1 is much larger than that expected for bulk silica ($\Delta v_B \approx 17$ MHz at $\lambda_p = 1.525$ μm). Other experiments have also shown similarly large bandwidths for silica-based fibers [17]–[19]. A part of the increase is due to the guided nature of acoustic modes in optical fibers [14]. Most of the increase, however, can be attributed to inhomogeneities in the fiber-core cross section along the fiber length. Because such inhomogeneities are specific to each fiber, Δv_B is generally different for different fibers and can be as large as 100 MHz in the 1.55-μm spectral region.

Equation (9.1.4) for the Brillouin gain is obtained under steady-state conditions and is valid for a CW or quasi-CW pump (pulse width $T_0 \gg T_B$), whose spectral width Δv_p is much smaller than Δv_B. For pump pulses of width $T_0 < T_B$, the Brillouin gain is substantially reduced [5] compared with that obtained from Eq. (9.1.5). Indeed, if the pulse width becomes much smaller than the phonon lifetime ($T_0 < 1$ ns), the Brillouin gain is reduced below the Raman gain; such a pump pulse generates a forward-propagating Raman pulse through SRS, as discussed in Section 8.4.

Even for a CW pump, the Brillouin gain is reduced considerably if the spectral width Δv_p of the pump exceeds Δv_B. This can happen when a multi-mode laser is used for pumping. It can also happen for a single-mode pump laser whose phase varies rapidly on a time scale shorter than the phonon lifetime T_B. Detailed calculations show that the Brillouin gain, under broadband pumping conditions, depends on the relative magnitudes of the pump-coherence length [21]–[23], defined by $L_{coh} = c/(n\Delta v_p)$, and the SBS-interaction length L_{int}, defined as the distance over which the Stokes amplitude varies appreciably. If $L_{coh} \gg L_{int}$, the SBS process is independent of the mode structure of the pump laser provided the longitudinal-mode spacing exceeds Δv_B, and the Brillouin gain is nearly the same as for a single-mode laser after a few interaction lengths [21]. In contrast, the Brillouin gain is reduced significantly if $L_{coh} \ll L_{int}$. The latter situation is generally applicable to optical fibers where the interaction length is typically comparable to the fiber length L. In the case of a pump laser with a Lorentzian spectral profile of width Δv_p, the gain spectrum is still given by Eq. (9.1.4) but the peak value of Brillouin gain is reduced by a factor $1 + \Delta v_p/\Delta v_B$ [23]. As a result, the SBS threshold increases by a large factor when $\Delta v_p \gg \Delta v_B$.

9.2 Quasi-CW SBS

Similar to the SRS case, the development of SBS in optical fibers requires consideration of mutual interaction between the pump and Stokes waves. In this section we develop a simple theory valid under CW or quasi-CW conditions and use it to discuss the concept of Brillouin threshold.

9.2.1   Coupled Intensity Equations

Under steady-state conditions, applicable for a CW or quasi-CW pump, SBS is governed by the two coupled equations similar to Eqs. (8.1.2) and (8.1.3). The only difference is that the sign of dI_s/dz should be changed to account for the counterpropagating nature of the Stokes wave with respect to the pump wave. Two simplifications can be made. First, owing to the relatively small values of the Brillouin shift, $\omega_p \approx \omega_s$. Second, for the same reason, fiber losses are nearly the same for the pump and Stokes waves, that is, $\alpha_p \approx \alpha_s \equiv \alpha$. With these changes, Eqs. (8.1.2) and (8.1.3) become

$$\frac{dI_p}{dz} = -g_B I_p I_s - \alpha I_p. \qquad (9.2.1)$$

$$\frac{dI_s}{dz} = -g_B I_p I_s + \alpha I_s, \qquad (9.2.2)$$

One can readily verify that in the absence of fiber losses ($\alpha = 0$),

$$\frac{d}{dz}(I_p - I_s) = 0, \qquad (9.2.3)$$

and $I_p - I_s$ remains constant along the fiber.

Equations (9.2.1) and (9.2.2) assume implicitly that the counterpropagating pump and Stokes waves are linearly polarized along the same direction and maintain their polarization along the fiber. This is the case when the two waves are polarized along a principal axis of a polarization-maintaining fiber. The relative polarization angle between the pump and Stokes waves varies randomly in conventional optical fibers. The Brillouin gain g_B is reduced in that case by a factor of 1.5 [24].

9.2.2   Brillouin Threshold

For the purpose of estimating the Brillouin threshold, pump depletion can be neglected. Using $I_p(z) = I_p(0)e^{-\alpha z}$ in Eq. (9.2.1) and integrating it over the fiber length L, the Stokes intensity is found to grow exponentially in the backward direction as

$$I_s(0) = I_s(L) \exp(g_B P_0 L_{\text{eff}}/A_{\text{eff}} - \alpha L), \qquad (9.2.4)$$

where $P_0 = I_p(0)A_{eff}$ is the input pump power, A_{eff} is the effective core area, and the effective interaction length is given by

$$L_{eff} = [1 - \exp(-\alpha L)]/\alpha. \tag{9.2.5}$$

Equation (9.2.4) shows how a Stokes signal incident at $z = L$ grows in the backward direction because of Brillouin amplification occurring as a result of SBS. In practice, no such signal is generally fed (unless the fiber is used as a Brillouin amplifier), and the Stokes wave grows from noise or spontaneous Brillouin scattering occurring throughout the fiber. Similar to the SRS case, this is equivalent to injecting a fictitious photon per mode at a distance where the gain exactly equals the fiber loss. Following a method similar to that of Section 8.1, the Brillouin threshold is found to occur at a critical pump power P_{cr} obtained from [7]

$$g_B P_{cr} L_{eff}/A_{eff} \approx 21, \tag{9.2.6}$$

where g_B is the peak value of the Brillouin gain given by Eq. (9.1.5). This equation should be compared with Eq. (8.1.15) obtained for the case of SRS. If we use typical values for fibers used in 1.55-μm optical communication systems, $A_{eff} = 50 \ \mu m^2$, $L_{eff} \approx 20$ km, and $g_B = 5 \times 10^{-11}$ m/W, Eq. (9.2.6) predicts $P_{cr} \sim 1$ mW. It is such a low Brillouin threshold that makes SBS a dominant nonlinear process in optical fibers.

The threshold level predicted by Eq. (9.2.6) is only approximate as the effective Brillouin gain can be reduced by many factors in practice. For example, the SBS threshold increases by 50% when the state of polarization of the pump field becomes completely scrambled [24]. Fiber inhomogeneities also affect the effective Brillouin gain in optical fibers. Variations in the doping level along the radial direction lead to slight changes in the acoustic velocity in that direction. As a result, the SBS threshold depends, to some extent, on various dopants used to make the fiber [25]. Similarly, longitudinal variations in the Brillouin shift Ω_B along the fiber length can reduce the effective Brillouin gain and increase the SBS threshold [26]. This feature can be used to suppress SBS by intentionally changing the core radius along fiber length because Ω_B depends on the core radius. In a variation of this idea, Ω_B was changed along the fiber length using nonuniform doping levels [27]. The SBS threshold for such a fiber exceeded 30 mW in the wavelength region near 1.55 μm.

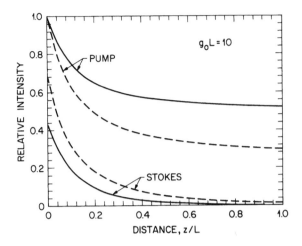

Figure 9.2 Variations in the pump and Stokes intensities (normalized to the input pump intensity) along fiber length. The Stokes power at $z = L$ relative to the pump power is 0.1% for solid curves and 1% for dashed curves.

9.2.3   Gain Saturation

Once Brillouin threshold is reached, a large part of the pump power is transferred to the Stokes wave. To account for pump depletion, it is necessary to solve Eqs. (9.2.1) and (9.2.2). Their general solution is somewhat complicated [28]. However, if fiber losses are neglected by setting $\alpha = 0$, we can make use of Eq. (9.2.3) and set $I_p = C + I_s$, where C is a constant. The resulting equation can be integrated along the fiber length to yield

$$\frac{I_s(z)}{I_s(0)} = \left(\frac{C + I_s(z)}{C + I_s(0)}\right) \exp(g_B C z). \tag{9.2.7}$$

Using $C = I_p(0) - I_s(0)$, the Stokes intensity $I_s(z)$ is given by [4]

$$I_s(z) = \frac{b_0(1 - b_0)}{G(z) - b_0} I_p(0), \tag{9.2.8}$$

where $G(z) = \exp[(1 - b_0)g_0 z]$ with

$$b_0 = I_s(0)/I_p(0), \qquad g_0 = g_B I_p(0). \tag{9.2.9}$$

The parameter b_0 is a measure of the SBS efficiency as it shows what fraction of the input pump power is converted to the Stokes power. The quantity g_0 is the small-signal gain associated with the SBS process.

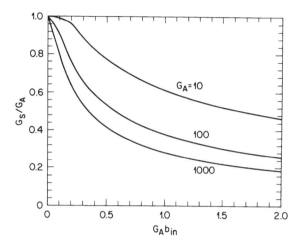

Figure 9.3 Gain saturation in Brillouin amplifiers for several values of G_A.

Equation (9.2.8) shows how the Stokes intensity varies along fiber length in a Brillouin amplifier when the input signal is launched at $z = L$ and the pump is incident at $z = 0$. Figure 9.2 shows this variation for two input signals such that $b_{in} = I_s(L)/I_p(0) = 0.001$ and 0.01. The value of $g_0L = 10$ corresponds to unsaturated amplifier gain of e^{10} or 43 dB. Because of pump depletion, the net gain is considerably smaller. Nonetheless, about 50 and 70% of the pump power is transferred to the Stokes for $b_{in} = 0.001$ and 0.01, respectively. Note also that most of the power transfer occurs within the first 20% of the fiber length.

The saturation characteristics of Brillouin amplifiers can be obtained from Eq. (9.2.8) if we define the saturated gain using

$$G_s = I_s(0)/I_s(L) = b_0/b_{in}, \tag{9.2.10}$$

and introduce the unsaturated gain as $G_A = \exp(g_0L)$. Figure 9.3 shows gain saturation by plotting G_s/G_A as a function of $G_A b_{in}$ for several values of G_A. It should be compared with Fig. 8.6 where the saturation characteristics of Raman amplifiers are shown. The saturated gain is reduced by a factor of two (by 3 dB) when $G_A b_{in} \approx 0.5$ for G_A in the range 20–30 dB. This condition is satisfied when the amplified signal power becomes about 50% of the input pump power. As typical pump powers are ~ 1 mW, the saturation power of Brillouin amplifiers is also ~ 1 mW.

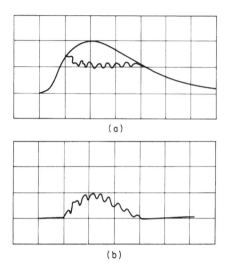

(b)

Figure 9.4 Oscilloscope traces of (a) input and transmitted pump pulses and (b) Stokes pulse for SBS occurring in a 5.8-m fiber. The horizontal scale is 200 ns per division while the vertical scale is arbitrary. (After Ref. [6].)

9.2.4   Experimental Results

In the 1972 demonstration of SBS in optical fibers, a xenon laser operating at 535.5 nm was used as a source of relatively wide (width $\sim$1 μs) pump pulses [6]. Because of large fiber losses (about 1300 dB/km), only small sections of fibers ($L = 5$–20 m) were used in the experiment. The measured Brillouin threshold was 2.3 W for a 5.8-m-long fiber and reduced to below 1 W for $L = 20$ m, in good agreement with Eq. (9.2.6) if we use the estimated value $A_{\text{eff}} = 13.5$ μm^2 for the effective core area. The Brillouin shift of $v_B = 32.2$ GHz also agreed with Eq. (9.1.3).

Figure 9.4 shows the oscilloscope traces of incident and transmitted pump pulses, as well as the corresponding Stokes pulse, for SBS occurring in a 5.8-m-long fiber. The oscillatory structure is due to relaxation oscillations whose origin is discussed in the next section. The oscillation period of 60 ns corresponds to the round-trip time within the fiber. The Stokes pulse is narrower than the pump pulse because SBS transfers energy only from the central part of the pump pulse where power is large enough to exceed the Brillouin threshold. As a result, peak power of the Stokes pulse can exceed the input peak power of the pump pulse. Indeed, SBS can damage the fiber permanently because of

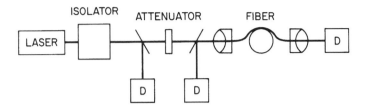

Figure 9.5 Schematic of the experimental setup used to observe SBS in optical fibers. (After Ref. [30].)

such an increase in the Stokes intensity [6].

In most of the early experiments, the SBS threshold was relatively high (>100 mW) because of extremely large fiber losses. As mentioned earlier, it can become as low as 1 mW if long lengths of low-loss fiber are employed. A threshold of 30 mW was measured when a ring dye laser, operating in a single-longitudinal mode at 0.71 μm, was used to couple CW light into a 4-km-long fiber having a loss of about 4 dB/km [29]. In a later experiment, the threshold power was reduced to only 5 mW [30] by using a Nd:YAG laser operating at 1.32 μm. Figure 9.5 shows a schematic of the experimental arrangement. The 1.6-MHz spectral width of the CW pump was considerably smaller than the Brillouin-gain bandwidth. The experiment employed a 13.6-km-long fiber with losses of only 0.41 dB/km, resulting in an effective length of 7.66 km. The isolator serves to block the Stokes light from entering the laser. The transmitted power and the reflected power are measured as a function of the input power launched into the fiber. Figure 9.6 shows the experimental data. At low input powers, the reflected signal is due to 4% reflection at the air–fiber interface. Brillouin threshold is reached at an input power of about 5 mW as manifested through a substantial increase in the reflected power through SBS. At the same time, the transmitted power decreases because of pump depletion, and reaches a saturation level of about 2 mW for input powers in excess of 10 mW. The SBS conversion efficiency is nearly 65% at that level.

In a 1987 experiment, SBS was observed by using a 1.3-μm semiconductor laser, which utilized the distributed-feedback mechanism to emit light in a single longitudinal mode with a spectral width $\sim$10 MHz [31]. The CW pump light was coupled into a 30-km-long fiber having a loss of 0.46 dB/km. The effective length was about 9 km from Eq. (9.2.5). The Brillouin threshold was reached at a pump power of 9 mW. To check if the bandwidth of the pump laser affected the SBS threshold, the measurements were also performed using

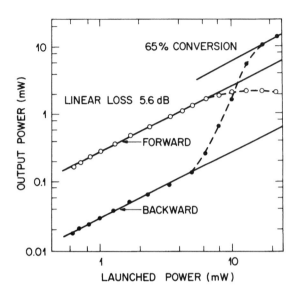

Figure 9.6 Transmitted (forward) and reflected (backward) powers as a function of the input power launched into a 13.6-km-long single-mode fiber. (After Ref. [30].)

a Nd:YAG laser having a bandwidth of only 20 kHz. The results were virtually identical with those obtained using the semiconductor laser, indicating that the Brillouin-gain bandwidth Δv_B was considerably larger than 15 MHz. As mentioned in Section 9.1, Δv_B is considerably enhanced from its value in bulk silica ($\approx$ 22 MHz at 1.3 μm) because of fiber inhomogeneities. The threshold power is considerably smaller ($\sim$1 mW) near 1.55 μm where the effective length is $\sim$20 km because of relatively low losses of optical fibers (0.2 dB/km) at that wavelength.

In most experiments on SBS, it is essential to use an optical isolator between the laser and the fiber to prevent the Stokes signal from reentering into the fiber after reflection at the laser-cavity mirror. In the absence of an isolator, a significant portion of the Stokes power can be fed back into the fiber. In one experiment, about 30% of the Stokes power was reflected back and appeared in the forward direction [32]. It was observed that several orders of Stokes and anti-Stokes lines were generated in the spectrum as a result of the feedback. Figure 9.7 shows the output spectra in the forward and backward directions for SBS occurring in a 53-m-long fiber. The 34-GHz spacing between adjacent spectral lines corresponds exactly to the Brillouin shift at $\lambda_p = 514.5$ nm. The anti-Stokes components are generated as a result of four-wave mix-

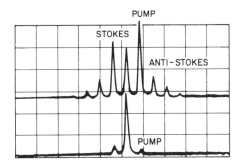

Figure 9.7 Output spectra in the forward (upper trace) and backward (lower trace) directions showing multiple orders of Stokes and anti-Stokes waves generated without isolation between laser and fiber. (After Ref. [32].)

ing between the copropagating pump and Stokes waves (see Section 10.1). The higher-order Stokes lines are generated when the power in the lower-order Stokes wave becomes large enough to satisfy the threshold condition given in Eq. (9.2.6). Even when optical isolators are used to prevent external feedback, Rayleigh backscattering, occurring within the fiber, can provide feedback to the SBS process. In a 1998 experiment, this internal feedback process in a high-loss fiber (length 300 m) was strong enough to produce lasing when the fiber was pumped using 1.06-μm CW radiation [33].

9.3 Dynamic Aspects

The dynamic aspects of SBS are more important than SRS as the medium response in the SBS case is governed by the phonon lifetime $T_B \sim 10$ ns. The quasi-CW regime is thus valid only for pump pulses of widths 100 ns or more. This section focuses on the time-dependent effects relevant to the SBS phenomenon in optical fibers.

9.3.1 Coupled Amplitude Equations

The time-dependent effects are described by using the coupled-amplitude equations similar to Eqs. (8.1.21) and (8.1.22). However, these equations should be modified to include the transient nature of SBS by adding a third equation for the material density [34]–[43]. The resulting set of equations can be written as

$$\frac{\partial A_p}{\partial z} + \frac{1}{v_g}\frac{\partial A_p}{\partial t} + \frac{i\beta_2}{2}\frac{\partial^2 A_p}{\partial t^2} + \frac{\alpha}{2}A_p$$

$$= i\gamma(|A_p|^2 + 2|A_s|^2)A_p - \frac{g_B}{2}A_s Q, \qquad (9.3.1)$$

$$-\frac{\partial A_s}{\partial z} + \frac{1}{v_g}\frac{\partial A_s}{\partial t} + \frac{i\beta_2}{2}\frac{\partial^2 A_s}{\partial t^2} + \frac{\alpha}{2}A_s$$

$$= i\gamma(|A_s|^2 + 2|A_p|^2)A_s + \frac{g_B}{2}A_p Q^*, \qquad (9.3.2)$$

$$T_B\frac{\partial Q}{\partial t} + (1 + i\delta)Q = A_p A_s^*, \qquad (9.3.3)$$

where the effects of GVD, SPM, and XPM have been included. The wavelength difference between the pump and Stokes waves is so small that numerical values of β_2, γ, and α are nearly the same for both waves. The material response to acoustic waves is governed by Eq. (9.3.3), where Q is related to the amplitude of density oscillations [10]. The detuning parameter δ is defined as $\delta = (\omega_p - \omega_s - \Omega_B)T_B$, where Ω_B is the Brillouin shift.

For pump pulses of widths $T_0 \gg 10$ ns, the preceding equations can be simplified considerably if we note that the GVD effects are negligible for such pulses. The SPM and XPM effects can also be neglected if the peak powers associated with the pump and Stokes pulses are relatively low. We can also set $\partial Q/\partial t = 0$ for such wide pulses. Defining $I_j = |A_j|^2$, where $j = p$ or s, and choosing $\delta = 0$, the temporal evolution of SBS is governed by the following two simple equations:

$$\frac{\partial I_p}{\partial z} + \frac{1}{v_g}\frac{\partial I_p}{\partial t} = -g_B I_p I_s - \alpha I_p, \qquad (9.3.4)$$

$$-\frac{\partial I_s}{\partial z} + \frac{1}{v_g}\frac{\partial I_s}{\partial t} = g_B I_p I_s - \alpha I_s. \qquad (9.3.5)$$

These equations reduce to Eqs. (9.2.1) and (9.2.2) under steady-state conditions in which I_p and I_s are time independent.

9.3.2 Relaxation Oscillations

The dynamic response of SBS has many interesting features even for pump pulses that are wider than T_B. In particular, the Stokes intensity does not approach its steady-state value monotonically but exhibits relaxation oscillations

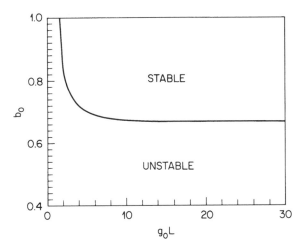

Figure 9.8 Stable and unstable regions of SBS in the presence of feedback. Solid line shows the critical value of the relative Stokes intensity $[b_0 = I_s(0)/I_p(0)]$ below which the steady state is unstable. (After Ref. [45].)

with a period $2T_r$, where $T_r = L/v_g$ is the transit time for a fiber of length L [44]. In the presence of external feedback, relaxation oscillations can turn into stable oscillations [45], that is, both the pump and Stokes waves can develop self-induced intensity modulation.

Even though the group velocity v_g is the same for the pump and the Stokes waves, their relative speed is $2v_g$ because of their counterpropagating nature. Relaxation oscillations occur as a result of this effective group-velocity mismatch. A simple way to obtain the frequency and the decay time of relaxation oscillations is to perform a linear stability analysis of the steady-state solution (9.2.8) by following a procedure similar to that of Section 5.1 used in the context of modulation instability. The effect of external feedback can be included by assuming that optical fiber is enclosed within a cavity and by applying the appropriate boundary conditions at fiber ends [45]. Such a linear stability analysis also provides the conditions under which the steady state becomes unstable.

Assuming that small perturbations from the steady state decay as $\exp(-ht)$, the complex parameter h can be determined by linearizing Eqs. (9.3.4) and (9.3.5). If the real part of h is positive, perturbations decay exponentially with time through relaxation oscillations whose frequency is given by $v_r = \text{Im}(h)/2\pi$. By contrast, if the real part of h becomes negative, perturbations

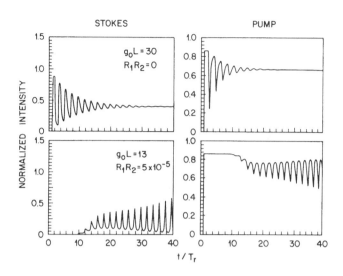

Figure 9.9 Temporal evolution of the Stokes (left column) and pump (right column) intensities without (top row) and with (bottom row) feedback. Fiber losses are such that $\alpha L = 0.15$. (After Ref. [45].)

grow with time, and the steady state is unstable. In that case, SBS leads to temporal modulation of the pump and Stokes intensities even for a CW pump. Figure 9.8 shows the stable and unstable regions in the case of feedback as a function of the gain factor $g_0 L$ that is related to the pump intensity through Eq. (9.2.9). The parameter b_0 is also defined there. It represents the fraction of the pump power converted to the Stokes power.

Figure 9.9 shows temporal evolution of the Stokes and pump intensities obtained by solving Eqs. (9.3.4) and (9.3.5) numerically. The top row for $g_0 L = 30$ shows relaxation oscillations occurring in the absence of feedback. The oscillation period is $2T_r$, where T_r is the transit time. Physically, the origin of relaxation oscillations can be understood as follows [44]. Rapid growth of the Stokes power near the input end of the fiber depletes the pump. This reduces the gain until the depleted portion of the pump passes out of the fiber. The gain then builds up and the process repeats.

The bottom row in Fig. 9.9 corresponds to the case of weak feedback such that $R_1 R_2 = 5 \times 10^{-5}$, where R_1 and R_2 are the reflectivities at the fiber ends. The gain factor of $g_0 L = 13$ is below the Brillouin threshold. The Stokes wave is nonetheless generated because of the reduction in Brillouin threshold as a result of the feedback. However, because of the instability indicated in Fig.

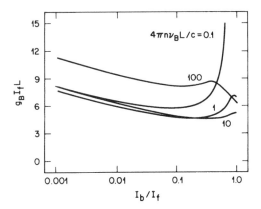

Figure 9.10 Threshold of SBS-induced modulation instability for counterpropagating pump waves of input intensities I_f and I_b. Normalized forward intensity is plotted as a function of I_b/I_f for $\Delta v_B/v_B = 0.06$ for several fiber lengths. (After Ref. [50].)

9.8, a steady state is not reached. Instead, the intensities of the pump output (at $z = L$) and the Stokes output (at $z = 0$) both exhibit steady oscillations. Interestingly enough, a steady state is reached if the feedback is increased such that $R_1 R_2 \geq 2 \times 10^{-2}$. This happens because b_0 for this amount of feedback lies in the stable regime of Fig. 9.8. All of these dynamic features have been observed experimentally [45].

9.3.3 Modulation Instability and Chaos

Another instability can occur when two counterpropagating pump waves are present simultaneously, even though none of them is intense enough to reach the Brillouin threshold [46]–[52]. The origin of this instability lies in the SBS-induced coupling between the counterpropagating pump waves through an acoustic wave at the frequency v_B. The instability manifests as the spontaneous growth of side modes in the pump spectrum at $v_p \pm v_B$, where v_p is the pump frequency [47]. In the time domain, both pump waves develop modulations at the frequency v_B. The SBS-induced modulation instability is analogous to the XPM-induced modulation instability discussed in Section 7.3 except that it occurs for waves propagating in the opposite directions. The instability threshold depends on the forward and backward input pump intensities I_f and I_b, fiber length L, and parameters g_B, v_B, and Δv_B.

Figure 9.10 shows the forward pump intensity I_f (in the normalized form) needed to reach the instability threshold as a function of the intensity ratio I_b/I_f for $\Delta v_B/v_B = 0.06$ and several values of the normalized fiber length $4\pi n v_B L/c$. The instability threshold is significantly smaller than the Brillouin threshold ($g_B I_f L = 21$) and can be as small as $g_B I_f L = 3$ for specific values of the parameters. Numerical results show that temporal evolution of pump intensities at the fiber output can become chaotic [50], following a period-doubling route, if the Brillouin-gain bandwidth Δv_B is comparable to the Brillouin shift v_B. Subharmonics of v_B appear in the spectrum of the scattered light with successive period-doubling bifurcations. The chaotic evolution is also predicted when the backward pump is not incident externally but is produced by the feedback of the forward pump at a reflector [48].

During the 1990s, considerable effort was devoted to observe and characterize SBS-induced chaos in optical fibers [53]–[64]. Irregular fluctuations in the Stokes intensity, occurring at a time scale ~ 0.1 μs, were observed in several experiments [53]–[55]. Figure 9.11 shows an example of fluctuations in the Stokes power observed for SBS occurring inside a 166-m-long fiber when the CW pump power was 50% above the SBS threshold ($P_0 = 1.5 P_{cr}$). Whether such fluctuations are stochastic or chaotic in nature is an issue that is not easy to resolve. Interpretation of the experimental results requires a careful consideration of both spontaneous Brillouin scattering and the effects of optical feedback. It was established by 1993 that fluctuations in the Stokes power, observed when care was taken to suppress optical feedback, were due to stochastic noise arising from spontaneous Brillouin scattering [62]. The mathematical description requires inclusion of spontaneous Brillouin scattering (through a Langevin noise source) in the coupled amplitude equations given earlier [59].

The SBS dynamics change drastically in the presence of optical feedback introduced either by using an external mirror or occurring naturally at the fiber ends because of a refractive-index discontinuity at the glass-air interface. As discussed earlier in this section, feedback destabilizes relaxation oscillations and leads to periodic output at a repetition rate $(2T_r)^{-1}$, where T_r is the transit time. Under certain conditions, the envelope of the pulse train exhibits irregular fluctuations occurring at a time scale much longer than T_r. In one experiment, such fluctuations were found to be stochastic in nature and were attributed to fluctuations of the relative phase between the pump light and the fiber resonator [62]. In another experiment, a quasi-periodic route to chaos was observed in a limited range of pump powers [63]. Fiber nonlinearity may

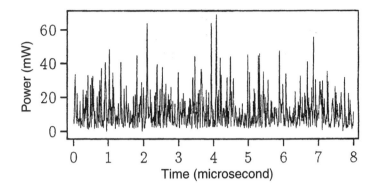

Figure 9.11 Fluctuations observed in Stokes power at a pump power $P_0 = 1.5P_{cr}$ when SBS occurred inside a 166-m-long fiber. (After Ref. [62].)

have played a role in this experiment because chaos was observed at pump powers ~0.8 W. Numerical simulations based on the coupled amplitude equations (9.3.1)–(9.3.3) predict that inclusion of SPM and XPM can lead to optical chaos, even in the absence of feedback, if the nonlinearity is large enough [56]–[58].

9.3.4 Transient Regime

Equations (9.3.1)–(9.3.3) can be used to study SBS in the transient regime applicable for pump pulses shorter than 10 ns. From a practical standpoint, two cases are of interest depending on the repetition rate of pump pulses. Both cases are discussed here.

In the case relevant for optical communication systems, the repetition rate of pump pulses is ~1 GHz, while the width of each pulse is ~100 ps. The pulse train is not uniform for lightwave signals as "1" and "0" bits form a pseudorandom sequence. Nevertheless, the effect of such a pulse train is similar to the quasi-CW case discussed earlier because of its high repetition rate. The time interval between pump pulses is short enough that successive pulses can pump the same acoustic wave in a coherent manner (except in rare instances of a long sequence of "0" bits). The main effect of a pseudorandom pulse train is that the Brillouin threshold is increased by a factor of two or so compared with the CW case; the exact factor depends on the bit rate as well as modulation format. The SBS threshold can be increased even more by modulating the phase of the CW beam at the optical transmitter (before information is encoded on it)

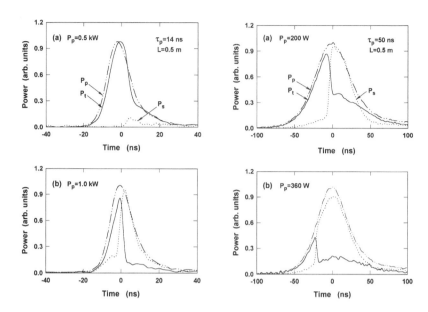

Figure 9.12 Stokes (dotted) and transmitted pump (solid) pulses observed for several peak powers when 14-ns (left column) and 50-ns (right column) input pump pulses (dot-dashed curve) were transmitted through a 0.5-m-long fiber. (After Ref. [43].)

at frequencies in excess of 100 MHz [36]. Such phase modulations reduce the effective Brillouin gain by increasing spectral bandwidth of the optical source. Phase modulations can also be used to convert the CW beam into a train of optical pulses whose width ($\sim$0.5 ns) is a small fraction of the phonon lifetime [38]–[41].

The case in which the repetition rate of pump pulses is relatively low ($<$10 MHz) is quite interesting. In this case, the acoustic wave created by each pump pulse decays almost completely before the next pump pulse arrives. This case has been studied by solving Eqs. (9.3.1)–(9.3.3) numerically [43]. The results show that the characteristics of the Stokes pulse depend not only on the width and peak power of the pump pulse but also on the fiber length. If the pump-pulse width T_p becomes much shorter than the phonon lifetime T_B, SBS ceases to occur. When the two are comparable, the Stokes pulse can become shorter than the pump pulse. Figure 9.12 shows the Stokes pulse (dotted curve) and the transmitted pump pulse (solid curve) for several peak powers when 14- and 50-ns pump pulses at the 10-Hz repetition rate (obtained from a Q-switched Nd:YAG laser) were transmitted through a 0.5-m-long optical fiber.

These experimental results agree quite well with the numerical predictions.

An interesting question is whether Eqs. (9.3.1)–(9.3.3) permit solitary-wave solutions such that each pump pulse generates the Stokes field A_s in the form of a backward-propagating soliton. It turns out that, under certain conditions, the pump and Stokes waves can support each other as a coupled bright–dark soliton pair [65]–[67], similar to the XPM-paired solitons discussed in Section 7.3. Such solitons exist even in the absence of GVD ($\beta_2 = 0$) and XPM ($\gamma = 0$) because they rely on the presence of a solitary acoustic wave. Such Brillouin solitons exist in spite of fiber losses and can move at a speed larger than the group velocity expected in the low-power limit.

9.4 Brillouin Fiber Lasers

Similar to the SRS case considered in Section 8.2, the Brillouin gain in optical fibers can be used to make lasers by placing the fiber inside a cavity. Such lasers were made as early as 1976 and have remained an active topic of study since then [68]–[86]. Both the ring-cavity and the Fabry–Perot geometries have been used for making Brillouin lasers, each having its own advantages. No mirrors are needed in the ring-cavity case as such a cavity can be made by using a directional fiber coupler.

9.4.1 CW Operation

The threshold pump power required for laser oscillations is considerably reduced from that given in Eq. (9.2.6) because of the feedback provided by the cavity. Consider a ring cavity. Using the boundary condition $I_s(L) = RI_s(0)$, the threshold condition can be written as

$$R\exp(g_B P_{th} L_{eff}/A_{eff} - \alpha L) = 1, \qquad (9.4.1)$$

where L is the ring-cavity length, R is the fraction of Stokes intensity fed back after each round trip and P_{th} is the threshold value of the pump power. Fiber losses can be neglected in most cases of practical interest because L is typically 100 m or less. A comparison with Eq. (9.2.6) shows that, for the same fiber length, the factor of 21 is typically replaced by a number in the range 0.1–1 depending on the value of R.

In a 1976 demonstration of a CW Brillouin laser, a ring cavity consisting of a 9.5-m-long fiber was pumped using an argon-ion laser [68]. The length

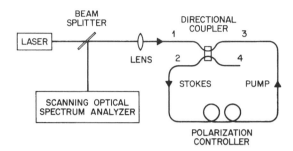

Figure 9.13 Schematic illustration of a Brillouin ring laser. (After Ref. [73].)

of fiber was chosen to be relatively small in view of its large losses (about 100 dB/km) at the pump wavelength of 514.5 nm. The round-trip losses (about 70%) were still so large that the threshold was more than 100 mW for this laser. By 1982, the threshold pump power was reduced to a mere 0.56 mW [73] by using an all-fiber ring resonator shown schematically in Fig. 9.13. The round-trip cavity losses were only 3.5% for this ring cavity. Such a low loss resulted in an enhancement of the input pump power by a factor of 30 inside the ring cavity. A He–Ne laser operating at 633 nm could be used as a pump source because of the low laser threshold. Semiconductor lasers were used for pumping of Brillouin lasers soon after, resulting in a compact device [74]. Such lasers are routinely used in high-performance laser gyroscopes for inertial rotation sensing.

Brillouin fiber lasers consisting of a Fabry–Perot cavity exhibit features that are qualitatively different from those making use of a ring-cavity config-uration. The difference arises from the simultaneous presence of the forward and backward propagating components associated with the pump and Stokes waves. Higher-order Stokes waves are generated through cascaded SBS, a pro-cess in which each successive Stokes component pumps the next-order Stokes component after its power becomes large enough to reach the Brillouin thresh-old. At the same time, anti-Stokes components are generated through four-wave mixing between copropagating pump and Stokes waves. The frequency spectrum of the laser output appears similar to that shown in Fig. 9.7. The number of Stokes and anti-Stokes lines depends on the pump power. In one experiment [68], up to 14 spectral lines were observed, with 10 lines appearing on the Stokes side, when a 20-m-long fiber was used inside a Fabry–Perot cav-ity, which was pumped by using a CW argon-ion laser operating at 514.5 nm.

The 34-GHz frequency spacing between adjacent lines corresponded to the expected Brillouin shift. The use of a Sagnac interferometer in 1998 resulted in generation of up to 34 spectral lines through cascaded SBS in an erbium-doped fiber laser [84].

Most Brillouin fiber lasers use a ring cavity to avoid generation of multiple Stokes lines through cascaded SBS. The performance of a Brillouin ring laser depends on the fiber length L used to make the cavity (see Fig. 9.13) as it determines the longitudinal-mode spacing through $\Delta v_L = c/(\bar{n}L)$, where $\bar{n}$ is the effective mode index. For short fibers such that $\Delta v_L > \Delta v_B$, where Δv_B is the Brillouin-gain bandwidth (typically 20 MHz), the ring laser operates stably in a single longitudinal mode. Such lasers can be designed to have low threshold [73] and to emit CW light with a narrow spectrum [76]. In contrast, a Brillouin ring laser operates in multiple longitudinal modes when $\Delta v_L \ll \Delta v_B$, and the number of modes increases with fiber length. As early as 1981, it was noted that such lasers required active intracavity stabilization to operate continuously [72]. In fact, their output can become periodic, and even chaotic, under some conditions. This issue is discussed later in this section.

An important application of CW Brillouin lasers consists of using them as a sensitive laser gyroscope [77]–[79]. Passive fiber gyroscopes have been discussed in Section 7.5.3. Laser gyroscopes differ from them both conceptually and operationally. Whereas passive fiber gyroscopes use a fiber ring as an interferometer, active laser gyroscopes use the fiber ring as a laser cavity. The rotation rate is determined by measuring the frequency difference between the counterpropagating laser beams. Similarly to the case of passive fiber gyroscopes, fiber nonlinearity affects the performance of a Brillouin-laser fiber gyroscope through XPM-induced nonreciprocity and constitutes a major error source [78].

9.4.2 Pulsed Operation

Brillouin fiber lasers with long cavity lengths can be forced to emit a pulse train using several different methods. The technique of active mode locking was used in a 1978 experiment by placing an amplitude modulator inside the laser cavity [70]. Laser output consisted of a train of pulses (width about 8 ns) at a repetition rate of 8 MHz, determined by the cavity length. These pulses result from locking of multiple longitudinal modes of the cavity.

Another type of mode locking can occur in Fabry–Perot cavities in which multiple Stokes lines are generated through cascaded SBS. Relaxation oscillations can seed the mode-locking process because their period is equal to the round-trip time in the cavity. Indeed, partial mode-locking of such a Brillouin laser by itself was observed [70], but the process was not very stable. The reason can be understood from Eq. (9.1.3) showing that the Brillouin shift depends on pump wavelength. In cascaded SBS, different Stokes waves act as pumps for the successive Stokes. As a result, multiple Stokes lines are not spaced equally but differ in frequencies by a small amount ~ 1 MHz. In a 1989 experiment, mode locking was achieved by using a multimode fiber [75]. As different modes have a slightly different effective index (modal dispersion), equally spaced Stokes lines can be generated by using different fiber modes.

An interesting technique for generating short Stokes pulses from a Brillouin laser makes use of synchronous pumping using a mode-locked train of pump pulses [80]. The idea is quite simple. The length of a ring cavity is adjusted such that the round-trip time exactly equals the spacing between pump pulses. Each pump wave is so short that it is unable to excite the acoustic wave significantly. However, if the next pump pulse arrives before the acoustic wave has decayed, the cumulative effect of multiple pump pulses can build up the acoustic wave to a large amplitude. After the build-up process is complete, a short Stokes pulse will be generated through transient SBS with the passage of each pump pulse. This technique has produced Stokes pulses of 200-ps duration when 300-ps pulses from a mode-locked Nd:YAG laser were used for pumping a Brillouin ring laser.

Brillouin ring lasers with long cavity lengths can produce pulse trains, even when pumped continuously, through a nonlinear self-pulsing mechanism. Typically, pulses have widths in the range 20 to 30 ns and are emitted with a repetition rate nearly equal to the longitudinal-mode spacing $\Delta v_L \equiv 1/t_r$, where t_r is the round-trip time. Physics behind such lasers attracted considerable attention during the 1990s [86]. Equations (9.3.1)–(9.3.3) describe the nonlinear dynamics in Brillouin ring lasers when supplemented with the boundary conditions appropriate for a ring cavity:

$$A_s(L,t) = \sqrt{R}A_s(0,t), \qquad A_p(0,t) = \sqrt{P_0} + \sqrt{R_p}A_p(L,t), \qquad (9.4.2)$$

where R and R_p are the feedback levels for the Stokes and pump fields after one round-trip inside the ring cavity. In modern Brillouin lasers, feedback of the pump on each round trip is avoided using either an optical isolator or an optical circulator in place of the directional coupler (see Fig. 9.11) so that $R_p = 0$.

Neither dispersive (GVD) nor nonlinear (SPM and XPM) effects play an important role in determining the self-pulsing threshold in Brillouin lasers. A linear stability analysis of Eqs. (9.3.1)–(9.3.3) [similar to that used for modulation instability] predicts that the stability of the CW state depends on the pumping parameter $g_0 \equiv g_B P_0 / A_{\text{eff}}$, defined as the small-signal gain at the pump power P_0. The linear stability analysis is quite complicated when the nonlinear effects and the finite medium response time T_B are taken into account. However, a simplified approach shows that the CW state is unstable whenever the pump power P_0 satisfies the inequality [86]

$$\ln\left(\frac{1}{R}\right) < g_0 L < 3 \ln\left(\frac{1+2R}{3R}\right). \qquad (9.4.3)$$

The laser threshold is reached when $g_0 L = \ln(1/R)$. However, the laser does not emit CW light until $g_0 L$ is large enough to be outside the instability domain predicted by Eq. (9.4.3). Noting that the instability domain shrinks as R approaches 1, it is easy to conclude that stable CW operation with a low threshold can be realized in low-loss ring cavities [73]. On the other hand, the pump-power level at which CW operation is possible becomes quite high when $R \ll 1$ (high-loss cavity). As an example, the laser threshold is reached at $g_0 L = 4.6$ when $R = 0.01$ but CW operation is possible only for $g_0 L > 10.6$.

Numerical solutions of Eqs. (9.3.1)–(9.3.3), performed with $\beta_2 = 0$ as GVD effects are negligible, show that the laser emits a pulse train in the unstable region except close to the instability boundary where the laser output exhibits periodic oscillations whose frequency depends on the number of longitudinal modes supported by the laser [86]. Moreover, in this transition region, the laser exhibits bistable behavior in the sense that the transition between the CW and periodic states occurs at different pump levels depending on whether P_0 is increasing or decreasing. The width of the hysteresis loop depends on the nonlinear parameter γ responsible for SPM and XPM, and bistability ceases to occur when $\gamma = 0$.

In the self-pulsing regime, the laser emits a train of optical pulses. As an example, Fig. 9.14 shows evolution of the Stokes and pump amplitudes over multiple round trips. The laser output (Stokes) exhibits transients (left traces) that last over hundreds of round trips. A regular pulse train is eventually formed (right traces) such that a pulse is emitted nearly once per round-trip time. The emitted pulse can be characterized as a Brillouin soliton. All of these features have been seen experimentally using two Brillouin lasers, one pumped at 514.5 nm by an argon-ion laser and another at 1319 nm by Nd:YAG laser [86]. The observed behavior was in agreement with the numerical solutions of Eqs. (9.3.1)–(9.3.3). Specifically, both lasers emitted a pulse train

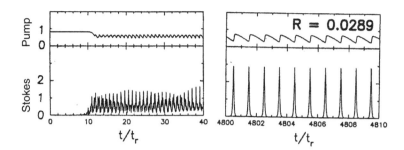

Figure 9.14 Evolution of Stokes (lower trace) and pump (upper trace) amplitudes over multiple round trips in the self-pulsing regime: (left) initial build-up from noise; (right) fully formed pulse train after 4800 round trips. (After Ref. [86].)

when the pumping level satisfied the inequality in Eq. (9.4.3). A Brillouin ring laser can thus be designed to emit a soliton train, with a pulse width $\sim$10 ns and a repetition rate ($\sim$1 MHz) determined by the round-trip time of the ring cavity.

9.5 SBS Applications

Stimulated Brillouin scattering has been used for several different kinds of applications. Its use for making lasers has already been discussed; it can also be used for making an optical amplifier. The SBS has also been exploited for making fiber sensors capable of sensing temperature and strain changes over relatively long distances. In this section we discuss some of these applications.

9.5.1 Brillouin Fiber Amplifiers

The SBS-produced gain in an optical fiber can be used to amplify a weak signal whose frequency is shifted from the pump frequency by an amount equal to the Brillouin shift. Such amplifiers were studied during the 1980s [87]–[99].

A semiconductor laser can be used to pump Brillouin amplifiers provided it operates in a single-longitudinal mode whose spectral width is considerably less than the Brillouin-gain bandwidth. Distributed-feedback or external-cavity semiconductor lasers [100] are most appropriate for pumping Brillouin amplifiers. In a 1986 experiment, two external-cavity semiconductor lasers, with line widths below 0.1 MHz, were used as pump and probe lasers [87]. Both lasers operated continuously and were tunable in the spectral region near

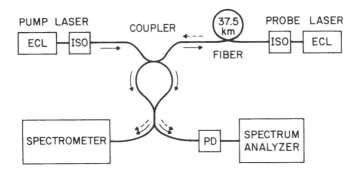

Figure 9.15 Schematic illustration of a Brillouin amplifier. ECL, ISO, and PD stand for external-cavity laser, isolator, and photodetectors, respectively. Solid and dashed arrows show the path of pump and probe lasers. (After Ref. [87].)

1.5 μm. Figure 9.15 shows the experimental setup schematically. Radiation from the pump laser was coupled into a 37.5-km-long fiber through a 3-dB coupler. The probe laser provided a weak input signal ($\sim$ 10 μW) at the other end of the fiber. Its wavelength was tuned in the vicinity of the Brillouin shift ($\nu_B = 11.3$ GHz) to maximize the Brillouin gain. The measured amplification factor increased exponentially with the pump power. This is expected from Eq. (9.2.10). If gain saturation is neglected, the amplification factor can be written as

$$G_A = I_s(0)/I_s(L) = \exp(g_B P_0 L_{\text{eff}}/A_{\text{eff}} - \alpha L). \qquad (9.5.1)$$

The amplifier gain was 16 dB ($G_A = 40$) at a pump power of only 3.7 mW because of a long fiber length used in the experiment.

An exponential increase in the signal power with increasing pump powers occur only if the amplified signal remains below the saturation level. The saturation characteristics of Brillouin amplifiers are shown in Fig. 9.3. The saturated gain G_s is reduced by 3 dB when

$$G_A(P_{\text{in}}/P_0) \approx 0.5, \qquad (9.5.2)$$

for G_A in the range 20–30 dB, where P_{in} is the incident power of the signal being amplified. As P_0 is typically $\sim$1 mW, the saturation power of Brillouin amplifiers is also $\sim$1 mW.

Brillouin amplifiers are capable of providing 30 dB gain at a pump power under 10 mW. However, because $\Delta \nu_B < 100$ MHz, the bandwidth of such amplifiers is also below 100 MHz, in sharp contrast with Raman amplifiers whose bandwidth exceeds 5 THz. In fact, the difference between the signal and pump frequencies should be matched to the Brillouin shift Ω_B (about 11 GHz in the 1.55-μm region) with an accuracy to better than 10 MHz. For this reason, Brillouin amplifiers are not suitable for amplifying signals in fiber-optic communication systems, which commonly use erbium-doped fiber amplifiers.

Brillouin amplifiers can be useful for applications requiring selective amplification [89]–[91]. One such application is based on a scheme in which receiver sensitivity is improved by selective amplification of the carrier while leaving modulation sidebands unamplified [101]. The underlying principle is similar to that of homodyne detection except that the amplified carrier acts as a reference signal. This feature eliminates the need of a local oscillator that must be phase locked to the transmitter, a difficult task in general. In a demonstration of this scheme, the carrier was amplified by 30 dB more than the modulation sidebands even at a modulation frequency as low as 80 MHz [89]. With a proper design, sensitivity improvements of up to 15 dB or more seem feasible at bit rates in excess of 100 Mb/s. The limiting factor is the nonlinear phase shift induced by the pump (a kind of cross-phase modulation) if the difference between the pump and carrier frequencies does not match the Brillouin shift exactly. The calculations show [90] that deviations from the Brillouin shift should be within 100 kHz for a phase stability of 0.1 rad. Nonlinear phase shifts can also lead to undesirable amplitude modulation of a frequency-modulated signal [93].

Another application of narrowband Brillouin amplifiers consists of using them as a tunable narrowband optical filter for channel selection in a densely packed multichannel communication system [91]. If channel spacing exceeds but the bit rate is smaller than the bandwidth Δv_B, the pump laser can be tuned to amplify a particular channel selectively. The concept was demonstrated using a tunable color-center laser as a pump [91]. Two 45-Mb/s channels were transmitted through a 10-km-long fiber. Each channel could be amplified by 20 to 25 dB by using 14 mW of pump power when pump frequency was tuned in the vicinity of the Brillouin shift associated with each channel. More importantly, each channel could be detected with a low bit-error rate ($< 10^{-8}$) when channel spacing exceeded 140 MHz. Because $\Delta v_B < 100$ MHz typically, channels can be packed as close as 1.5 Δv_B without introducing crosstalk from neighboring channels. Brillouin gain has been used as a narrowband amplifier to simultaneously amplify and demodulate FSK signals at bit rates of up to 250 Mb/s by using commercially available semiconductor lasers [94].

9.5.2   Fiber Sensors

SBS can be used for making distributed fiber sensors capable of sensing temperature and strain changes over relatively long distances [102]–[110]. The basic idea behind the use of SBS for fiber sensors is quite simple and can be easily understood from Eq. (9.1.3). As the Brillouin shift depends on the effective refractive index of the fiber mode, it changes whenever the refractive index of silica changes in response to local environmental variations. Both temperature and strain can change the refractive index of silica. By monitoring changes in the Brillouin shift along fiber length, it is possible to map out the distribution of temperature or strain over long distances over which the SBS signal can be detected with a good signal-to-noise ratio.

The basic idea has been implemented in several experiments to demonstrate distributed sensing over distances as long as 32 km. A tunable CW probe laser and a pulsed pump laser inject light at the opposite ends of a fiber. The CW signal is amplified through SBS only when the pump-probe frequency difference coincides exactly with the Brillouin shift. The time delay between the launch of the pump pulse and increase in the received probe signal indicates the exact location where Brillouin amplification occurs. By tuning the probe frequency and measuring time delays, one can map the distribution of temperature or strain over the entire fiber length. In one experiment [104], two diode-pumped, 1.319-μm, Nd:YAG lasers were used for pump and probe signals. Frequency difference between the two lasers was adjusted by temperature tuning the probe-laser cavity. A Bragg cell was used as an optical switch to generate pump pulses of widths in the range 0.1–1 μs. A temperature resolution of 1°C and a spatial resolution of 10 m were realized for a 22-km-long fiber. In a later experiment [105], spatial resolution was improved to 5 m, and fiber length was increased to 32 km. Similar performance is achieved for sensing of distributed strain. A resolution of 20 microstrain with a spatial resolution of 5 m has been demonstrated by using Brillouin loss [105]. It is even possible to combine the temperature and strain measurements in a single fiber sensor [110].

Problems

9.1 What is meant by Brillouin scattering? Explain its origin. What is the difference between spontaneous and stimulated Brillouin scattering?

9.2 Why does SBS occur only in the backward direction in single-mode fibers?

9.3 What are the main differences between SBS and SRS? What is the origin of these differences and how do they manifest in practice?

9.4 Estimate SBS threshold at 1.55 μm for a 40-km-long fiber with 8-μm core diameter. How much does it change at 1.3 μm? Use $g_B = 5 \times 10^{-11}$ m/W and loss values of 0.5 and 0.2 dB/km at 1.3 and 1.55 μm, respectively.

9.5 Solve Eqs. (9.2.1) and (9.2.2) neglecting pump deletion. Use the solution to derive the threshold condition for SBS.

9.6 Solve Eqs. (9.2.1) and (9.2.2) including pump deletion. Neglect fiber losses by setting $\alpha = 0$.

9.7 Solve Eqs. (9.2.1) and (9.2.2) numerically for a 20-km-long fiber using $g_B I_p(0) = 1$ km^{-1} and $\alpha = 0.2$ dB/km. Plot I_p and I_s along the fiber length assuming $I_p(0) = 2$ MW/cm^2 and $I_s(L) = 1$ kW/cm^2.

9.8 Solve Eqs. (9.3.4) and (9.3.5) numerically assuming that both pump and Stokes pulses are Gaussian in shape initially with a FWHM of 1 μs. Plot output pulse shapes when SBS occurs in a 10-m-long fiber assuming $g_B I_p(0) = 1$ m^{-1}.

9.9 Follow the analysis of Reference [86] and derive the inequality given in Eq. (9.4.3).

9.10 How can SBS be used for temperature sensing? Design a SBS-based fiber sensor for this purpose. Sketch the experimental setup and identify all components.

References

[1] R. Y. Chiao, C. H. Townes, and B. P. Stoicheff, *Phys. Rev. Lett.* **12**, 592 (1964).

[2] E. Garmire and C. H. Townes, *Appl. Phys. Lett.* **5**, 84 (1964).

[3] N. M. Kroll, *J. Appl. Phys.* **36**, 34 (1965).

[4] C. L. Tang, *J. Appl. Phys.* **37**, 2945 (1966).

[5] W. Kaiser and M. Maier, in *Laser Handbook*, Vol. 2, F. T. Arecchi and E. O. Schulz-Dubois, Eds. (North-Holland, Amsterdam, 1972), Chap. E2.

[6] E. P. Ippen and R. H. Stolen, *Appl. Phys. Lett.* **21**, 539 (1972).

[7] R. G. Smith, *Appl. Opt.* **11**, 2489 (1972).

[8] D. Cotter, *J. Opt. Commun.* **4**, 10 (1983).

[9] Y. R. Shen, *The Principles of Nonlinear Optics* (Wiley, New York, 1984), Chap. 11.

[10] R. W. Boyd, *Nonlinear Optics* (Academic Press, San Diego, 1992), Chap. 8.

[11] R. M. Shelby, M. D. Levenson, and P. W. Bayer, *Phys. Rev. Lett.* **54**, 939 (1985); *Phys. Rev. B* **31**, 5244 (1985).

[12] R. S. Krishnan, *Nature* **165**, 933 (1950).

[13] D. Heiman, D. S. Hamilton, and R. W. Hellwarth, *Phys. Rev. B* **19**, 6583 (1979).

[14] P. J. Thomas, N. L. Rowell, H. M. van Driel, and G. I. Stegeman, *Phys. Rev. B* **19**, 4986 (1979).

[15] J. Stone and A. R. Chraplyvy, *Electron. Lett.* **19**, 275 (1983).

[16] R. W. Tkach, A. R. Chraplyvy, and R. M. Derosier, *Electron. Lett.* **22**, 1011 (1986).

[17] N. Shibata, R. G. Waarts, and R. P. Braun, *Opt. Lett.* **12**, 269 (1987).

[18] Y. Azuma, N. Shibata, T. Horiguchi, and M. Tateda, *Electron. Lett.* **24**, 250 (1988).

[19] N. Shibata, K. Okamoto, and Y. Azuma, *J. Opt. Soc. Am. B* **6**, 1167 (1989).

[20] T.-O. Sun, A. Wada, T. Sakai, and R. Yamuchi, *Electron. Lett.* **28**, 247 (1992).

[21] G. C. Valley, *IEEE J. Quantum Electron.* **QE-22**, 704 (1986).

[22] P. Narum, M. D. Skeldon, and R. W. Boyd, *IEEE J. Quantum Electron.* **QE-22**, 2161 (1986).

[23] E. Lichtman, A. A. Friesem, R. G. Waarts, and H. H. Yaffe, *J. Opt. Soc. Am. B* **4**, 1397 (1987).

[24] M. O. van Deventer and A. J. Boot, *J. Lightwave Technol.* **12**, 585 (1994).

[25] J. Botineau, E. Picholle, and D. Bahloul, *Electron. Lett.* **31**, 2032 (1995).

[26] K. Shiraki, M. Ohashi, and M. Tateda, *Electron. Lett.* **31**, 668 (1995); *J. Lightwave Technol.* **14**, 50 (1996).

[27] K. Shiraki, M. Ohashi, and M. Tateda, *J. Lightwave Technol.* **14**, 549 (1996).

[28] L. Chen and X. Bao, *Opt. Commun.* **152**, 65 (1998).

[29] N. Uesugi, M. Ikeda, and Y. Sasaki, *Electron. Lett.* **17**, 379 (1981).

[30] D. Cotter, *Electron. Lett.* **18**, 495 (1982).

[31] Y. Aoki, K. Tajima, and I. Mito, *Opt. Quantum Electron.* **19**, 141 (1987).

[32] P. Labudde, P. Anliker, and H. P. Weber, *Opt. Commun.* **32**, 385 (1980).

[33] A. A. Fotiadi and R. V. Kiyan, *Opt. Lett.* **23**, 1805 (1998).

[34] G. N. Burlak, V. V. Grimal'skii, and Y. N. Taranenko, *Sov. Tech. Phys. Lett.* **32**, 259 (1986).

[35] J. Costes and C. Montes, *Phys. Rev. A* **34**, 3940 (1986).

[36] E. M. Dianov, B. Y. Zeldovich, A. Y. Karasik, and A. N. Pilipetskii, *Sov. J. Quantum Electron.* **19**, 1051 (1989).

[37] E. Lichtman, R. G. Waarts, and A. A. Friesem, *J. Lightwave Technol.* **7**, 171 (1989).

[38] A. Höök, A. Bolle, G. Grosso, and M. Martinelli, *Electron. Lett.* **26**, 470 (1990).

[39] A. Höök, *J. Opt. Soc. Am. B* **8**, 1284 (1991).

[40] A. Höök and A. Bolle, *J. Lightwave Technol.* **10**, 493 (1992).

[41] G. Grosso and A. Höök, *J. Opt. Soc. Am. B* **10**, 946 (1993).

[42] S. Rae, I. Bennion, and M. J. Carswell, *Opt. Commun.* **123**, 611 (1996).

[43] H. Li and K. Ogusu, *Jpn. J. Appl. Phys.* **38**, 6309 (1999).

[44] R. V. Johnson and J. H. Marburger, *Phys. Rev. A* **4**, 1175 (1971).

[45] I. Bar-Joseph, A. A. Friesem, E. Lichtman, and R. G. Waarts, *J. Opt. Soc. Am. B* **2**, 1606 (1985).

[46] B. Y. Zeldovich and V. V. Shkunov, *Sov. J. Quantum Electron.* **12**, 223 (1982).

[47] N. F. Andreev, V. I. Besapalov, A. M. Kiselev, G. A. Pasmanik, and A. A. Shilov, *Sov. Phys. JETP* **55**, 612 (1982).

[48] C. J. Randall and J. R. Albritton, *Phys. Rev. Lett.* **52**, 1887 (1984).

[49] P. Narum and R. W. Boyd, *IEEE J. Quantum Electron.* **QE-23**, 1216 (1987).

[50] P. Narum, A. L. Gaeta, M. D. Skeldon, and R. W. Boyd, *J. Opt. Soc. Am. B* **5**, 623 (1988).

[51] Y. Takushima and K. Kikuchi, *Opt. Lett.* **20**, 34 (1995).

[52] K. Ogusu, *J. Opt. Soc. Am. B* **17**, 769 (2000).

[53] R. G. Harrison, J. S. Uppal, A. Johnstone, and J. V. Moloney, *Phys. Rev. Lett.* **65**, 167 (1990).

[54] A. L. Gaeta and R. W. Boyd, *Phys. Rev. A* **44**, 3205 (1991).

[55] M. Dämmig, C. Boden, and F. Mitschke, *Appl. Phys. B* **55**, 121 (1992).

[56] A. Johnstone, W. Lu, J. S. Uppal, and R. G. Harrison, *Opt. Commun.* **81**, 122 (1991).

[57] W. Lu and R. G. Harrison, *Europhys. Lett.* **16**, 655 (1991).

[58] W. Lu and, A. Johnstone, and R. G. Harrison, *Phys. Rev. A* **46**, 4114 (1992).

[59] A. L. Gaeta and R. W. Boyd, *Int. J. Nonlinear Opt. Phys.* **1**, 581 (1992).

[60] C. Chow and A. Bers, *Phys. Rev. A* **47**, 5144 (1993).

[61] R. G. Harrison, W. Lu, D. S. Lim, D. Yu, and P. M. Ripley, *Proc. SPIE* **2039**, 91 (1993).

[62] M. Dämmig, G. Zimmer, F. Mitschke, and H. Welling, *Phys. Rev. A* **48**, 3301 (1993).

[63] R. G. Harrison, P. M. Ripley, and W. Lu, *Phys. Rev. A* **49**, R24 (1994).

[64] Y. Imai and H. Aso, *Opt. Rev.* **4**, 476 (1997).

[65] E. Picholle, C. Montes, C. Leycuras, O. Legrand, and J. Botineau, *Phys. Rev. Lett.* **66**, 1454 (1991).

[66] Y. N. Taranenko and L. G. Kazovsky, *IEEE Photon. Technol. Lett.* **4**, 494 (1992).

[67] C. Montes, A. Mikhailov, A. Picozii, and F. Ginovart, *Phys. Rev. E* **55**, 1086 (1997).

[68] K. O. Hill, B. S. Kawasaki, and D. C. Johnson, *Appl. Phys. Lett.* **28**, 608 (1976).

[69] K. O. Hill, D. C. Johnson, and B. S. Kawasaki, *Appl. Phys. Lett.* **29**, 185 (1976).

[70] B. S. Kawasaki, D. C. Johnson, Y. Fujii, and K. O Hill, *Appl. Phys. Lett.* **32**, 429 (1978).

[71] R. H. Stolen, *IEEE J. Quantum Electron.* **QE-15**, 1157 (1979).

[72] D. R. Ponikvar and S. Ezekiel, *Opt. Lett.* **6**, 398 (1981).

[73] L. F. Stokes, M. Chodorow, and H. J. Shaw, *Opt. Lett.* **7**, 509 (1982).

[74] P. Bayvel and I. P. Giles, *Electron. Lett.* **25**, 260 (1989); *Opt. Lett.* **14**, 581 (1989).

[75] E. M. Dianov, S. K. Isaev, L. S. Kornienko, V. V. Firsov, and Y. P. Yatsenko, *Sov. J. Quantum Electron.* **19**, 1 (1989).

[76] S. P. Smith, F. Zarinetchi, and S. Ezekiel, *Opt. Lett.* **16**, 393 (1991).

[77] F. Zarinetchi, S. P. Smith, and S. Ezekiel, *Opt. Lett.* **16**, 229 (1991).

[78] S. Huang, K. Toyama, P.-A. Nicati, L. Thévenaz, B. Y. Kim, and H. J. Shaw, *Proc. SPIE* **1795**, 48 (1993).

[79] S. Huang, L. Thévenaz, K. Toyama, B. Y. Kim, and H. J. Shaw, *IEEE Photon. Technol. Lett.* **5**, 365 (1993).

[80] T. P. Mirtchev and N. I. Minkovski, *IEEE Photon. Technol. Lett.* **5**, 158 (1993).

[81] P.-A. Nicati, K. Toyama, S. Huang, and H. J. Shaw, *Opt. Lett.* **18**, 2123 (1993); *IEEE Photon. Technol. Lett.* **6**, 801 (1994).

[82] C. Montes, A. Mahmhoud, and E. Picholle, *Phys. Rev. A* **49**, 1344 (1994).

[83] S. Randoux, V. Lecoueche, B. Ségrad, and J. Zemmouri, *Phys. Rev. A* **51**, R4345 (1995); *Phys. Rev. A* **52**, 221 (1995).

[84] D. S. Lim, H. K. Lee, K. H. Kim, S. B. Kang, J. T. Ahn, and M. Y. Jeon, *Opt. Lett.* **23**, 1671 (1998).

[85] S. Randoux and J. Zemmouri, *Phys. Rev. A* **59**, 1644 (1999).

[86] C. Montes, D. Bahloul, I. Bongrand, J. Botineau, G. Cheval, A. Mahmhoud, E. Picholle, and A. Picozzi, *J. Opt. Soc. Am. B* **16**, 932 (1999).

[87] N. A. Olsson and J. P. van der Ziel, *Appl. Phys. Lett.* **48**, 1329 (1986).

[88] N. A. Olsson and J. P. van der Ziel, *Electron. Lett.* **22**, 488 (1986).

[89] C. G. Atkins, D. Cotter, D. W. Smith, and R. Wyatt, *Electron. Lett.* **22**, 556 (1986).

[90] D. Cotter, D. W. Smith, C. G. Atkins, and R. Wyatt, *Electron. Lett.* **22**, 671 (1986).

[91] A. R. Chraplyvy and R. W. Tkach, *Electron. Lett.* **22**, 1084 (1986).

[92] N. A. Olsson and J. P. van der Ziel, *J. Lightwave Technol.* **LT-5**, 147 (1987).

[93] R. G. Waarts, A. A. Friesem, and Y. Hefetz, *Opt. Lett.* **13**, 152 (1988).

[94] R. W. Tkach, A. R. Chraplyvy, R. M. Derosier, and H. T. Shang, *Electron. Lett.* **24**, 260 (1988); *IEEE Photon. Technol. Lett.* **1**, 111 (1989).

[95] A. A. Fotiadi, E. A. Kuzin, M. P. Petrov, and A. A. Ganichev, *Sov. Tech. Phys. Lett.* **15**, 434 (1989).

[96] R. W. Tkach and A. R. Chraplyvy, *Opt. Quantum Electron.* **21**, S105 (1989).

[97] M. Tsubokawa and Y. Sasaki, *J. Opt. Commun.* **10**, 42 (1989).

[98] A. S. Siddiqui and G. G. Vienne, *J. Opt. Commun.* **13**, 33 (1992).

[99] G. P. Agrawal, *Fiber-Optic Communication Systems*, 2nd ed. (Wiley, New York, 1997).

[100] G. P. Agrawal and N. K. Dutta, *Semiconductor Lasers*, 2nd ed. (Van Nostrand Reinhold, New York, 1993).

[101] J. A. Arnaud, *IEEE J. Quantum Electron.* **QE-4**, 893 (1968).

[102] C. Culverhouse, F. Farahi, C. N. Pannell, and D. A. Jackson, *Electron. Lett.* **25**, 914 (1989).

[103] T. Kurashima, T. Horiguchi, and M. Tateda, *Opt. Lett.* **15**, 1038 (1990).

[104] X. Bao, D. J. Webb, and D. A. Jackson, *Opt. Lett.* **18**, 552 (1993); *Opt. Lett.* **18**, 1561 (1993).

[105] X. Bao, D. J. Webb, and D. A. Jackson, *Opt. Commun.* **104**, 298 (1994); *Opt. Lett.* **19**, 141 (1994).

[106] A. H. Hartog, in *Optical Fiber Sensor Technology*, K. T. V. Grattan and B. T. Meggitt, Eds. (Chapman & Hall, New York, 1995).

[107] T. R. Parker, M. Farhadiroushan, R. Feced, V. A. Handerek, and A. J. Rogers, *IEEE J. Quantum Electron.* **34**, 645 (1998).

[108] L. Thévenaz, M. Nikles, A. Fellay, N. Facchini, and P. Robert, *Proc. SPIE* **3407**, 374 (1998).

[109] J. Smith, A. Brown, M. DeMerchant, and X. Bao, *Proc. SPIE* **3670**, 366 (1999).

[110] H. H. Kee, G. P. Lees, and T. P. Newson, *Opt. Lett.* **25**, 695 (2000).

Chapter 10

Parametric Processes

In the stimulated scattering processes covered in Chapters 8 and 9, optical fibers play an active role in the sense that the process depends on molecular vibrations or density variations of silica. In a separate class of nonlinear phenomena, optical fibers play a passive role except for mediating interaction among several optical waves. Such nonlinear processes are referred to as parametric processes because they involve modulation of a medium parameter such as the refractive index. Among others, parametric processes include four-wave mixing (FWM) and harmonic generation, both of which are described in this chapter. In Section 10.1 we consider the origin of FWM and discuss its theory in Section 10.2, with emphasis on the parametric gain. The techniques used for phase-matching are covered in Section 10.3. In Section 10.4 we focus on parametric amplification; other applications of FWM are covered in Section 10.5. The last section is devoted to second-harmonic generation in optical fibers.

10.1 Origin of Four-Wave Mixing

The origin of parametric processes lies in the nonlinear response of bound electrons of a material to an applied optical field. More specifically, the polarization induced in the medium is not linear in the applied field but contains nonlinear terms whose magnitude is governed by the nonlinear susceptibilities [1]–[5]. The parametric processes can be classified as second- or third-order processes depending on whether the second-order susceptibility $\chi^{(2)}$ or the third-order susceptibility $\chi^{(3)}$ is responsible for them. The second-order susceptibility $\chi^{(2)}$ vanishes for an isotropic medium in the dipole approximation. For this reason, the second-order parametric processes, such as second-harmonic generation and sum-frequency generation, should not occur in silica fibers. In practice, these processes do occur because of quadrupole and magnetic-dipole effects, but with a relatively low conversion efficiency. Unexpectedly high conversion efficiencies ($\sim 1\%$) for second-harmonic generation have been observed in optical fibers under specific conditions. This topic is discussed in Section 10.6.

The third-order parametric processes involve, in general, nonlinear interaction among four optical waves and include the phenomena such as third-harmonic generation, FWM, and parametric amplification [1]–[5]. FWM in optical fibers has been studied extensively because it can be quite efficient for generating new waves [6]–[57]. Its main features can be understood by considering the third-order polarization term in Eq. (1.3.1) given as

$$\mathbf{P}_{NL} = \varepsilon_0 \chi^{(3)} \vdots \mathbf{EEE}, \qquad (10.1.1)$$

where $\mathbf{E}$ is the electric field, $\mathbf{P}_{NL}$ is the induced nonlinear polarization, and ε_0 is the vacuum permittivity.

Consider four optical waves oscillating at frequencies ω_1, ω_2, ω_3, and ω_4 and linearly polarized along the same axis x. The total electric field can be written as

$$\mathbf{E} = \frac{1}{2}\hat{x} \sum_{j=1}^{4} E_j \exp[i(k_j z - \omega_j t)] + \text{c.c.}, \qquad (10.1.2)$$

where the propagation constant $k_j = n_j \omega_j / c$, n_j is the refractive index, and all four waves are assumed to be propagating in the same direction. If we substitute Eq. (10.1.2) in Eq. (10.1.1) and express $\mathbf{P}_{NL}$ in the same form as $\mathbf{E}$ using

$$\mathbf{P}_{NL} = \frac{1}{2}\hat{x} \sum_{j=1}^{4} P_j \exp[i(k_j z - \omega_j t)] + \text{c.c.}, \qquad (10.1.3)$$

we find that P_j (j =1 to 4) consists of a large number of terms involving the products of three electric fields. For example, P_4 can be expressed as

$$P_4 = \frac{3\varepsilon_0}{4} \chi_{xxxx}^{(3)} [|E_4|^2 E_4 + 2(|E_1|^2 + |E_2|^2 + |E_3|^2)E_4$$
$$+ 2E_1 E_2 E_3 \exp(i\theta_+) + 2E_1 E_2 E_3^* \exp(i\theta_-) + \cdots], \qquad (10.1.4)$$

where θ_+ and θ_- are defined as

$$\theta_+ = (k_1 + k_2 + k_3 - k_4)z - (\omega_1 + \omega_2 + \omega_3 - \omega_4)t, \qquad (10.1.5)$$
$$\theta_- = (k_1 + k_2 - k_3 - k_4)z - (\omega_1 + \omega_2 - \omega_3 - \omega_4)t. \qquad (10.1.6)$$

The first four terms containing E_4 in Eq. (10.1.4) are responsible for the SPM and XPM effects. The remaining terms result from FWM. How many of these are effective in producing a parametric coupling depends on the phase mismatch between E_4 and P_4 governed by θ_+, θ_-, or a similar quantity.

Significant FWM occurs only if the phase mismatch nearly vanishes. This requires matching of the frequencies as well as of the wave vectors. The latter requirement is often referred to as phase matching. In quantum-mechanical terms, FWM occurs when photons from one or more waves are annihilated and new photons are created at different frequencies such that the net energy and momentum are conserved during the parametric interaction. The main difference between the parametric processes and the stimulated scattering processes discussed in Chapters 8 and 9 is that the phase-matching condition is automatically satisfied in the case of stimulated Raman or Brillouin scattering as a result of the active participation of the nonlinear medium. By contrast, the phase-matching condition requires a specific choice of the frequencies and the refractive indices for parametric processes to occur.

There are two types of FWM terms in Eq. (10.1.4). The term containing θ_+ corresponds to the case in which three photons transfer their energy to a single photon at the frequency $\omega_4 = \omega_1 + \omega_2 + \omega_3$. This term is responsible for the phenomena such as third-harmonic generation ($\omega_1 = \omega_2 = \omega_3$), or frequency conversion when $\omega_1 = \omega_2 \neq \omega_3$. In general, it is difficult to satisfy the phase-matching condition for such processes to occur in optical fibers with high efficiencies. The term containing θ_- in Eq. (10.1.4) corresponds to the case in which two photons at frequencies ω_1 and ω_2 are annihilated with simultaneous creation of two photons at frequencies ω_3 and ω_4 such that

$$\omega_3 + \omega_4 = \omega_1 + \omega_2. \qquad (10.1.7)$$

The phase-matching requirement for this process to occur is

$$\Delta k = k_3 + k_4 - k_1 - k_2$$
$$= (n_3\omega_3 + n_4\omega_4 - n_1\omega_1 - n_2\omega_2)/c = 0. \qquad (10.1.8)$$

It is relatively easy to satisfy $\Delta k = 0$ in the specific case $\omega_1 = \omega_2$. This partially degenerate case is most relevant for optical fibers. Physically, it manifests in a way similar to SRS. A strong pump wave at ω_1 creates two sidebands located symmetrically at frequencies ω_3 and ω_4 with a frequency shift

$$\Omega_s = \omega_1 - \omega_3 = \omega_4 - \omega_1, \qquad (10.1.9)$$

where we assumed for definiteness $\omega_3 < \omega_4$. The low-frequency sideband at ω_3 and the high-frequency sideband at ω_4 are referred to as the Stokes and anti-Stokes bands in direct analogy with SRS. The partially degenerate FWM was originally called three-wave mixing as only three distinct frequencies are involved in the nonlinear process [6]. In this chapter, the term three-wave mixing is reserved for the processes mediated by $\chi^{(2)}$. The name four-photon mixing is also used for FWM synonymously [7]. Note also that the Stokes and anti-Stokes bands are often called the signal and idler bands, borrowing the terminology from the field of microwaves, when an input signal at ω_3 is amplified through the process of FWM.

10.2 Theory of Four-Wave Mixing

Four-wave mixing transfers energy from a strong pump wave to two waves, upshifted and downshifted in frequency from the pump frequency ω_1 by an amount Ω_s given in Eq. (10.1.9). If only the pump wave is incident at the fiber, and the phase-matching condition is satisfied, the Stokes and anti-Stokes waves at the frequencies ω_3 and ω_4 can be generated from noise, similarly to the stimulated scattering processes discussed in Chapters 8 and 9. On the other hand, if a weak signal at ω_3 is also launched into the fiber together with the pump, the signal is amplified while a new wave at ω_4 is generated simultaneously. The gain responsible for such amplification is called the parametric gain. In this section, we consider the FWM mixing process in detail and drive an expression for the parametric gain. The nondegenerate case ($\omega_1 \neq \omega_2$) is considered for generality.

10.2.1 Coupled Amplitude Equations

The starting point is, as usual, the wave equation (2.3.1) for the total electric field $\mathbf{E}(\mathbf{r},t)$ with $\mathbf{P}_{NL}$ given in Eq. (10.1.1). We substitute Eqs. (10.1.2) and (10.1.3) in the wave equation, together with a similar expression for the linear part of the polarization, and neglect the time dependence of the field components E_j ($j = 1$ to 4) assuming quasi-CW conditions. Their spatial dependence is, however, included using

$$E_j(\mathbf{r}) = F_j(x,y)A_j(z), \tag{10.2.1}$$

where $F_j(x,y)$ is the spatial distribution of the fiber mode in which the jth field propagates inside the fiber [12]. Evolution of the amplitude $A_j(z)$ inside a multimode fiber is governed by a set of four coupled equations which, in the paraxial approximation, can be written as

$$\frac{dA_1}{dz} = \frac{in_2'\omega_1}{c}[(f_{11}|A_1|^2 + 2\sum_{k \neq 1} f_{1k}|A_k|^2)A_1 + 2f_{1234}A_2^*A_3A_4e^{i\Delta kz}], \tag{10.2.2}$$

$$\frac{dA_2}{dz} = \frac{in_2'\omega_2}{c}[(f_{22}|A_2|^2 + 2\sum_{k \neq 2} f_{2k}|A_k|^2)A_2 + 2f_{2134}A_1^*A_3A_4e^{i\Delta kz}], \tag{10.2.3}$$

$$\frac{dA_3}{dz} = \frac{in_2'\omega_3}{c}[(f_{33}|A_3|^2 + 2\sum_{k \neq 3} f_{3k}|A_k|^2)A_3 + 2f_{3412}A_1A_2A_4^*e^{-i\Delta kz}], \tag{10.2.4}$$

$$\frac{dA_4}{dz} = \frac{in_2'\omega_4}{c}[(f_{44}|A_4|^2 + 2\sum_{k \neq 4} f_{4k}|A_k|^2)A_4 + 2f_{4312}A_1A_2A_3^*e^{-i\Delta kz}], \tag{10.2.5}$$

where the wave-vector mismatch Δk is given by [see Eq. (10.1.8)]

$$\Delta k = (\tilde{n}_3\omega_3 + \tilde{n}_4\omega_4 - \tilde{n}_1\omega_1 - \tilde{n}_2\omega_2)/c. \tag{10.2.6}$$

The refractive indices $\tilde{n}_1$ to $\tilde{n}_4$ stand for the effective indices of the fiber modes. Note that $\tilde{n}_1$ and $\tilde{n}_2$ can differ from each other when the pump waves A_1 and A_2 propagate in different fiber modes even if they are degenerate in frequencies. The overlap integral f_{jk} is defined in Eq. (7.1.14) of Section 7.1. The new overlap integral f_{ijkl} is given by [12]

$$f_{ijkl} = \frac{\langle F_i^* F_j^* F_k F_l \rangle}{[\langle|F_i|^2\rangle\langle|F_j|^2\rangle\langle|F_k|^2\rangle\langle|F_l|^2\rangle]^{1/2}}, \tag{10.2.7}$$

where angle brackets denote integration over the transverse coordinates x and y. In deriving Eqs. (10.2.2)–(10.2.5), we kept only nearly phase-matched terms [see Eq. (10.1.4)] and neglected frequency dependence of $\chi^{(3)}$. The parameter n_2' is the nonlinear parameter defined by Eq. (2.3.13) [the prime distinguishes it from n_2 appearing in Eq. (10.1.8)].

10.2.2   Approximate Solution

Equations (10.2.2)–(10.2.5) are quite general in the sense that they include the effects of SPM, XPM, and pump depletion on the FWM process; a numerical approach is necessary to solve them exactly. Considerable physical insight is gained if the pump waves are assumed to be much intense compared with the Stokes and anti-Stokes waves and to remain undepleted during the parametric interaction. As a further simplification, we assume that all overlap integrals are nearly the same, that is,

$$f_{ijkl} \approx f_{ij} \approx 1/A_{\text{eff}} \quad (i,j = 1,2,3,4), \qquad (10.2.8)$$

where A_{eff} is the effective core area introduced in Section 2.3. This assumption is valid for single-mode fibers. The analysis can easily be extended to include differences in the overlap integrals [12].

We can now introduce the nonlinear parameter γ_j using the definition

$$\gamma_j = n_2' \omega_j/(cA_{\text{eff}}) \approx \gamma, \qquad (10.2.9)$$

where γ is an average value if we ignore relatively small differences in optical frequencies of four waves. Equations (10.2.2) and (10.2.3) for the pump fields are easily solved to obtain

$$A_1(z) = \sqrt{P_1} \exp[i\gamma(P_1 + 2P_2)z], \qquad (10.2.10)$$
$$A_2(z) = \sqrt{P_2} \exp[i\gamma(P_2 + 2P_1)z], \qquad (10.2.11)$$

where $P_j = |A_j(0)|^2$, and P_1 and P_2 are the incident pump powers at $z = 0$. This solution shows that, in the undepleted-pump approximation, the pump waves only acquire a phase shift occurring as a result of SPM and XPM.

Substituting Eqs. (10.2.10) and (10.2.11) in Eqs. (10.2.4) and (10.2.5), we obtain two linear coupled equations for the signal and idler fields:

$$\frac{dA_3}{dz} = 2i\gamma[(P_1 + P_2)A_3 + \sqrt{P_1P_2}e^{-i\theta}A_4^*], \qquad (10.2.12)$$

$$\frac{dA_4^*}{dz} = -2i\gamma[(P_1 + P_2)A_4^* + \sqrt{P_1P_2}e^{i\theta}A_3], \qquad (10.2.13)$$

where

$$\theta = [\Delta k - 3\gamma(P_1 + P_2)]z. \qquad (10.2.14)$$

To solve these equations, we introduce

$$B_j = A_j \exp[-2i\gamma(P_1 + P_2)z], \quad (j = 3, 4).$$ (10.2.15)

Using Eqs. (10.2.12)–(10.2.15), we then obtain

$$\frac{dB_3}{dz} = 2i\gamma\sqrt{P_1 P_2} \exp(-i\kappa z) B_4^*,$$ (10.2.16)

$$\frac{dB_4^*}{dz} = -2i\gamma\sqrt{P_1 P_2} \exp(i\kappa z) B_3,$$ (10.2.17)

where the net phase mismatch is given by

$$\kappa = \Delta k + \gamma(P_1 + P_2).$$ (10.2.18)

Equations (10.2.16) and (10.2.17) govern growth of the signal and idler waves occurring as a result of FWM. Their general solution is of the form [12]

$$B_3(z) = (a_3 e^{gz} + b_3 e^{-gz}) \exp(-i\kappa z/2),$$ (10.2.19)
$$B_4^*(z) = (a_4 e^{gz} + b_4 e^{-gz}) \exp(i\kappa z/2),$$ (10.2.20)

where a_3, b_3, a_4, and b_4 are determined from the boundary conditions. The parametric gain g depends on the pump power and is defined as

$$g = \sqrt{(\gamma P_0 r)^2 - (\kappa/2)^2},$$ (10.2.21)

where we have introduced the parameters r and P_0 as

$$r = 2(P_1 P_2)^{1/2}/P_0, \qquad P_0 = P_1 + P_2.$$ (10.2.22)

The solution given by Eqs. (10.2.19) and (10.2.20) is valid only when the conversion efficiency of the FWM process is relatively small so that the pump waves remain largely undepleted. Pump depletion can be included by solving the complete set of four equations, Eqs. (10.2.2)–(10.2.5). Such a solution can be obtained in terms of elliptic functions [29] but is not discussed here because of its complexity.

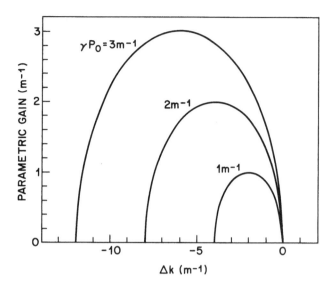

Figure 10.1 Variation of parametric gain with wave-vector mismatch Δk for several pump powers P_0. The shift of the gain peak from $\Delta k = 0$ is due to a combination of the SPM and XPM effects.

10.2.3 Effect of Phase Matching

The derivation of the parametric gain has assumed that the two pump waves are distinct. If the pump fields were indistinguishable on the basis of their frequency, polarization, or spatial mode, the preceding procedure should be carried out with only three terms in Eq. (10.1.2). The parametric gain is still given by Eq. (10.2.21) if we choose $P_1 = P_2 = P_0$ and replace κ with

$$\kappa = \Delta k + 2\gamma P_0. \tag{10.2.23}$$

Figure 10.1 shows variations of g with Δk in this specific case for several values of γP_0. The maximum gain ($g_{max} = \gamma P_0$) occurs at $\kappa = 0$, or at $\Delta k = -2\gamma P_0$. The range over which the gain exists is given by $0 > \Delta k > -4\gamma P_0$. All of these features can be understood from Eqs. (10.2.21) and (10.2.23). The shift of the gain peak from $\Delta k = 0$ is due to the contribution of SPM and XPM to the phase mismatch as apparent from Eq. (10.2.23).

It is interesting to compare the peak value of the parametric gain with that of the Raman gain [7]. From Eq. (10.2.21) the maximum gain is given by (assuming $r = 1$)

$$g_{max} = \gamma P_0 = g_P(P_0/A_{eff}), \tag{10.2.24}$$

where γ is used from Eq. (10.2.9) and g_P is defined as $g_P = 2\pi n_2'/\lambda_1$ at the pump wavelength λ_1. Using $\lambda_1 = 1\ \mu$m and $n_2' \approx 3 \times 10^{-20}$ m^2/W, we obtain $g_P \approx 2 \times 10^{-13}$ m/W. This value should be compared with the peak value of the Raman gain g_R in Fig. 8.1. The parametric gain is larger by about a factor of 2 compared with g_R. As a result, the threshold pump power for the FWM process is expected to be lower than the Raman threshold if phase matching is achieved. In practice, however, SRS dominates for long fibers. This is so because it is difficult to maintain phase matching over long fiber lengths as a result of variations in the core diameter.

One can define a length scale, known as the coherence length, by using

$$L_{\mathrm{coh}} = 2\pi/|\kappa|, \tag{10.2.25}$$

where $\Delta\kappa$ is the maximum value of the wave-vector mismatch that can be tolerated. Significant FWM occurs if $L < L_{\mathrm{coh}}$. Even when this condition is satisfied, SRS can influence the FWM process significantly when the frequency shift Ω_s lies within the Raman-gain bandwidth (see Chapter 8). The interplay between SRS and FWM has been studied extensively [36]–[41]. The main effect in practice is that the Stokes component gets amplified through SRS, resulting in an asymmetric sideband spectrum. This feature is discussed further in the next section where experimental results are presented.

10.2.4 Ultrafast FWM

The simplified analysis of this section is based on Eqs. (10.2.12) and (10.2.13) which assume, among other things, CW or quasi-CW conditions so that group-velocity dispersion (GVD) can be neglected. The effects of both GVD and fiber losses can be included following the analysis of Section 2.3 and allowing $A_j(z)$ in Eq. (10.2.1) for $j = 1$ to 4 to be a slowly varying function of time. If polarization effects are neglected assuming that all four waves are polarized along a principle axis of a birefringent fiber, the inclusion of GVD effects in Eqs. (10.2.3)–(10.2.6) amounts to replacing the derivative dA_j/dz with

$$\frac{dA_j}{dz} \rightarrow \frac{\partial A_j}{\partial z} + \beta_{1j}\frac{\partial A_j}{\partial t} + \frac{i}{2}\beta_{2j}\frac{\partial^2 A_j}{\partial t^2} + \frac{1}{2}\alpha_j A_j \tag{10.2.26}$$

for all four waves ($j = 1$–4) in analogy with Eq. (2.3.27). The resulting four coupled nonlinear Schrödinger (NLS) equations describe FWM of picosecond optical pulses and include the effects of GVD, SPM, and XPM. It is difficult to solve the four coupled NLS equations analytically under general conditions, and a numerical approach is used in practice. The group velocity of four pulses participating in the FWM process can be quite different. As a result, efficient FWM requires not only phase matching but also matching of the group velocities.

A natural question is whether the four coupled NLS equations have solutions in the form of optical solitons that support each other in the same way as the nonlinear phenomenon of XPM allows pairing of two solitons. Such solitons do exist for specific combination of parameters and are sometimes called parametric or FWM solitons. They have been investigated for both three- and four-wave interactions [58]–[61]. As an example, if we assume that the four waves satisfy both the phase-matching and group-velocity-matching conditions and, at the same time $|\beta_2|$ is the same for all waves, a solitary-wave solution in the form of two bright and two dark solitons has been found for a specific choice of the signs of GVD parameters [60].

In the case of intense CW pumping and a relatively small conversion efficiency so that the pump remains nearly undepleted, the pump equations can be solved analytically. Assuming that a single pump beam of power P_0 is incident at $z = 0$, the signal and idler fields are found to satisfy the following set of two coupled NLS equations:

$$
\frac{\partial A_3}{\partial z} + \beta_{13}\frac{\partial A_3}{\partial t} + \frac{i}{2}\beta_{23}\frac{\partial^2 A_3}{\partial t^2} + \frac{1}{2}\alpha_3 A_3
$$
$$
= i\gamma(|A_3|^2 + 2|A_4|^2 + 2P_0)A_3 + i\gamma P_0 A_4^* e^{-i\theta}, \qquad (10.2.27)
$$
$$
\frac{\partial A_4}{\partial z} + \beta_{14}\frac{\partial A_4}{\partial t} + \frac{i}{2}\beta_{24}\frac{\partial^2 A_4}{\partial t^2} + \frac{1}{2}\alpha_4 A_4
$$
$$
= i\gamma(|A_4|^2 + 2|A_3|^2 + 2P_0)A_4 + i\gamma P_0 A_3^* e^{i\theta}, \qquad (10.2.28)
$$

where the net phase mismatch $\theta = \Delta k + 2\gamma P_0 z$ takes into account the SPM-induced phase shift of the pump. Numerical results show that these equations can support "symbiotic" soliton pairs [62], similar to those discussed in Section 7.3, if the pump wavelength nearly coincides with the zero-dispersion wavelength of the fiber and the signal and idler wavelengths are equally spaced from it such that $\beta_{13} = \beta_{14}$ and $\beta_{23} = -\beta_{24}$ (opposite GVD but same group velocities). Such solitons require a balance between the parametric gain and fiber losses, similar to the case of Brillouin solitons, and are referred to as dissipative solitons. Both members of the soliton pair are bright solitons even though one pulse travels in the normal-GVD region of the fiber.

The use of multiple NLS equations is necessary when carrier frequencies of four pulses are widely separated (>10 THz). In the case of smaller frequency spacings (<1 THz), it is more practical to use a single NLS equation of the form given in Eq. (2.3.33) or (2.3.39) and solve it with an initial amplitude of the form

$$A(0,t) = A_1(0,t) + A_3(0,t) \exp(-i\Omega_s t) + A_4(0,t) \exp(i\Omega_s t), \qquad (10.2.29)$$

where two pump waves are assumed to be degenerate in frequency (a common case in practice) and Ω_s is the frequency shift given by Eq. (10.1.9). Such an approach includes SPM, XPM and FWM effects automatically and is routinely used for modeling of WDM lightwave systems. The only requirement is that the time step used in numerical simulations should be much shorter than $2\pi/\Omega_s$. This approach also permits inclusion of the Raman and birefringence effects and provides a unified treatment of various nonlinear phenomena for pulses propagating inside optical fibers [63].

10.3 Phase-Matching Techniques

The parametric gain responsible for FWM peaks when the phase mismatch $\kappa = 0$, where κ is given by Eq. (10.2.18). This section discusses several different methods used for realizing phase matching in practice.

10.3.1 Physical Mechanisms

The phase-matching condition $\kappa = 0$ can be written in the form

$$\kappa = \Delta k_M + \Delta k_W + \Delta k_{NL} = 0, \qquad (10.3.1)$$

where Δk_M, Δk_W, and Δk_{NL} represent the mismatch occurring as a result of material dispersion, waveguide dispersion, and the nonlinear effects, respectively. The contributions Δk_M and Δk_W can be obtained from Eq. (10.2.6) if the effective indices are written as

$$\tilde{n}_j = n_j + \Delta n_j, \qquad (10.3.2)$$

where Δn_j is the change in the material index n_j due to waveguiding. In the partially degenerate case ($\omega_1 = \omega_2$), the three contributions in Eq. (10.3.1) are

$$\Delta k_M = [n_3 \omega_3 + n_4 \omega_4 - 2n_1 \omega_1]/c, \qquad (10.3.3)$$

$$\Delta k_W = [\Delta n_3 \omega_3 + \Delta n_4 \omega_4 - (\Delta n_1 + \Delta n_2)\omega_1]/c, \qquad (10.3.4)$$

$$\Delta k_{NL} = \gamma(P_1 + P_2). \qquad (10.3.5)$$

To realize phase matching, at least one of them should be negative.

The material contribution Δk_M can be expressed in terms of the frequency shift Ω_s [see Eq. (10.1.9)] if we use the expansion (2.3.23) and note that $\beta_j = n_j \omega_j/c$ ($j = 1$–4). Retaining up to terms quadratic in Ω_s in this expansion,

$$\Delta k_M \approx \beta_2 \Omega_s^2, \qquad (10.3.6)$$

where β_2 is the GVD coefficient at the pump frequency ω_1. Equation (10.3.6) is valid if the pump wavelength ($\lambda_1 = 2\pi c/\omega_1$) is not too close to the zero-dispersion wavelength λ_D of the fiber. As $\beta_2 > 0$ for $\lambda_1 < \lambda_D$, Δk_M is positive in the visible or near-infrared region. Phase matching for $\lambda_1 < 1.3$ μm can be realized making Δk_W negative by propagating different waves in different modes of a multimode fiber. Most of the early experiments used this method of phase matching [6]–[11].

In the case of a single-mode fiber, $\Delta k_W = 0$ because Δn is nearly the same for all waves. Three techniques can be used to achieve phase matching in single-mode fibers. If the pump wavelength exceeds λ_D, Δk_M becomes negative. This allows us to achieve phase matching for λ_1 in the vicinity of λ_D. For $\lambda_1 > \lambda_D$, phase matching can also be obtained by adjusting Δk_{NL} through the pump power. For $\lambda_1 < \lambda_D$, modal birefringence in polarization-preserving fibers makes it possible to achieve phase matching by polarizing different waves differently with respect to a principal axis of the fiber. All of these techniques are discussed in this section.

10.3.2 Phase Matching in Multimode Fibers

Multimode fibers allow phase matching when the waveguide contribution Δk_W is negative and exactly compensates the positive contribution $\Delta k_M + \Delta k_{NL}$ in Eq. (10.3.1). The magnitude of Δk_W depends on the choice of fiber modes in which four waves participating in the FWM process propagate. The eigenvalue equation (2.2.9) of Section 2.2 can be used to calculate Δn_j ($j = 1$–4) for each mode. Equation (10.3.4) is then used to calculate Δk_W.

Figure 10.2 shows the calculated value of Δk_W as a function of the frequency shift ($v_s = \Omega_s/2\pi$) for a fiber with 5-μm-core radius and a core-cladding index difference of 0.006. The dashed line shows the quadratic variation of

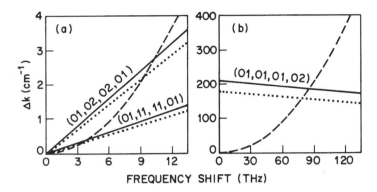

Figure 10.2 Phase-matching diagrams for (a) mixed-mode and (b) single-mode pump propagation. Solid and dashed lines show variations of $|\Delta k_W|$ and Δk_M with frequency shift. Dotted lines illustrate the effect of increasing core radius by 10%. Fiber modes are indicated using the LP_{mn} terminology. (After Ref. [7] ©1975 IEEE.)

Δk_M from Eq. (10.3.6). The frequency shift v_s is determined by the intersection of the solid and dashed curves (assuming that Δk_{NL} is negligible). Two cases are shown in Fig. 10.2 corresponding to whether the pump wave propagates with its power divided in two different fiber modes or whether it propagates in a single fiber mode. In the former case, frequency shifts are in the range 1–10 THz while in the latter case $v_s \sim 100$ THz. The exact value of frequency shifts is sensitive to several fiber parameters. The dotted lines in Fig. 10.2 show how v_s changes with a 10% increase in the core radius. In general, the phase-matching condition can be satisfied for several combinations of the fiber modes.

In the 1974 demonstration of phase-matched FWM in silica fibers, pump pulses at 532 nm with peak powers ~ 100 W were launched in a 9-cm-long fiber, together with a CW signal (power ~ 10 mW) obtained from a dye laser tunable in the range of 565 to 640 nm [6]. FWM generated a new wave in the blue region ($\omega_4 = 2\omega_1 - \omega_3$), called the idler wave in the parametric-amplifier configuration used for the experiment. Figure 10.3 shows the observed idler spectrum obtained by varying the signal frequency ω_3. The five different peaks correspond to different combinations of fiber modes for which phase matching is achieved. Different far-field patterns for the two dominant peaks clearly indicate that the idler wave is generated in different fiber modes. In this experiment, the pump propagated in a single fiber mode. As expected from Fig. 10.2, phase-matching occurred for relatively large frequency shifts in the range of 50

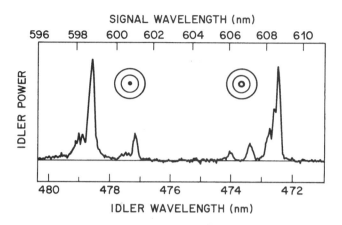

Figure 10.3 Idler power as a function of wavelength obtained by tuning the signal wavelength (upper scale). Far-field patterns corresponding to the two dominant peaks are also shown as an inset. (After Ref. [6].)

to 60 THz. In another experiment, the frequency shift was as large as 130 THz, a value that corresponds to 23% change in the pump frequency [11].

FWM with much smaller frequency shifts ($v_s = 1$–10 THz) can occur if the pump power is divided between two different fiber modes (see Fig. 10.2). This configuration is also relatively insensitive to variations in the core diameter [7] and results in coherence lengths ~ 10 m. For $v_s \sim 10$ THz, the Raman process can interfere with the FWM process as the generated Stokes line falls near the Raman-gain peak and can be amplified by SRS. In an experiment in which 532-nm pump pulses with peak powers ~ 100 W were transmitted through a multimode fiber, the Stokes line was always more intense than the anti-Stokes line as a result of Raman amplification [7].

When picosecond pump pulses are propagated through a multimode fiber, the FWM process is affected not only by SRS but also by SPM, XPM, and GVD. In a 1987 experiment [28], 25-ps pump pulses were transmitted through a 15-m-long fiber, supporting four modes at the pump wavelength of 532 nm. Figure 10.4 shows the observed spectra at the fiber output as the pump peak intensity is increased above the FWM threshold occurring near 500 MW/cm^2. Only the pump line is observed below threshold (trace a). Three pairs of Stokes and anti-Stokes lines with frequency shifts in the range 1–8 THz are observed just above threshold (trace b). All of these lines have nearly the same amplitude, indicating that SRS does not play a significant role at this pump power.

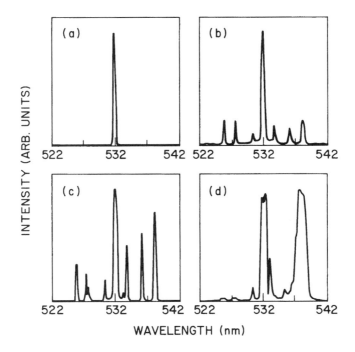

Figure 10.4 Spectra of 25-ps pump pulses at fiber output. The peak intensity is increased progressively beyond the FWM threshold (≈ 500 MW/cm^2) in going from (a) to (d). (After Ref. [28] ©1987 IEEE.)

As pump power is slightly increased, the Stokes lines become much more intense than the anti-Stokes lines as a result of Raman amplification (trace c). With a further increase in pump power, the Stokes line closest to the Raman-gain peak becomes as intense as the pump line itself whereas the anti-Stokes lines are nearly depleted (trace d). At the same time, the pump and the dominant Stokes line exhibit spectral broadening and splitting that are characteristic of SPM and XPM (see Section 7.4). As the pump power is increased further, higher-order Stokes lines are generated through cascaded SRS. At a pump intensity of 1.5 GW/cm^2, the broadened multiple Stokes lines merge, and a supercontinuum extending from 530 to 580 nm is generated as a result of the combined effects of SPM, XPM, SRS, and FWM. Figure 10.5 shows the spectra observed at the fiber output under such conditions. Under certain conditions, supercontinuum can extend over a wide range ($\sim$200 nm) [64]. Supercontinuum generation in single-mode fibers is discussed in Section 10.5.

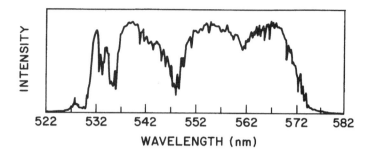

Figure 10.5   Supercontinuum observed when pump intensity is increased to 1.5 GW/cm². (After Ref. [28] ©1987 IEEE.)

10.3.3   Phase Matching in Single-Mode Fibers

In single-mode fibers the waveguide contribution Δk_W in Eq. (10.3.1) is very small compared with the material contribution Δk_M for identically polarized waves except near the zero-dispersion wavelength λ_D where the two become comparable. The three possibilities for approximate phase matching consist of: (i) reducing Δk_M and Δk_{NL} by using small frequency shifts and low pump powers; (ii) operating near the zero-dispersion wavelength so that Δk_W nearly cancels $\Delta k_M + \Delta k_{NL}$; and (iii) working in the anomalous GVD regime so that Δk_M is negative and can be cancelled by $\Delta k_{NL} + \Delta k_W$.

Nearly Phase-Matched Four-Wave Mixing

The gain spectrum shown in Fig. 10.1 indicates that significant FWM can occur even if phase matching is not perfect to make $\kappa = 0$ in Eq. (10.3.1). The amount of tolerable wave-vector mismatch depends on how long the fiber is compared with the coherence length L_{coh}. Assuming that the contribution Δk_M dominates in Eq. (10.3.1), the coherence length can be related to the frequency shift Ω_s by using Eqs. (10.2.25) and (10.3.6) and is given by

$$L_{coh} = \frac{2\pi}{|\Delta k_M|} = \frac{2\pi}{|\beta_2|\Omega_s^2}. \tag{10.3.7}$$

In the visible region, $\beta_2 \sim 50$ ps²/km, resulting in $L_{coh} \sim 1$ km for frequency shifts $v_s = \Omega_s/2\pi \sim 100$ GHz. Such large coherence lengths indicate that significant FWM can occur if the fiber length $L \leq L_{coh}$.

    In an early experiment, three CW waves with a frequency separation in the range 1–10 GHz were propagated through a 150-m-long fiber whose 4-μm

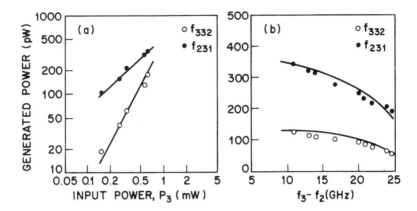

Figure 10.6 Variation of FWM-generated power at the output of a 3.5-km-long fiber with (a) input power P_3 and (b) frequency separation. (After Ref. [27] ©1987 IEEE.)

core diameter ensured single-mode operation near the argon-ion laser wavelength of 514.5 nm [8]. The FWM generated nine new frequencies such that $\omega_4 = \omega_i + \omega_j - \omega_k$, where $i, j, k = 1$, 2, or 3 with $j \neq k$. The experiment also showed that FWM can lead to spectral broadening whose magnitude increases with an increase in the incident power. The 3.9-GHz linewidth of the CW input from a multimode argon laser increased to 15.8 GHz at an input power of 1.63 W after passing through the fiber. The spectral components within the incident light generate new frequency components through FWM as the light propagates through the fiber. In fact, SPM-induced spectral broadening discussed in Section 4.1 can be interpreted in terms of such a FWM process [65].

From a practical standpoint FWM can lead to crosstalk in multichannel (WDM) communication systems where the channel spacing is typically in the range of 10 to 100 GHz. This issue attracted considerable attention during the 1990s because of the advent of WDM systems [42]–[47]. In an early experiment [27], three CW waves with a frequency separation ~10 GHz were propagated through a 3.5-km-long fiber and the amount of power generated in the nine frequency components was measured by varying the frequency separation and the input power levels. Figure 10.6 shows measured variations for two frequency components f_{332} and f_{231} using the notation

$$f_{ijk} = f_i + f_j - f_k, \qquad f_j = \omega_j/2\pi. \tag{10.3.8}$$

In the left-hand part, $f_3 - f_1 = 11$ GHz, $f_2 - f_1 = 17.2$ GHz, $P_1 = 0.43$ mW, and $P_2 = 0.14$ mW while P_3 was varied from 0.15 to 0.60 mW. In the right-

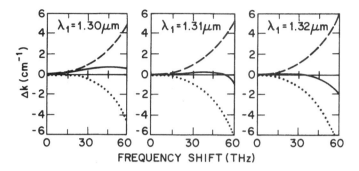

Figure 10.7 Phase-matching diagrams near the zero-dispersion wavelength for three values of the pump wavelength λ_1. Dotted, dashed, and solid lines show respectively Δk_M, Δk_W, and their sum. (After Ref. [15].)

hand part, $f_3 - f_2$ was varied from 10 to 25 GHz with $P_3 = 0.55$ mW while all other parameters were the same.

The generated power P_4 varies with P_3 linearly for the frequency component f_{231} but quadratically for the frequency component f_{332}. This is expected from the theory of Section 10.2 by noting that f_{231} results from nondegenerate pump waves but the pump waves are degenerate in frequency for f_{332}. More power is generated in the frequency component f_{231} because f_{231} and f_{321} are degenerate, and the measured power is the sum of powers generated through two FWM processes. Finally, P_4 decreases with increasing frequency separation because of a larger phase mismatch. A noteworthy feature of Fig. 10.6 is that up to 0.5 nW of power is generated for input powers < 1 mW. This can be a source of significant performance degradation in coherent communication systems [45]. Even in the case of direct detection, input channel powers should typically be kept below 1 mW to avoid degradation induced by FWM [35].

Phase Matching near the Zero-Dispersion Wavelength

The material contribution Δk_M to the wave-vector mismatch becomes quite small near the zero-dispersion wavelength of the fiber as it changes from positive to negative values around 1.28 μm. The waveguide contribution Δk_W depends on the fiber design, but is generally positive near 1.3 μm. In a limited range of pump wavelengths Δk_M can cancel $\Delta k_W + \Delta k_{NL}$ for specific values of frequency shifts ν_s. Figure 10.7 shows such cancelation, assuming Δk_{NL} to be negligible, for a fiber of 7-μm core diameter and a core-cladding index differ-

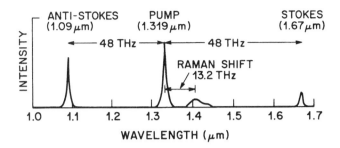

Figure 10.8 Output spectra showing Stokes and anti-Stokes bands generated by FWM. The Raman band is also shown. (After Ref. [15].)

ence of 0.006. The frequency shift depends on the pump wavelength λ_1 and can vary over a wide range 1–100 THz. It is also sensitive to the values of core diameter and index difference. These two parameters can be used to tailor the frequency shift at a given pump wavelength [16].

In a 1980 experiment on FWM near 1.3 μm, a 30-m-long fiber was used as a parametric amplifier, pumped by 1.319-μm pulses from a Q-switched Nd:YAG laser [14]. A signal at 1.338 μm ($v_s = 3.3$ THz) was amplified by up to 46 dB while, at the same time, three pairs of Stokes and anti-Stokes lines were observed at the fiber output. These equally spaced lines (spacing 3.3 THz) originate from a cascade of FWM processes in which successive lines interact with each other to generate new frequencies. In a later experiment, FWM occurred spontaneously without an input signal [15]. Mode-locked input pulses at 1.319 μm, with peak powers ~1 kW, were propagated through a 50-m-long fiber. Their peak power was large enough to exceed the Raman threshold. Figure 10.8 shows the spectrum observed at the fiber output. The Stokes and anti-Stokes lines at 1.67 and 1.09 μm, respectively, originate from FWM. The huge frequency shift ($v_s \approx 48$ THz) is comparable to that achieved in multimode fibers. Similar experiments show that v_s can be varied in the range 3–50 THz by changing the core diameter from 7.2–8.2 μm [16]. This scheme is useful for realizing new optical sources pumped by a 1.319-μm Nd:YAG laser.

Phase Matching due to Self-Phase Modulation

When the pump wavelength lies in the anomalous-GVD regime and deviates considerably from λ_D, Δk_M exceeds significantly from Δk_W and it becomes difficult to achieve phase matching (see Fig. 10.7). However, because $\Delta k_M + \Delta k_W$ is negative, it is possible to compensate it by the nonlinear contribution Δk_{NL} in Eq. (10.3.1). The frequency shift Ω_s in that case depends on the input pump power. In fact, if we use Eq. (10.2.23) with $\Delta k \approx \Delta k_M = \beta_2 \Omega_s^2$ from Eq. (10.3.6), phase-matching occurs ($\kappa = 0$) when

$$\Omega_s = (2\gamma P_0/|\beta_2|)^{1/2}, \tag{10.3.9}$$

where P_0 is the input pump power. Thus, a pump wave propagating in the anomalous-GVD regime would develop sidebands located at $\omega_1 \pm \Omega_s$ as a result of FWM that is phase-matched by the nonlinear process of self-phase modulation. This case has been discussed in Section 5.1 in the context of modulation instability. As was indicated there, modulation instability can be interpreted in terms of FWM in the frequency domain, whereas in the time domain it results from an unstable growth of weak perturbations from the steady state. In fact, the modulation frequency given by Eq. (5.1.10) is identical to Ω_s of Eq. (10.3.9). The output spectrum shown in Fig. 5.2 provides an experimental evidence of phase matching occurring as a result of self-phase modulation. The frequency shifts are in the range 1–10 THz for pump powers P_0 ranging from 1–100 W. This phenomenon has been used to convert the wavelength of femtosecond pulses from 1.5- to 1.3-μm spectral region [48].

10.3.4 Phase Matching in Birefringent Fibers

An important phase-matching technique in single-mode fibers takes advantage of the modal birefringence, resulting from different effective indices for the waves propagating with orthogonal polarizations. The index difference

$$\delta n = \Delta n_x - \Delta n_y, \tag{10.3.10}$$

where Δn_x and Δn_y represent changes in the refractive indices (from the material value) for the optical fields polarized along the slow and fast axes of the fiber, respectively. A complete description of the parametric gain in birefringent fibers should generalize the formalism of Section 10.2 by following an approach similar to that used in Section 6.1. Assuming that each of the four waves is polarized along the slow or the fast axis, one still obtains the parametric gain in the form of Eq. (10.2.21) with minor changes in the definitions of the parameters γ and κ. In particular, γ is reduced by a factor of 3 compared with the value obtained from Eq. (10.2.9) if the electronic contribution to $\chi^{(3)}$ dominates in Eq. (6.1.5). The wave-vector mismatch κ still has three contributions as in Eq. (10.3.1). However, the waveguide contribution Δk_W is now dominated by δn. The nonlinear contribution Δk_{NL} is also different than that given by Eq. (10.3.5). In the following discussion Δk_{NL} is assumed to be negligible compared with Δk_M and Δk_W.

As before, phase matching occurs when Δk_M and Δk_W cancel each other. Both of them can be positive or negative. For $\lambda_1 < \lambda_D$, a range that covers the visible region, Δk_M is positive because β_2 is positive in Eq. (10.3.6). The waveguide contribution Δk_W can be made negative if the pump wave is polarized along the slow axis, while the Stokes and anti-Stokes waves are polarized along the fast axis. This can be seen from Eq. (10.3.4). As $\Delta n_3 = \Delta n_4 = \Delta n_y$ and $\Delta n_1 = \Delta n_2 = \Delta n_x$, the waveguide contribution becomes

$$\Delta k_W = [\Delta n_y(\omega_3 + \omega_4) - 2\Delta n_x\omega_1]/c = -2\omega_1(\delta n)/c, \qquad (10.3.11)$$

where Eq. (10.3.10) was used together with $\omega_3 + \omega_4 = 2\omega_1$. From Eqs. (10.3.6) and (10.3.11), Δk_M and Δk_W compensate each other for a frequency shift Ω_s given by [17]

$$\Omega_s = \left(\frac{4\pi\delta n}{\beta_2\lambda_1}\right)^{1/2}, \qquad (10.3.12)$$

where $\lambda_1 = 2\pi c/\omega_1$. At a pump wavelength $\lambda_1 = 0.532\ \mu m$, $\beta_2 \approx 60\ ps^2/km$, and a typical value $\delta n = 1 \times 10^{-5}$ for the fiber birefringence, the frequency shift v_s is ~ 10 THz. In a 1981 experiment on FWM in birefringent fibers, the frequency shift was in the range of 10 to 30 THz [17]. Furthermore, the measured values of v_s agreed well with those estimated from Eq. (10.3.12).

Equation (10.3.12) for the frequency shift is derived for a specific choice of field polarizations, namely, that the pump fields A_1 and A_2 are polarized along the slow axis while A_3 and A_4 are polarized along the fast axis. Several other combinations can be used for phase matching depending on whether β_2 is positive or negative. The corresponding frequency shifts Ω_s are obtained by evaluating Δk_W from Eq. (10.3.4) with Δn_j ($j = 1$ to 4) replaced by Δn_x or Δn_y depending on the wave polarization and by using Eq. (10.3.6). Table 10.1 lists the four phase-matching processes that can occur in birefringent fibers together with the corresponding frequency shifts [39]. The frequency shifts of the first two processes are smaller by more than one order of magnitude compared with the other two processes. All frequency shifts in Table 10.1 are approximate because frequency dependence of δn has been ignored; its inclusion can reduce them by about 10%. Several other phase-matched processes

Table 10.1 Phase-Matched FWM processes in Birefringent Fibers[a]

Process	A_1	A_2	A_3	A_4	Frequency shift Ω_s	Condition
I	s	f	s	f	$\delta n/(|\beta_2|c)$	$\beta_2 > 0$
II	s	f	f	s	$\delta n/(|\beta_2|c)$	$\beta_2 < 0$
III	s	s	f	f	$(4\pi\delta n/|\beta_2|\lambda_1)^{1/2}$	$\beta_2 > 0$
IV	f	f	s	s	$(4\pi\delta n/|\beta_2|\lambda_1)^{1/2}$	$\beta_2 < 0$

[a]The symbols s and f denote the direction of polarization along the slow and fast axes, respectively.

have been identified [20] but are not generally observed in a silica fiber because of its predominantly isotropic nature.

From a practical standpoint, the four processes shown in Table 10.1 can be divided into two categories. The first two correspond to the case in which pump power is divided between the slow and fast modes. In contrast, the pump field is polarized along a principal axis for the remaining two processes. In the first category, the parametric gain is maximum when the pump power is divided equally by choosing $\theta = 45°$, where θ is the polarization angle measured from the slow axis. Even then, different processes compete with each other because the parametric gain is nearly the same for all of them. In one experiment, FWM occurring as a result of Process I was observed by using 15-ps pump pulses from a mode-locked dye laser operating at 585.3 nm [21]. Because of a relatively small group-velocity mismatch among the four waves in this case, Process I became dominant compared with the others.

Figure 10.9 shows the spectrum observed at the output of a 20-m-long fiber for an input peak power $\sim$1 kW and a pump-polarization angle $\theta = 44°$. The Stokes and anti-Stokes bands located near ± 4 THz are due to FWM phase-matched by Process I. As expected, the Stokes band is polarized along the slow axis, while the anti-Stokes band is polarized along the fast axis. Asymmetric broadening of the pump and Stokes bands results from the combined effects of SPM and XPM (see Section 7.4). Selective enhancement of the Stokes band is due to the Raman gain. The peak near 13 THz is also due to SRS. It is polarized along the slow axis because the pump component along that axis is slightly more intense for $\theta = 44°$. An increase in θ by 2° flips the polarization of the Raman peak along the fast axis. The small peak near 10 THz results from a nondegenerate FWM process in which both the pump and the Stokes bands act as pump waves ($\omega_1 \neq \omega_2$) and the Raman band provides

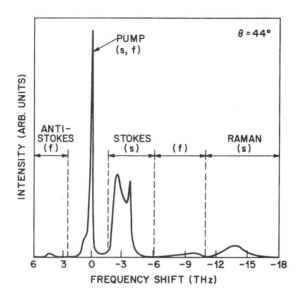

Figure 10.9 Output spectra showing Stokes and anti-Stokes bands generated in a 20-m-long birefringent fiber when 15-ps pump pulses with a peak power ~1 kW are incident at the fiber. Pump is polarized at 44° from the slow axis. (After Ref. [21].)

a weak signal for the parametric process to occur. Phase matching can occur only if the Raman band is polarized along the slow axis. Indeed, the peak near 10 THz disappeared when θ was increased beyond 45° to flip the polarization of the Raman band.

The use of birefringence for phase matching in single-mode fibers has an added advantage in that the frequency shift ν_s can be tuned over a considerable range (~4 THz). Such a tuning is possible because birefringence can be changed through external factors such as stress and temperature. In one experiment, the fiber was pressed with a flat plate to apply the stress [18]. The frequency shift ν_s could be tuned over 4 THz for a stress of 0.3 kg/cm. In a similar experiment, stress was applied by wrapping the fiber around a cylindrical rod [19]. The frequency shift ν_s was tuned over 3 THz by changing the rod diameter. Tuning is also possible by varying temperature as the built-in stress in birefringent fibers is temperature dependent. A tuning range of 2.4 THz was demonstrated by heating the fiber up to 700°C [22]. In general, FWM provides a convenient way of measuring the net birefringence of a fiber because the frequency shift depends on δn [23].

The frequency shift associated with the FWM process depends on the pump power through the nonlinear contribution Δk_{NL} in Eq. (10.3.1). This contribution has been neglected in obtaining Eq. (10.3.12) and other expressions of Ω_s in Table 10.1, but can be included in a straightforward manner. In general Ω_s decreases with increasing pump power. In one experiment, Ω_s decreased with pump power at a rate of 1.4% W^{-1} [24]. The nonlinear contribution Δk_{NL} can also be used to satisfy the phase-matching condition. This feature is related to modulation instability in birefringent fibers (see Section 6.4).

10.4 Parametric Amplification

Similar to the cases of Raman and Brillouin gains, the parametric gain in optical fibers can be used for making parametric amplifiers and lasers. Such devices have attracted considerable attention [66]–[75]. This section describes their characteristics and applications.

10.4.1 Gain and Bandwidth

A complete description of parametric amplification often requires a numerical solution of Eqs. (10.2.2)–(10.2.5). However, considerable physical insight is gained by first considering the approximate analytic solution, given by Eqs. (10.2.19) and (10.2.20), that neglects pump depletion. The constants a_3, b_3, a_4, and b_4 in these equations are determined from the boundary conditions. If we assume that only the signal and pump waves are incident at the fiber input, the signal and idler powers at the fiber output ($z = L$) are given by [12]

$$P_3(L) = P_3(0)[1 + (1 + \kappa^2/4g^2)\sinh^2(gL)], \qquad (10.4.1)$$
$$P_4(L) = P_3(0)(1 + \kappa^2/4g^2)\sinh^2(gL), \qquad (10.4.2)$$

where the parametric gain g is given by Eq. (10.2.21). As one would expect, the signal is amplified and, at the same time, the idler wave is generated. Thus, the same FWM process can be used to amplify a weak signal *and* to generate simultaneously a new wave at the idler frequency. Here we focus on signal amplification. The amplification factor is obtained from Eq. (10.4.1) and can be written using Eq. (10.2.21) as

$$G_p = P_3(L)/P_3(0) = 1 + (\gamma P_0 r/g)^2 \sinh^2(gL). \qquad (10.4.3)$$

The parameter r is given in Eq. (10.2.22); $r = 1$ when a single pump beam is used for parametric amplification.

The gain expression (10.4.3) should be compared with Eq. (8.2.5) obtained for a Raman amplifier. The main difference is that the parametric gain depends on the phase mismatch κ and can become quite small if phase matching is not achieved. In the limit $\kappa \gg \gamma P_0 r$, Eqs. (10.2.21) and (10.4.3) yield

$$G_p \approx 1 + (\gamma P_0 rL)^2 \frac{\sin^2(\kappa L/2)}{(\kappa L/2)^2}. \tag{10.4.4}$$

The parametric gain is relatively small and increases with pump power as P_0^2 if phase mismatch is relatively large. On the other hand, if phase matching is perfect ($\kappa = 0$) and $gL \gg 1$, the amplifier gain increases exponentially with P_0 as

$$G_p \approx \tfrac{1}{4} \exp(2\gamma P_0 rL). \tag{10.4.5}$$

The amplifier bandwidth $\Delta\Omega_A$ can be determined from Eq. (10.4.3) and depends on both the fiber length L and the pump power P_0. Consider first the limit $\kappa \gg \gamma P_0$ in which G_p is given by Eq. (10.4.4). The gain decreases by a factor of $\pi^2/4$ for $\kappa L = \pm \pi$. A convenient definition of $\Delta\Omega_A$ corresponds to a wave-vector mismatch $\Delta\kappa = 2\pi/L$; it provides a bandwidth slightly larger than the full width at half maximum [12]. Noting that κ is dominated by the material-dispersion contribution given by Eq. (10.3.6), $\Delta\kappa \approx 2|\beta_2|\Omega_s\Delta\Omega_A$, and the bandwidth becomes

$$\Delta\Omega_A = \frac{\Delta\kappa}{2|\beta_2|\Omega_s} = \frac{\pi}{|\beta_2|\Omega_s L}, \tag{10.4.6}$$

where Ω_s is the frequency shift between the pump and signal waves corresponding to the phase-matching condition $\kappa = 0$.

Pump-induced broadening of the parametric gain (see Fig. 10.1) increases the bandwidth over that given in Eq. (10.4.6). If Eq. (10.4.3) is used to determine $\Delta\Omega_A$, the amplifier bandwidth becomes [12]

$$\Delta\Omega_A = \frac{1}{|\beta_2|\Omega_s} \left[\left(\frac{\pi}{L}\right)^2 + (\gamma P_0 r)^2 \right]^{1/2}. \tag{10.4.7}$$

At high pump powers, the bandwidth can be approximated by

$$\Delta\Omega_A \approx \gamma P_0 / |\beta_2\Omega_s| \equiv (|\beta_2|\Omega_s L_{\text{NL}})^{-1}, \tag{10.4.8}$$

where the nonlinear length $L_{NL} = (\gamma P_0)^{-1}$ and $r = 1$ has been assumed. Equation (10.4.8) is valid for $L_{NL} \ll L$. As a rough estimate, the bandwidth is ~ 1 THz for $|\beta_2| \sim 10$ ps^2/km, $\Omega_s/2\pi \sim 10$ THz, and $L_{NL} \sim 1$ m. As a result, the bandwidth of a parametric amplifier is typically smaller than that of Raman amplifiers (~ 5 THz) but much larger than that of Brillouin amplifiers (~ 10 MHz). It can be increased considerably by matching the pump wavelength to the zero-dispersion wavelength of the fiber so that $|\beta_2|$ is reduced. In the limit β_2 approaches zero, and the bandwidth is determined by the fourth-order dispersion parameter β_4 and can exceed 5 THz or 50 nm [72].

The amplifier gain is quite different when both the signal and idler waves are incident at the fiber input together with the pump. In particular, the signal and the idler waves may be amplified or attenuated depending on the relative input phase between them. This dependence on the relative phase has been seen in an experiment in which the signal and idler waves, shifted from the pump frequency by 130 MHz, were propagated through a 350-m-long fiber and their relative phase $\Delta\phi$ was varied using a delay line [68]. The signal power was minimum and maximum for $\Delta\phi = \pi/2$ and $3\pi/2$, respectively, while it remained nearly unchanged for $\Delta\phi = 0$ and π. The amplifier gain varies linearly with pump power and fiber length under such conditions.

10.4.2 Pump Depletion

Inclusion of pump depletion modifies the characteristics of parametric amplifiers considerably compared with those expected from Eqs. (10.4.1) and (10.4.2). It is generally necessary to solve Eqs. (10.2.2)–(10.2.5) numerically although an analytic solution in terms of the elliptic functions can be obtained under specific conditions [29]. Whether the signal and idler waves are amplified or attenuated depends on the relative phase θ defined by

$$\theta = \phi_1 + \phi_2 - \phi_3 - \phi_4, \tag{10.4.9}$$

where ϕ_j is the phase of the amplitude A_j for $j = 1$ to 4. Maximum amplification occurs for $\theta = -\pi/2$. The relative phase θ, however, changes during parametric interaction. As a result, even if $\theta = -\pi/2$ at the fiber input, eventually θ lies in the range 0 to $\pi/2$, and both the signal and idler waves experience deamplification. This behavior is shown in Fig. 10.10 where evolution of the relative phase θ, the idler power P_4, and the pump power P_1 is shown along the fiber for an input pump power of 70 W after assuming perfect phase matching,

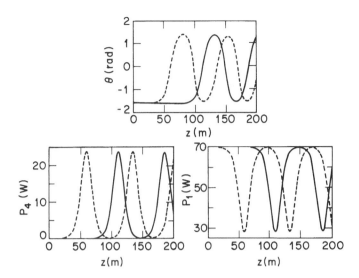

Figure 10.10 Evolution of relative phase θ, idler power P_4, and pump power P_1 along fiber length when $\kappa = 0$. Input parameters correspond to $P_1 = P_2 = 70$ W, $P_4 = 0.1$ μW, and $\theta = \pi/2$. Input signal power $P_3 = 0.1$ μW for solid curves and 6 mW for dashed curves. (After Ref. [70] ©1987 IEEE.)

$P_1 = P_2$, and $\theta = -\pi/2$ at the fiber input; $P_3(0) = P_4(0) = 0.1$ μW for solid lines while $P_3(0) = 6$ mW with $P_4(0) = 0.1$ μW for dashed lines. The latter case corresponds to a parametric amplifier; both waves grow from noise in the former case. In both cases, the signal and idler waves amplify and deamplify periodically. This behavior can be understood by noting that pump depletion changes the relative phase θ from its initial value of $-\pi/2$. The main point is that a parametric amplifier requires a careful control of the fiber length even in the case of perfect phase matching.

The preceding discussion is based on the CW theory of FWM. In the case of a fiber pumped by short pump pulses, two effects can reduce parametric interaction among the four waves participating in FWM. First, the pump spectrum is broadened by SPM as the pump wave propagates through the fiber. If the spectral width of the pump exceeds the amplifier bandwidth $\Delta\Omega_A$, the parametric gain would decrease, similar to the case of Brillouin gain discussed in Section 9.1. Second, the group-velocity mismatch among the pump, signal, and idler pulses manifests through a separation of pulses from each other. Both of these effects reduce the effective length over which FWM can occur. For ultrashort pulses, one must also include the GVD effects using Eq. (10.2.26).

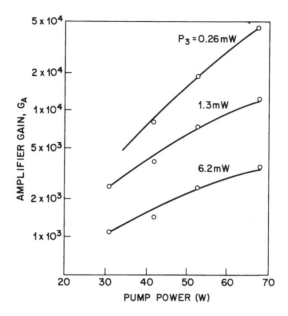

Figure 10.11 Measured gain G_p of a parametric amplifier as a function of pump power for three values of the input signal power. (After Ref. [14].)

10.4.3 Parametric Amplifiers

All of the phase-matching techniques discussed in Section 10.3 have been used for making parametric amplifiers. The main difference between the FWM experiments and parametric amplifiers is whether or not a signal at the phase-matched frequency is copropagated with the pump. In the absence of a signal, the signal and idler waves are generated from amplification of noise.

In the 1974 experiment on parametric amplification in optical fibers, phase-matching was achieved using a multimode fiber [6]. The peak power of 532-nm pump pulses was $\sim$100 W whereas the CW signal with a power $\sim$10 mW was tunable near 600 nm. The amplifier gain was quite small because of a short fiber length (9 cm) used in the experiment. In a later experiment [14], phase matching was achieved using a pump at 1.319 μm, a wavelength close to the zero-dispersion wavelength (see Fig. 10.7). The peak power of the pump pulses was varied in the range of 30 to 70 W and the amplified signal power was measured propagating a CW 1.338-μm signal through a 30-m-long fiber.

Figure 10.11 shows the amplifier gain G_p as a function of pump power P_0 for three values of the input signal power P_3. Departure from the exponential increase in G_p with P_0 is due to gain saturation occurring as a result of pump depletion. Note also that G_p is considerably reduced as P_3 is increased from 0.26 to 6.2 mW. For the 0.26-mW input signal, the single-pass amplifier gain was as large as 46 dB for $P_0 = 70$ W. Such large values clearly indicate the potential use of optical fibers as parametric amplifiers provided phase matching can be achieved. This requirement puts stringent limits on the control of frequency shift Ω_s between the pump and signal waves.

The use of fiber birefringence for phase matching is attractive as birefringence can be adjusted to match Ω_s by applying external stress or by bending the fiber. Parametric amplifiers with such schemes have been demonstrated. In one experiment, the 1.292-μm signal from a semiconductor laser was amplified by 38 dB when Ω_s was tuned by applying external stress to the fiber [66]. In another, the 1.57-μm signal from a distributed feedback semiconductor laser was amplified by 37 dB using a 1.319-μm pump [67].

A common technique for realizing broadband parametric amplifiers consists of choosing a pump laser whose wavelength λ_p is close to the zero-dispersion wavelength λ_0 of the optical fiber. In one implementation of this idea [72], a DFB semiconductor laser operating near $\lambda_p \approx 1.54$ μm was used for pumping while a tunable external-cavity semiconductor laser provided the signal. The length of the dispersion-shifted fiber ($\lambda_0 = 1.5393$) used for parametric amplification was 200 m. The amplifier bandwidth changed considerably as pump wavelength was varied in the vicinity of λ_0 and was larger when pump wavelength was detuned such that $\lambda_p - \lambda_0 = 0.8$ nm. These results are readily understood when the effects of fourth-order dispersion are included. The use of dispersion management and high-nonlinearity fibers can provide higher gain and larger bandwidths for parametric amplifiers [74].

10.4.4 Parametric Oscillators

The parametric gain can be used to make a laser by placing the fiber inside a Fabry–Perot or a ring cavity. Such lasers are called parametric oscillators; the term four-photon fiber laser is also used. In a 1987 experiment [69], a parametric oscillator was pumped by using 100-ps pulses obtained from a 1.06-μm, mode-locked, Q-switched, Nd:YAG laser. It emitted 1.15-μm pulses of about 65-ps duration. Synchronous pumping was realized by adjusting the cavity length such that each laser pulse, after a round trip, overlapped with a successive pump pulse. The bandwidth of the parametric laser was about 100 GHz, in agreement with Eq. (10.4.7).

In the anomalous-GVD region of optical fibers, a new kind of parametric oscillator, called the modulation-instability laser, has been made. As discussed in Section 10.3.2, modulation instability can be interpreted in terms of a FWM process phase-matched by the nonlinear index change responsible for SPM. The modulation-instability laser was first made in 1988 by pumping a fiber-ring cavity (length ~ 100 m) synchronously with a mode-locked color-center laser (pulse width ~ 10 ps) operating in the 1.5-μm wavelength region [71]. For a 250-m ring, the laser reached threshold at an average pump power of 17.5 mW (peak power of 13.5 W). The laser generated signal and idler bands with a frequency shift of about 2 THz, and this value agreed with the theory of Section 5.1. No attempt was made to separate the pump, signal, and idler waves, resulting in a triply resonant parametric oscillator. In some sense, a modulation-instability laser is different from a conventional parametric oscillator as the objective is to convert a CW pump into a train of short optical pulses rather than generating a tunable CW signal. This objective was realized in 1999 by pumping a 115-m-long ring cavity with a CW laser (a DFB fiber laser). The SBS process was suppressed by modulating the phase of pump light at a frequency (>80 MHz) larger than the Brillouin-gain bandwidth. The laser reached threshold at a pump power of about 80 mW and emitted a pulse train at the 58-GHz repetition rate when pumped harder [75]. The spectrum exhibited multiple peaks, separated by 58 GHz and generated through a cascaded FWM process.

A parametric oscillator, tunable over a 40-nm range and centered at the pump wavelength of 1539 nm, has been made [73]. This laser used a nonlinear Sagnac interferometer (loop length 105 m) for a parametric amplifier that was pumped by using 7.7-ps mode-locked pulses from a color-center laser operating at 1539 nm. Such an interferometer separates pump light from the signal and idler waves while amplifying both of them. In effect, it acts as a mirror of a Fabry–Perot cavity with internal gain. A grating at the other end of the cavity separates the idler and signal waves so that the Fabry–Perot cavity is resonant for the signal only. The grating is also used for tuning laser wavelength. The laser emitted 1.7-ps pulses at the 100-MHz repetition rate of the pump laser.

10.5 FWM Applications

FWM in optical fibers can be both harmful and useful depending on the application. It can induce crosstalk in WDM communication systems and limit the performance of such systems. However, FWM can be avoided using unequal channel spacings or using fibers with large enough GVD that the the FWM process is not phase matched over long fiber lengths. This issue is covered in Chapter B.7. Here we focus on applications in which FWM plays a useful role.

10.5.1 Wavelength Conversion

Similar to the case of Raman amplifiers, parametric amplifiers are useful for signal amplification. However, as such amplifiers also generate a idler wave at a frequency $2\omega_p - \omega_s$, where ω_p and ω_p are pump and signal frequencies, they can be used for wavelength conversion. Equation (10.4.2) provides the following measure of the conversion efficiency:

$$\eta_c = P_4(L)/P_3(0) = (\gamma P_0 r/g)^2 \sinh^2(gL). \tag{10.5.1}$$

The FWM process becomes quite efficient if the phase-matching condition is satisfied in the sense that η_c can exceed 1. From a practical standpoint, more power appears at the new wavelength compared with the signal power incident at the input end. This is not surprising if we note that the pump beam supplies energy to both the signal and idler waves simultaneously.

The use of FWM for wavelength conversion attracted considerable attention during the 1990s because of its potential application in WDM lightwave systems [76]–[80]. If a CW pump beam is injected together with a signal pulse train consisting of a pseudorandom sequence of "1" and "0" bits inside a parametric amplifier, the idler wave is generated through FWM only when the pump and signal are present simultaneously. As a result, the idler wave appears in the form of a pulse train consisting of same sequence of "1" and "0" bits as the signal. In effect, FWM transfers the signal data to the idler at a new wavelength with perfect fidelity. It can even improve the signal quality by reducing intensity noise [80].

A related application uses FWM for demultiplexing a time-division-multi-plexed (TDM) signal [81]. In a TDM signal, bits from different channels are packed together such that the bits belonging to a specific channel are separated by $T_B = 1/B_{ch}$, where B_{ch} is the bit rate of each channel, whereas individual bits are spaced apart only T_B/N_{ch} if N_{ch} channels are multiplexed together in the time domain. A specific channel can be demultiplexed if an optical pulse train at the repetition rate B_{ch} is used for pumping (referred to as an optical clock). During FWM inside an optical fiber, the idler wave is generated only when the pump and signal pulses overlap in the time domain. As a result, the idler will consist of a replica of the bit pattern associated with a single channel. FWM for demultiplexing a TDM signal was used as early as 1991 [82]. By 1997, 10-Gb/s channels were demultiplexed with 22-dB conversion efficiency associated with the FWM process [83]. The same idea can be used for all-optical picosecond sampling as the intensity of the idler pulse is proportional to the signal pulse. In a demonstration of this concept 20-ps resolution was achieved by using actively mode-locked semiconductor lasers [84].

10.5.2 Phase Conjugation

An interesting feature of the idler wave that has considerable practical importance is apparent from Eqs. (10.2.12) and (10.2.13). These equations show that A_3 is coupled to A_4^* rather than A_4. Because the idler field is "complex conjugate" of the signal field, such a FWM process is called phase conjugation. An important application of phase conjugation, first proposed in 1979 [85], consists of using it for dispersion compensation in optical communication systems [86]–[90].

To understand how phase conjugation leads to dispersion compensation, consider what would happen if a phase conjugator is placed in the middle of a fiber link. Then, during the first half of the fiber link, the signal $A(z,t)$ would satisfy Eq. (3.1.1) or

$$\frac{\partial A}{\partial z} + \frac{i\beta_2}{2}\frac{\partial^2 A}{\partial dT^2} - \frac{\beta_3}{6}\frac{\partial^3 A}{\partial dT^3} = i\gamma|A|^2 A - \frac{1}{2}\alpha A, \tag{10.5.2}$$

where effects of third-order dispersion have also been included. After phase conjugation, A becomes A^*. As a result, in the second half of the fiber link, signal propagation is governed by an equation obtained by taking the complex conjugate of Eq. (10.5.2):

$$\frac{\partial A^*}{\partial z} - \frac{i\beta_2}{2}\frac{\partial^2 A^*}{\partial dT^2} - \frac{\beta_3}{6}\frac{\partial^3 A^*}{\partial dT^3} = -i\gamma|A|^2 A^* - \frac{1}{2}\alpha A^*. \tag{10.5.3}$$

To simplify the following discussion, consider first the linear case and assume that the nonlinear effects can be neglected. Because the sign of the β_2 term is reversed for the idler wave, the net GVD becomes zero when the phase conjugator is placed exactly in the middle of the fiber link. However, as the sign of the β_3 term remains unchanged, phase conjugation has no effect on third-order dispersion. As a result, such a system is equivalent to operating at the zero-dispersion wavelength even though β_2 may be quite large.

Noting that the sign of the nonlinear term is also reversed in Eq. (10.5.3), one may be tempted to conclude that the nonlinear effects can also be compensated by this technique. This conclusion is correct if $\alpha = 0$ but becomes invalid in the presence of fiber losses because the SPM effects—governed by the γ term—become weaker with propagation when $\alpha \neq 0$. This can be seen more clearly by using the transformation $A = B\exp(-\alpha z/2)$ in Eq. (10.5.2), which then becomes

$$i\frac{\partial B}{\partial z} - \frac{\beta_2}{2}\frac{\partial^2 B}{\partial T^2} + \gamma_z|B|^2 B = 0, \tag{10.5.4}$$

where $\gamma_z = \gamma\exp(-\alpha z)$ and the β_3 term has been neglected for simplicity. The effect of fiber loss is mathematically equivalent to the loss-free case but with a z-dependent nonlinear parameter. By taking the complex conjugate of Eq. (10.5.4) and changing z to $-z$, it is easy to see that perfect SPM compensation can occur only if $\gamma(z) = \gamma(L - z)$. This condition cannot be satisfied for $\alpha \neq 0$.

Perfect compensation of both GVD and SPM can be realized by using dispersion-decreasing fibers . To see how such a scheme can be implemented, assume that β_2 in Eq. (10.5.4) is a function of z. Making the transformation,

$$\xi = \int_0^z \gamma_z(z)\,dz, \tag{10.5.5}$$

Eq. (10.5.4 can be written as [90]

$$i\frac{\partial B}{\partial \xi} - \frac{d(\xi)}{2}\frac{\partial^2 B}{\partial T^2} + |B|^2 B = 0, \tag{10.5.6}$$

where $d(\xi) = \beta_2(\xi)/\gamma_z(\xi)$. Both GVD and SPM are compensated if $d(\xi) = d(\xi_L - \xi)$, where ξ_L is the value of ξ at $z = L$. A simple solution is provided by the case in which the dispersion is tailored in exactly the same way as γ_z. Because fiber losses cause γ_z to vary exponentially, both GVD and SPM can be compensated exactly in a dispersion-decreasing fiber whose GVD decreases exponentially. Even in the absence of such fibers, GVD and SPM can be compensated by controlling β_2 and γ along the fiber link. This approach is quite general and applies even when optical amplifiers are used [90].

FWM in fibers has been used to generate phase-conjugated signals in the middle of a fiber link and compensate the GVD effects. In a 1993 experiment [86], a 1546-nm signal was phase-conjugated using FWM in a 23-km fiber while pumping it with a CW semiconductor laser operating at 1549 nm. The 6-Gb/s signal could be transmitted over 152 km of standard fiber because of GVD compensation. In another experiment, a 10-Gb/s signal was transmitted over 360 km [87]. FWM was performed in a 21-km-long fiber using a pump laser whose wavelength was tuned exactly to the zero-dispersion wavelength of the fiber. The pump and signal wavelengths differed by 3.8 nm.

Several factors need to be considered while implementing the midspan phase-conjugation technique in practice. First, as the signal wavelength changes from ω_3 to $\omega_4 \equiv 2\omega_1 - \omega_3$ at the phase conjugator, the GVD parameter β_2 becomes different in the the second-half section. As a result, perfect compensation occurs only if the phase conjugator is slightly offset from the midpoint of the fiber link. The exact location L_p can be determined by using the condition $\beta_2(\omega_3)L_p = \beta_2(\omega_4)(L - L_p)$, where L is the total link length. For a typical wavelength shift of 6 nm, the phase-conjugator location changes by about 1%. The effect of residual dispersion and SPM in the phase-conjugation fiber itself can also affect the placement of phase conjugator [89].

Phase conjugation has several other potential applications [91]–[98]. It can be used to determine the phase of optical signals without performing homodyne or heterodyne detection [91]. Its judicious use can cancel the frequency shifts of solitons induced by intrapulse Raman scattering [92] and reduce the timing jitter induced by amplifiers and soliton collisions [93]–[96]. Phase conjugation has been used for all-optical storage of picosecond-pulse packets [97]. It can also be used for reducing the noise below the shot-noise level. This topic is discussed next.

10.5.3 Squeezing

An interesting application of FWM has led to the reduction of quantum noise through a phenomenon called squeezing [99]–[101]. Squeezing refers to the process of generating the special states of an electromagnetic field for which noise fluctuations in some frequency range are reduced below the quantum-noise level. FWM can be used for squeezing as noise components at the signal and idler frequencies are coupled through the fiber nonlinearity. An accurate description of squeezing in optical fibers requires a quantum-mechanical approach in which the signal and idler amplitudes B_3 and B_4 are replaced by annihilation operators [102]. Furthermore, quantum noise should be included by adding a fluctuating source term (known as the Langevin force) on the right-hand side of Eqs. (10.2.16) and (10.2.17). It is also necessary to account for fiber losses.

From a physical standpoint, squeezing can be understood as deamplification of signal and idler waves for certain values of the relative phase between the two waves [68]. Spontaneous emission at the signal and idler frequencies generates photons with random phases. FWM increases or decreases the number of specific signal-idler photon pairs depending on their relative phases. A phase-sensitive (homodyne or heterodyne) detection scheme would show noise reduced below the quantum-noise level when the phase of the local oscillator is adjusted to match the relative phase corresponding to the photon pair whose number was reduced as a result of FWM.

The observation of squeezing in optical fibers is hindered by the competing processes such as spontaneous and stimulated Brillouin scattering (SBS). A particularly important noise process turned out to be Brillouin scattering caused by guided acoustic waves [102]. If noise generated by this phenomenon exceeds the reduction expected from FWM, squeezing is washed out. Several techniques have been developed to reduce the impact of this noise source [101]. A simple method consists of immersing the fiber in a liquid-helium bath. Indeed, a 12.5% reduction in the quantum-noise level was observed in a 1986 experiment in which a 647-nm CW pump beam was propagated through a 114-m-long fiber [103]. SBS was suppressed by modulating the pump beam at 748 MHz, a frequency much larger than the Brillouin-gain bandwidth. Thermal Brillouin scattering from guided acoustic waves was the most limiting factor in the experiment. Figure 10.12 shows the observed noise spectrum when the local oscillator phase is set to obtain the minimum noise. The large peaks are due to guided-acoustic-wave Brillouin scattering. Squeezing occurs in the spectral bands located around 45 and 55 MHz.

Squeezing in optical fibers has continued to attract attention since its first observation in 1986 [104]–[115]. In a 1987 experiment, two CW pump beams at wavelengths 647 and 676 nm were propagated through the same fiber [104]. A 20% reduction in the quantum-noise level was measured by using a dual-frequency homodyne detection scheme. This scheme is referred to as four-mode squeezing as the two signal-idler pairs associated with the two pump beams are responsible for squeezing. Both SPM and XPM effects play an important role in this kind of squeezing. By 1991, quadrature squeezing was observed using intense ultrashort pulses propagating as a soliton inside a balanced Sagnac interferometer. In one experiment, 200-fs pulses were propa-

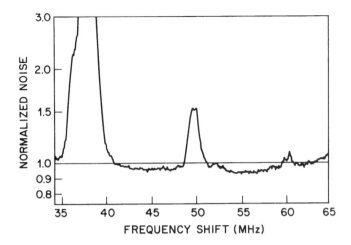

Figure 10.12 Noise spectrum under minimum-noise conditions. Horizontal line shows the quantum-noise level. Reduced noise around 45 and 55 MHz is a manifestation of squeezing in optical fibers occurring as a result of FWM. (After Ref. [103].)

gated through 5 m of fiber at a repetition rate of 168 MHz [108]. A 32% reduction in the shot noise was observed in the frequency range 3–30 MHz. In a similar experiment, 100-ps optical pulses centered at the zero-dispersion wavelength of the fiber were used, and noise reduction by as much as 5 dB was observed [109].

Starting in 1996, attention focused on amplitude or photon-number squeezing that can be observed directly without requiring a local oscillator for homodyne (or heterodyne) detection [111]–[115]. Initial experiments used spectral filtering of optical solitons. More recent experiments have used interferometric techniques. Amplitude squeezing by more than 6 dB was observed when 182-fs pulses at a repetition rate of 82 MHz were propagated as solitons in a 3.5-m asymmetric fiber loop acting as a Sagnac interferometer [113]. This value dropped to 2.5 dB when pulses were propagated in the normal-GVD regime of the fiber loop [114]. The reader is referred to a 1999 review for further details [101].

10.5.4 Supercontinuum Generation

When ultrashort optical pulses propagate through an optical fiber, the FWM process is accompanied by a multitude of other nonlinear effects, such as SPM,

XPM, and SRS, together with the effects of dispersion. All of these nonlinear processes are capable of generating new frequencies within the pulse spectrum. It turns out that, for sufficiently intense pulses, the pulse spectrum can become so broad that it extends over a frequency range exceeding 10 THz. This broadening is referred to as supercontinuum generation and was initially studied in solid and gaseous nonlinear media [64]. In the case of optical fibers, supercontinuum was first observed using multimode fibers (see Section 10.3.1). As seen in Fig. 10.5, the spectrum of a 532-nm picosecond pulse extended over 50 nm after it propagated through a 15-m-long four-mode fiber because of the combined effects of SPM, XPM, SRS, and FWM. Similar features are expected when single-mode fibers are used [116].

Starting in 1993, supercontinuum generation in single-mode fibers was used as a practical tool for obtaining picosecond pulses at multiple wavelengths simultaneously, acting as an ideal source for WDM communication systems [117]–[126]. In a 1994 experiment [118], 6-ps pulses with a peak power of 3.8 W at a 6.3-GHz repetition rate (obtained by amplifying the output of a gain-switched semiconductor laser operating at 1553 nm) were propagated in the anomalous-GVD region of a 4.9-km-long fiber ($|\beta_2| < 0.1$ ps^2/km). The supercontinuum generated at the fiber output was wide enough (>40 nm) that it could be used to produce 40 WDM channels by using a periodic optical filter. The 6.3-GHz pulse train in different channels consisted of nearly transform-limited pulses whose width varied in the range 5–12 ps. By 1995, this technique produced a 200-nm-wide supercontinuum, resulting in a 200-channel WDM source [119]. The 6.3-GHz pulse train had a low jitter (<0.3 ps) and high frequency stability. The same technique can also produce femtosecond pulses by enlarging the bandwidth of the optical filter. In fact, pulse widths in the range 0.37 to 11.3 ps were generated by using an array-waveguide grating filter having variable bandwidth in the range 0.3–10.5 nm [120]. A supercontinuum source was used in 1997 to demonstrate data transmission at a bit rate of 1.4 Tb/s using seven WDM channels with 600-GHz spacing while time-division multiplexing was used to operate each channel at 200 Gb/s [121].

GVD plays an important role in the formation of a supercontinuum in optical fibers. Indeed, because of the large spectral bandwidth associated with a supercontinuum, β_2 cannot be treated as constant and its wavelength dependence should be considered. Numerical simulations show that uniformity or flatness of the supercontinuum can be improved considerably if β_2 increases along the fiber length such that the optical pulse experiences anomalous GVD

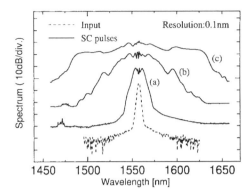

Figure 10.13 Supercontinuum (SC) pulse spectra measured at average power levels of (a) 45 mW, (b) 140 mW, and (c) 210 mW. Dashed curve shows the spectrum of input pulses. (After Ref. [124] ©1998 IEEE.)

initially and normal GVD after some fiber length [122]. However, the role of anomalous GVD is to simply compress the pulse that propagates as a higher-order soliton.

A fiber with normal GVD ($\beta_2 > 0$) along its entire length can also be used if the input pulse in unchirped and a dispersion-flattened fiber is used [124]. Such a fiber has β_3 nearly zero so that β_2 is almost constant over the entire bandwidth of the supercontinuum. In fact, a 280-nm-wide (10-dB bandwidth) flat supercontinuum was generated when 0.5-ps chirp-free pulses, obtained from a mode-locked fiber laser, were propagated in a dispersion-flattened fiber with a small positive value of β_2. Even when 2.2-ps pulses from a mode-locked semiconductor laser were used, supercontinuum as broad as 140 nm could be generated provided pulses were initially compressed to make them nearly chirp-free. Figure 10.13 shows the pulse spectra obtained at several power levels when a 1.7-km-long dispersion-flattened fiber with $\beta_2 = 0.1$ ps^2/km at 1569 nm was used for supercontinuum generation. Input spectrum is also shown as a dashed curve for comparison. The nearly symmetric nature of the spectra indicates that SRS played a minor role. The combination of SPM, XPM, and FWM is responsible for most of spectral broadening seen in Fig. 10.13. Small fluctuations near the pump wavelength originate from residual chirp. Such a supercontinuum spectrum has been used to produce 10-GHz pulse trains at 20 different wavelengths, with nearly the same pulse width in each channel [125].

10.6   Second-Harmonic Generation

FWM, by far the dominant parametric process in optical fibers, generates spectral sidebands separated from the pump frequency by up to ~ 100 THz. In several experiments, the spectrum at the output of an optical fiber, pumped by intense 1.06-μm pump pulses, was found to extend into the visible and ultraviolet regions [127]–[134]. Spectral components in these regions can be generated through mixing of two or more waves such that the generated frequency is a sum of the frequencies of the participating waves. Thus, two waves with frequencies ω_1 and ω_2 can generate second-harmonic frequencies $2\omega_1$ and $2\omega_2$, third-harmonic frequencies $3\omega_1$ and $3\omega_2$, and sum frequencies of the form $\omega_1 + \omega_2$, $2\omega_1 + \omega_2$, and $2\omega_2 + \omega_1$. Frequencies such as $3\omega_1$, $3\omega_2$, $2\omega_1 + \omega_2$, and $2\omega_2 + \omega_1$ are expected to be generated in optical fibers because they are due to third-order parametric processes controlled by $\chi^{(3)}$. In general, efficiency is quite low as phase matching is difficult to achieve for such parametric processes. In a 1983 experiment [131], the third harmonic and sum frequencies $2\omega_1 + \omega_2$ and $2\omega_2 + \omega_1$ were generated as a result of mixing between the input pump wave and the Raman-generated Stokes waves near 1.06-μm and 1.12-μm wavelengths. Third-harmonic generation, phase matched through Čerenkov radiation, has also been observed in both standard and erbium-doped fibers [134]. This section focuses on second-harmonic generation in silica fibers.

10.6.1   Experimental Results

Several early experiments showed that the second harmonic as well as the sum frequencies of the form $\omega_1 + \omega_2$ are generated when intense 1.06-μm pump pulses from a mode-locked, Q-switched, Nd:YAG laser are propagated through optical fibers [128]–[130]. Conversion efficiencies as high as $\sim 0.1\%$ were achieved both for the sum-frequency and the second-harmonic processes. Such high efficiencies for second-order parametric processes are unexpected as the second-order susceptibility $\chi^{(2)}$, resulting from the nonlinear response of electric dipoles, vanishes for centerosymmetric materials such as silica. There are several higher-order nonlinearities that can provide an effective $\chi^{(2)}$ for such processes to occur; the most important ones among them are surface nonlinearities at the core-cladding interface and nonlinearities resulting from quadrupole and magnetic-dipole moments. However, detailed calculations show that these nonlinearities should give rise to a maximum conversion ef-

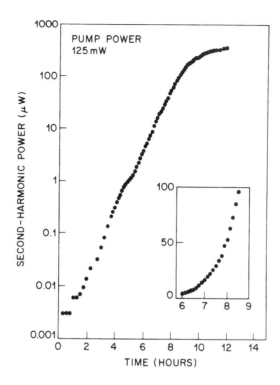

Figure 10.14 Average second-harmonic power generated in a silica fiber as a function of time. Average power of mode-locked, Q-switched, 1.06-μm pump pulses was 125 mW. The inset shows exponential growth on a linear scale. (After Ref. [136].)

ficiency $\sim 10^{-5}$ under phase-matching conditions [135]. It appears that some other mechanism leads to an enhancement of the second-order parametric processes in optical fibers.

A clue to the origin of such a mechanism came when it was discovered in 1986 that the second-harmonic power grows considerably if the fiber is illuminated with pump radiation for several hours [136]. Figure 10.14 shows the average second-harmonic power as a function of time when 1.06-μm pump pulses of duration 100 to 130 ps and average power 125 mW, obtained from a mode-locked and Q-switched Nd:YAG laser, were propagated through a 1-m-long fiber. The second-harmonic power grew almost exponentially with time and began to saturate after 10 hours. The maximum conversion efficiency was about 3%. The 0.53-μm pulses at the fiber output had a width of about 55 ps and were intense enough to pump a dye laser. This experiment stimulated further work and led to extensive research on second-harmonic generation in optical fibers [137]–[194]. It turned out that optical fibers exhibit photosensitivity and can change their optical properties permanently when exposed to intense radiation of certain wavelengths.

Efficient generation of the second harmonic in optical fibers requires an incubation period during which the fiber is prepared by a seeding process. In the 1986 work [136], fibers were prepared by sending intense 1.06-μm pump pulses of duration $\sim$100 ps. The preparation time depended on the pump power and was several hours at peak powers $\sim$10 kW. The fiber core had both germanium and phosphorus as dopants. The presence of phosphorus appeared to be necessary for the preparation process. In a later experiment, in which a mode-locked Kr-ion laser operating at 647.1 nm was used to provide 100-ps pump pulses, a fiber with Ge-doped core could be prepared in about 20 minutes with a peak power of only 720 W [141]. In an important development [138], it was found that a fiber could be prepared in only a few minutes, even by 1.06-μm pulses, provided a weak second-harmonic signal, acting as a seed, is propagated together with pump pulses. The same fiber could not be prepared even after 12 hours without the seeding beam.

10.6.2 Physical Mechanism

Several physical mechanisms were proposed as early as 1987 to explain the second-harmonic generation in optical fibers [137]–[140]. They all relied on a periodic ordering of some entity, such as color centers or defects, along the fiber in such a way that the phase-matching condition was automatically satisfied. In one model, ordering occurs through a third-order parametric process in which the pump and the second-harmonic light (internally generated or externally applied) mix together to create a static or dc polarization (at zero frequency) given by [138]

$$P_{dc} = (3\varepsilon_0/4)\text{Re}[\chi^{(3)} E_p^* E_p^* E_{SH} \exp(i\Delta k_p z)], \qquad (10.6.1)$$

where E_p is the pump field at the frequency ω_p, E_{SH} is the second-harmonic seed field at $2\omega_p$, and the wave-vector mismatch Δk_p is given by

$$\Delta k_p = [n(2\omega_p) - 2n(\omega_p)]\omega_p/c. \qquad (10.6.2)$$

The polarization P_{dc} induces a dc electric field E_{dc} whose polarity changes periodically along the fiber with the phase-matching period $2\pi/\Delta k_p$ ($\sim$30 μm for a 1.06-μm pump). This electric field redistributes electric charges and creates a periodic array of dipoles. The physical entity participating in the dipole formation could be defects, traps, or color centers. The main point is that such a redistribution of charges breaks the inversion symmetry and is also periodic with the right periodicity required for phase matching. In effect, the fiber has organized itself to produce second-harmonic light. Mathematically, the dipoles can respond to an applied optical field with an effective value of $\chi^{(2)}$. In the simplest case, $\chi^{(2)}$ is assumed to be proportional to P_{dc}, that is,

$$\chi^{(2)} \equiv \alpha_{SH}P_{dc} = (3\alpha_{SH}/4)\varepsilon_0\chi^{(3)}|E_p|^2|E_{SH}|\cos(\Delta k_p z + \phi_p), \qquad (10.6.3)$$

where α_{SH} is a constant whose magnitude depends on the microscopic process responsible for $\chi^{(2)}$ and ϕ_p is a phase shift that depends on the initial phases of the pump and the second-harmonic seed. Because of the periodic nature of $\chi^{(2)}$, the preparation process is said to create a $\chi^{(2)}$ grating.

This simple model in which a dc electric field E_{dc} is generated through a $\chi^{(3)}$ process suffers from a major drawback [149]. Under typical experimental conditions, Eq. (10.6.1) leads to $E_{dc} \sim 1$ V/cm for pump powers ~ 1 kW and second-harmonic seed powers ~ 10 W if we use $\chi^{(3)} \approx 10^{-22}$ (m/V)^2. This value is too small to orient defects and generate a $\chi^{(2)}$ grating.

Several alternative mechanisms have been proposed to solve this discrepancy. In one model [166] a charge-delocalization process enhances $\chi^{(3)}$ by several orders of magnitudes, resulting in a corresponding enhancement in E_{dc}. In another model [167], free electrons are generated through photoionization of defects, and a strong electric field ($E_{dc} \sim 10^5$ V/cm) is created through a coherent photovoltaic effect. In a third model [170], ionization occurs through three multiphoton processes (four pump photons, two second-harmonic photons, or two pump photons and one second-harmonic photon). In this model, the $\chi^{(2)}$ grating is created through quantum-interference effects that cause the electron-injection process to depend on the relative phase between the pump and second-harmonic fields. This charge-transport model is in qualitative agreement with most of the observed features. In the next subsection, Eq. (10.6.3) is used to study second-harmonic generation in optical fibers because the results are qualitatively valid irrespective of the exact physical mechanism involved in the creation of the $\chi^{(2)}$ grating.

Conversion efficiencies achieved in the second-harmonic experiments are typically $\sim 1\%$, with a maximum value of about 5% [136]. A natural question is what limits the conversion efficiency? As seen in Fig. 10.14, the second-harmonic power grows exponentially during initial stages of the preparation process but then saturates. One possibility is that the generated second-harmonic interferes with formation of the $\chi^{(2)}$ grating. It has been pointed out that the $\chi^{(2)}$ grating formed by the generated harmonic would be out of phase with the original grating [138]. If this is the case, it should be possible to erase the grating by sending just the second-harmonic light through the fiber without the pump. Indeed, such an erasure has been observed [146]. The erasing rate depends on the second-harmonic power launched into the fiber. In one experiment [148], conversion efficiency decreased from its initial value by a factor of 10 in about five minutes for average second-harmonic powers ~ 2 mW. The decay was not exponential but followed a time dependence of the form $(1 + Ct)^{-1}$, where C is a constant. Furthermore, erasure was reversible, that is, the fiber could be reprepared to recover its original conversion efficiency. These observations are consistent with the model in which the $\chi^{(2)}$ grating is formed by ordering of charged entities such as color centers, defects, or traps.

10.6.3 Simple Theory

One can follow a standard procedure [1]–[5] to study second-harmonic generation from $\chi^{(2)}$ given by Eq. (10.6.3). Assume that a pump wave at the frequency ω_1 is incident on such a prepared fiber. The frequency ω_1 can be different from ω_p in general. The pump field E_1 and the second-harmonic field E_2 then satisfy the coupled amplitude equations of the form [137]

$$\frac{dA_1}{dz} = i\gamma_1(|A_1|^2 + 2|A_2|^2)A_1 + \frac{i}{2}\gamma_{SH}^* A_2 A_1^* \exp(-i\kappa z), \qquad (10.6.4)$$

$$\frac{dA_2}{dz} = i\gamma_2(|A_2|^2 + 2|A_1|^2)A_2 + i\gamma_{SH}A_1^2 \exp(i\kappa z), \qquad (10.6.5)$$

where γ_1 and γ_2 are defined similarly to Eq. (2.3.28),

$$\gamma_{SH} = (3\omega_1/4n_1 c)\varepsilon_0^2 \alpha_{SH} f_{112} \chi^{(3)} |E_p|^2 |E_{SH}|, \qquad (10.6.6)$$

f_{112} is an overlap integral (see Section 10.2), $\kappa = \Delta k_p - \Delta k$, and Δk given by Eq. (10.6.2) after replacing ω_p with ω_1. The parameter κ is the residual wave-vector mismatch occurring when $\omega_1 \neq \omega_p$. The terms proportional to γ_1 and γ_2 are due to SPM and XPM and must be included in general.

Equations (10.6.4) and (10.6.5) can be solved using the procedure of Section 10.2. If we assume that the pump field remains undepleted ($|A_2|^2 \ll |A_1|^2$), Eq. (10.6.4) has the solution

$$A_1(z) = \sqrt{P_1} \exp(i\gamma_1 P_1 z), \qquad (10.6.7)$$

where P_1 is the incident pump power. Introducing $A_2 = B_2 \exp(2i\gamma_1 P_1 z)$ in Eq. (10.6.5), we obtain

$$\frac{dB_2}{dz} = i\gamma_{\text{SH}} P_1 \exp(i\kappa z) + 2i(\gamma_2 - \gamma_1)P_1 B_2. \tag{10.6.8}$$

Equation (10.6.8) is readily solved to obtain the second-harmonic power as

$$P_2(L) = |B_2(L)|^2 = |\gamma_{\text{SH}} P_1 L|^2 \frac{\sin^2(\kappa' L/2)}{(\kappa' L/2)^2}, \tag{10.6.9}$$

where $\kappa' = \kappa - 2(\gamma_2 - \gamma_1)P_1$. The difference between κ' and κ depends on $\gamma_2 - \gamma_1$ and could be significant if the effective mode areas are about the same for the pump and second-harmonic beams because $\gamma_2 = 2\gamma_1$ in that case. Physically, SPM and XPM modify κ as they contribute to the phase-matching condition.

In the simple theory presented here Eqs. (10.6.4) and (10.6.5) are solved approximately assuming that the pump remains undepleted. This approximation begins to break down for conversion efficiencies $> 1\%$. It turns out that these equations can be solved analytically even when the pump is allowed to deplete because the sum of pump and second-harmonic powers remains constant during the generation process. The solution is in the form of elliptic functions similar to that obtained as early as 1962 [1]. The periodic nature of the elliptic functions implies that the conversion process is periodic along the fiber length, and energy is transferred back to the pump after the second harmonic attains its maximum value. The analysis also predicts the existence of a parametric mixing instability induced by SPM and XPM [184]. The instability occurs when $\gamma_1 \sqrt{P_1}$ becomes comparable to γ_{SH} and manifests as doubling of the spatial period associated with the frequency-conversion process. As pump depletion is generally negligible in most experimental situations, these effects are not yet observed, and the following discussion is based on Eq. (10.6.9).

The derivation of Eq. (10.6.9) assumes that the $\chi^{(2)}$ grating is created coherently throughout the fiber. This would be the case if the pump used during the preparation process were a CW beam of narrow spectral width. In practice, mode-locked pulses of duration ~ 100 ps are used. The use of such short pulses affects grating formation in two ways. First, the group-velocity mismatch between the pump and second-harmonic pulses leads to their separation within a few walk-off lengths L_W. If we use $T_0 \approx 80$ ps and $|d_{12}| \approx 80$ ps/m in Eq. (1.2.14), the values appropriate for 1.06-μm experiments, $L_W \approx 1$ m. Thus, the $\chi^{(2)}$ grating stops to form within a distance ~ 1 m for pump pulses of duration ~ 100 ps. Second, SPM-induced spectral broadening reduces the coherence length L_{coh} over which the $\chi^{(2)}$ grating can generate the second harmonic coherently. It turns out that L_{coh} sets the ultimate limit because $L_{coh} < L_W$ under typical experimental conditions. This can be seen by noting that each pump frequency creates its own grating with a slightly different period $2\pi/\Delta k_p$, where Δk_p is given by Eq. (10.6.2). Mathematically, Eq. (10.6.1) for P_{dc} should be integrated over the pump-spectral range to include the contribution of each grating. Assuming Gaussian spectra for both pump and second-harmonic waves, the effective dc polarization becomes [147]

$$P_{dc}^{eff} = P_{dc} \exp[-(z/L_{coh})^2], \qquad L_{coh} = 2/|d_{12}\delta\omega_p|, \qquad (10.6.10)$$

where d_{12} is defined in Eq. (1.2.13) and $\delta\omega_p$ is the spectral half-width (at $1/e$ point). For $|d_{12}| = 80$ ps/m and 10-GHz spectral width (FWHM), $L_{coh} \approx 60$ cm.

In most experiments performed with 1.06-μm pump pulses the spectral width at the fiber input is ~ 10 GHz. However, SPM broadens the pump spectrum as the pump pulse travels down the fiber [see Eqs. (4.1.6) and (4.1.11)]. This broadening reduces the coherence length considerably, and $L_{coh} \sim 10$ cm is expected. Equation (10.6.9) should be modified if $L_{coh} < L$. In a simple approximation [137], L is replaced by L_{coh} in Eq. (10.6.9). This amounts to assuming that $P_{dc}^{eff} = P_{dc}$ for $z \leq L_{coh}$ and zero for $z > L_{coh}$. One can improve over this approximation using Eq. (10.6.10). Its use requires that γ_{SH} in Eq. (10.6.8) be multiplied by the exponential factor $\exp[-(z/L_{coh})^2]$. If $L_{coh} \ll L$, Eq. (10.6.8) can be integrated with the result [147]

$$P_2(\kappa) = (\pi/4)|\gamma_{SH}P_1 L_{coh}|^2 \exp(-\tfrac{1}{2}\kappa^2 L_{coh}^2). \qquad (10.6.11)$$

This expression is also approximate as it is based on Eqs. (10.6.4) and (10.6.5) that are valid only under quasi-CW conditions. These equations can be generalized to include time-dependent features by adding the first and second derivatives with respect to time [156].

Frequency dependence of the second-harmonic power has been measured in several experiments. Figure 10.15 shows the data for an experiment in which the pump wavelength was tuned over 4 nm to vary κ [137]. The solid curve is a theoretical fit based on Eq. (10.6.9) with $L = L_{coh}$. The best fit occurs for $L_{coh} = 12$ cm. In this experiment the peak power of pump pulses during the preparation process was about 10 kW. A short coherence length is expected because of SPM-induced spectral broadening of such intense pump pulses. In

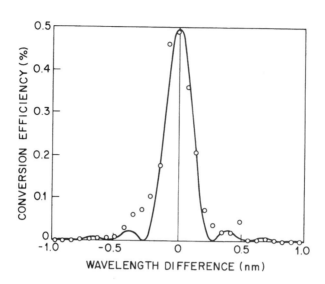

Figure 10.15 Second-harmonic conversion efficiency as a function of pump-wavelength deviation from 1.064 μm. Solid curve is the theoretical fit. (After Ref. [137].)

a 1988 experiment [147], the coherence length could be increased to 35 cm using less intense pump pulses (230-W peak power).

For ultrashort pump pulses, it is important to include the effects of fiber dispersion. The GVD effects can be included in Eqs. (10.6.4) and (10.6.5) by replacing the spatial derivatives by a sum of partial derivatives as indicated in Eq. (10.2.26). The resulting equations, under certain conditions, are found to have solitary-wave solutions [189]–[194], similar to the XPM-paired solitons discussed in Section 7.3.

10.6.4 Quasi-Phase-Matching Technique

Because of a limited conversion efficiency ($<10\%$) associated with photosensitive fibers, the technique of quasi-phase matching has been used to make fibers suitable for harmonic generation. This technique was proposed in 1962 [1] but has been developed only during the 1990s, mostly for ferroelectric materials such as $LiNbO_3$. It was used for optical fibers as early as 1989 and was referred to as electric-field-induced second-harmonic generation [195]. The basic idea is quite simple. Rather than inducing an internal dc electric field optically [see Eq. (10.6.1)], the dc field is applied externally, resulting in an effective $\chi^{(2)}$.

However, a constant value of $\chi^{(2)}$ along the fiber length is not very useful in practice because of the large phase mismatch governed by Δk_p in Eq. (10.6.2).

The technique of quasi-phase matching enhances the second-harmonic efficiency by reversing the sign of $\chi^{(2)}$ along the sample periodically [196]. When this technique is applied to optical fibers, one needs to reverse the polarity of electric field periodically along the fiber length. The period should be chosen such that the sign of $\chi^{(2)}$ is reversed before the second-harmonic power reverts back to the pump; its numerical value is just $2\pi/\Delta k_p$. Quasi-phase matching is commonly used in combination with thermal poling, a technique that can produce relatively large values of $\chi^{(2)}$ of permanent nature in silica glasses and fibers [197]–[200]. Although thermal poling has been used since 1991, the exact physical mechanism responsible for producing $\chi^{(2)}$ was being debated even in 1999 [201]–[205].

Thermal poling requires that a large dc electric field be applied across the fiber core, at an elevated temperature in the range of 250 to 300°, for a duration ranging anywhere from 10 minutes to several hours. If one wants to establish a constant value of $\chi^{(2)}$ across the entire fiber length, electrodes can be inserted through two holes within the cladding of a fiber. The positive electrode should pass quite close to the fiber core because formation of a negatively charged layer close to this electrode plays an important role in the charge migration and ionization process thought to be responsible for inducing $\chi^{(2)}$ [200].

The technique of quasi-phase matching requires reversal of the polarity of the electric field periodically along fiber length during the thermal-poling process [206]. To make such periodically poled silica fibers, the cladding on one side of the fiber is etched away to produce a D-shaped fiber. The flat surface of the fiber should be quite close to the core, 5 μm being a typical distance [207]. A patterned aluminum contact is fabricated on the flat surface using a standard lithographic technique. The required period is close to 56.5 μm for a pump wavelength of 1.54 μm. Thermal poling of the fiber then produces a $\chi^{(2)}$ grating with the same period.

Such a quasi-phase-matched fiber (length 7.5 cm) was pumped by using 2-ns pulses with peak powers of up to 30 kW at a wavelength of 1.532 μm [207]. Second-harmonic light at 766 nm was generated with an average efficiency of 21%. Further improvement is possible using longer fibers and optimizing the thermal-poling process to provide higher $\chi^{(2)}$ values. The experiment shows that quasi-phase-matched fibers represent a viable and potentially useful non-linear medium for the second-order parametric processes. The main drawback of quasi-phase matching is that the the period of $\chi^{(2)}$ grating depends on the pump wavelength and should be matched precisely. A mismatch of even 1 nm in the pump wavelength can reduce the efficiency by 50%.

Problems

10.1 Find an expression similar to that given in Eq. (10.1.4) for P_3 by using Eqs. (10.1.1)–(10.1.3).

10.2 Consider the FWM geometry in which a single CW pump beam produces signal and idler waves. Starting from Eq. (10.1.1), derive the three nonlinear equations, similar to those appearing in Eqs. (10.2.2)–(10.2.5), governing the FWM process.

10.3 Solve the set of three equations obtained in Problem 10.2 assuming that the pump beam remains undepleted. Find the parametric gain for the signal and idler waves as a function of the pump power and the phase mismatch Δk.

10.4 Explain how self-phase modulation can satisfy the phase-matching condition for FWM to occur in a single-mode fiber. What should be the pump power when pump and signal wavelengths are 1.50 and 1.51 μm, respectively? Assume $\gamma = 5$ W^{-1}/km and $\beta_2 = -20$ ps^2/km.

10.5 FWM is observed to occur in a birefringent fiber when a 1.5-μm pump beam is launched such that it is polarized at 40° from the slow axis. What are the wavelengths and the directions of polarization for the spectral sidebands generated through FWM?

10.6 Derive Eqs. (10.4.1) and (10.4.2) by solving Eqs. (10.2.16) and (10.2.17).

10.7 Use Eq. (10.4.1) to find the gain and bandwidth of a parametric amplifier assuming that the FWM process is nearly phase matched.

10.8 How can you use FWM for wavelength conversion in WDM systems? Derive an expression for the conversion efficiency.

10.9 Explain how the second harmonic is generated in an optical fiber. Describe the physical processes involved in detail.

10.10 Derive an expression for the second-harmonic power by solving Eqs. (10.6.4) and (10.6.5). Assume that the conversion efficiency is low enough that the pump remains mostly undepleted.

References

[1] J. A. Armstrong, N. Bloembergen, J. Ducuing, and P. S. Pershan, *Phys. Rev.* **127**, 1918 (1962).

[2] Y. R. Shen, *The Principles of Nonlinear Optics* (Wiley, New York, 1984).

[3] M. Schubert and B. Wilhelmi, *Nonlinear Optics and Quantum Electronics* (Wiley, New York, 1986).

[4] P. N. Butcher and D. Cotter, *Elements of Nonlinear Optics* (Cambridge University Press, Cambridge, UK, 1990).

[5] R. W. Boyd, *Nonlinear Optics* (Academic Press, San Diego, CA, 1992).

[6] R. H. Stolen, J. E. Bjorkholm, and A. Ashkin, *Appl. Phys. Lett.* **24**, 308 (1974).

[7] R. H. Stolen, *IEEE J. Quantum Electron.* **QE-11**, 100 (1975).

[8] K. O. Hill, D. C. Johnson, B. S. Kawasaki, and R.I. MacDonald, *J. Appl. Phys.* **49**, 5098 (1978).

[9] A. Säisy, J. Botineau, A. A. Azéma and F. Gires, *Appl. Opt.* **19**, 1639 (1980).

[10] K. O. Hill, D. C. Johnson, and B. S. Kawasaki, *Appl. Opt.* **20**, 1075 (1981).

[11] C. Lin and M. A. Bösch, *Appl. Phys. Lett.* **38**, 479 (1981).

[12] R. H. Stolen and J. E. Bjorkholm, *IEEE J. Quantum Electron.* **QE-18**, 1062 (1982).

[13] C. Lin, *J. Opt. Commun.* **4**, 2 (1983).

[14] K. Washio, K. Inoue, and S. Kishida, *Electron. Lett.* **16**, 658 (1980).

[15] C. Lin, W. A. Reed, A. D. Pearson, and H. T. Shang, *Opt. Lett.* **6**, 493 (1981).

[16] C. Lin, W. A. Reed, A. D. Pearson, H. T. Shang, and P. F. Glodis, *Electron. Lett.* **18**, 87 (1982).

[17] R. H. Stolen, M. A. Bösch, and C. Lin, *Opt. Lett.* **6**, 213 (1981).

[18] K. Kitayama, S. Seikai, and N. Uchida, *Appl. Phys. Lett.* **41**, 322 (1982).

[19] K. Kitayama and M. Ohashi, *Appl. Phys. Lett.* **41**, 619 (1982).

[20] R. K. Jain and K. Stenersen, *Appl. Phys. B* **35**, 49 (1984).

[21] K. Stenersen and R. K. Jain, *Opt. Commun.* **51**, 121 (1984).

[22] M. Ohashi, K. Kitayama, N. Shibata, and S. Seikai, *Opt. Lett.* **10**, 77 (1985).

[23] N. Shibata, M. Ohashi, K. Kitayama, and S. Seikai, *Opt. Lett.* **10**, 154 (1985).

[24] H. G. Park, J. D. Park, and S. S. Lee, *Appl. Opt.* **26**, 2974 (1987).

[25] S. J. Garth and C. Pask, *Opt. Lett.* **11**, 380 (1986).

[26] N. Shibata, R. P. Braun, and R. G. Waarts, *Electron. Lett.* **22**, 675 (1986); R. G. Waarts and R. P. Braun, *Electron. Lett.* **22**, 873 (1986).

[27] N. Shibata, R. P. Braun, and R. G. Waarts, *IEEE J. Quantum Electron.* **QE-23**, 1205 (1987).

[28] P. L. Baldeck and R. R. Alfano, *J. Lightwave Technol.* **LT-5**, 1712 (1987).

[29] Y. Chen and A. W. Snyder, *Opt. Lett.* **14**, 87 (1989).

[30] S. Trillo and S. Wabnitz, *J. Opt. Soc. Am. B* **6**, 238 (1989).

[31] P. Bayvel and I. P. Giles, *Opt. Commun.* **75**, 57 (1990).

[32] Y. Chen, *J. Opt. Soc. Am. B* **7**, 43 (1990).

[33] C. J. McKinstrie, G. G. Luther, and S. H. Batha, *J. Opt. Soc. Am. B* **7**, 340 (1990).

[34] M. W. Maeda, W. B. Sessa, W. I. Way, A. Yi-Yan, L. Curtis, R. Spicer, and R. I. Laming, *J. Lightwave Technol.* **8**, 1402 (1990).

[35] A. R. Chraplyvy, *J. Lightwave Technol.* **8**, 1548 (1990).

[36] J. K. Chee and J. M. Liu, *Opt. Lett.* **14**, 820 (1989); *IEEE J. Quantum Electron.* **26**, 541 (1990).

[37] Y. Chen, *J. Opt. Soc. Am. B* **7** , 43 (1990).

[38] E. A. Golovchenko, P. V. Mamyshev, A. N. Pilipetskii, and E. M. Dianov, *IEEE J. Quantum Electron.* **26**, 1815 (1990); *J. Opt. Soc. Am. B* **8**, 1626 (1991).

[39] P. N. Morgon and J. M. Liu, *IEEE J. Quantum Electron.* **27**, 1011 (1991).

[40] G. Cappellini and S. Trillo, *Phys. Rev. A* **44**, 7509 (1991).

[41] S. Trillo and S. Wabnitz, *J. Opt. Soc. Am. B* **9**, 1061 (1992).

[42] N. Minkovski, I. Mirtchev, and L. Ivanov, *Appl. Phys. B* **55**, 430 (1992).

[43] G. R. Walker, D. M. Spirit, P. J. Chidgey, E. G. Bryant, and C. R. Batchellor, *Electron. Lett.* **28**, 989 (1992).

[44] K. Inoue, *IEEE J. Quantum Electron.* **28**, 883 (1992); *Opt. Lett.* **17** , 801 (1992); *J. Lightwave Technol.* **10**, 1553 (1992).

[45] K. Inoue and H. Toba, *IEEE Photon. Technol. Lett.* **3**, 77 (1991); *J. Lightwave Technol.* **10**, 361 (1992).

[46] K. Inoue, H. Toba, and K. Oda, *J. Lightwave Technol.* **10**, 350 (1992).

[47] D. G. Schadt and I. D. Stephens, *J. Lightwave Technol.* **10**, 1715 (1992).

[48] Z. Su, X. Zhu, and W. Sibbett, *J. Opt. Soc. Am. B* **10**, 1053 (1993).

[49] C. J. McKinstrie, X. D. Cao, and J. S. Li, *J. Opt. Soc. Am. B* **10**, 1856 (1993).

[50] T. Yamamoto and M. Nakazawa, *IEEE Photon. Technol. Lett.* **9**, 327 (1997).

[51] S. Trillo, G. Millot, E. Seve, and S. Wabnitz, *Appl. Phys. Lett.* **72**, 150 (1998).

[52] K. P. Lor and K. S. Chiang, *Opt. Commun.* **152**, 26 (1998).

[53] M. Karlsson, *J. Opt. Soc. Am. B* **15**, 2269 (1998).

[54] K. Nakajima, M. Ohashi, K. Shiraki, T. Horiguchi, K. Kurokawa, and Y. Miyajima, *J. Lightwave Technol.* **17**, 1814 (1999).

[55] M. Eiselt, *J. Lightwave Technol.* **17**, 2261 (1999).

[56] S. Song, C. T. Allen, K. R. Demarest, and R. Hui, *J. Lightwave Technol.* **17**, 2285 (1999).

[57] F. Matera, M. Settembre, M. Tamburrini, M. Zitelli, and S. Turitsyn, *Opt. Commun.* **181**, 407 (2000).

[58] Y. Inoue, *J. Phys. Soc. Jp.* **39**, 1092 (1975).

[59] D. J. Kaup, A. Reiman, and A. Bers, *Rev. Mod. Phys.* **51**, 275 (1979).

[60] L. M. Kovachek and V. N. Serkin, *Sov. J. Quantum Electron.* **19**, 1211 (1989).

[61] A. A. Zobolotskii, *Sov. Phys. JETP* **70**, 71 (1990).

[62] S. Wabnitz and J. M. Soto-Crespo, *Opt. Lett.* **23**, 265 (1998).

[63] E. A. Golovchenko and A. N. Pilipetskii, *J. Opt. Soc. Am. B* **11**, 92 (1994).

[64] R. R. Alfano, Ed., *Supercontinuum Laser Source* (Springer-Verlag, New York, 1989).

[65] J. Botineau and R. H. Stolen, *J. Opt. Soc. Am.* **72**, 1592 (1982).

[66] M. Ohashi, K. Kitayama, Y. Ishida, and N. Uchida, *Appl. Phys. Lett.* **41**, 1111 (1982).

[67] J. P. Pocholle, J. Raffy, M. Papuchon, and E. Desurvire, *Opt. Exp.* **24**, 600 (1985).

[68] I. Bar-Joseph, A. A. Friesem, R. G. Waarts, and H. H. Yaffe, *Opt. Lett.* **11**, 534 (1986).

[69] W. Margulis and U. Österberg, *Opt. Lett.* **12**, 519 (1987).

[70] A. Vatarescu, *J. Lightwave Technol.* **LT-5**, 1652 (1987).

[71] M. Nakazawa, K. Suzuki, and H. A. Haus, *Phys. Rev. A* **38**, 5193 (1988).

[72] M. E. Marhic, N. Kagi, T. K. Chiang, and L. G. Kazovsky, *Opt. Lett.* **21**, 573 (1996).

[73] D. K. Serkland and P. Kumar, *Opt. Lett.* **24**, 92 (1999).

[74] M. E. Marhic, F. S. Yang, M. C. Ho, and L. G. Kazovsky, *J. Lightwave Technol.* **17**, 210 (1999).

[75] S. Coen, M. Haelterman, P. Emplit, L. Delage, L. M. Simohamed, and F. Reynaud, *J. Opt. Soc. Am. B* **15**, 2283 (1998); *J. Opt. B* **1**, 36 (1999).

[76] S. Watanabe and T. Chikama, *Electron. Lett.* **30**, 163 (1994).

[77] K. Inoue, *Opt. Lett.* **19**, 1189 (1994).

[78] X. Zhang and B. F. Jogersen, *Opt. Fiber Technol.* **3**, 28 (1997).

[79] I. Zacharopoulos, I. Tomkos, D. Syvridis, F. Girardin, L. Occhi, and G. Guekos, *IEEE Photon. Technol. Lett.* **11**, 430 (1999).

[80] E. Ciaramella and S. Trillo, *IEEE Photon. Technol. Lett.* **12**, 849 (2000).

[81] G. P. Agrawal, *Fiber-Optic Communication Systems*, 2nd. ed. (Wiley, New York, 1997).

[82] P. A. Andrekson, N. A. Olsson, J. R. Simpson, T. Tanbun-Ek, R. A. Logan, and M. Haner, *Electron. Lett.* **27**, 922 (1991).

[83] P. O. Hedekvist, M. Karlsson, and P. A. Andrekson, *J. Lightwave Technol.* **15**, 2051 (1997).

[84] P. A. Andrekson, *Electron. Lett.* **27**, 1440 (1991).

[85] A. Yariv, D. Fekete, and D. M. Pepper, *Opt. Lett.* **4**, 52 (1979).

[86] S. Watanabe, N. Saito, and T. Chikama, *IEEE Photon. Technol. Lett.* **5**, 92 (1993).

[87] R. M. Jopson, A. H. Gnauck, and R. M. Derosier, *IEEE Photon. Technol. Lett.* **5**, 663 (1993).

[88] K. Kikuchi and C. Lorattanasane, *IEEE Photon. Technol. Lett.* **6**, 104 (1994); *IEEE Photon. Technol. Lett.* **7**, 1375 (1995).

[89] M. Yu, G. P. Agrawal, and C. J. McKinstrie, *IEEE Photon. Technol. Lett.* **7**, 932 (1995).

[90] S. Watanabe and M. Shirasaki, *J. Lightwave Technol.* **14**, 243 (1996).

[91] G. P. Agrawal, *J. Europ. Opt. Soc. B* **8**, 383 (1996).

[92] S. Chi and S. Wen, *Opt. Lett.* **19**, 1705 (1994).

[93] S. Wen and S. Chi, *Electron. Lett.* **30**, 663 (1994); *Opt. Lett.* **20**, 976 (1995).

[94] W. Forysiak and N. J. Doran, *J. Lightwave Technol.* **13**, 850 (1995).

[95] R. J. Essiambre and G. P. Agrawal, *J. Opt. Soc. Am. B* **14**, 323 (1997).

[96] C. G. Goedde, W. L. Kath, and P. Kumar, *Opt. Lett.* **20**, 1365 (1995); *J. Opt. Soc. Am. B* **14**, 1371 (1997).

[97] G. D. Bartolini, D. K. Serkland, P. Kumar, and W. L. Kath, *IEEE Photon. Technol. Lett.* **9**, 1020 (1997).

[98] P. O. Hedekvist and P. A. Andrekson, *J. Lightwave Technol.* **17**, 74 (1999).

[99] D. F. Walls and G. J. Milburn, *Quantum Optics* (Springer, New York, 1994).

[100] M. O. Scully and M. S. Zubairy, *Quantum Optics* (Cambridge University Press, New York, 1997).

[101] A. Sizmann and G. Leuchs, in *Progress in Optics*, Vol. 39, E. Wolf, Ed. (Elsevier, New York, 1999), Chap. 5.

[102] M. D. Levenson, R. M. Shelby, A. Aspect, M. Reid, and D. F. Walls, *Phys. Rev. A* **32**, 1550 (1985).

[103] R. M. Shelby, M. D. Levenson, S. H. Perlmutter, R. G. DeVoe, and D. F. Walls, *Phys. Rev. Lett.* **57**, 691 (1986).

[104] B. L. Schumaker, S. H. Perlmutter, R. M. Shelby, and M. D. Levenson, *Phys. Rev. Lett.* **58**, 357 (1987).

[105] G. J. Milburn, M. D. Levenson, R. M. Shelby, S. H. Perlmutter, R. G. DeVoe, and D. F. Walls, *J. Opt. Soc. Am. B* **4**, 1476 (1987).

[106] H. A. Haus and Y. Lai, *J. Opt. Soc. Am. B* **7**, 386 (1990).

[107] R. M. Shelby, P. D. Drummond, and S. J. Carter, *Phys. Rev. A* **42**, 2966 (1990).

[108] M. Rosenbluh and R. M. Shelby, *Phys. Rev. Lett.* **66**, 153 (1991).

[109] K. Bergman and H. A. Haus, *Opt. Lett.* **16**, 663 (1991).

[110] K. Bergman, C R. Doerr, H. A. Haus, and M. Shirasaki, *Opt. Lett.* **18**, 643 (1993).

[111] S. R. Friberg, S. Machida, M. J. Werner, A. Levanon, and T. Mukai, *Phys. Rev. Lett.* **77**, 3775 (1996).

[112] S. Schmitt, J. Ficker, M. Wolff, F. Konig, A. Sizmann, and G. Leuchs, *Phys. Rev. Lett.* **81**, 2446 (1998).

[113] D. Krylov and K. Bergman, *Opt. Lett.* **24**, 1390 (1998).

[114] D. Krylov, K. Bergman and Y. Lai, *Opt. Lett.* **24**, 774 (1999).

[115] D. Levandovsky, M. Vasilyev, and P. Kumar, *Opt. Lett.* **24**, 89 (1999); *Opt. Lett.* **24**, 984 (1999).

[116] B. Gross and J. T. Manassah, *Phys. Lett. A* **160**, 261 (1991); *J. Opt. Soc. Am. B* **9**, 1813 (1992).

[117] T. Morioka, K. Mori, and M. Saruwatari, *Electron. Lett.* **29**, 862 (1993).

[118] T. Morioka, K. Mori, S. Kawanishi, and M. Saruwatari, *IEEE Photon. Technol. Lett.* **6**, 365 (1994).

[119] T. Morioka, K. Uchiyama, S. Kawanishi, S. Suzuki, and M. Saruwatari, *Electron. Lett.* **31**, 1064 (1995).

[120] T. Morioka, K. Okamoto, M. Ishiiki, and M. Saruwatari, *Electron. Lett.* **32**, 836 (1997).

[121] S. Kawanishi, H. Takara, K. Uchiyama, I. Shake, O. Kamatani, and H. Takahashi, *Electron. Lett.* **33**, 1716 (1997).

[122] K. Mori, H. Takara, S. Kawanishi, M. Saruwatari, and T. Morioka, *Electron. Lett.* **33**, 1806 (1997).

[123] T. Okuno, M. Onishi, and M. Nishimura, *IEEE Photon. Technol. Lett.* **10**, 72 (1998).

[124] Y. Takushima, F. Futami, and K. Kikuchi, *IEEE Photon. Technol. Lett.* **10**, 1560 (1998).

[125] Y. Takushima and K. Kikuchi, *IEEE Photon. Technol. Lett.* **11**, 324 (1999).

[126] M. Prabhu, N. S. Kim, and K. Ueda, *Jpn. J. Appl. Phys.* **39**, L291 (2000).

[127] C. Lin, V. T. Nguyen, and W. G. French, *Electron. Lett.* **14** , 822 (1978).

[128] Y. Fujii, B. S. Kawasaki, K. O. Hill, and D. C. Johnson, *Opt. Lett.* **5**, 48 (1980).

[129] Y. Sasaki and Y. Ohmori, *Appl. Phys. Lett.* **39**, 466 (1981); *J. Opt. Commun.* **4**, 83 (1983).

[130] Y. Ohmori and Y. Sasaki, *IEEE J. Quantum Electron.* **QE-18**, 758 (1982).

[131] J. M. Gabriagues, *Opt. Lett.* **8**, 183 (1983).

[132] M. Nakazawa, T. Nakashima, and S. Seikai, *Appl. Phys. Lett.* **45**, 823 (1984).

[133] M. V. D. Vermelho, D. L. Nicácio, E. A. Gouveia, A. S. Gouveia-Neto, I. C. S. Carvalho, J. P. von der Weid, and W. Margulis, *J. Opt. Soc. Am. B* **10**, 1820 (1993).

[134] J. Thogersen and J. Mark, *Opt. Commun.* **110**, 435 (1994).

[135] R. W. Terhune and D. A. Weinberger, *J. Opt. Soc. Am. B* **4**, 661 (1987).

[136] U. Österberg and W. Margulis, *Opt. Lett.* **11**, 516 (1986); *Opt. Lett.* **12**, 57 (1987).

[137] M. C. Farries, P. S. J. Russel, M. E. Fermann, and D.N. Payne, *Electron. Lett.* **23**, 322 (1987).

[138] R. H. Stolen and H. W. K. Tom, *Opt. Lett.* **12**, 585 (1987).

[139] J. M. Gabriagues and H. Février, *Opt. Lett.* **12**, 720 (1987).

[140] N. B. Baranova and B. Y. Zeldovitch, *JETP Lett.* **45**, 12 (1987).

[141] B. Valk, E. M. Kim, and M. M. Salour, *Appl. Phys. Lett.* **51**, 722 (1987).

[142] F. P. Payne, *Electron. Lett.* **23**, 1215 (1987).

[143] M. C. Farries and M. E. Fermann, *Electron. Lett.* **24**, 294 (1988).

[144] V. Mizrahi, U. Österberg, J. E. Sipe, and G. I. Stegeman, *Opt. Lett.* **13**, 279 (1988).

[145] M. E. Fermann, M. C. Farries, P. St. J. Russell, and L. Poyntz-Wright, *Opt. Lett.* **13**, 282 (1988).

[146] A. Krotkus and W. Margulis, *Appl. Phys. Lett.* **52**, 1942 (1988).

[147] H. W. K. Tom, R. H. Stolen, G. D. Aumiller, and W. Pleibel, *Opt. Lett.* **13**, 512 (1988).

[148] F. Ouellette, K. O. Hill, and D. C. Johnson, *Opt. Lett.* **13**, 515 (1988).

[149] M.-V. Bergot, M. C. Farries, M. E. Fermann, L. Li, L. J. Poyntz-Wright, P. S. J. Russell, and A. Smithson, *Opt. Lett.* **13**, 592 (1988).

[150] M. E. Fermann, L. Li, M. C. Farries, and D. N. Payne, *Electron. Lett.* **24**, 894 (1988).

[151] P. Chmela, *Opt. Lett.* **13**, 669 (1988).

[152] V. Mizrahi, U. Österberg, C. Krautschik, G. I. Stegeman, J. E. Sipe, and T. F. Morse, *Appl. Phys. Lett.* **53**, 557 (1988).

[153] M. A. Saifi and M. J. Andrejco, *Opt. Lett.* **13**, 773 (1988).

[154] L. J. Poyntz-Wright, M. E. Fermann, and P. S. J. Russell, *Opt. Lett.* **13**, 1023 (1988).

[155] J. T. Manassah and O. R. Cockings, *Opt. Lett.* **12**, 1005 (1987).

[156] N. C. Kothari and X. Carlotti, *J. Opt. Soc. Am. B* **5**, 756 (1988).

[157] M. E. Fermann, L. Li, M. C. Farries, L. J. Poyntz-Wright, and L. Dong, *Opt. Lett.* **14**, 748 (1989).

[158] F. Ouellette, *Opt. Lett.* **14**, 964 (1989).

[159] T. E. Tsai, M. A. Saifi, E. J. Friebele, D. L. Griscom, and U. Österberg, *Opt. Lett.* **14**, 1023 (1989).

[160] E. M. Dianov, P. G. Kazansky, and D. Y. Stepanov, *Sov. J. Quantum Electron.* **19**, 575 (1989); *Sov. J. Quantum Electron.* **20**, 849 (1990).

[161] E. V. Anoikin, E. M. Dianov, P. G. Kazansky, and D. Yu. Stepanov, *Opt. Lett.* **15**, 834 (1990).

[162] A. Kamal, D. A. Weinberger, and W. H. Weber, *Opt. Lett.* **15**, 613 (1990).

[163] Y. E. Kapitzky and B. Ya. Zeldovich, *Opt. Commun.* **78**, 227 (1990).

[164] N. B. Baranova, A. N. Chudinov, and B. Y. Zeldovich, *Opt. Commun.* **79**, 116 (1990).

[165] V. Mizrahi, Y. Hibino, and G. I. Stegeman, *Opt. Commun.* **78**, 283 (1990).

[166] N. M. Lawandy, *Opt. Commun.* **74**, 180 (1989); *Phys. Rev. Lett.* **65**, 1745 (1990).

[167] E. M. Dianov, P. G. Kazansky, and D. Y. Stepanov, *Sov. Lightwave Commun.* **1**, 247 (1991); *Proc. SPIE* **1516**, 81 (1991).

[168] D. M. Krol, M. M. Broer, K. T. Nelson, R. H. Stolen, H. W. K. Tom, and W. Pleibel, *Opt. Lett.* **16**, 211 (1991).

[169] P. Chmela, *J. Mod. Opt.* **37**, 327 (1990); *Opt. Lett.* **16**, 443 (1991).

[170] D. Z. Anderson, *Proc. SPIE* **1148**, 186 (1989).

[171] J. K. Lucek, R. Kashyap, S. T. Davey, and D. L. Williams, *J. Mod. Opt.* **37**, 533 (1990).

[172] D. Z. Anderson, V. Mizrahi, and J. E. Sipe, *Opt. Lett.* **16**, 796 (1991).

[173] I. C. S. Carvalho, W. Margulis, and B. Lesche, *Opt. Lett.* **16**, 1487 (1991).

[174] D. M. Krol and J. R. Simpson, *Opt. Lett.* **16**, 1650 (1991).

[175] M. D. Selker and N. M. Lawandy, *Opt. Commun.* **81**, 38 (1991).

[176] R. Kashyap, *Appl. Phys. Lett.* **58**, 1233 (1991).

[177] M. I. Dyakonov and A. S. Furman, *Electron. Lett.* **27**, 1429 (1991).

[178] E. M. Dianov, P. G. Kazansky, D. S. Stardubov, and D. Y. Stepanov, *Sov. Lightwave Commun.* **2**, 83 (1992).

[179] V. M. Churikov, Y. E. Kapitzky, V. N. Lukyanov, and B. Ya. Zeldovich, *Sov. Lightwave Commun.* **2**, 389 (1992).

[180] T. J. Driscoll and N. M. Lawandy, *Opt. Lett.* **17**, 571 (1992).

[181] V. Dominic and J. Feinberg, *Opt. Lett.* **17**, 1761 (1992); *Opt. Lett.* **18**, 784 (1993).

[182] V. Y. Glushchenko and V. B. Smirnov, *Opt. Spectroscopy* **72**, 538 (1992).

[183] S. Trillo and S. Wabnitz, *Opt. Lett.* **17**, 1572 (1992).

[184] R. I. MacDonald and N. M. Lawandy, *Opt. Lett.* **18**, 595 (1993).

[185] P. G. Kazansky, A. Kamal, and P. S. J. Russel, *Opt. Lett.* **18**, 693 (1993); *Opt. Lett.* **18**, 1141 (1993); *Opt. Lett.* **19**, 701 (1994).

[186] M. A. Bolshtyansky, V. M. Churikov, Y. E. Kapitzky, A. Y. Savchenko, and B. Y. Zeldovich, *Opt. Lett.* **18**, 1217 (1993).

[187] P. S. Weitzman and U. Österberg, *IEEE J. Quantum Electron.* **29**, 1437 (1993).

[188] G. Demouchy and G. R. Boyer, *Opt. Commun.* **101**, 385 (1993).

[189] Q. Guo, *Quantum Opt.* **5**, 133 (1993).

[190] R. Schiek, *J. Opt. Soc. Am. B* **10**, 1848 (1993).

[191] R. Hayata and M. Koshiba, *Phys. Rev. Lett.* **71**, 3275 (1993).

[192] M. J. Werner and P. D. Drummond, *Opt. Lett.* **19**, 613 (1994).

[193] M. M. Lacerda, I. C. S. Carvalho, W. Margulis, and B. Lesche, *Electron. Lett.* **30**, 732 (1994).

[194] A. G. Kalocsai and J. W. Haus, *Phys. Rev. A* **49**, 574 (1994).

[195] R. Kashyap, *J. Opt. Soc. Am. B* **6**, 313 (1989); *Appl. Phys. Lett.* **58**, 1233 (1991).

[196] L. E. Myers and W. R. Bosenberg, *IEEE J. Quantum Electron.* **33** 1663 (1997).

[197] R. A. Myers, N. Mukherjee, and S. R. J. Brueck, *Opt. Lett.* **16**, 1732 (1991).

[198] A. Okada, K. Ishii, K. Mito, and K. Sasaki, *Appl. Phys. Lett.* **60**, 2853 (1992).

[199] N. Mukherjee, R. A. Myers, and S. R. J. Brueck, *J. Opt. Soc. Am. B* **11**, 665 (1994).

[200] P. G. Kazansky and P. S. J. Russell, *Opt. Commun.* **110**, 611 (1994).

[201] A. L. Calvez, E. Freysz, and A. Ducasse, *Opt. Lett.* **22**, 1547 (1997).

[202] P. G. Kazansky, P. S. J. Russell, and H. Takabe, *J. Lightwave Technol.* **15**, 1484 (1997).

[203] N. Wada and K. Morinaga, *J. Jpn. Inst. Metals* **62**, 932 (1998).

[204] V. Pruneri, F. Samoggia, G. Bonfrate, P. G. Kazansky, and G. M. Yang, *Appl. Phys. Lett.* **74**, 2423 (1999).

[205] W. Xu, J. Arentoft, and S. Glemming, *IEEE Photon. Technol. Lett.* **11**, 1265 (1999).

[206] V. Pruneri and P. G. Kazansky, *Electron. Lett.* **33**, 318 (1997).

[207] V. Pruneri, G. Bonfrate, P. G. Kazansky, D. J. Richardson, N. G. Broderick, J. P. de Sandro, C. Simonneau, P. Vidakovic, and J. A. Levenson, *Opt. Lett.* **24**, 208 (1999).

Appendix A

Decibel Units

In both linear and nonlinear fiber optics it is common to make use of decibel units, abbreviated as dB and used by engineers in many different fields. Any ratio R can be converted into decibels using the general definition

$$R \text{ (in dB)} = 10 \log_{10} R. \qquad (A.1)$$

The logarithmic nature of the decibel scale allows a large ratio to be expressed as a much smaller number. For example, 10^9 and 10^{-9} correspond to 90 dB and -90 dB, respectively. As $R = 1$ corresponds to 0 dB, ratios smaller than 1 are negative on the decibel scale. Furthermore, negative ratios cannot be expressed using decibel units.

The most common use of the decibel scale occurs for power ratios. For instance, the fiber-loss parameter α appearing in Eq. (1.2.3) can be expressed in decibel units by noting that fiber losses decrease the optical power launched into an optical fiber from its value at the input end, and thus can be written as a power ratio. Equation (1.2.4) shows how fiber losses can be expressed in units of dB/km. If a 1-mW signal reduces to 1 μW after transmission over 100 km of fiber, the power reduction by a factor of 1000 translates into a 30-dB loss from Eq. (A.1). Spreading this loss over the 100-km fiber length produces a loss of 0.3 dB/km. The same technique can be used to define the insertion loss of any component. For instance, a 1-dB loss of a fiber connector implies that the optical power is reduced by 1 dB ($\approx 20\%$) when the signal passes through the connector. Examples of other quantities that are often quoted using the decibel scale include the signal-to-noise ratio and the amplification factor of an optical amplifier.

445

If optical losses of all components in a fiber-optic communication system are expressed in decibel units, it is useful to express the transmitted and received powers also by using a decibel scale. This is achieved by using a derived unit, denoted as dBm and defined as

$$\text{power (in dBm)} = 10 \log_{10} \left(\frac{\text{power}}{1 \text{ mW}} \right), \qquad (A.2)$$

where the reference level of 1 mW is chosen for convenience; the letter m in dBm is a reminder of the 1-mW reference level. In this decibel scale for the absolute power, 1 mW corresponds to 0 dBm, whereas powers < 1 mW are expressed as negative numbers. For example, a power of 10 μW corresponds to -20 dBm. By contrast, large peak powers of intense pulses commonly used in experiments on nonlinear fiber optics are represented by positive numbers. Thus, a peak power of 10 W corresponds to 40 dBm. It is possible to introduce other decibel scales for representing optical powers by changing the reference level. For example, if the reference power in Eq. (A.2) is 1 μW, we obtain powers in dBμ units. When 1 W is used as a reference level, the power unit is known as dBW.

Appendix B

Nonlinear Refractive Index

Since the nonlinear part of the refractive index $\delta n_{NL} = n_2 |E|^2$ in Eq. (1.3.2) governs a large number of the nonlinear effects in optical fibers, an accurate measurement of the nonlinear-index coefficient n_2 is necessary. However, before discussing the measurements of n_2, it is important to clarify the units used to express its numerical values [1].

In the standard metric system of units (called SI, short for the French expression Système international d'unités), the electric field has units of V/m. As δn_{NL} is dimensionless, the units of n_2 are m²/V². In practice, it is more convenient to write the nonlinear index in the form $\delta n_{NL} = n_2' I$, where I is the intensity of the optical field and is related to E as

$$I = \tfrac{1}{2}\varepsilon_0 cn |E|^2. \tag{B.1}$$

Here ε_0 is the vacuum permittivity ($\varepsilon_0 = 8.8542 \times 10^{-12}$ F/m), c is the velocity of light in vacuum ($c = 2.998 \times 10^8$ m/s), and n is the linear part of the refractive index ($n \approx 1.45$ for silica fibers). The parameter n_2' has units of m²/W and is related to n_2 as

$$n_2' = 2n_2/(\varepsilon_0 cn). \tag{B.2}$$

It is common to quote n_2 in units of m²/W. Equation (B.2) shows the conversion factor explicitly. In earlier measurements of n_2, electrostatic units (esu) have sometimes been used. One can convert from esu to SI units by using the relation [1]

$$n_2' = (80\pi/cn)n_2(\text{esu}) \approx 5.78 \times 10^{-7} n_2(\text{esu}), \tag{B.3}$$

if we use $n = 1.45$ for silica fibers.

The measurements of n_2 for bulk silica glasses yield a value $n_2 = 2.7 \times 10^{-20}$ m^2/W at the 1.06-μm wavelength [2]. This value should decrease by 3–4% for wavelengths near 1.5 μm. The earliest measurements of n_2 in silica fibers were carried out in 1978 [3] using spectral broadening of 90-ps optical pulses (obtained from an argon-ion laser operating near 515 nm) that was produced by self-phase modulation (SPM). The estimated value of 3.2×10^{-20} m^2/W from this experiment was used almost exclusively in most studies of the nonlinear effects in optical fibers in spite of the fact that n_2 normally varies from fiber to fiber.

The growing importance of the nonlinear effects in optical communication systems revived interest in the measurements of n_2 during the 1990s, especially because fiber manufacturers are often required to specify its numerical value for their fibers [4]. Several different techniques have been used to measure n_2 for various kinds of fibers [5]–[24]: Measured values are found to vary in the range 2.2–3.9×10^{-20} m^2/W [19]. To understand the origin of such a large range of variations in n_2 values for silica fibers, note that the core of silica fibers is doped with other materials such as GeO$_2$ and Al$_2$O$_3$. These dopants affect the measured value of n_2. As a result, n_2 is expected to be different for various dispersion-shifted fibers made using different amounts of dopants inside the fiber core. Moreover, optical fibers do not maintain the state of polarization during propagation of light. As the polarization state changes randomly along fiber length, one measures an average value of n_2 that is reduced by a factor of $\frac{8}{9}$ compared with the value expected for bulk samples that maintain the linear polarization of the incident light [14].

The measured value of n_2 is also affected by the experimental technique used to measure it. The reason is that two other mechanisms, related to molecular motion (the Raman effect) and excitation of acoustic waves through electrostriction (Brillouin scattering), also contribute to n_2. However, their relative contributions depend on whether the pulse width is longer or shorter than the response time associated with the corresponding process. The electrostrictive contribution vanishes for pulses shorter than 100 ps but attains its maximum value ($\sim 16\%$ of total n_2) for pulse widths > 10 ns [13]. In contrast, the Raman contribution does not vanish until pulse widths become < 50 fs and is $\sim 18\%$ for pulse widths > 10 ps. One should be careful when comparing measurements made using different pulse widths. The largest value of n_2 is expected for measurements performed under quasi-CW conditions using pulse widths > 10 ns (or CW beams).

Table B.1 Nonlinear-Index Coefficient for Different Fibers

Method used	λ (μm)	Fiber type	Measured n_2 (10^{-20}m^2/W)	Comments
SPM	1.319	silica core	2.36	110-ps pulses [7]
	1.319	DSF	2.62	110-ps pulses [7]
	1.548	DSF	2.31	34-ps pulses [8]
	1.550	DSF	2.50	5-ps pulses [12]
	1.550	standard	2.20	$\sim$50-GHz modulation [15]
	1.550	DSF	2.32	$\sim$50-GHz modulation [15]
	1.550	DCF	2.57	$\sim$50-GHz modulation [15]
XPM	1.550	silica core	2.48	7.4-MHz modulation [9]
	1.550	standard	2.63	7.4-MHz modulation [9]
	1.550	DSF	2.98	7.4-MHz modulation [9]
	1.550	DCF	3.95	7.4-MHz modulation [9]
	1.548	standard	2.73	10-MHz modulation [18]
	1.548	standard	2.23	2.3-GHz modulation [18]
FWM	1.555	DSF	2.25	CW lasers [6]
	1.553	DSF	2.64	modulation instability [11]

Several different nonlinear effects have been used to measure n_2 for silica fibers; the list includes SPM, cross-phase modulation (XPM), four-wave mixing (FWM), and modulation instability. The SPM technique makes use of broadening of the pulse spectrum (see Section 4.1) and was first used in 1978 [3]. The uncertainty in n_2 values measured depends on how accurately one can estimate the effective core area A_{eff} from the mode-field diameter and how well one can characterize input pulses. The SPM technique is used extensively. In one set of measurements, mode-locked pulses of 110-ps duration, obtained from a Nd:YAG laser operating at 1.319 μm, were used [7]. For a silica-core fiber (no dopants inside the core), the measured value of n_2 was 2.36×10^{-20} m^2/W. The measured n_2 values were larger for dispersion-shifted fibers (DSFs) (average value 2.62×10^{-20} m^2/W) because of the contribution of dopants. The measured values of n_2 in the 1.55-μm wavelength region are somewhat smaller. Table B.1 summarizes the results obtained in several experiments. The uncertainty is estimated to be $\sim$ 5% for these measurements.

Perhaps the simplest SPM-based technique is the one used in Reference [15]. It makes use of two CW DFB semiconductor lasers whose wavelength difference (0.3–0.5 nm) is stabilized by controlling the laser temperature. The optical signal entering the fiber oscillates sinusoidally at the beat frequency ($\sim$ 50 GHz). The optical spectrum at the fiber output exhibits peaks at multiples of this beat frequency because of the SPM-induced phase modulation (FM sidebands). The ratio of the peak heights depends only on the nonlinear phase shift ϕ_{NL} given in Eq. (4.1.6) and can be used to deduce n_2. For standard telecommunication fibers, the measured value was 2.2×10^{-20} m^2/W. The technique was used to measure n_2 for many DSFs and dispersion-compensating fibers (DCFs) having different amounts of dopants (see Table B.1). A self-aligned interferometric technique, allowing a direct measurement of the SPM-induced phase shift, has also been used [24].

The XPM-induced phase shift was used as early as 1987 to measure n_2 [5]. A 1995 experiment used a pump-probe configuration in which both pump and probe signals were obtained from CW sources [9]. When the pump light was modulated at a low frequency (< 10 MHz), the probe signal developed FM sidebands because of the XPM-induced phase modulation (see Section 7.1). To ensure that the relative polarization between pump and probe varied randomly, pump light was depolarized before entering the fiber. The average values of n_2 for several kinds of fibers are listed in Table B.1. The measured n_2 values in this experiment were consistently larger than those obtained using the SPM-based techniques. The most likely explanation for this discrepancy is related to the electrostrictive contribution to n_2 that occurs for pulse widths > 1 ns (or modulation frequencies < 1 GHz). The 7.4-MHz modulation frequency used for the pump beam is small enough that the measured values of n_2 include the full electrostrictive contribution. The XPM technique has also been used to study the frequency dependence of the electrostrictive contribution by changing the pump-modulation frequency from 10 MHz to > 1 GHz [17]–[19].

The nonlinear phenomenon of FWM can also be used to estimate n_2. As discussed in Chapter 10, FWM produces sidebands whose amplitude and frequency depend on n_2. In one experiment, the CW output of two DFB lasers, operating near 1.55 μm with a wavelength separation of 0.8 nm, was amplified by using a fiber amplifier and then injected into 12.5 km of a DSF [6]. The powers in the FWM sidebands were fitted numerically and used to estimate a value of $n_2 = 2.25 \times 10^{-20}$ m^2/W for the test fiber. This value agrees with other measurements if we assume that electrostriction did not contribute to n_2. This is a reasonable assumption if we note that the spectral bandwidth of the laser was broadened to > 600 MHz through current modulation to avoid the onset of stimulated Brillouin scattering inside the test fiber.

The modulation instability can be thought of as a special case of FWM (see Section 10.3.2). The main difference is that only a single pump beam is needed at the fiber input. As discussed in Section 5.1, two spectral sidebands centered at the frequencies $\omega_0 \pm \Omega$ appear at the fiber output, where ω_0 is the pump frequency. The frequency shift Ω as well as the sideband amplitudes depend among other things on n_2 and can be used to deduce it. In a 1995 experiment [11], a DFB laser operating at 1.553 μm was modulated externally to produce 25 ns pulses at a repetition rate of 4 MHz. Such pulses were amplified using two cascaded fiber amplifiers and then launched into a 10.1 km of DSF. The amplitude of modulation instability sidebands was used to deduce a value $n_2 = 2.64 \times 10^{-20}$ m^2/W for the test fiber. This value includes the electrostrictive contribution because of the relatively wide pump pulses used in the experiment. Indeed, if the 16% electrostrictive contribution is removed from the measured value, we recover $n_2 = 2.22 \times 10^{-20}$ m^2/W, a value close to other measurements in which the electrostrictive contribution was absent.

What nominal value of n_2 should be used for estimating the nonlinear parameter $\gamma \equiv 2\pi n_2/(\lambda A_{\text{eff}})$ used in this book? An appropriate recommendation is $n_2 = 2.6 \times 10^{-20}$ m^2/W for pulse wider than 1 ns and $n_2 = 2.2 \times 10^{-20}$ m^2/W for pulses as short as 1 ps. For femtosecond pulses, the Raman contribution should be modified as discussed in Section 2.3. The n_2 value should be multiplied by the factor of $\frac{9}{8}$ in experiments that use polarization-maintaining fibers. Also, the effective core area should be estimated properly for the fiber used in a specific experiment. This parameter can vary from a value as small as 20 μm^2 for some DCFs to values > 100 μm^2 for a fiber known as the large-effective-area fiber (LEAF).

Because of a relatively low value of n_2 for silica fibers, several other kinds of glasses with larger nonlinearities have been used to make optical fibers [25]–[30]. For a lead-silicate fiber n_2 was measured to be $\sim 2 \times 10^{-19}$ m^2/W [25]. In chalcogenide As$_2$S$_3$-based fibers, the measured value $n_2 = 4.2 \times 10^{-18}$ m^2/W can be larger by more than two orders of magnitude compared with the value for silica fibers [26]. Such fibers are attracting increasing attention for applications related to nonlinear fiber optics in spite of their relatively high losses. Their use for making fiber gratings and nonlinear switches has reduced power requirements considerably [27].

References

[1] P. N. Butcher and D. N. Cotter, *The Elements of Nonlinear Optics* (Cambridge University Press, Cambridge, UK, 1990), Appendix 5.

[2] D. Milam and M. J. Weber, *J. Appl. Phys.* **47**, 2497 (1976).

[3] R. H. Stolen and C. Lin, *Phys. Rev. A* **17**, 1448 (1978).

[4] G. P. Agrawal, in *Properties of Glass and Rare-Earth Doped Glasses for Optical Fibers*, D. Hewak, Ed. (INSPEC, IEE, London, UK, 1998), pp. 17–21.

[5] M. Monerie and Y. Durteste, *Electron. Lett.* **23**, 961 (1987).

[6] L. Prigent and J. P. Hamaide, *IEEE Photon. Technol. Lett.* **5**, 1092 (1993).

[7] K. S. Kim, R. H. Stolen, W. A. Reed, and K. W. Quoi, *Opt. Lett.* **19**, 257 (1994).

[8] Y. Namihira, A. Miyata, and N. Tanahashi, *Electron. Lett.* **30**, 1171 (1994).

[9] T. Kato, Y. Suetsugu, M. Takagi, E. Sasaoka, and M. Nishimura, *Opt. Lett.* **20**, 988 (1995).

[10] T. Kato, Y. Suetsugu, and M. Nishimura, *Opt. Lett.* **20**, 2279] (1995).

[11] M. Artiglia, E. Ciaramella, and B. Sordo, *Electron. Lett.* **31**, 1012 (1995).

[12] M. Artiglia, R. Caponi, F. Cisternino, C. Naddeo, and D. Roccato, *Opt. Fiber Technol.* **2**, 75 (1996).

[13] E. L. Buckland and R. W. Boyd, *Opt. Lett.* **21**, 1117 (1996).

[14] S. V. Chernikov and J. R. Taylor, *Opt. Lett.* **21**, 1559 (1996).

[15] A. Boskovic, S. V. Chernikov, J. R. Taylor, L. Gruner-Nielsen, and O. A. Levring, *Opt. Lett.* **21**, 1966 (1996).

[16] K. Li, Z. Xiong, G. D. Peng, and P. L. Chu, *Opt. Commun.* **136**, 223 (1997).

[17] E. L. Buckland and R. W. Boyd, *Opt. Lett.* **22**, 676 (1997).

[18] A. Fellegara, A. Melloni, and M. Martinelli, *Opt. Lett.* **22**, 1615 (1997).

[19] A. Fellegara, M. Artiglia, S. B. Andereasen, A. Melloni, F. P. Espunes, and M. Martinelli, *Electron. Lett.* **33**, 1168 (1997).

[20] A. Melloni, M. Martinelli, and A. Fellegara, *Fiber Integ. Opt.* **18**, 1 (1999).

[21] C. Mazzali, D. F. Grosz, and H. L. Fragnito, *IEEE Photon. Technol. Lett.* **11**, 251 (1999).

[22] A. Fellegara, A. Melloni, and P. Sacchetto, *Opt. Commun.* **162**, 333 (1999).

[23] S. Smolorz, F. Wise, and N. F. Borrelli, *Opt. Lett.* **24**, 1103 (1999).

[24] C. Vinegoni, M. Wegumuller, and N. Gisin, *Electron. Lett.* **36**, 886 (2000).

[25] M. A. Newhouse, D. L. Weidman, and D. W. Hall, *Opt. Lett.* **15**, 1185 (1990).

[26] M. Asobe, T. Kanamori, and K. Kubodera, *IEEE J. Quantum Electron.* **29**, 2325 (1993).

[27] M. Asobe, *Opt. Fiber Technol.* **3**, 142 (1997).

[28] S. Smolorz, I. Kang, F. Wise, B. G. Aitken, and N. F. Borrelli, *J. Non-Crystal. Solids* **256**, 310 (1999).

[29] T. Cardinal, K. A. Richardson, H. Shim, A. Schulte, R. Beatty, K. Le Foulgoc, C. Meneghini, J. F. Viens, and A. Villeneuve, *J. Non-Crystal. Solids* **256**, 353 (1999).

[30] G. Lenz, J. Zimmermann, T. Katsufuji, M. E. Lines, H. Y. Hwang, S. Spalter, R. E. Slusher, S. W. Cheong, J. S. Sanghera, and I. D. Aggarwal, *Opt. Lett.* **25**, 254 (2000).

Appendix C

Acronyms

Each scientific field has its own jargon, and the field of nonlinear fiber optics is not an exception. Although an attempt was made to avoid extensive use of acronyms, many still appear throughout the book. Each acronym is defined the first time it appears in a chapter so that the reader does not have to search the entire text to find its meaning. As a further help, all acronyms are listed here in alphabetical order.

AM	amplitude modulation
ASE	amplified spontaneous emission
ASK	amplitude-shift keying
BER	bit-error rate
CVD	chemical vapor deposition
CW	continuous wave
DCF	dispersion-compensating fiber
DFB	distributed feedback
DSF	dispersion-shifted fiber
EDFA	erbium-doped fiber amplifier
EDFL	erbium-doped fiber laser
FDM	frequency-division multiplexing
FDTD	finite-difference time domain
FFT	fast Fourier transform
FM	frequency modulation
FROG	frequency-resolved optical gating
FWHM	full width at half maximum
FWM	four-wave mixing

GVD	group-velocity dispersion
LCM	liquid-crystal modulator
LEAF	large-effective-area fiber
MCVD	modified chemical vapor deposition
MI	modulation instability
MQW	multiquantum well
MZ	Mach–Zehnder
NLS	nonlinear Schrödinger
NOLM	nonlinear optical-loop mirror
NRZ	nonreturn to zero
OOK	on-off keying
OPC	optical phase conjugation
OTDM	optical time-division multiplexing
OVD	outside-vapor deposition
PCM	pulse-code modulation
PDM	polarization-division multiplexing
PM	phase modulation
PMD	polarization-mode dispersion
PSK	phase-shift keying
RDF	reverse-dispersion fiber
RIN	relative intensity noise
RMS	root mean square
RZ	return to zero
SBS	stimulated Brillouin scattering
SCM	subcarrier multiplexing
SDH	synchronous digital hierarchy
SI	Système international d'unités
SLA	semiconductor laser amplifier
SNR	signal-to-noise ratio
SPM	self-phase modulation
SRS	stimulated Raman scattering
TDM	time-division multiplexing
TOD	third-order dispersion
TROG	time-resolved optical gating
VAD	vapor-axial deposition
VCSEL	vertical-cavity surface-emitting laser
VPE	vapor-phase epitaxy

WDM	wavelength-division multiplexing
XPM	cross-phase modulation
YAG	yttrium aluminum garnet
ZDWL	zero-dispersion wavelength

Index

Optics and Photonics
(formerly Quantum Electronics)

Editors: Paul L. Kelly, Tufts University, Medford, Massachusetts
Ivan Kaminow, Lucent Technologies, Holmdel, New Jersey
Govind P. Agrawal, University of Rochester, Rochester, New York

Yoh-Han Pao, Case Western Reserve University, Cleveland, Ohio, Founding Editor 1972-1979